The *Blueprint for Problem Solving* is a general outline that you can use to help you solve application problems.

BLUEPRINT FOR PROBLEM SOLVING

STEP 1: **Read** the problem, and then mentally **list** the items that are known and the items that are unknown.

STEP 2: **Assign a variable** to one of the unknown items. (In most cases this will amount to letting x = the item that is asked for in the problem.) Then **translate** the other **information** in the problem to expressions involving the variable.

STEP 3: **Reread** the problem, and then **write an equation,** using the items and variables listed in steps 1 and 2, that describes the situation.

STEP 4: **Solve the equation** found in step 3.

STEP 5: **Write** your **answer** using a complete sentence.

STEP 6: **Reread** the problem, and **check** your solution with the original words in the problem.

Here is a *Blueprint for Problem Solving* that you can use specifically for problems about systems of equations (beginning in Chapter 3).

BLUEPRINT FOR PROBLEM SOLVING USING A SYSTEM OF EQUATIONS

STEP 1: **Read** the problem, and then mentally **list** the items that are known and the items that are unknown.

STEP 2: **Assign variables** to each of the unknown items; that is, let x = one of the unknown items and y = the other unknown item. Then **translate** the other **information** in the problem to expressions involving the two variables.

STEP 3: **Reread** the problem, and then **write a system of equations,** using the items and variables listed in steps 1 and 2, that describes the situation.

STEP 4: **Solve the system** found in step 3.

STEP 5: **Write** your **answers** using complete sentences.

STEP 6: **Reread** the problem, and **check** your solution with the original words in the problem.

Resources

By gathering resources early in the term, before you need help, the information about these resources will be available to you when they are needed.

INSTRUCTOR

Knowing the contact information for your instructor is very important. You may already have this information from the course syllabus. It is a good idea to write it down again.

Name _____ Office Location _____

Available Hours: M _____ T _____ W _____ TH _____ F _____

Phone Number _____ ext. _____ E-mail Address _____

TUTORING CENTER

Many schools offer tutoring, free of charge to their students. If this is the case at your school, find out when and where tutoring is offered.

Tutoring Location _____ Phone Number _____ ext. _____

Available Hours: M _____ T _____ W _____ TH _____ F _____

COMPUTER LAB

Many schools offer a computer lab where students can use the online resources and software available with their textbook. Other students using the same software and websites as you can be very helpful. Find out where the computer lab at your school is located.

Computer Lab Location _____ Phone Number _____ ext. _____

Available Hours: M _____ T _____ W _____ TH _____ F _____

VIDEO LESSONS

A complete set of video lessons is available to your school. These videos feature the author of your textbook presenting full-length, 15- to 20-minute lessons from every section of your textbook. If you miss class, or find yourself behind, these lessons will prove very useful.

Video Location _____ Phone Number _____ ext. _____

Available Hours: M _____ T _____ W _____ TH _____ F _____

CLASSMATES

Form a study group and meet on a regular basis. When you meet try to speak to each other using proper mathematical language. That is, use the words that you see in the definition and property boxes in your textbook.

Name _____ Phone _____ E-mail _____

Name _____ Phone _____ E-mail _____

Name _____ Phone _____ E-mail _____

ELEMENTARY AND INTERMEDIATE ALGEBRA

A COMBINED COURSE

THIRD EDITION

Charles P. McKeague

CUESTA COLLEGE

BROOKS/COLE
CENGAGE Learning™

Australia • Brazil • Japan • Korea • Mexico • Singapore • Spain • United Kingdom • United States

BROOKS/COLE
CENGAGE Learning™

Elementary and Intermediate Algebra, Third Edition

Charles P. McKeague

Executive Editor: Charlie Van Wagner

Development Editor: Donald Gecewicz

Assistant Editor: Laura Localio

Editorial Assistant: Lisa Lee

Technology Project Manager: Rebecca Subity

Marketing Manager: Greta Kleinert

Marketing Assistant: Cassandra Cummings

Marketing Communications Manager: Darlene Amidon-Brent

Project Manager, Editorial Production: Cheryll Linthicum

Creative Director: Rob Hugel

Art Director: Vernon T. Boes

Print Buyer: Karen Hunt

Permissions Editor: Roberta Broyer

Production Service: Graphic World Publishing Services

Text Designer: Diane Beasley

Photo Researcher: Kathleen Olson

Illustrator: Graphic World Illustration Studio

Cover Designer: Diane Beasley

Cover Image: Triton Bridge, Nagoya, Honshu, Japan, Photodisc Green/Getty Images

Compositor: Graphic World Inc.

For product information and technology assistance, contact us at **Cengage Learning Customer & Sales Support, 1-800-354-9706**
For permission to use material from this text or product, submit all requests online at **cengage.com/permissions**
Further permissions questions can be emailed to **permissionrequest@cengage.com**

ExamView® and ExamView Pro® are registered trademarks of FSCreations, Inc. Windows is a registered trademark of the Microsoft Corporation used herein under license. Macintosh and Power Macintosh are registered trademarks of Apple Computer, Inc. Used herein under license.

Library of Congress Control Number: 2006940283
ISBN-13: 978-0-495-10851-1
ISBN-10: 0-495-10851-0

Brooks/Cole
10 Davis Drive
Belmont, CA 94002-3098
USA

Cengage Learning is a leading provider of customized learning solutions with office locations around the globe, including Singapore, the United Kingdom, Australia, Mexico, Brazil, and Japan. Locate your local office at: **international.cengage.com/region**

Cengage Learning products are represented in Canada by Nelson Education, Ltd.

For your course and learning solutions, visit **academic.cengage.com**
Purchase any of our products at your local college store or at our preferred online store **www.ichapters.com**

Printed in China
2   3   4   5   6   7   11   10   09   08

Brief Contents

Contents

CHAPTER 9 Rational Exponents and Roots 535

CHAPTER 10 Quadratic Functions 595

CHAPTER 11 Exponential and Logarithmic Functions 657

CHAPTER 12   Conic Sections   711

CHAPTER 13   Sequences and Series   743

Preface to the Instructor

I have a passion for teaching mathematics. That passion carries through to my textbooks. My goal is a textbook that is user-friendly for both students and instructors. For the instructor, I build features into the text that reinforce the habits and study skills we know will bring success to our students. For students, this combined book forms a bridge between the elementary and intermediate algebra courses. Students benefit, too, from clear, concise writing; continuous review; and foreshadowing of topics to come.

The third edition of *Elementary and Intermediate Algebra: A Combined Course* builds upon these strengths. In addition, the combined book offers the required course topics in a more efficient physical format. Many instructors like the convenience of having two courses in one book.

In this edition, renewal of the problem sets, along with the continued emphasis on foreshadowing of later topics, a program of continuous cumulative review, and the focus on applications, make this the best edition of *Elementary and Intermediate Algebra* yet.

Renewal and Reorganization of Exercise Sets

This edition of *Elementary and Intermediate Algebra* contains over 2,000 new problems in the reorganized problem sets, roughly 30 percent of the problems. Most of these new problems have been added to enhance the midrange of exercises. Doing so gives our problem sets a better bridge from easy problems to more difficult problems. Many of these midrange problems cover more than one concept or technique in a new and slightly more challenging way. In short, they start students thinking mathematically and working with algebra productively.

This enhanced midrange of exercises also underscores our series' appeal to the middle level of rigor for the course. We think instructors will use many of these midrange problems as classroom examples, so we have labeled some of them as *chalkboard problems*.

As part of our revamping of the problem sets in this edition, we have also reordered the categories of problems to make a more logical bridge between sections.

A Better Progression of Categories of Problems

The categories in our problem sets now appear in the following order.

General (Undesignated) Exercises These "starter" exercises normally do not have labels. They involve a certain amount of the drill necessary to master basic techniques. These problems then progress in difficulty so that students can begin to put together more than one concept or idea. It is here that you will find the foreshadowing problems. Instead of drill for the sake of drill, we have students work the problems that they will need later in the course—hence the description *foreshadowing*. This category is also where you will find the midrange of problems discussed earlier. As in previous editions, we have kept the odd-even similarity of the problems in this part of the problem set.

Applying the Concepts Students are always curious about how the algebra they are learning can be applied, so we have included applied problems in most of the problem sets in the book and have labeled them to show students the array of uses of mathematics. These applied problems are written in an inviting way, to help students overcome some of the apprehension associated with application problems. We have a number of new applications under the heading *Improving Your Quantitative Literacy* that are particularly accessible and inviting.

Maintaining Your Skills One of the major themes of our book is continuous review. We strive to continuously hone techniques learned earlier by keeping the important concepts in the forefront of the course. The *Maintaining Your Skills* problems review material from the previous chapter, or they review problems that form the foundation of the course—the problems that you expect students to be able to solve when they get to the next course.

Getting Ready for the Next Section Many students think of mathematics as a collection of discrete, unrelated topics. Their instructors know that this is not the case. The new *Getting Ready for the Next Section* problems reinforce the cumulative, connected nature of this course by showing how the concepts and techniques flow one from another throughout the course. These problems review all of the material that students will need in order to be successful, forming a bridge to the next section and gently preparing students to move forward.

The Combined Course: A Natural Bridge

Elementary algebra is a bridge course. One of its main goals is to get students ready to make a successful start in intermediate algebra. Intermediate algebra, too, is a bridge. The course and its syllabus bring the student to the level of ability to do quantitative work in his or her major. Many students will go on after the intermediate algebra course to precalculus, calculus, and majors requiring quantitative literacy. At many schools, a large number of students enroll in the elementary–intermediate sequence. Instructors at those schools want some flexibility regarding which topics they cover and in what order.

Chapters 1 through 6 cover concepts and techniques traditionally associated with the elementary course. Chapters 8 through 13 would normally be in an intermediate algebra course. Chapter 7, the transitional chapter, allows greater flexibility in the design of the courses and in teaching the sequence.

The combined book is ideal for adapting the developmental math sequence to a school's specific needs. The comprehensive coverage that a combined book offers gives instructors a chance to end the elementary course at a point suitable for their students or to begin the intermediate course at a conceptual level more appropriate for the kinds of students who take intermediate algebra at their schools.

Our Proven Commitment to Student Success

After two successful editions, we have developed several interlocking, proven features that will improve students' chances of success in the course. We placed practical, easily understood study skills in the first six chapters (look for them on the page after the chapter opener). Here are some of the other, important success features of the book.

Getting Ready for Class Just before each problem set is a list of four questions under the heading *Getting Ready for Class.* These problems require written responses from students and are to be done before students come to class. The answers can be found by reading the preceding section. These questions reinforce the importance of reading the section before coming to class.

Linking Objectives and Examples At the end of each section we place a small box that shows which examples support which learning objectives. This feature helps students to understand how the section and its examples are built around objectives. We think that this feature helps to make the structure of exposition of concepts clearer.

Blueprint for Problem Solving Found in the main text, the *Blueprint for Problem Solving* is a detailed outline of steps needed to successfully attempt application problems. Intended as a guide to problem solving in general, the blueprint takes the student through the solution process to various kinds of applications.

Maintaining Your Skills We believe that students who consistently work review problems will be much better prepared for class than students who do not engage in continuous review. The *Maintaining Your Skills* problems cumulatively review the most important concepts from the previous chapter as well as concepts that form the foundation of the course.

Getting Ready for the Next Section At the ends of section problem sets, you will find *Getting Ready for the Next Section,* a category of problems that students can work to prepare themselves to navigate the next section successfully. These problems polish techniques and reinforce the idea that all topics in the course are built on previous topics.

End-of-Chapter Summary, Review, and Assessment

We have learned that students are more comfortable with a chapter that sums up what they have learned thoroughly and accessibly through a well-organized presentation that reinforces concepts and techniques well. To help students grasp concepts and get more practice, each chapter ends with the following features that together give a comprehensive reexamination of the chapter.

Chapter Summary The chapter summary recaps all main points from the chapter in a visually appealing grid. In a column next to each topic is an example that illustrates the type of problem associated with the topic being reviewed. Our way of summarizing shows students that concepts in mathematics do relate—and that mastering one concept is a bridge to the next. When students prepare for a test, they can use the chapter summary as a guide to the main concepts of the chapter.

Chapter Review Test Following the chapter summary in each chapter is the chapter review test. It contains an extensive set of problems that review all the main topics in the chapter. This feature can be used flexibly—as assigned review, as a recommended self-test for students as they prepare for examinations, or as an in-class quiz or test.

Chapter Projects Each chapter closes with a pair of projects. One is a group project, suitable for students to work on in class. Group projects list details about number of participants, equipment, and time, so that instructors can determine how well the project fits into their classroom. The second project is a research project for students to do outside of class and tends to be open ended.

Additional Features of the Book

Facts from Geometry Many of the important facts from geometry are listed under this heading. In most cases, an example or two accompanies each of the facts to give students a chance to see how topics from geometry are related to the algebra they are learning.

Unit Analysis Chapter 6 contains problems requiring students to convert from one unit of measure to another. The method used to accomplish the conversions is the method they will use if they take a chemistry class. Since this method is similar to the method we use to multiply rational expressions, unit analysis is covered in Section 6.2 as an application of multiplying rational expressions.

Chapter Openings Each chapter opens with an introduction in which a real-world application is used to spark interest in the chapter. We expand on each of these opening applications later in the chapter.

Supplements

Enhanced WebAssign (0534495826) Instant feedback and ease of use are just two reasons why WebAssign is the most widely used homework system in higher education. WebAssign's Homework Delivery System lets you deliver, collect, grade, and record assignments using the web. This proven system has been enhanced to include end-of-chapter problems from McKeague's *Elementary and Intermediate Algebra,* Third Edition—incorporating figures, videos, examples, PDF pages of the text, and quizzes to promote active learning and provide the immediate, relevant feedback students want.

Test Bank (0495382973) Drawing from hundreds of text-specific questions, an instructor can easily create tests that target specific course objectives. The Test Bank includes multiple tests per chapter as well as final exams. The tests are made up of a combination of multiple-choice, free-response, true/false, and fill-in-the-blank questions.

ExamView® (0495383066) This computerized algorithmic test bank on CD allows instructors to create exams using a CD.

Text-Specific Videos (0495382981) These text-specific DVD sets completed by Pat McKeague are available at no charge to qualified adopters of the text. They feature 10- to 20-minute problem-solving lessons that cover each section of every chapter.

Student Solutions Manual (0495383007) The *Student Solutions Manual* provides worked-out solutions to the odd-numbered problems in the text.

Annotated Instructor's Edition (0495383031) The Instructor's Edition provides the complete student text with answers next to each respective exercise.

Complete Solutions Manual (0495383023) The *Complete Solutions Manual* provides worked-out solutions to all of the problems in the text.

CENGAGENOW™ CengageNOW, a powerful and fully integrated teaching and learning system, provides instructors and students with unsurpassed control, variety, and all-in-one utility. CengageNOW ties together the fundamental learning activities: diagnostics, tutorials, homework, personalized study, quizzing, and testing. Personalized Study is a learning companion that helps students gauge their unique study needs and makes the most of their study time by building focused Personalized Study plans that reinforce key concepts. **Pre-Tests** give students an initial assessment of their knowledge. **Personalized Study** plans, based on the students' answers to the pre-test questions, outline key elements for review. **Post-Tests** assess student mastery of core chapter concepts. Results can be e-mailed to the instructor!

To package a printed access card with each new student text, order using ISBN 0495392944. Students without a new copy of the text can purchase an instant access code by using ISBN 0495392952.

JoinIn™ on TurningPoint® (0495383058) Brooks/Cole, a part of Cengage Learning, is pleased to offer book-specific JoinIn™ content from the Third Edition of McKeague's *Elementary and Intermediate Algebra*. This content for student classroom response systems allows instructors to transform the classroom and assess students' progress with instant in-class quizzes and polls. Our agreement to offer TurningPoint® software lets an instructor pose book-specific questions and display students' answers seamlessly within the Microsoft® PowerPoint® slides of your own lecture, in conjunction with the "clicker" hardware of your choice. Enhance how your students interact with you, your lecture, and each other. For college and university adopters only. Contact your local Cengage Learning representative to learn more.

Blackboard® CengageNOW Integration (049538352X) It is easier than ever to integrate electronic course-management tools with content from this text's rich companion website. This supplement is ready to use as soon as you log on, or you can customize WebTutor ToolBox with web links, images, and other resources.

WebCT® CengageNOW Integration (0495383538) Another option is available for integrating easy-to-use course management tools with content from this text's rich companion website. Ready to use as soon as you log on, or you can customize WebTutor ToolBox with web links, images, and other resources.

Website academic.cengage.com/math/mckeague
The book's website offers instant access to the Student Resource Center, a rich array of teaching and learning resources that offers videos, chapter-by-chapter online tutorial quizzes, a final exam, chapter outlines, chapter reviews, chapter-by-chapter web links, flash cards, and more interactive options.

Acknowledgments

I would like to thank my editor at Brooks/Cole, Charlie Van Wagner, for his help and encouragement with this project. Many thanks also to Don Gecewicz, my developmental editor, for his suggestions on content, his proofreading, and his availability for consulting. This is a better book because of Don. Patrick McKeague, Tammy Fisher-Vasta, and Devin Christ assisted me with all parts of this revision, from manuscript preparation to proofreading page proofs and preparing the index. They are a fantastic team to work with, and this project could not have been completed without them. Susan Caire and Jeff Brouwer did an excellent job of proofreading the entire book in page proofs. Mary Gentilucci, Shane Wilwand, and Annie Stephens assisted with error checking and proofreading. Thanks to Rebecca Subity and Laura Localio for handling the media and ancillary packages on this project, and to Lisa Lee for her administrative expertise. Cheryll Linthicum of Brooks/Cole and Carol O'Connell of Graphic World Publishing Services turned the manuscript into a book. Ross Rueger produced the excellent solutions manuals that accompany the book.

Thanks also to Diane McKeague and Amy Jacobs for their encouragement with all my writing endeavors.

Finally, I am grateful to the following instructors for their suggestions and comments: Jess L. Collins, McLennan Community College; Richard Drey, Northampton Community College; Peg Hovde, Grossmont College; Sarah Jackman, Richland College; Carol Juncker, Delgado Community College; Joanne Kendall, College of the Mainland; Harriet Kiser, Floyd College; Domingo Javier Litong, South Texas Community College; Cindy Lucas, College of the Mainland; Jan MacInnes, Florida Community College of Jacksonville; Rudolfo Maglio, Oakton Community College; Janice McFatter, Gulf Coast Community College; Nancy Olson, Johnson County Community College; John H. Pleasants, Orange County Community College; Barbara Jane Sparks, Camden County College; Jim Stewart, Jefferson Community College; David J. Walker, Hinds Community College; and Deborah Woods, University of Cincinnati.

Charles P. McKeague
January 2007

The Basics

Caryl Bryer Fallert

Much of what we do in mathematics is concerned with recognizing patterns. If you recognize the patterns in the following two sequences, then you can easily extend each sequence.

Sequence of odd numbers = 1, 3, 5, 7, 9, . . .
Sequence of squares = 1, 4, 9, 16, 25, . . .

Once we have classified groups of numbers as to the characteristics they share, we sometimes discover that a relationship exists between the groups. Although it may not be obvious at first, there is a relationship that exists *between* the two sequences shown. The introduction to *The Book of Squares,* written in 1225 by the mathematician known as Fibonacci, begins this way:

> I thought about the origin of all square numbers and discovered that they arise out of the increasing sequence of odd numbers.

The relationship that Fibonacci refers to is shown visually in Figure 1.

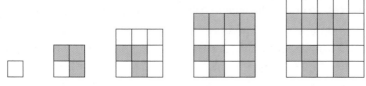

FIGURE 1

Many times we describe a relationship or pattern in a number of different ways. Figure 1 is a visual description of a relationship. In this chapter we will work on describing relationships numerically and verbally (in writing).

Some of the students enrolled in my algebra classes develop difficulties early in the course. Their difficulties are not associated with their ability to learn mathematics; they all have the potential to pass the course. Students who get off to a poor start do so because they have not developed the study skills necessary to be successful in algebra. Here is a list of things you can do to begin to develop effective study skills.

1 Put Yourself on a Schedule

The general rule is that you spend 2 hours on homework for every hour you are in class. Make a schedule for yourself in which you set aside 2 hours each day to work on algebra. Once you make the schedule, stick to it. Don't just complete your assignments and stop. Use all the time you have set aside. If you complete an assignment and have time left over, read the next section in the book, and then work more problems.

2 Find Your Mistakes and Correct Them

There is more to studying algebra than just working problems. You must always check your answers with the answers in the back of the book. When you have made a mistake, find out what it is and correct it. Making mistakes is part of the process of learning mathematics. In the prologue to *The Book of Squares,* Leonardo Fibonacci (ca. 1170–ca. 1250) had this to say about the content of his book:

> I have come to request indulgence if in any place it contains something more or less than right or necessary; for to remember everything and be mistaken in nothing is divine rather than human . . .

Fibonacci knew, as you know, that human beings make mistakes. You cannot learn algebra without making mistakes.

3 Gather Information on Available Resources

You need to anticipate that you will need extra help sometime during the course. There is a form to fill out in Appendix A to help you gather information on resources available to you. One resource is your instructor; you need to know your instructor's office hours and where the office is located. Another resource is the math lab or study center, if they are available at your school. It also helps to have the phone numbers of other students in the class, in case you miss class. You want to anticipate that you will need these resources, so now is the time to gather them together.

Notation and Symbols

OBJECTIVES

A Translate between phrases written in English and expressions written in symbols.

B Simplify expressions containing exponents.

C Simplify expressions using the rule for order of operations.

D Read tables and charts.

E Recognize the pattern in a sequence of numbers.

Suppose you have a checking account that costs you $15 a month, plus $0.05 for each check you write. If you write 10 checks in a month, then the monthly charge for your checking account will be

$$15 + 10(0.05)$$

Do you add 15 and 10 first and then multiply by 0.05? Or do you multiply 10 and 0.05 first and then add 15? If you don't know the answer to this question, you will after you have read through this section.

Because much of what we do in algebra involves comparison of quantities, we will begin by listing some symbols used to compare mathematical quantities. The comparison symbols fall into two major groups: equality symbols and inequality symbols.

We will let the letters a and b stand for (represent) any two mathematical quantities. When we use letters to represent numbers, as we are doing here, we call the letters *variables*.

Variables: An Intuitive Look

When you filled out the application for the school you are attending, there was a space to fill in your first name. "First name" is a variable quantity because the value it takes depends on who is filling out the application. For example, if your first name is Manuel, then the value of "First Name" is Manuel. However, if your first name is Christa, then the value of "First Name" is Christa.

If we denote "First Name" as *FN*, "Last Name" as *LN*, and "Whole Name" as *WN*, then we take the concept of a variable further and write the relationship between the names this way:

$$FN + LN = WN$$

(We use the + symbol loosely here to represent writing the names together with a space between them.) This relationship we have written holds for all people who have only a first name and a last name. For those people who have a middle name, the relationship between the names is

$$FN + MN + LN = WN$$

A similar situation exists in algebra when we let a letter stand for a number or a group of numbers. For instance, if we say "let a and b represent numbers," then a and b are called *variables* because the values they take on vary. We use the variables a and b in the following lists so that the relationships shown there are true for all numbers that we will encounter in this book. By using variables, the following statements are general statements about all numbers, rather than specific statements about only a few numbers.

> **Comparison Symbols**
>
> | | | |
> |---|---|---|
> | *Equality:* | $a = b$ | a is equal to b (a and b represent the same number) |
> | | $a \neq b$ | a is not equal to b |
> | *Inequality:* | $a < b$ | a is less than b |
> | | $a \not< b$ | a is not less than b |
> | | $a > b$ | a is greater than b |
> | | $a \not> b$ | a is not greater than b |
> | | $a \geq b$ | a is greater than or equal to b |
> | | $a \leq b$ | a is less than or equal to b |

The symbols for inequality, $<$ and $>$, always point to the smaller of the two quantities being compared. For example, $3 < x$ means 3 is smaller than x. In this case we can say "3 is less than x" or "x is greater than 3"; both statements are correct. Similarly, the expression $5 > y$ can be read as "5 is greater than y" or as "y is less than 5" because the inequality symbol is pointing to y, meaning y is the smaller of the two quantities.

Next, we consider the symbols used to represent the four basic operations: addition, subtraction, multiplication, and division.

> **Operation Symbols**
>
> | | | |
> |---|---|---|
> | *Addition:* | $a + b$ | The *sum* of a and b |
> | *Subtraction:* | $a - b$ | The *difference* of a and b |
> | *Multiplication:* | $a \cdot b,\ (a)(b),\ a(b),\ (a)b,\ ab$ | The *product* of a and b |
> | *Division:* | $a \div b,\ a/b,\ \dfrac{a}{b},\ b\overline{)a}$ | The *quotient* of a and b |

When we encounter the word *sum,* the implied operation is addition. To find the sum of two numbers, we simply add them. *Difference* implies subtraction, *product* implies multiplication, and *quotient* implies division. Notice also that there is more than one way to write the product or quotient of two numbers.

> **Grouping Symbols** Parentheses ( ) and brackets [ ] are the symbols used for grouping numbers together. (Occasionally, braces { } are also used for grouping, although they are usually reserved for set notation, as we shall see.)

The following examples illustrate the relationship between the symbols for comparing, operating, and grouping and the English language.

EXAMPLES

Mathematical Expression	*English Equivalent*
1. $4 + 1 = 5$	The sum of 4 and 1 is 5.
2. $8 - 1 < 10$	The difference of 8 and 1 is less than 10.
3. $2(3 + 4) = 14$	Twice the sum of 3 and 4 is 14.

4. $3x \geq 15$ The product of 3 and x is greater than or equal to 15.

5. $\dfrac{y}{2} = y - 2$ The quotient of y and 2 is equal to the difference of y and 2.

The last type of notation we need to discuss is the notation that allows us to write repeated multiplications in a more compact form—*exponents*. In the expression 2^3, the 2 is called the *base* and the 3 is called the *exponent*. The exponent 3 tells us the number of times the base appears in the product; that is,

$$2^3 = 2 \cdot 2 \cdot 2 = 8$$

The expression 2^3 is said to be in exponential form, whereas $2 \cdot 2 \cdot 2$ is said to be in expanded form. Here are some additional examples of expressions involving exponents.

 EXAMPLES Expand and multiply.

6. $5^2 = 5 \cdot 5 = 25$ Base 5, exponent 2

7. $2^5 = 2 \cdot 2 \cdot 2 \cdot 2 \cdot 2 = 32$ Base 2, exponent 5

8. $10^3 = 10 \cdot 10 \cdot 10 = 1{,}000$ Base 10, exponent 3

Notation and Vocabulary Here is how we read expressions containing exponents.

Mathematical Expression	Written Equivalent
5^2	five to the second power
5^3	five to the third power
5^4	five to the fourth power
5^5	five to the fifth power
5^6	five to the sixth power

We have a shorthand vocabulary for second and third powers because the area of a square with a side of 5 is 5^2, and the volume of a cube with a side of 5 is 5^3.

5^2 can be read "five squared." 5^3 can be read "five cubed."

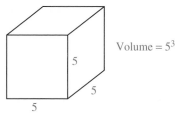

The symbols for comparing, operating, and grouping are to mathematics what punctuation symbols are to English. These symbols are the punctuation symbols for mathematics.

Consider the following sentence:

Paul said John is tall.

It can have two different meanings, depending on how it is punctuated.

1. "Paul," said John, "is tall."

2. Paul said, "John is tall."

Let's take a look at a similar situation in mathematics. Consider the following mathematical statement:

$$5 + 2 \cdot 7$$

If we add the 5 and 2 first and then multiply by 7, we get an answer of 49. However, if we multiply the 2 and the 7 first and then add 5, we are left with 19. We have a problem that seems to have two different answers, depending on whether we add first or multiply first. We would like to avoid this type of situation. Every problem like $5 + 2 \cdot 7$ should have only one answer. Therefore, we have developed the following rule for the order of operations.

> **Rule (Order of Operations)**   When evaluating a mathematical expression, we will perform the operations in the following order, beginning with the expression in the innermost parentheses or brackets first and working our way out.
> 1. Simplify all numbers with exponents, working from left to right if more than one of these expressions is present.
> 2. Then do all multiplications and divisions left to right.
> 3. Perform all additions and subtractions left to right.

 EXAMPLES   Simplify each expression using the rule for order of operations.

9. $5 + 8 \cdot 2 = 5 + 16$ Multiply $8 \cdot 2$ first
$\qquad = 21$

10. $12 \div 4 \cdot 2 = 3 \cdot 2$ Work left to right
$\qquad = 6$

11. $2[5 + 2(6 + 3 \cdot 4)] = 2[5 + 2(6 + 12)]$ Simplify within the innermost
$\qquad = 2[5 + 2(18)]$ grouping symbols first
$\qquad = 2[5 + 36]$ Next, simplify inside
$\qquad = 2[41]$ the brackets
$\qquad = 82$ Multiply

12. $10 + 12 \div 4 + 2 \cdot 3 = 10 + 3 + 6$ Multiply and divide left to right
$\qquad = 19$ Add left to right

13. $2^4 + 3^3 \div 9 - 4^2 = 16 + 27 \div 9 - 16$ Simplify numbers with exponents
$\qquad = 16 + 3 - 16$ Then, divide
$\qquad = 19 - 16$ Finally, add and subtract
$\qquad = 3$ left to right

Reading Tables and Bar Charts

The following table shows the average amount of caffeine in a number of beverages. The diagram in Figure 1 is a bar chart. It is a visual presentation of the information in Table 1. The table gives information in numerical form, whereas the chart gives the same information in a geometric way. In mathematics, it is important to be able to move back and forth between the two forms.

TABLE 1
Caffeine Content of Hot Drinks

Drink (6-ounce cup)	Caffeine (milligrams)
Brewed coffee	100
Instant coffee	70
Tea	50
Cocoa	5
Decaffeinated coffee	4

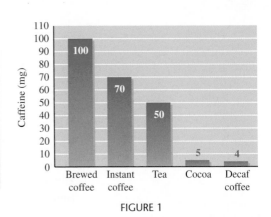

FIGURE 1

 EXAMPLE 14　Referring to Table 1 and Figure 1, suppose you have 3 cups of brewed coffee, 1 cup of tea, and 2 cups of decaf in one day. Write an expression that will give the total amount of caffeine in these six drinks, then simplify the expression.

SOLUTION　From the table or the bar chart, we find the number of milligrams of caffeine in each drink; then we write an expression for the total amount of caffeine:

$$3(100) + 50 + 2(4)$$

Using the rule for order of operations, we get 358 total milligrams of caffeine.

Number Sequences and Inductive Reasoning

Suppose someone asks you to give the next number in the sequence of numbers below. (The dots mean that the sequence continues in the same pattern forever.)

$$2, 5, 8, 11, \ldots$$

If you notice that each number is 3 more than the number before it, you would say the next number in the sequence is 14 because $11 + 3 = 14$. When we reason in this way, we are using what is called *inductive reasoning.* In mathematics we use inductive reasoning when we notice a pattern to a sequence of numbers and then use the pattern to extend the sequence.

 EXAMPLE 15　Find the next number in each sequence.
  a. $3, 8, 13, 18, \ldots$
  b. $2, 10, 50, 250, \ldots$
  c. $2, 4, 7, 11, \ldots$

SOLUTION　To find the next number in each sequence, we need to look for a pattern or relationship.

  a. For the first sequence, each number is 5 more than the number before it; therefore, the next number will be $18 + 5 = 23$.
  b. For the sequence in part (b), each number is 5 times the number before it; therefore, the next number in the sequence will be $5 \cdot 250 = 1{,}250$.

c. For the sequence in part (c), there is no number to add each time or multiply by each time. However, the pattern becomes apparent when we look at the differences between the numbers:

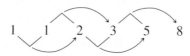

Proceeding in the same manner, we would add 5 to get the next term, giving us $11 + 5 = 16$.

In the introduction to this chapter we mentioned the mathematician known as Fibonacci. There is a special sequence in mathematics named for Fibonacci. Here it is.

Fibonacci sequence = 1, 1, 2, 3, 5, 8, . . .

Can you see the relationship among the numbers in this sequence? Start with two 1's, then add two consecutive members of the sequence to get the next number. Here is a diagram.

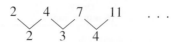

Sometimes we refer to the numbers in a sequence as *terms* of the sequence.

 EXAMPLE 16 Write the first 10 terms of the Fibonacci sequence.

SOLUTION The first six terms are given above. We extend the sequence by adding 5 and 8 to obtain the seventh term, 13. Then we add 8 and 13 to obtain 21. Continuing in this manner, the first 10 terms in the Fibonacci sequence are

1, 1, 2, 3, 5, 8, 13, 21, 34, 55

LINKING OBJECTIVES AND EXAMPLES

Next to each **objective** we have listed the examples that are best described by that objective. Connecting the examples to the objectives gives us a more complete understanding of the overall structure of the section.

A	1–5
B	6–8
C	9–13
D	14
E	15, 16

GETTING READY FOR CLASS

Each section of the book will end with some problems and questions like the ones below. They are for you to answer after you have read through the section but before you go to class. All of them require that you give written responses in complete sentences. Writing about mathematics is a valuable exercise. If you write with the intention of explaining and communicating what you know to someone else, you will find that you understand the topic you are writing about even better than you did before you started writing. As with all problems in this course, you want to approach these writing exercises with a positive point of view. You will get better at giving written responses to questions as you progress through the course. Even if you never feel comfortable writing about mathematics, just the process of attempting to do so will increase your understanding and ability in mathematics.

After reading through the preceding section, respond in your own words and in complete sentences.

1. What is a variable?
2. Write the first step in the rule for order of operations.
3. What is inductive reasoning?
4. Explain the relationship between an exponent and its base.

For each sentence below, write an equivalent expression in symbols.

1. The sum of x and 5 is 14.

2. The difference of x and 4 is 8.

▶ **3.** The product of 5 and y is less than 30.

4. The product of 8 and y is greater than 16.

5. The product of 3 and y is less than or equal to the sum of y and 6.

6. The product of 5 and y is greater than or equal to the difference of y and 16.

7. The quotient of x and 3 is equal to the sum of x and 2.

8. The quotient of x and 2 is equal to the difference of x and 4.

Expand and multiply.

9. 3^2 **10.** 4^2

11. 7^2 **12.** 9^2

13. 2^3 **14.** 3^3

▶ **15.** 4^3 **16.** 5^3

17. 2^4 **18.** 3^4

19. 10^2 **20.** 10^4

21. 11^2 **22.** 111^2

Use the rule for order of operations to simplify each expression as much as possible.

23. a. $2 \cdot 3 + 5$ **24. a.** $8 \cdot 7 + 1$
    b. $2(3 + 5)$ **b.** $8(7 + 1)$

▶ **25. a.** $5 + 2 \cdot 6$ **26. a.** $8 + 9 \cdot 4$
    b. $(5 + 2) \cdot 6$ **b.** $(8 + 9) \cdot 4$

27. a. $5 \cdot 4 + 5 \cdot 2$ **28. a.** $6 \cdot 8 + 6 \cdot 3$
    b. $5(4 + 2)$ **b.** $6(8 + 3)$

29. a. $8 + 2(5 + 3)$ **30. a.** $7 + 3(8 - 2)$
    b. $(8 + 2)(5 + 3)$ **b.** $(7 + 3)(8 - 2)$

31. $20 + 2(8 - 5) + 1$ **32.** $10 + 3(7 + 1) + 2$

33. $5 + 2(3 \cdot 4 - 1) + 8$ **34.** $11 - 2(5 \cdot 3 - 10) + 2$

▶ **35.** $4 + 8 \div 4 - 2$ **36.** $6 + 9 \div 3 + 2$

37. $3 \cdot 8 + 10 \div 2 + 4 \cdot 2$ **38.** $5 \cdot 9 + 10 \div 2 + 3 \cdot 3$

39. a. $(5 + 3)(5 - 3)$ **40. a.** $(7 + 2)(7 - 2)$
    b. $5^2 - 3^2$ **b.** $7^2 - 2^2$

41. a. $(4 + 5)^2$ **42. a.** $(6 + 3)^2$
    b. $4^2 + 5^2$ **b.** $6^2 + 3^2$

43. $2 \cdot 10^3 + 3 \cdot 10^2 + 4 \cdot 10 + 5$

44. $5 \cdot 10^3 + 6 \cdot 10^2 + 7 \cdot 10 + 8$

▶ **45.** $10 - 2(4 \cdot 5 - 16)$ **46.** $15 - 5(3 \cdot 2 - 4)$

47. $4[7 + 3(2 \cdot 9 - 8)]$

48. $5[10 + 2(3 \cdot 6 - 10)]$

49. $3(4 \cdot 5 - 12) + 6(7 \cdot 6 - 40)$

50. $6(8 \cdot 3 - 4) + 5(7 \cdot 3 - 1)$

▶ **51.** $3^4 + 4^2 \div 2^3 - 5^2$ **52.** $2^5 + 6^2 \div 2^2 - 3^2$

53. $5^2 + 3^4 \div 9^2 + 6^2$ **54.** $6^2 + 2^5 \div 4^2 + 7^2$

Simplify each expression.

55. $20 \div 2 \cdot 10$ **56.** $40 \div 4 \cdot 5$

57. $24 \div 8 \cdot 3$ **58.** $24 \div 4 \cdot 6$

59. $36 \div 6 \cdot 3$ **60.** $36 \div 9 \cdot 2$

61. $16 - 8 + 4$ **62.** $16 - 4 + 8$

63. $24 - 14 + 8$ **64.** $24 - 16 + 6$

65. $36 - 6 + 12$ **66.** $36 - 9 + 20$

We are assuming that you know how to do arithmetic with decimals. Here are some problems to practice. Simplify each expression.

▶ **67.** $0.08 + 0.09$ ▶ **68.** $0.06 + 0.04$

▶ **69.** $0.10 + 0.12$ ▶ **70.** $0.08 + 0.06$

▶ **71.** $4.8 - 2.5$ ▶ **72.** $6.3 - 4.8$

▶ **73.** $2.07 + 3.48$ ▶ **74.** $4.89 + 2.31$

▶ **75.** $0.12(2,000)$ ▶ **76.** $0.09(3,000)$

▶ **77.** $0.25(40)$ ▶ **78.** $0.75(40)$

▶ **79.** $510 \div 0.17$ ▶ **80.** $400 \div 0.1$

▶ **81.** $240 \div 0.12$ ▶ **82.** $360 \div 0.12$

Use a calculator to find the following quotients. Round your answers to the nearest hundredth, if necessary.

83. $37.80 \div 1.07$ **84.** $85.46 \div 4.88$

85. $555 \div 740$ **86.** $740 \div 108$

87. $70 \div 210$ **88.** $15 \div 80$

89. $6,000 \div 22$ **90.** $51,000 \div 17$

Applying the Concepts

Food Labels In 1993 the government standardized the way in which nutrition information was presented on the labels of most packaged food products. Figure 2 shows a standardized food label from a package of cookies that I ate at lunch the day I was writing the problems for this problem set. Use the information in Figure 2 to answer the following questions.

Nutrition Facts	Amount/serving	%DV*	Amount/serving	%DV*
Serving Size 5 Cookies (about 43 g) Servings Per Container 2	**Total Fat** 9 g	**15%**	**Total Carb.** 30 g	**10%**
	Sat. Fat 2.5 g	**12%**	Fiber 1 g	**2%**
Calories 210	**Cholest.** less than 5 mg	**2%**	Sugars 14 g	
Fat Calories 90	**Sodium** 110 mg	**5%**	**Protein** 3 g	
* Percent Daily Values (DV) are based on a 2,000 calorie diet.	Vitamin A 0% • Vitamin C 0% • Calcium 2% • Iron 8%			

Sandwich cremes

FIGURE 2

91. How many cookies are in the package?

92. If I paid $0.50 for the package of cookies, how much did each cookie cost?

93. If the "calories" category stands for calories per serving, how many calories did I consume by eating the whole package of cookies?

94. Suppose that, while swimming, I burn 11 calories each minute. If I swim for 20 minutes, will I burn enough calories to cancel out the calories I added by eating 5 cookies?

95. Reading Tables and Charts The following table and bar chart give the amount of caffeine in five different soft drinks. How much caffeine is in each of the following?
   a. A 6-pack of Jolt
   b. 2 Coca-Colas plus 3 Tabs

Caffeine Content in Soft Drinks

Drink	Caffeine (milligrams)
Jolt	100
Tab	47
Coca-Cola	45
Diet Pepsi	36
7UP	0

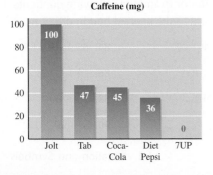

Caffeine (mg)

96. Reading Tables and Charts The following table and bar chart give the amount of caffeine in five different nonprescription drugs. How much caffeine is in each of the following?
   a. A box of 12 Excedrin
   b. 1 Dexatrim plus 4 Excedrin

Caffeine Content in Nonprescription Drugs

Nonprescription Drug	Caffeine (milligrams)
Dexatrim	200
NoDoz	100
Excedrin	65
Triaminicin tablets	30
Dristan tablets	16

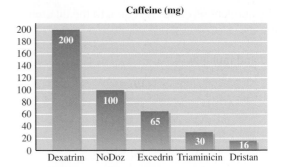

Caffeine (mg)

97. Reading Tables and Charts The following bar chart gives the number of calories burned by a 150-pound person during 1 hour of various exercises. The accompanying table should display the same information. Use the bar chart to complete the table.

Calories Burned by 150-Pound Person

Activity	Calories Burned in 1 Hour
Bicycling	374
Bowling	
Handball	
Jogging	
Skiing	

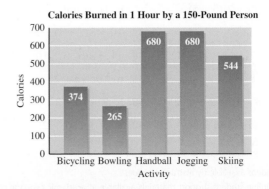

Calories Burned in 1 Hour by a 150-Pound Person

98. Reading Tables and Charts The following bar chart gives the number of calories consumed by eating some popular fast foods. The accompanying table should display the same information. Use the bar chart to complete the table.

Calories in Fast Food

Food	Calories
McDonald's Hamburger	270
Burger King Hamburger	
Jack in the Box Hamburger	
McDonald's Big Mac	
Burger King Whopper	

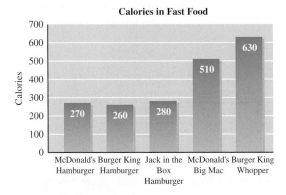

Calories in Fast Food

Find the next number in each sequence.

99. 1, 2, 3, 4, . . . (The sequence of counting numbers)

100. 0, 1, 2, 3, . . . (The sequence of whole numbers)

101. 2, 4, 6, 8, . . . (The sequence of even numbers)

102. 1, 3, 5, 7, . . . (The sequence of odd numbers)

103. 1, 4, 9, 16, . . . (The sequence of squares)

104. 1, 8, 27, 64, . . . (The sequence of cubes)

105. 2, 2, 4, 6, . . . (A Fibonacci-like sequence)

106. 5, 5, 10, 15, . . . (A Fibonacci-like sequence)

1.2 Real Numbers

OBJECTIVES

A Locate and label points on the number line.

B Change a fraction to an equivalent fraction with a new denominator.

C Simplify expressions containing absolute value.

D Identify the opposite of a number.

E Multiply fractions.

F Identify the reciprocal of a number.

G Find the value of an expression.

H Find the perimeter and area of squares, rectangles, and triangles.

Table 1 and Figure 1 give the record low temperature, in degrees Fahrenheit, for each month of the year in the city of Jackson, Wyoming. Notice that some of these temperatures are represented by negative numbers.

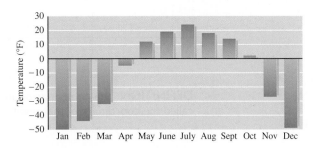

FIGURE 1

In this section we start our work with negative numbers. To represent negative numbers in algebra, we use what is called the *real number line*. Here is how we construct a real number line: We first draw a straight line and label a conve-

TABLE 1
Record Low Temperatures for Jackson, Wyoming

Month	Temperature (Degrees Fahrenheit)	Month	Temperature (Degrees Fahrenheit)
January	−50	July	24
February	−44	August	18
March	−32	September	14
April	−5	October	2
May	12	November	−27
June	19	December	−49

Note

If there is no sign (+ or −) in front of a number, the number is assumed to be positive (+).

Note

There are other numbers on the number line that you may not be as familiar with. They are irrational numbers such as π, $\sqrt{2}$, $\sqrt{3}$. We will introduce these numbers later in the chapter.

nient point on the line with 0. Then we mark off equally spaced distances in both directions from 0. Label the points to the right of 0 with the numbers 1, 2, 3, . . . (the dots mean "and so on"). The points to the left of 0 we label in order, −1, −2, −3, ... Here is what it looks like.

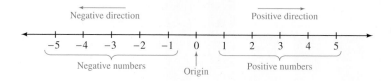

The numbers increase in value going from left to right. If we "move" to the right, we are moving in the positive direction. If we move to the left, we are moving in the negative direction. When we compare two numbers on the number line, the number on the left is always smaller than the number on the right. For instance, −3 is smaller than −1 because it is to the left of −1 on the number line.

 EXAMPLE 1 Locate and label the points on the real number line associated with the numbers $-3.5, -1\frac{1}{4}, \frac{1}{2}, \frac{3}{4}, 2.5$.

SOLUTION We draw a real number line from −4 to 4 and label the points in question.

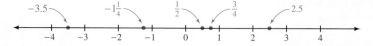

DEFINITION The number associated with a point on the real number line is called the **coordinate** of that point.

In the preceding example, the numbers $\frac{1}{2}, \frac{3}{4}, 2.5, -3.5$, and $-1\frac{1}{4}$ are the coordinates of the points they represent.

DEFINITION The numbers that can be represented with points on the real number line are called **real numbers**.

Real numbers include whole numbers, fractions, decimals, and other numbers that are not as familiar to us as these.

Fractions on the Number Line

As we proceed through Chapter 1, from time to time we will review some of the major concepts associated with fractions. To begin, here is the formal definition of a fraction.

> **DEFINITION** If a and b are real numbers, then the expression
> $$\frac{a}{b} \qquad b \neq 0$$
> is called a **fraction.** The top number a is called the **numerator,** and the bottom number b is called the **denominator.** The restriction $b \neq 0$ keeps us from writing an expression that is undefined. (As you will see, division by zero is not allowed.)

The number line can be used to visualize fractions. Recall that for the fraction $\frac{a}{b}$, a is called the numerator and b is called the denominator. The denominator indicates the number of equal parts in the interval from 0 to 1 on the number line. The numerator indicates how many of those parts we have. If we take that part of the number line from 0 to 1 and divide it into *three equal parts,* we say that we have divided it into *thirds* (Figure 2). Each of the three segments is $\frac{1}{3}$ (one third) of the whole segment from 0 to 1.

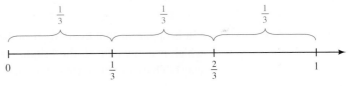

FIGURE 2

Two of these smaller segments together are $\frac{2}{3}$ (two thirds) of the whole segment. And three of them would be $\frac{3}{3}$ (three thirds), or the whole segment.

Let's do the same thing again with six equal divisions of the segment from 0 to 1 (Figure 3). In this case we say each of the smaller segments has a length of $\frac{1}{6}$ (one sixth).

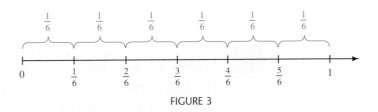

FIGURE 3

The same point we labeled with $\frac{1}{3}$ in Figure 2 is now labeled with $\frac{2}{6}$. Likewise, the point we labeled earlier with $\frac{2}{3}$ is now labeled $\frac{4}{6}$. It must be true then that

$$\frac{2}{6} = \frac{1}{3} \qquad \text{and} \qquad \frac{4}{6} = \frac{2}{3}$$

Actually, there are many fractions that name the same point as $\frac{1}{3}$. If we were to divide the segment between 0 and 1 into 12 equal parts, 4 of these 12 equal parts ($\frac{4}{12}$) would be the same as $\frac{2}{6}$ or $\frac{1}{3}$; that is,

$$\frac{4}{12} = \frac{2}{6} = \frac{1}{3}$$

Even though these three fractions look different, each names the same point on the number line, as shown in Figure 4. All three fractions have the same *value* because they all represent the same number.

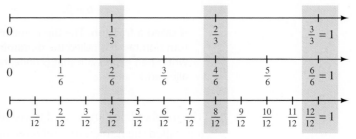

FIGURE 4

> **DEFINITION** Fractions that represent the same number are said to be **equivalent.** Equivalent fractions may look different, but they must have the same value.

It is apparent that every fraction has many different representations, each of which is equivalent to the original fraction. The next two properties give us a way of changing the terms of a fraction without changing its value.

> **Property 1** Multiplying the numerator and denominator of a fraction by the same nonzero number never changes the value of the fraction.

> **Property 2** Dividing the numerator and denominator of a fraction by the same nonzero number never changes the value of the fraction.

EXAMPLE 2   Write $\frac{3}{4}$ as an equivalent fraction with denominator 20.

SOLUTION   The denominator of the original fraction is 4. The fraction we are trying to find must have a denominator of 20. We know that if we multiply 4 by 5, we get 20. Property 1 indicates that we are free to multiply the denominator by 5 as long as we do the same to the numerator.

$$\frac{3}{4} = \frac{3 \cdot 5}{4 \cdot 5} = \frac{15}{20}$$

The fraction $\frac{15}{20}$ is equivalent to the fraction $\frac{3}{4}$.

Absolute Values and Opposites

Representing numbers on the number line lets us give each number two important properties: a direction from zero and a distance from zero. The direction from zero is represented by the sign in front of the number. (A number without a sign is understood to be positive.) The distance from zero is called the absolute value of the number, as the following definition indicates.

> **DEFINITION** The **absolute value** of a real number is its distance from zero on the number line. If x represents a real number, then the absolute value of x is written $|x|$.

 EXAMPLES   Write each expression without absolute value symbols.

3. $|5| = 5$ The number 5 is 5 units from zero

4. $|-5| = 5$ The number -5 is 5 units from zero

5. $\left|-\dfrac{1}{2}\right| = \dfrac{1}{2}$ The number $-\dfrac{1}{2}$ is $\dfrac{1}{2}$ units from zero

The absolute value of a number is *never* negative. It is the distance the number is from zero without regard to which direction it is from zero. When working with the absolute value of sums and differences, we must simplify the expression inside the absolute value symbols first and then find the absolute value of the simplified expression.

 EXAMPLES   Simplify each expression.

6. $|8 - 3| = |5| = 5$

7. $|3 \cdot 2^3 + 2 \cdot 3^2| = |3 \cdot 8 + 2 \cdot 9| = |24 + 18| = |42| = 42$

8. $|9 - 2| - |8 - 6| = |7| - |2| = 7 - 2 = 5$

Another important concept associated with numbers on the number line is that of opposites. Here is the definition.

> **DEFINITION** Numbers the same distance from zero but in opposite directions from zero are called **opposites.**

 EXAMPLES   Give the opposite of each number.

	Number	Opposite	
9.	5	-5	5 and -5 are opposites
10.	-3	3	-3 and 3 are opposites
11.	$\dfrac{1}{4}$	$-\dfrac{1}{4}$	$\dfrac{1}{4}$ and $-\dfrac{1}{4}$ are opposites
12.	-2.3	2.3	-2.3 and 2.3 are opposites

Each negative number is the opposite of some positive number, and each positive number is the opposite of some negative number. The opposite of a negative number is a positive number. In symbols, if a represents a positive number, then

$$-(-a) = a$$

Opposites always have the same absolute value. And, when you add any two opposites, the result is always zero:

$$a + (-a) = 0$$

Reciprocals and Multiplication with Fractions

The last concept we want to cover in this section is the concept of reciprocals. Understanding reciprocals requires some knowledge of multiplication with fractions. To multiply two fractions, we simply multiply numerators and multiply denominators.

 EXAMPLE 13   Multiply $\frac{3}{4} \cdot \frac{5}{7}$.

SOLUTION   The product of the numerators is 15, and the product of the denominators is 28:

$$\frac{3}{4} \cdot \frac{5}{7} = \frac{3 \cdot 5}{4 \cdot 7} = \frac{15}{28}$$

Note

In past math classes you may have written fractions like $\frac{7}{3}$ (improper fractions) as mixed numbers, such as $2\frac{1}{3}$. In algebra it is usually better to write them as improper fractions rather than mixed numbers.

 EXAMPLE 14   Multiply $7\left(\frac{1}{3}\right)$.

SOLUTION   The number 7 can be thought of as the fraction $\frac{7}{1}$:

$$7\left(\frac{1}{3}\right) = \frac{7}{1}\left(\frac{1}{3}\right) = \frac{7 \cdot 1}{1 \cdot 3} = \frac{7}{3}$$

 EXAMPLE 15   Expand and multiply $\left(\frac{2}{3}\right)^3$.

SOLUTION   Using the definition of exponents from the previous section, we have

$$\left(\frac{2}{3}\right)^3 = \frac{2}{3} \cdot \frac{2}{3} \cdot \frac{2}{3} = \frac{8}{27}$$

We are now ready for the definition of reciprocals.

> **DEFINITION** Two numbers whose product is 1 are called **reciprocals**.

 EXAMPLES   Give the reciprocal of each number.

	Number	Reciprocal	
16.	5	$\frac{1}{5}$	Because $5\left(\frac{1}{5}\right) = \frac{5}{1}\left(\frac{1}{5}\right) = \frac{5}{5} = 1$
17.	2	$\frac{1}{2}$	Because $2\left(\frac{1}{2}\right) = \frac{2}{1}\left(\frac{1}{2}\right) = \frac{2}{2} = 1$
18.	$\frac{1}{3}$	3	Because $\frac{1}{3}(3) = \frac{1}{3}\left(\frac{3}{1}\right) = \frac{3}{3} = 1$
19.	$\frac{3}{4}$	$\frac{4}{3}$	Because $\frac{3}{4}\left(\frac{4}{3}\right) = \frac{12}{12} = 1$

Although we will not develop multiplication with negative numbers until later in the chapter, you should know that the reciprocal of a negative number is also a negative number. For example, the reciprocal of -4 is $-\frac{1}{4}$.

Previously we mentioned that a variable is a letter used to represent a number or a group of numbers. An expression that contains any combination of numbers, variables, operation symbols, and grouping symbols is called an *algebraic expression* (sometimes referred to as just an *expression*). This definition includes the use of exponents and fractions. Each of the following is an algebraic expression.

$$3x + 5 \qquad 4t^2 - 9 \qquad x^2 - 6xy + y^2 \qquad -15x^2y^4z^5 \qquad \frac{a^2 - 9}{a - 3} \qquad \frac{(x - 3)(x + 2)}{4x}$$

In the last two expressions, the fraction bar separates the numerator from the denominator and is treated the same as a pair of grouping symbols; it groups the numerator and denominator separately.

The Value of an Algebraic Expression

An expression such as $3x + 5$ will take on different values depending on what x is. If we were to let x equal 2, the expression $3x + 5$ would become 11. On the other hand, if x is 10, the same expression has a value of 35:

When	$x = 2$	When	$x = 10$
the expression	$3x + 5$	the expression	$3x + 5$
becomes	$3(2) + 5$	becomes	$3(10) + 5$
	$= 6 + 5$		$= 30 + 5$
	$= 11$		$= 35$

Table 2 lists some other algebraic expressions, along with specific values for the variables and the corresponding value of the expression after the variable has been replaced with the given number.

TABLE 2

Original Expression	Value of the Variable	Value of the Expression
$5x + 2$	$x = 4$	$5(4) + 2 = 20 + 2$
		$= 22$
$3x - 9$	$x = 2$	$3(2) - 9 = 6 - 9$
		$= -3$
$4t^2 - 9$	$t = 5$	$4(5^2) - 9 = 4(25) - 9$
		$= 100 - 9$
		$= 91$
$\dfrac{a^2 - 9}{a - 3}$	$a = 8$	$\dfrac{8^2 - 9}{8 - 3} = \dfrac{64 - 9}{8 - 3}$
		$= \dfrac{55}{5}$
		$= 11$

FACTS FROM GEOMETRY

Formulas for Area and Perimeter

A square, rectangle, and triangle are shown in the following figures. Note that we have labeled the dimensions of each with variables. The formulas for the perimeter and area of each object are given in terms of its dimensions.

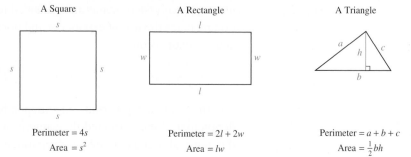

A Square	A Rectangle	A Triangle
Perimeter $= 4s$	Perimeter $= 2l + 2w$	Perimeter $= a + b + c$
Area $= s^2$	Area $= lw$	Area $= \frac{1}{2}bh$

The formula for perimeter gives us the distance around the outside of the object along its sides, whereas the formula for area gives us a measure of the amount of surface the object has.

EXAMPLE 20   Find the perimeter and area of each figure.

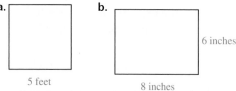

a.   5 feet

b.   6 inches   8 inches

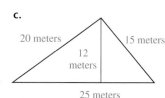

c.   20 meters   15 meters   12 meters   25 meters

SOLUTION   We use the preceding formulas to find the perimeter and the area. In each case, the units for perimeter are linear units, whereas the units for area are square units.

a.   Perimeter $= 4s = 4 \cdot 5$ feet $= 20$ feet
  Area $= s^2 = (5 \text{ feet})^2 = 25$ square feet

b.   Perimeter $= 2l + 2w = 2(8 \text{ inches}) + 2(6 \text{ inches}) = 28$ inches
  Area $= lw = (8 \text{ inches})(6 \text{ inches}) = 48$ square inches

c.   Perimeter $= a + b + c = (20 \text{ meters}) + (25 \text{ meters}) + (15 \text{ meters})$
    $= 60$ meters

  Area $= \frac{1}{2}bh = \frac{1}{2}(25 \text{ meters})(12 \text{ meters}) = 150$ square meters

GETTING READY FOR CLASS

After reading through the preceding section, respond in your own words and in complete sentences.

1. What is a real number?
2. Explain multiplication with fractions.
3. How do you find the opposite of a number?
4. Explain how you find the perimeter and the area of a rectangle.

Draw a number line that extends from -5 to $+5$. Label the points with the following coordinates.

1. 5

2. -2

3. -4

4. -3

5. 1.5

6. -1.5

7. $\dfrac{9}{4}$

8. $\dfrac{8}{3}$

Write each of the following fractions as an equivalent fraction with denominator 24.

9. $\dfrac{3}{4}$

10. $\dfrac{5}{6}$

11. $\dfrac{1}{2}$

12. $\dfrac{1}{8}$

13. $\dfrac{5}{8}$

14. $\dfrac{7}{12}$

Write each fraction as an equivalent fraction with denominator 60.

15. $\dfrac{3}{5}$

16. $\dfrac{5}{12}$

17. $\dfrac{11}{30}$

18. $\dfrac{9}{10}$

Fill in the missing numerator so the fractions are equal.

19. $\dfrac{1}{2} = \dfrac{}{4}$

20. $\dfrac{1}{5} = \dfrac{}{10}$

21. $\dfrac{5}{9} = \dfrac{}{45}$

22. $\dfrac{2}{5} = \dfrac{}{45}$

23. $\dfrac{3}{4} = \dfrac{}{8}$

24. $\dfrac{1}{2} = \dfrac{}{8}$

For each of the following numbers, give the opposite, the reciprocal, and the absolute value. (Assume all variables are nonzero.)

25. 10

26. $\dfrac{3}{4}$

27. -3

28. $-\dfrac{2}{5}$

29. x

30. a

Place one of the symbols $<$ or $>$ between each of the following to make the resulting statement true.

31. $-5 \quad -3$

32. $-8 \quad -1$

33. $-3 \quad -7$

34. $-6 \quad 5$

35. $|-4| \quad -|-4|$

36. $3 \quad -|-3|$

37. $7 \quad -|-7|$

38. $-7 \quad |-7|$

39. $-\dfrac{3}{4} \quad -\dfrac{1}{4}$

40. $-\dfrac{2}{3} \quad -\dfrac{1}{3}$

41. $-\dfrac{3}{2} \quad -\dfrac{3}{4}$

42. $-\dfrac{8}{3} \quad -\dfrac{17}{3}$

Simplify each expression.

43. $|8 - 2|$

44. $|6 - 1|$

45. $|5 \cdot 2^3 - 2 \cdot 3^2|$

46. $|2 \cdot 10^2 + 3 \cdot 10|$

47. $|7 - 2| - |4 - 2|$

48. $|10 - 3| - |4 - 1|$

49. $10 - |7 - 2(5 - 3)|$

50. $12 - |9 - 3(7 - 5)|$

51. $15 - |8 - 2(3 \cdot 4 - 9)| - 10$

52. $25 - |9 - 3(4 \cdot 5 - 18)| - 20$

Multiply the following.

53. $\dfrac{2}{3} \cdot \dfrac{4}{5}$

54. $\dfrac{1}{4} \cdot \dfrac{3}{5}$

55. $\dfrac{1}{2}(3)$

56. $\dfrac{1}{5}(4)$

57. $\dfrac{4}{3} \cdot \dfrac{3}{4}$

58. $\dfrac{5}{7} \cdot \dfrac{7}{5}$

59. $3 \cdot \dfrac{1}{3}$

60. $4 \cdot \dfrac{1}{4}$

61. Multiply.

   a. $\dfrac{1}{2}(4)$

   b. $\dfrac{1}{2}(8)$

   c. $\dfrac{1}{2}(16)$

   d. $\dfrac{1}{2}(0.06)$

62. Multiply.

   a. $\dfrac{1}{4}(8)$

   b. $\dfrac{1}{4}(24)$

   c. $\dfrac{1}{4}(16)$

   d. $\dfrac{1}{4}(0.20)$

63. Multiply.

   a. $\dfrac{3}{2}(4)$

   b. $\dfrac{3}{2}(8)$

   c. $\dfrac{3}{2}(16)$

   d. $\dfrac{3}{2}(0.06)$

64. Multiply.

   a. $\dfrac{3}{4}(8)$

   b. $\dfrac{3}{4}(24)$

   c. $\dfrac{3}{4}(16)$

   d. $\dfrac{3}{4}(0.20)$

= Videos available by instructor request

▶ = Online student support materials available at academic.cengage.com/login

Expand and multiply.

65. $\left(\dfrac{3}{4}\right)^2$ **66.** $\left(\dfrac{5}{6}\right)^2$

67. $\left(\dfrac{2}{3}\right)^3$ **68.** $\left(\dfrac{1}{2}\right)^3$

69. Find the value of $2x - 6$ when
   a. $x = 5$
   b. $x = 10$
   c. $x = 15$
   d. $x = 20$

70. Find the value of $2(x - 3)$ when
   a. $x = 5$
   b. $x = 10$
   c. $x = 15$
   d. $x = 20$

71. Find the value of each expression when x is 10.
   a. $x + 2$
   b. $2x$
   c. x^2
   d. 2^x

72. Find the value of each expression when x is 3.
   a. $x + 3$
   b. $3x$
   c. x^2
   d. 3^x

73. Find the value of each expression when x is 4.
   a. $x^2 + 1$
   b. $(x + 1)^2$
   c. $x^2 + 2x + 1$

74. Find the value of $b^2 - 4ac$ when
   a. $a = 2, b = 6, c = 3$
   b. $a = 1, b = 5, c = 6$
   c. $a = 1, b = 2, c = 1$

Find the perimeter and area of each figure.

75.

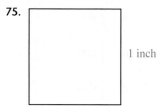

1 inch

1 inch

76.

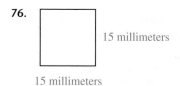

15 millimeters

15 millimeters

77.

0.75 inches

1.5 inches

78.

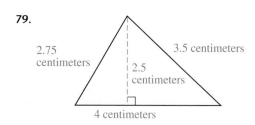

1.5 centimeters

4.5 centimeters

79.

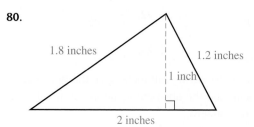

2.75 centimeters   3.5 centimeters

2.5 centimeters

4 centimeters

80.

1.8 inches   1.2 inches

1 inch

2 inches

Applying the Concepts

81. Football Yardage A football team gains 6 yards on one play and then loses 8 yards on the next play. To what number on the number line does a loss of 8 yards correspond? The total yards gained or lost on the two plays corresponds to what negative number?

82. Checking Account Balance A woman has a balance of $20 in her checking account. If she writes a check for $30, what negative number can be used to represent the new balance in her checking account?

Temperature In the United States, temperature is measured on the Fahrenheit temperature scale. On this scale, water boils at 212 degrees and freezes at 32 degrees. To denote a temperature of 32 degrees on the Fahrenheit scale, we write

32°F, which is read "32 degrees Fahrenheit"

Use this information for Problems 83 and 84.

83. Temperature and Altitude Marilyn is flying from Seattle to San Francisco on a Boeing 737 jet. When the plane reaches an altitude of 35,000 feet, the temperature outside the plane is 64 degrees below zero Fahrenheit. Represent the temperature with a negative number. If the temperature outside the plane gets warmer by 10 degrees, what will the new temperature be?

84. Temperature Change At 10:00 in the morning in White Bear Lake, Minnesota, John notices the temperature outside is 10 degrees below zero Fahrenheit. Write the temperature as a negative number. An hour later it has warmed up by 6 degrees. What is the temperature at 11:00 that morning?

85. Scuba Diving Steve is scuba diving near his home in Maui. At one point he is 100 feet below the surface. Represent this number with a negative number. If he descends another 5 feet, what negative number will represent his new position?

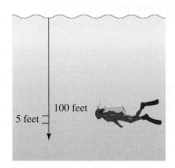

86. Reading a Chart The chart shows yields for certificates of deposit during one week in 2005. Write a mathematical statement using one of the symbols < or > to compare the following:

   a. 6-month yield a year ago to 1-year yield last week

   b. 2½-year yield this week to 5-year yield a year ago

   c. 5-year yield last week to 6-month yield this week

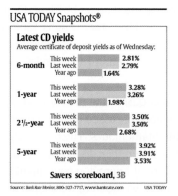

From *USA Today.* Copyright 2005. Reprinted with permission.

87. Geometry Find the area and perimeter of an $8\frac{1}{2}$-by-11-inch piece of notebook paper.

88. Geometry Find the area and perimeter of an $8\frac{1}{2}$-by-$5\frac{1}{2}$-inch piece of paper.

Calories and Exercise Table 3 gives the amount of energy expended per hour for various activities for a person weighing 120, 150, or 180 pounds. Use Table 3 to answer questions 89–92.

TABLE 3
Energy Expended from Exercising

Activity	Calories per Hour 120 lb	150 lb	180 lb
Bicycling	299	374	449
Bowling	212	265	318
Handball	544	680	816
Horseback trotting	278	347	416
Jazzercise	272	340	408
Jogging	544	680	816
Skiing (downhill)	435	544	653

89. Suppose you weigh 120 pounds. How many calories will you burn if you play handball for 2 hours and then ride your bicycle for an hour?

90. How many calories are burned by a person weighing 150 pounds who jogs for $\frac{1}{2}$ hour and then goes bicycling for 2 hours?

91. Two people go skiing. One weighs 180 pounds and the other weighs 120 pounds. If they ski for 3 hours, how many more calories are burned by the person weighing 180 pounds?

92. Two people spend 3 hours bowling. If one weighs 120 pounds and the other weighs 150 pounds, how many more calories are burned during the evening by the person weighing 150 pounds?

93. Improving Your Quantitative Literacy Quantitative literacy is a subject discussed by many people involved in teaching mathematics. The person they are concerned with when they discuss it is you. We are going to work at improving your quantitative literacy; but before we do that, we should answer the question: What is quantitative literacy? Lynn Arthur Steen, a noted mathematics educator, has stated that quantitative literacy is "the capacity to deal effectively with the quantitative aspects of life."
   a. Give a definition for the word *quantitative*.
   b. Give a definition for the word *literacy*.
   c. Are there situations that occur in your life that you find distasteful, or that you try to avoid, because they involve numbers and mathematics? If so, list some of them here. (For example, some people find the process of buying a car particu-

larly difficult because they feel that the numbers and details of the financing are beyond them.)

94. Improving Your Quantitative Literacy Use the chart shown here to answer the following questions.
   a. How many millions of camera phones were sold in 2004?
   b. True or false? The chart shows projected sales in 2005 to be more than 155 million camera phones.
   c. True or false? The chart shows projected sales in 2007 to be less than 310 million camera phones.

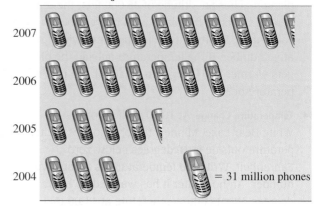

Camera Phone Growth:
Projected sales in millions of units

= 31 million phones

1.3 Addition of Real Numbers

OBJECTIVES

A Add any combination of positive and negative numbers.

B Simplify expressions using the rule for order of operations.

C Extend an arithmetic sequence.

Suppose that you are playing a friendly game of poker with some friends, and you lose $3 on the first hand and $4 on the second hand. If you represent winning with positive numbers and losing with negative numbers, how can you translate this situation into symbols? Because you lost $3 and $4 for a total of $7, one way to represent this situation is with addition of negative numbers:

$$(-\$3) + (-\$4) = -\$7$$

From this equation, we see that the sum of two negative numbers is a negative number. To generalize addition with positive and negative numbers, we use the number line.

Because real numbers have both a distance from zero (absolute value) and a direction from zero (sign), we can think of addition of two numbers in terms of distance and direction from zero.

Let's look at a problem for which we know the answer. Suppose we want to add the numbers 3 and 4. The problem is written 3 + 4. To put it on the number line, we read the problem as follows:

1. The 3 tells us to "start at the origin and move 3 units in the positive direction."
2. The + sign is read "and then move."
3. The 4 means "4 units in the positive direction."

To summarize, 3 + 4 means to start at the origin, move 3 units in the *positive* direction, and then move 4 units in the *positive* direction.

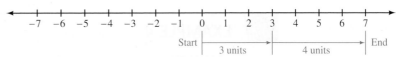

We end up at 7, which is the answer to our problem: 3 + 4 = 7.

Let's try other combinations of positive and negative 3 and 4 on the number line.

 EXAMPLE 1   Add 3 + (−4).

SOLUTION   Starting at the origin, move 3 units in the *positive* direction and then 4 units in the *negative* direction.

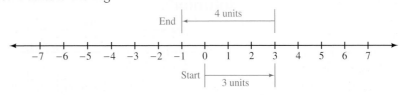

We end up at −1; therefore, 3 + (−4) = −1.

 EXAMPLE 2   Add −3 + 4.

SOLUTION   Starting at the origin, move 3 units in the *negative* direction and then 4 units in the *positive* direction.

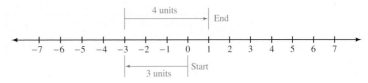

We end up at +1; therefore, −3 + 4 = 1.

 EXAMPLE 3   Add −3 + (−4).

SOLUTION   Starting at the origin, move 3 units in the *negative* direction and then 4 units in the *negative* direction.

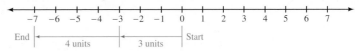

We end up at −7; therefore, −3 + (−4) = −7.

Here is a summary of what we have just completed:

$$3 + 4 = 7$$
$$3 + (-4) = -1$$
$$-3 + 4 = 1$$
$$-3 + (-4) = -7$$

Let's do four more problems on the number line and then summarize our results into a rule we can use to add any two real numbers.

 EXAMPLE 4   Show that $5 + 7 = 12$.

SOLUTION

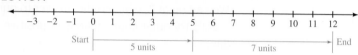

 EXAMPLE 5   Show that $5 + (-7) = -2$.

SOLUTION

 EXAMPLE 6   Show that $-5 + 7 = 2$.

SOLUTION

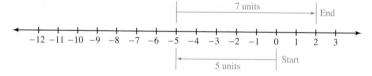

EXAMPLE 7   Show that $-5 + (-7) = -12$.

SOLUTION

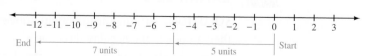

If we look closely at the results of the preceding addition problems, we can see that they support (or justify) the following rule.

1.4 Subtraction of Real Numbers

OBJECTIVES

A Subtract any combination of positive and negative numbers.

B Simplify expressions using the rule for order of operations.

C Translate sentences from English into symbols and then simplify.

D Find the complement and the supplement of an angle.

Suppose that the temperature at noon is 20° Fahrenheit and 12 hours later, at midnight, it has dropped to −15° Fahrenheit. What is the difference between the temperature at noon and the temperature at midnight? Intuitively, we know the difference in the two temperatures is 35°. We also know that the word difference indicates subtraction. The difference between 20 and −15 is written

$$20 - (-15)$$

It must be true that $20 - (-15) = 35$. In this section we will see how our definition for subtraction confirms that this last statement is in fact correct.

In the previous section we spent some time developing the rule for addition of real numbers. Because we want to make as few rules as possible, we can define subtraction in terms of addition. By doing so, we can then use the rule for addition to solve our subtraction problems.

Rule To subtract one real number from another, simply add its opposite.

Algebraically, the rule is written like this: If a and b represent two real numbers, then it is always true that

$$\underbrace{a - b}_{\text{To subtract } b} \quad = \quad \underbrace{a + (-b)}_{\text{add the opposite of } b}$$

This is how subtraction is defined in algebra. This definition of subtraction will not conflict with what you already know about subtraction, but it will allow you to do subtraction using negative numbers.

 EXAMPLE 1 Subtract all possible combinations of positive and negative 7 and 2.

SOLUTION

$$\left.\begin{array}{l} 7 - 2 = 7 + (-2) = 5 \\ -7 - 2 = -7 + (-2) = -9 \end{array}\right\} \quad \begin{array}{l}\textbf{Subtracting 2 is the same}\\ \textbf{as adding } -\textbf{2}\end{array}$$

$$\left.\begin{array}{l} 7 - (-2) = 7 + 2 = 9 \\ -7 - (-2) = -7 + 2 = -5 \end{array}\right\} \quad \begin{array}{l}\textbf{Subtracting } -\textbf{2 is the same}\\ \textbf{as adding 2}\end{array}$$

Notice that each subtraction problem is first changed to an addition problem. The rule for addition is then used to arrive at the answer.

We have defined subtraction in terms of addition, and we still obtain answers consistent with the answers we are used to getting with subtraction. Moreover, we now can do subtraction problems involving both positive and negative numbers.

As you proceed through the following examples and the problem set, you will begin to notice shortcuts you can use in working the problems. You will not always have to change subtraction to addition of the opposite to be able to get answers quickly. Use all the shortcuts you wish as long as you consistently get the correct answers.

EXAMPLE 2   Subtract all combinations of positive and negative 8 and 13.

SOLUTION

$$8 - 13 = 8 + (-13) = -5$$ **Subtracting +13 is the**
$$-8 - 13 = -8 + (-13) = -21$$ **same as adding −13**

$$8 - (-13) = 8 + 13 = 21$$ **Subtracting −13 is the**
$$-8 - (-13) = -8 + 13 = 5$$ **same as adding +13**

 EXAMPLES   Simplify each expression as much as possible.

3. $7 + (-3) - 5 = 7 + (-3) + (-5)$ **Begin by changing all**
$\qquad\qquad\quad = 4 + (-5)$ **subtractions to additions**
$\qquad\qquad\quad = -1$ **Then add left to right**

4. $8 - (-2) - 6 = 8 + 2 + (-6)$ **Begin by changing all**
$\qquad\qquad\quad = 10 + (-6)$ **subtractions to additions**
$\qquad\qquad\quad = 4$ **Then add left to right**

5. $-2 - (-3 + 1) - 5 = -2 - (-2) - 5$ **Do what is in the**
$\qquad\qquad\qquad\quad = -2 + 2 + (-5)$ **parentheses first**
$\qquad\qquad\qquad\quad = -5$

The next two examples involve multiplication and exponents as well as subtraction. Remember, according to the rule for order of operations, we evaluate the numbers containing exponents and multiply before we subtract.

 EXAMPLE 6   Simplify $2 \cdot 5 - 3 \cdot 8 - 4 \cdot 9$.

SOLUTION   First, we multiply left to right, and then we subtract.

$$2 \cdot 5 - 3 \cdot 8 - 4 \cdot 9 = 10 - 24 - 36$$
$$= -14 - 36$$
$$= -50$$

 EXAMPLE 7   Simplify $3 \cdot 2^3 - 2 \cdot 4^2$.

SOLUTION   We begin by evaluating each number that contains an exponent. Then we multiply before we subtract:

$$3 \cdot 2^3 - 2 \cdot 4^2 = 3 \cdot 8 - 2 \cdot 16$$
$$= 24 - 32$$
$$= -8$$

 EXAMPLE 8   Subtract 7 from −3.

SOLUTION   First, we write the problem in terms of subtraction. We then change to addition of the opposite:

$$-3 - 7 = -3 + (-7)$$
$$= -10$$

EXAMPLE 9   Subtract −5 from 2.

SOLUTION   Subtracting −5 is the same as adding +5:

$$2 - (-5) = 2 + 5$$
$$= 7$$

EXAMPLE 10   Find the difference of 9 and 2.

SOLUTION   Written in symbols, the problem looks like this:

$$9 - 2 = 7$$

The difference of 9 and 2 is 7.

EXAMPLE 11   Find the difference of 3 and −5.

SOLUTION   Subtracting −5 from 3 we have

$$3 - (-5) = 3 + 5$$
$$= 8$$

In the sport of drag racing, two cars at the starting line race to the finish line $\frac{1}{4}$ mile away. The car that crosses the finish line first wins the race.

Jim Rizzoli owns and races an alcohol dragster. On board the dragster is a computer that records data during each of Jim's races. Table 1 gives some of the data from a race Jim was in. Figure 1 gives the same information visually.

TABLE 1
Speed of a Race Car

Time in Seconds	Speed in Miles/Hour
0	0
1	72.7
2	129.9
3	162.8
4	192.2
5	212.4
6	228.1

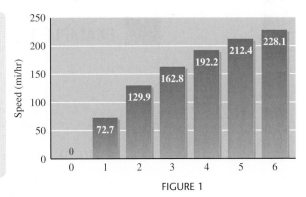

FIGURE 1

EXAMPLE 12   Use Table 1 to find the difference in speed after 5 seconds and after 2 seconds have elapsed during the race.

SOLUTION   We know the word *difference* implies subtraction. The speed at 2 seconds is 129.9 miles per hour, whereas the speed at 5 seconds is 212.4 miles per hour. Therefore, the expression that represents the solution to our problem looks like this:

$$212.4 - 129.9 = 82.5 \text{ miles per hour}$$

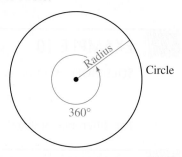

FACTS FROM GEOMETRY

Complementary and Supplementary Angles

If you have studied geometry at all, you know that there are 360° in a full rotation—the number of degrees swept out by the radius of a circle as it rotates once around the circle.

Radius

Circle

360°

We can apply our knowledge of algebra to help solve some simple geometry problems. Before we do, however, we need to review some of the vocabulary associated with angles.

Complementary angles: $x + y = 90°$

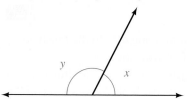

Supplementary angles: $x + y = 180°$

DEFINITION In geometry, two angles that add to 90° are called **complementary angles.** In a similar manner, two angles that add to 180° are called **supplementary angles.** The diagrams at the left illustrate the relationships between angles that are complementary and between angles that are supplementary.

EXAMPLE 13 Find x in each of the following diagrams.

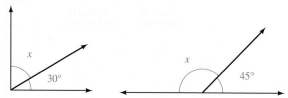

SOLUTION We use subtraction to find each angle.

a. Because the two angles are complementary, we can find x by subtracting 30° from 90°:

$x = 90° - 30° = 60°$

We say 30° and 60° are complementary angles. The complement of 30° is 60°.

b. The two angles in the diagram are supplementary. To find x, we subtract 45° from 180°:

$x = 180° - 45° = 135°$

We say 45° and 135° are supplementary angles. The supplement of 45° is 135°.

Subtraction and Taking Away

For some people taking algebra for the first time, subtraction of positive and negative numbers can be a problem. These people may believe that $-5 - 9$ should be -4 or 4, not -14. If this is happening to you, you probably are thinking of subtraction in terms of taking one number away from another. Thinking of subtraction in this way works well with positive numbers if you always subtract the smaller number from the larger. In algebra, however, we encounter many situations other than this. The definition of subtraction, that $a - b = a + (-b)$, clearly indicates the correct way to use subtraction; that is, when working subtraction problems, you should think "addition of the opposite," not "take one number away from another." To be successful in algebra, you need to apply properties and definitions exactly as they are presented here.

LINKING OBJECTIVES AND EXAMPLES

Next to each **objective** we have listed the examples that are best described by that objective.

A	1, 2
B	3–7
C	8–12
D	13

GETTING READY FOR CLASS

After reading through the preceding section, respond in your own words and in complete sentences.

1. Why do we define subtraction in terms of addition?
2. Write the definition for $a - b$.
3. Explain in words how you would subtract 3 from -7.
4. What are complementary angles?

Problem Set 1.4

Online support materials can be found at academic.cengage.com/login

The following problems are intended to give you practice with subtraction of positive and negative numbers. Remember, in algebra subtraction is not taking one number away from another. Instead, subtracting a number is equivalent to adding its opposite.

Subtract.

1. $5 - 8$
2. $6 - 7$
3. $3 - 9$
4. $2 - 7$
5. $5 - 5$
6. $8 - 8$
▶ 7. $-8 - 2$
8. $-6 - 3$
9. $-4 - 12$
10. $-3 - 15$
11. $-6 - 6$
12. $-3 - 3$
13. $-8 - (-1)$
14. $-6 - (-2)$
▶ 15. $15 - (-20)$
16. $20 - (-5)$
▶ 17. $-4 - (-4)$
18. $-5 - (-5)$

Simplify each expression by applying the rule for order of operations.

19. $3 - 2 - 5$
20. $4 - 8 - 6$
▶ 21. $9 - 2 - 3$
22. $8 - 7 - 12$
23. $-6 + 8 - 10$
24. $-5 - 7 - 9$
25. $-22 + 4 - 10$
26. $-13 + 6 - 5$
27. $10 - (-20) - 5$
28. $15 - (-3) - 20$
29. $8 - (2 - 3) - 5$
30. $10 - (4 - 6) - 8$
▶ 31. $7 - (3 - 9) - 6$
32. $4 - (3 - 7) - 8$
33. $5 - (-8 - 6) - 2$
34. $4 - (-3 - 2) - 1$
35. $-(5 - 7) - (2 - 8)$
36. $-(4 - 8) - (2 - 5)$
37. $-(3 - 10) - (6 - 3)$
38. $-(3 - 7) - (1 - 2)$
39. $16 - [(4 - 5) - 1]$
40. $15 - [(4 - 2) - 3]$
41. $5 - [(2 - 3) - 4]$
42. $6 - [(4 - 1) - 9]$
43. $21 - [-(3 - 4) - 2] - 5$
44. $30 - [-(10 - 5) - 15] - 25$

■ = Videos available by instructor request
▶ = Online student support materials available at academic.cengage.com/login

The following problems involve multiplication and exponents. Use the rule for order of operations to simplify each expression as much as possible.

45. $2 \cdot 8 - 3 \cdot 5$

46. $3 \cdot 4 - 6 \cdot 7$

47. $3 \cdot 5 - 2 \cdot 7$

48. $6 \cdot 10 - 5 \cdot 20$

49. $5 \cdot 9 - 2 \cdot 3 - 6 \cdot 2$

50. $4 \cdot 3 - 7 \cdot 1 - 9 \cdot 4$

51. $3 \cdot 8 - 2 \cdot 4 - 6 \cdot 7$

52. $5 \cdot 9 - 3 \cdot 8 - 4 \cdot 5$

53. $2 \cdot 3^2 - 5 \cdot 2^2$

54. $3 \cdot 7^2 - 2 \cdot 8^2$

55. $4 \cdot 3^3 - 5 \cdot 2^3$

56. $3 \cdot 6^2 - 2 \cdot 3^2 - 8 \cdot 6^2$

Subtract.

57. $-3.4 - 7.9$

58. $-3.5 - 2.3$

59. $3.3 - 6.9$

60. $2.2 - 7.5$

61. Find the value of $x + y - 4$ when
  a. $x = -3$ and $y = -2$
  b. $x = -9$ and $y = 3$
  c. $x = -\frac{3}{5}$ and $y = \frac{8}{5}$

62. Find the value of $x - y - 3$ when
  a. $x = -4$ and $y = 1$
  b. $x = 2$ and $y = -1$
  c. $x = -5$ and $y = -2$

Rewrite each of the following phrases as an equivalent expression in symbols, and then simplify.

63. Subtract 4 from -7.

64. Subtract 5 from -19.

65. Subtract -8 from 12.

66. Subtract -2 from 10.

67. Subtract -7 from -5.

68. Subtract -9 from -3.

69. Subtract 17 from the sum of 4 and -5.

70. Subtract -6 from the sum of 6 and -3.

Recall that the word *difference* indicates subtraction. The difference of a and b is $a - b$, in that order. Write a numerical expression that is equivalent to each of the following phrases, and then simplify.

71. The difference of 8 and 5

72. The difference of 5 and 8

73. The difference of -8 and 5

74. The difference of -5 and 8

75. The difference of 8 and -5

76. The difference of 5 and -8

Answer the following questions.

77. What number do you subtract from 8 to get -2?

78. What number do you subtract from 1 to get -5?

79. What number do you subtract from 8 to get 10?

80. What number do you subtract from 1 to get 5?

Applying the Concepts

81. **Savings Account Balance** A man with $1,500 in a savings account makes a withdrawal of $730. Write an expression using subtraction that describes this situation.

First Bank Account No. 12345			
Date	Withdrawals	Deposits	Balance
1/1/99			1,500
2/2/99	730		

82. **Temperature Change** The temperature inside a Space Shuttle is 73°F before reentry. During reentry the temperature inside the craft increases 10°. On landing it drops 8°F. Write an expression using the numbers 73, 10, and 8 to describe this situation. What is the temperature inside the shuttle on landing?

83. **Gambling** A man who has lost $35 playing roulette in Las Vegas wins $15 playing blackjack. He then loses $20 playing the wheel of fortune. Write an expression using the numbers -35, 15, and 20 to describe this situation and then simplify it.

84. **Altitude Change** An airplane flying at 10,000 feet lowers its altitude by 1,500 feet to avoid other air traffic. Then it increases its altitude by 3,000 feet to clear a mountain range. Write an expression that describes this situation and then simplify it.

85. **Checkbook Balance** Bob has $98 in his checking account when he writes a check for $65 and then another check for $53. Write a subtraction problem that gives the new balance in Bob's checkbook. What is his new balance?

86. Temperature Change The temperature at noon is 23°F. Six hours later it has dropped 19°F, and by midnight it has dropped another 10°F. Write a subtraction problem that gives the temperature at midnight. What is the temperature at midnight?

87. Depreciation Stacey buys a used car for $4,500. With each year that passes, the car drops $550 in value. Write a sequence of numbers that gives the value of the car at the beginning of each of the first 5 years she owns it. Can this sequence be considered an arithmetic sequence?

88. Depreciation Wade buys a computer system for $6,575. Each year after that he finds that the system is worth $1,250 less than it was the year before. Write a sequence of numbers that gives the value of the computer system at the beginning of each of the first four years he owns it. Can this sequence be considered an arithmetic sequence?

Drag Racing The table shown here extends the information given in Table 1 of this section. In addition to showing the time and speed of Jim Rizzoli's dragster during a race, it also shows the distance past the starting line that his dragster has traveled. Use the information in the table shown here to answer the following questions.

Speed and Distance for a Race Car

Time in Seconds	Speed in Miles/Hour	Distance Traveled in Feet
0	0	0
1	72.7	69
2	129.9	231
3	162.8	439
4	192.2	728
5	212.4	1,000
6	228.1	1,373

89. Find the difference in the distance traveled by the dragster after 5 seconds and after 2 seconds.

90. How much faster is he traveling after 4 seconds than he is after 2 seconds?

91. How far from the starting line is he after 3 seconds?

92. How far from the starting line is he when his speed is 192.2 miles per hour?

93. How many seconds have gone by between the time his speed is 162.8 miles per hour and the time at which he has traveled 1,000 feet?

94. How many seconds have gone by between the time at which he has traveled 231 feet and the time at which his speed is 228.1 miles per hour?

Find x in each of the following diagrams.

95.

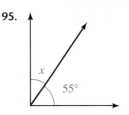

96.

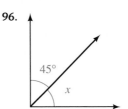

97.

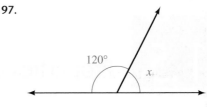

98.

99. Grass Growth The bar chart below shows the growth of a certain species of grass over a period of 10 days.
  a. Use the chart to fill in the missing entries in the table.
  b. How much higher is the grass after 8 days than after 3 days?

Day	Plant Height (inches)	Day	Plant Height (inches)
0	0	6	
1	0.5	7	
2			13
	1.5	9	18
4		10	
5	4		

Overall Plant Height

Height (in.) vs. Days after germination

Values shown on bars: 0, 0.5, 1, 1.5, 3, 4, 6, 9, 13, 18, 23

100. Improving Your Quantitative Literacy The chart shown here appeared in *USA Today* during the first week of June 2004. Use the chart to answer the following questions.

a. Do you think the numbers in the chart have been rounded? If so, to which place were they rounded?

b. How many more participants were there in 2003 than in 2000?

c. If the trend shown in the table continues, estimate how many participants there will be in 2006.

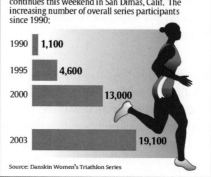

USA TODAY Snapshots

Triathlon appeals to more women
The 15th annual Danskin Women's Triathlon Series continues this weekend in San Dimas, Calif. The increasing number of overall series participants since 1990:

1990 1,100
1995 4,600
2000 13,000
2003 19,100

Source: Danskin Women's Triathlon Series

By Ellen J. Horrow and Keith Simmons, USA TODAY

1.5 Properties of Real Numbers

OBJECTIVES

A Rewrite expressions using the commutative and associative properties.

B Multiply using the distributive property.

C Identify properties used to rewrite an expression.

In this section we will list all the facts (properties) that you know from past experience are true about numbers in general. We will give each property a name so we can refer to it later in this book. Mathematics is very much like a game. The game involves numbers. The rules of the game are the properties and rules we are developing in this chapter. The goal of the game is to extend the basic rules to as many situations as possible.

You know from past experience with numbers that it makes no difference in which order you add two numbers; that is, $3 + 5$ is the same as $5 + 3$. This fact about numbers is called the *commutative property of addition*. We say addition is a commutative operation. Changing the order of the numbers does not change the answer.

There is one other basic operation that is commutative. Because $3(5)$ is the same as $5(3)$, we say multiplication is a commutative operation. Changing the order of the two numbers you are multiplying does not change the answer.

For all properties listed in this section, a, b, and c represent real numbers.

> **Commutative Property of Addition**
>
> *In symbols:* $a + b = b + a$
>
> *In words:*   Changing the *order* of the numbers in a sum will not change the result.

> **Commutative Property of Multiplication**
>
> *In symbols:* $a \cdot b = b \cdot a$
>
> *In words:* Changing the *order* of the numbers in a product will not change the result.

EXAMPLES

1. The statement $5 + 8 = 8 + 5$ is an example of the commutative property of addition.

2. The statement $2 \cdot y = y \cdot 2$ is an example of the commutative property of multiplication.

3. The expression $5 + x + 3$ can be simplified using the commutative property of addition:

$$5 + x + 3 = x + 5 + 3 \qquad \textbf{Commutative property of addition}$$
$$= x + 8 \qquad \textbf{Addition}$$

> **Note**
>
> At this point, some students are confused by the expression $x + 8$; they feel that there is more to do, but they don't know what. At this point, there isn't any more that can be done with $x + 8$ unless we know what x is. So $x + 8$ is as far as we can go with this problem.

The other two basic operations, subtraction and division, are not commutative. The order in which we subtract or divide two numbers makes a difference in the answer.

Another property of numbers that you have used many times has to do with grouping. You know that when we add three numbers it makes no difference which two we add first. When adding $3 + 5 + 7$, we can add the 3 and 5 first and then the 7, or we can add the 5 and 7 first and then the 3. Mathematically, it looks like this: $(3 + 5) + 7 = 3 + (5 + 7)$. This property is true of multiplication as well. Operations that behave in this manner are called *associative* operations. The answer will not change when we change the association (or grouping) of the numbers.

> **Associative Property of Addition**
>
> *In symbols:* $a + (b + c) = (a + b) + c$
>
> *In words:* Changing the *grouping* of the numbers in a sum will not change the result.

> **Associative Property of Multiplication**
>
> *In symbols:* $a(bc) = (ab)c$
>
> *In words:* Changing the *grouping* of the numbers in a product will not change the result.

The following examples illustrate how the associative properties can be used to simplify expressions that involve both numbers and variables.

EXAMPLES   Simplify.

4. $4 + (5 + x) = (4 + 5) + x \qquad$ **Associative property of addition**
$$= 9 + x \qquad\qquad\quad \textbf{Addition}$$

5. $5(2x) = (5 \cdot 2)x \qquad\qquad$ **Associative property of multiplication**
$$= 10x \qquad\qquad\qquad \textbf{Multiplication}$$

6. $\dfrac{1}{5}(5x) = \left(\dfrac{1}{5} \cdot 5\right)x$ **Associative property of multiplication**

$ = 1x$ **Multiplication**

$ = x$

7. $3\left(\dfrac{1}{3}x\right) = \left(3 \cdot \dfrac{1}{3}\right)x$ **Associative property of multiplication**

$ = 1x$ **Multiplication**

$ = x$

8. $12\left(\dfrac{2}{3}x\right) = \left(12 \cdot \dfrac{2}{3}\right)x$ **Associative property of multiplication**

$ = 8x$ **Multiplication**

The associative and commutative properties apply to problems that are either all multiplication or all addition. There is a third basic property that involves both addition and multiplication. It is called the *distributive property* and looks like this.

> **Distributive Property**
>
> *In symbols:* $a(b + c) = ab + ac$
>
> *In words:*   Multiplication *distributes* over addition.

Note

Because subtraction is defined in terms of addition, it is also true that the distributive property applies to subtraction as well as addition; that is, $a(b - c) = ab - ac$ for any three real numbers a, b, and c.

You will see as we progress through the book that the distributive property is used very frequently in algebra. We can give a visual justification to the distributive property by finding the areas of rectangles. Figure 1 shows a large rectangle that is made up of two smaller rectangles. We can find the area of the large rectangle two different ways.

Method 1 We can calculate the area of the large rectangle directly by finding its length and width. The width is 5 inches, and the length is $(3 + 4)$ inches.

$$\text{Area of large rectangle} = 5(3 + 4)$$
$$= 5(7)$$
$$= 35 \text{ square inches}$$

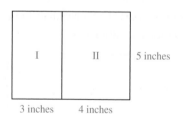

I   II   5 inches

3 inches   4 inches

FIGURE 1

Method 2 Because the area of the large rectangle is the sum of the areas of the two smaller rectangles, we find the area of each small rectangle and then add to find the area of the large rectangle.

$$\text{Area of large rectangle} = \text{Area of rectangle I} + \text{Area of rectangle II}$$
$$= 5(3) + 5(4)$$
$$= 15 + 20$$
$$= 35 \text{ square inches}$$

In both cases the result is 35 square inches. Because the results are the same, the two original expressions must be equal. Stated mathematically, $5(3 + 4) = 5(3) + 5(4)$. We can either add the 3 and 4 first and then multiply that sum by 5, or we can multiply the 3 and the 4 separately by 5 and then add the products. In either case we get the same answer.

Here are some examples that illustrate how we use the distributive property.

EXAMPLES   Apply the distributive property to each expression, and then simplify the result.

9. $2(x + 3) = 2(x) + 2(3)$ **Distributive property**

$ = 2x + 6$ **Multiplication**

10. $5(2x - 8) = 5(2x) - 5(8)$ **Distributive property**

$\qquad\qquad\quad = 10x - 40$ **Multiplication**

Notice in this example that multiplication distributes over subtraction as well as addition.

11. $4(x + y) = 4x + 4y$ **Distributive property**

12. $5(2x + 4y) = 5(2x) + 5(4y)$ **Distributive property**

$\qquad\qquad\quad = 10x + 20y$ **Multiplication**

13. $\dfrac{1}{2}(3x + 6) = \dfrac{1}{2}(3x) + \dfrac{1}{2}(6)$ **Distributive property**

$\qquad\qquad\quad = \dfrac{3}{2}x + 3$ **Multiplication**

14. $4(2a + 3) + 8 = 4(2a) + 4(3) + 8$ **Distributive property**

$\qquad\qquad\qquad = 8a + 12 + 8$ **Multiplication**

$\qquad\qquad\qquad = 8a + 20$ **Addition**

15. $a\left(1 + \dfrac{1}{a}\right) = a \cdot 1 + a \cdot \dfrac{1}{a}$ **Distributive property**

$\qquad\qquad\quad = a + 1$ **Multiplication**

16. $3\left(\dfrac{1}{3}x + 5\right) = 3 \cdot \dfrac{1}{3}x + 3 \cdot 5$ **Distributive property**

$\qquad\qquad\quad = x + 15$ **Multiplication**

17. $12\left(\dfrac{2}{3}x + \dfrac{1}{2}y\right) = 12 \cdot \dfrac{2}{3}x + 12 \cdot \dfrac{1}{2}y$ **Distributive property**

$\qquad\qquad\quad = 8x + 6y$ **Multiplication**

Special Numbers

In addition to the three properties mentioned so far, we want to include in our list two special numbers that have unique properties. They are the numbers zero and one.

Additive Identity Property There exists a unique number 0 such that

In symbols: $a + 0 = a$ and $0 + a = a$

In words: Zero preserves identities under addition. (The identity of the number is unchanged after addition with 0.)

Multiplicative Identity Property There exists a unique number 1 such that

In symbols: $a(1) = a$ and $1(a) = a$

In words: The number 1 preserves identities under multiplication. (The identity of the number is unchanged after multiplication by 1.)

Additive Inverse Property For each real number a, there exists a unique number $-a$ such that

In symbols: $a + (-a) = 0$

In words: Opposites add to 0.

> **Multiplicative Inverse Property** For every real number a, except 0, there exists a unique real number $\frac{1}{a}$ such that
>
> *In symbols:* $\quad a\left(\frac{1}{a}\right) = 1$
>
> *In words:* $\qquad$ Reciprocals multiply to 1.

Of all the basic properties listed, the commutative, associative, and distributive properties are the ones we will use most often. They are important because they will be used as justifications or reasons for many of the things we will do.

The following examples illustrate how we use the preceding properties. Each one contains an algebraic expression that has been changed in some way. The property that justifies the change is written to the right.

EXAMPLES $\quad$ State the property that justifies the given statement.

18. $x + 5 = 5 + x$ $\qquad$ **Commutative property of addition**

19. $(2 + x) + y = 2 + (x + y)$ $\qquad$ **Associative property of addition**

20. $6(x + 3) = 6x + 18$ $\qquad$ **Distributive property**

21. $2 + (-2) = 0$ $\qquad$ **Additive inverse property**

22. $3\left(\dfrac{1}{3}\right) = 1$ $\qquad$ **Multiplicative inverse property**

23. $(2 + 0) + 3 = 2 + 3$ $\qquad$ **Additive identity property**

24. $(2 + 3) + 4 = 3 + (2 + 4)$ $\qquad$ **Commutative and associative properties of addition**

25. $(x + 2) + y = (x + y) + 2$ $\qquad$ **Commutative and associative properties of addition**

As a final note on the properties of real numbers, we should mention that although some of the properties are stated for only two or three real numbers, they hold for as many numbers as needed. For example, the distributive property holds for expressions like $3(x + y + z + 5 + 2)$; that is,

$$3(x + y + z + 5 + 2) = 3x + 3y + 3z + 15 + 6$$

It is not important how many numbers are contained in the sum, only that it is a sum. Multiplication, you see, distributes over addition, whether there are two numbers in the sum or 200.

LINKING OBJECTIVES AND EXAMPLES

Next to each **objective** we have listed the examples that are best described by that objective.

A $\quad$ 1–8

B $\quad$ 9–17

C $\quad$ 18–25

GETTING READY FOR CLASS

After reading through the preceding section, respond in your own words and in complete sentences.

1. What is the commutative property of addition?
2. Do you know from your experience with numbers that the commutative property of addition is true? Explain why.
3. Write the commutative property of multiplication in symbols and words.
4. How do you rewrite expressions using the distributive property?

State the property or properties that justify the following.

1. $3 + 2 = 2 + 3$

2. $5 + 0 = 5$

3. $4\left(\dfrac{1}{4}\right) = 1$

4. $10(0.1) = 1$

5. $4 + x = x + 4$

6. $3(x - 10) = 3x - 30$

7. $2(y + 8) = 2y + 16$

8. $3 + (4 + 5) = (3 + 4) + 5$

9. $(3 + 1) + 2 = 1 + (3 + 2)$

10. $(5 + 2) + 9 = (2 + 5) + 9$

11. $(8 + 9) + 10 = (8 + 10) + 9$

12. $(7 + 6) + 5 = (5 + 6) + 7$

13. $3(x + 2) = 3(2 + x)$

14. $2(7y) = (7 \cdot 2)y$

15. $x(3y) = 3(xy)$

16. $a(5b) = 5(ab)$

17. $4(xy) = 4(yx)$

18. $3[2 + (-2)] = 3(0)$

19. $8[7 + (-7)] = 8(0)$

20. $7(1) = 7$

Each of the following problems has a mistake in it. Correct the right-hand side.

21. $3(x + 2) = 3x + 2$

22. $5(4 + x) = 4 + 5x$

23. $9(a + b) = 9a + b$

24. $2(y + 1) = 2y + 1$

25. $3(0) = 3$

26. $5\left(\dfrac{1}{5}\right) = 5$

27. $3 + (-3) = 1$

28. $8(0) = 8$

29. $10(1) = 0$

30. $3 \cdot \dfrac{1}{3} = 0$

Use the associative property to rewrite each of the following expressions, and then simplify the result. (See Examples 4, 5, and 6.)

31. $4 + (2 + x)$

32. $5 + (6 + x)$

33. $(x + 2) + 7$

34. $(x + 8) + 2$

35. $3(5x)$

36. $5(3x)$

37. $9(6y)$

38. $6(9y)$

39. $\dfrac{1}{2}(3a)$

40. $\dfrac{1}{3}(2a)$

41. $\dfrac{1}{3}(3x)$

42. $\dfrac{1}{4}(4x)$

43. $\dfrac{1}{2}(2y)$

44. $\dfrac{1}{7}(7y)$

45. $\dfrac{3}{4}\left(\dfrac{4}{3}x\right)$

46. $\dfrac{3}{2}\left(\dfrac{2}{3}x\right)$

47. $\dfrac{6}{5}\left(\dfrac{5}{6}a\right)$

48. $\dfrac{2}{5}\left(\dfrac{5}{2}a\right)$

Apply the distributive property to each of the following expressions. Simplify when possible.

49. $8(x + 2)$

50. $5(x + 3)$

51. $8(x - 2)$

52. $5(x - 3)$

53. $4(y + 1)$

54. $4(y - 1)$

55. $3(6x + 5)$

56. $3(5x + 6)$

57. $2(3a + 7)$

58. $5(3a + 2)$

59. $9(6y - 8)$

60. $2(7y - 4)$

Apply the distributive property to each of the following expressions. Simplify when possible.

61. $\dfrac{1}{2}(3x - 6)$

62. $\dfrac{1}{3}(2x - 6)$

63. $\dfrac{1}{3}(3x + 6)$

64. $\dfrac{1}{2}(2x + 4)$

65. $3(x + y)$

66. $2(x - y)$

67. $8(a - b)$

68. $7(a + b)$

69. $6(2x + 3y)$

70. $8(3x + 2y)$

71. $4(3a - 2b)$

72. $5(4a - 8b)$

73. $\dfrac{1}{2}(6x + 4y)$

74. $\dfrac{1}{3}(6x + 9y)$

75. $4(a + 4) + 9$

76. $6(a + 2) + 8$

77. $2(3x + 5) + 2$

78. $7(2x + 1) + 3$

79. $7(2x + 4) + 10$

80. $3(5x + 6) + 20$

Here are some problems you will see later in the book. Apply the distributive property and simplify, if possible.

▸ **81.** $\frac{1}{2}(4x + 2)$ ▸ **82.** $\frac{1}{3}(6x + 3)$

▸ **83.** $\frac{3}{4}(8x - 4)$ ▸ **84.** $\frac{2}{5}(5x + 10)$

▸ **85.** $\frac{5}{6}(6x + 12)$ ▸ **86.** $\frac{2}{3}(9x - 3)$

▸ **87.** $10\left(\frac{3}{5}x + \frac{1}{2}\right)$ ▸ **88.** $8\left(\frac{1}{4}x - \frac{5}{8}\right)$

89. $15\left(\frac{1}{3}x + \frac{2}{5}\right)$ **90.** $12\left(\frac{1}{12}m + \frac{1}{6}\right)$

91. $12\left(\frac{1}{2}m - \frac{5}{12}\right)$ **92.** $8\left(\frac{1}{8} + \frac{1}{2}m\right)$

93. $21\left(\frac{1}{3} + \frac{1}{7}x\right)$ **94.** $6\left(\frac{3}{2}y + \frac{1}{3}\right)$

95. $6\left(\frac{1}{2}x - \frac{1}{3}y\right)$ **96.** $12\left(\frac{1}{4}x + \frac{2}{3}y\right)$

97. $0.09(x + 2,000)$

98. $0.04(x + 7,000)$

99. $0.12(x + 500)$

100. $0.06(x + 800)$

101. $a\left(1 + \frac{1}{a}\right)$ **102.** $a\left(1 - \frac{1}{a}\right)$

103. $a\left(\frac{1}{a} - 1\right)$ **104.** $a\left(\frac{1}{a} + 1\right)$

Applying the Concepts

105. Getting Dressed While getting dressed for work, a man puts on his socks and puts on his shoes. Are the two statements "put on your socks" and "put on your shoes" commutative?

106. Getting Dressed Are the statements "put on your left shoe" and "put on your right shoe" commutative?

107. Skydiving A skydiver flying over the jump area is about to do two things: jump out of the plane and pull the rip cord. Are the two events "jump out of the plane" and "pull the rip cord" commutative? That is, will changing the order of the events always produce the same result?

108. Commutative Property Give an example of two events in your daily life that are commutative.

109. Division Give an example that shows that division is not a commutative operation; that is, find two numbers for which changing the order of division gives two different answers.

110. Subtraction Simplify the expression $10 - (5 - 2)$ and the expression $(10 - 5) - 2$ to show that subtraction is not an associative operation.

111. Take-Home Pay Jose works at a winery. His monthly salary is $2,400. To cover his taxes and retirement, the winery withholds $480 from each check. Calculate his yearly "take-home" pay using the numbers 2,400, 480, and 12. Do the calculation two different ways so that the results give further justification for the distributive property.

112. Hours Worked Carlo works as a waiter. He works double shifts 4 days a week. The lunch shift is 2 hours and the dinner shift is 3 hours. Find the total number of hours he works per week using the numbers 2, 3, and 4. Do the calculation two different ways so that the results give further justification for the distributive property.

113. College Expenses Maria is estimating her expenses for attending college for a year. Tuition is $650 per academic quarter. She estimates she will spend $225 on books each quarter. If she plans on attending 3 academic quarters during the year, how much can she expect to spend? Do the calculation two different ways so that the results give further justification for the distributive property.

114. Improving Your Quantitative Literacy Although everything you do in this course will help improve your quantitative literacy, these problems will extend the type of reasoning and thinking you are using in the classroom to situations you will find outside of the classroom. Here is what the *Mathematical*

The next examples continue the work we did previously with finding the value of an algebraic expression for given values of the variable or variables.

 EXAMPLE 27   Find the value of $-\dfrac{2}{3}x - 4$ when

a. $x = 0$ **b.** $x = 3$ **c.** $x = -\dfrac{9}{2}$

SOLUTION   Substituting the values of x into our expression one at a time we have

a. $-\dfrac{2}{3}(0) - 4 = 0 - 4 = -4$

b. $-\dfrac{2}{3}(3) - 4 = -2 - 4 = -6$

c. $-\dfrac{2}{3}\left(-\dfrac{9}{2}\right) - 4 = 3 - 4 = -1$

 EXAMPLE 28   Find the value of $5x - 4y$ when
a. $x = 4$ and $y = 0$
b. $x = 0$ and $y = -8$
c. $x = -2$ and $y = 3$

SOLUTION   We substitute the given values for x and y and then simplify

a. $5(4) - 4(0) = 20 - 0 = 20$
b. $5(0) - 4(-8) = 0 + 32 = 32$
c. $5(-2) - 4(3) = -10 - 12 = -22$

Geometric Sequences

A *geometric sequence* is a sequence of numbers in which each number (after the first number) comes from the number before it by multiplying by the same amount each time. For example, the sequence

$$2, 6, 18, 54, \ldots$$

is a geometric sequence because each number is obtained by multiplying the number before it by 3.

EXAMPLE 29   Each sequence below is a geometric sequence. Find the next number in each sequence.
a. $5, 10, 20, \ldots$ **b.** $3, -15, 75, \ldots$ **c.** $\dfrac{1}{8}, \dfrac{1}{4}, \dfrac{1}{2}, \ldots$

SOLUTION   Because each sequence is a geometric sequence, we know that each term is obtained from the previous term by multiplying by the same number each time.

a. $5, 10, 20, \ldots$: Starting with 5, each number is obtained from the previous number by multiplying by 2 each time. The next number will be $20 \cdot 2 = 40$.

b. $3, -15, 75, \ldots$: The sequence starts with 3. After that, each number is obtained by multiplying by -5 each time. The next number will be $75(-5) = -375$.

**LINKING OBJECTIVES
AND EXAMPLES**

Next to each **objective** we
have listed the examples
that are best described by
that objective.

A	1–10
B	11–15
C	16–18
D	20–22
E	23–26
F	29

c. $\frac{1}{8}, \frac{1}{4}, \frac{1}{2}, \ldots$: This sequence starts with $\frac{1}{8}$. Multiplying each number in the sequence by 2 produces the next number in the sequence. To extend the sequence, we multiply $\frac{1}{2}$ by 2: $\frac{1}{2} \cdot 2 = 1$ The next number in the sequence is 1.

GETTING READY FOR CLASS

*After reading through the preceding section, respond in your own words
and in complete sentences.*

1. How do you multiply two negative numbers?
2. How do you multiply two numbers with different signs?
3. Explain how some multiplication problems can be thought of as repeated addition.
4. What is a geometric sequence?

Problem Set 1.6

Online support materials can be found at academic.cengage.com/login

Use the rule for multiplying two real numbers to find
each of the following products.

▶ **1.** $7(-6)$ **2.** $8(-4)$

3. $-8(2)$ **4.** $-16(3)$

▶ **5.** $-3(-1)$ **6.** $-7(-1)$

7. $-11(-11)$ **8.** $-12(-12)$

Use the rule for order of operations to simplify each
expression as much as possible.

9. $-3(2)(-1)$ **10.** $-2(3)(-4)$

▶ **11.** $-3(-4)(-5)$ **12.** $-5(-6)(-7)$

13. $-2(-4)(-3)(-1)$ **14.** $-1(-3)(-2)(-1)$

15. $(-7)^2$ **16.** $(-8)^2$

17. $(-3)^3$ **18.** $(-2)^4$

19. $-2(2 - 5)$ **20.** $-3(3 - 7)$

21. $-5(8 - 10)$ **22.** $-4(6 - 12)$

▶ **23.** $(4 - 7)(6 - 9)$ **24.** $(3 - 10)(2 - 6)$

25. $(-3 - 2)(-5 - 4)$ **26.** $(-3 - 6)(-2 - 8)$

27. $-3(-6) + 4(-1)$ **28.** $-4(-5) + 8(-2)$

29. $2(3) - 3(-4) + 4(-5)$ **30.** $5(4) - 2(-1) + 5(6)$

31. $4(-3)^2 + 5(-6)^2$ **32.** $2(-5)^2 + 4(-3)^2$

33. $7(-2)^3 - 2(-3)^3$ **34.** $10(-2)^3 - 5(-2)^4$

35. $6 - 4(8 - 2)$ **36.** $7 - 2(6 - 3)$

37. $9 - 4(3 - 8)$ **38.** $8 - 5(2 - 7)$

39. $-4(3 - 8) - 6(2 - 5)$ **40.** $-8(2 - 7) - 9(3 - 5)$

41. $7 - 2[-6 - 4(-3)]$ **42.** $6 - 3[-5 - 3(-1)]$

43. $7 - 3[2(-4 - 4) - 3(-1 - 1)]$

44. $5 - 3[7(-2 - 2) - 3(-3 + 1)]$

45. Simplify each expression.
     a. $5(-4)(-3)$
     b. $5(-4) - 3$
     c. $5 - 4(-3)$
     d. $5 - 4 - 3$

46. Simplify each expression.
     a. $-2(-3)(-5)$
     b. $-2(-3) - 5$
     c. $-2 - 3(-5)$
     d. $-2 - 3 - 5$

= Videos available by instructor request

▶ = Online student support materials available at academic.cengage.com/login

Multiply the following fractions.

47. $-\dfrac{2}{3} \cdot \dfrac{5}{7}$

48. $-\dfrac{6}{5} \cdot \dfrac{2}{7}$

49. $-8\left(\dfrac{1}{2}\right)$

50. $-12\left(\dfrac{1}{3}\right)$

51. $\left(-\dfrac{3}{4}\right)^2$

52. $\left(-\dfrac{2}{5}\right)^2$

53. Simplify each expression.

   a. $\dfrac{5}{8}(24) + \dfrac{3}{7}(28)$

   b. $\dfrac{5}{8}(24) - \dfrac{3}{7}(28)$

   c. $\dfrac{5}{8}(-24) + \dfrac{3}{7}(-28)$

   d. $-\dfrac{5}{8}(24) - \dfrac{3}{7}(28)$

54. Simplify each expression.

   a. $\dfrac{5}{6}(18) + \dfrac{3}{5}(15)$

   b. $\dfrac{5}{6}(18) - \dfrac{3}{5}(15)$

   c. $\dfrac{5}{6}(-18) + \dfrac{3}{5}(-15)$

   d. $-\dfrac{5}{6}(18) - \dfrac{3}{5}(15)$

Simplify.

▶ **55.** $\left(\dfrac{1}{2} \cdot 6\right)^2$

▶ **56.** $\left(\dfrac{1}{2} \cdot 10\right)^2$

▶ **57.** $\left(\dfrac{1}{2} \cdot 5\right)^2$

▶ **58.** $\left[\dfrac{1}{2}(0.8)\right]^2$

▶ **59.** $\left[\dfrac{1}{2}(-4)\right]^2$

▶ **60.** $\left[\dfrac{1}{2}(-12)\right]^2$

▶ **61.** $\left[\dfrac{1}{2}(-3)\right]^2$

▶ **62.** $\left[\dfrac{1}{2}(-0.8)\right]^2$

Find the following products.

63. $-2(4x)$

64. $-8(7x)$

65. $-7(-6x)$

66. $-8(-9x)$

67. $-\dfrac{1}{3}(-3x)$

68. $-\dfrac{1}{5}(-5x)$

Apply the distributive property to each expression, and then simplify the result.

69. $-\dfrac{1}{2}(3x - 6)$

70. $-\dfrac{1}{4}(2x - 4)$

▶ **71.** $-3(2x - 5) - 7$

72. $-4(3x - 1) - 8$

73. $-5(3x + 4) - 10$

74. $-3(4x + 5) - 20$

75. $-4(3x + 5y)$

76. $5(5x + 4y)$

77. $-2(3x + 5y)$

78. $-2(2x - y)$

▶ **79.** $\dfrac{1}{2}(-3x + 6)$

▶ **80.** $\dfrac{1}{4}(5x - 20)$

▶ **81.** $\dfrac{1}{3}(-2x + 6)$

▶ **82.** $\dfrac{1}{5}(-4x + 20)$

83. $-\dfrac{1}{3}(-2x + 6)$

84. $-\dfrac{1}{2}(-2x + 6)$

85. $8\left(-\dfrac{1}{4}x + \dfrac{1}{8}y\right)$

86. $9\left(-\dfrac{1}{9}x + \dfrac{1}{3}y\right)$

87. Find the value of $-\dfrac{1}{3}x + 2$ when

   a. $x = 0$

   b. $x = 3$

   c. $x = -3$

88. Find the value of $-\dfrac{2}{3}x + 1$ when

   a. $x = 0$

   b. $x = 3$

   c. $x = -3$

89. Find the value of $2x + y$ when

   a. $x = 2$ and $y = -1$

   b. $x = 0$ and $y = 3$

   c. $x = \dfrac{3}{2}$ and $y = -7$

90. Find the value of $2x - 5y$ when

   a. $x = 2$ and $y = 3$

   b. $x = 0$ and $y = -2$

   c. $x = \dfrac{5}{2}$ and $y = 1$

91. Find the value of $2x^2 - 5x$ when

   a. $x = 4$

   b. $x = -\dfrac{3}{2}$

92. Find the value of $49a^2 - 16$ when

   a. $a = \dfrac{4}{7}$

   b. $a = -\dfrac{4}{7}$

93. Find the value of $y(2y + 3)$ when

   a. $y = 4$

   b. $y = -\dfrac{11}{2}$

94. Find the value of $x(13 - x)$ when
   a. $x = 5$
   b. $x = 8$

95. Five added to the product of 3 and -10 is what number?

96. If the product of -8 and -2 is decreased by 4, what number results?

97. Write an expression for twice the product of -4 and x, and then simplify it.

98. Write an expression for twice the product of -2 and $3x$, and then simplify it.

99. What number results if 8 is subtracted from the product of -9 and 2?

100. What number results if -8 is subtracted from the product of -9 and 2?

Each of the following is a geometric sequence. In each case, find the next number in the sequence.

101. $1, 2, 4, \ldots$ **102.** $1, 5, 25, \ldots$

103. $10, -20, 40, \ldots$ **104.** $10, -30, 90, \ldots$

105. $1, \dfrac{1}{2}, \dfrac{1}{4}, \ldots$ **106.** $1, \dfrac{1}{3}, \dfrac{1}{9}, \ldots$

107. $3, -6, 12, \ldots$ **108.** $-3, 6, -12, \ldots$

Applying the Concepts

109. Stock Value Suppose you own 20 shares of a stock. If the price per share drops $3, how much money have you lost?

110. Stock Value Imagine that you purchase 50 shares of a stock at a price of $18 per share. If the stock is selling for $11 a share a week after you purchased it, how much money have you lost?

111. Temperature Change The temperature is 25°F at 5:00 in the afternoon. If the temperature drops 6°F every hour after that, what is the temperature at 9:00 in the evening?

112. Temperature Change The temperature is -5°F at 6:00 in the evening. If the temperature drops 3°F every hour after that, what is the temperature at midnight?

113. Improving Your Quantitative Literacy This table appeared as part of an article on tuition costs. Although tuition costs rose considerably between 1998 and 2003, the average cost per student declined.

   a. Find the difference of student grants in 2003 and student grants in 1998.

   b. Find the difference of actual cost in 2003 and actual cost in 1998.

   c. How does the information in the table explain how students are paying less to go to college when tuition costs have increased?

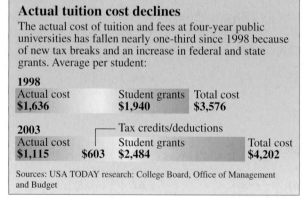

Actual tuition cost declines

The actual cost of tuition and fees at four-year public universities has fallen nearly one-third since 1998 because of new tax breaks and an increase in federal and state grants. Average per student:

1998		
Actual cost $1,636	Student grants $1,940	Total cost $3,576

2003	Tax credits/deductions		
Actual cost $1,115	$603	Student grants $2,484	Total cost $4,202

Sources: USA TODAY research: College Board, Office of Management and Budget

By Karl Geties. USA TODAY

Division of Real Numbers

OBJECTIVES

A Divide any combination of positive and negative numbers.

B Divide fractions.

C Simplify expressions using the rule for order of operations.

Suppose that you and four friends bought equal shares of an investment for a total of $15,000 and then sold it later for only $13,000. How much did each person lose? Because the total amount of money that was lost can be represented by −$2,000, and there are 5 people with equal shares, we can represent each person's loss with division:

$$\frac{-\$2,000}{5} = -\$400$$

From this discussion it seems reasonable to say that a negative number divided by a positive number is a negative number. Here is a more detailed discussion of division with positive and negative numbers.

The last of the four basic operations is division. We will use the same approach to define division as we used for subtraction; that is, we will define division in terms of rules we already know.

Recall that we developed the rule for subtraction of real numbers by defining subtraction in terms of addition. We changed our subtraction problems to addition problems and then added to get our answers. Because we already have a rule for multiplication of real numbers, and division is the inverse operation of multiplication, we will simply define division in terms of multiplication.

We know that division by the number 2 is the same as multiplication by $\frac{1}{2}$; that is, 6 divided by 2 is 3, which is the same as 6 times $\frac{1}{2}$. Similarly, dividing a number by 5 gives the same result as multiplying by $\frac{1}{5}$. We can extend this idea to all real numbers with the following rule.

Note

We are defining division this way simply so that we can use what we already know about multiplication to do division problems. We actually want as few rules as possible. Defining division in terms of multiplication allows us to avoid writing a separate rule for division.

Rule If a and b represent any two real numbers (b cannot be 0), then it is always true that

$$a \div b = \frac{a}{b} = a\left(\frac{1}{b}\right)$$

Division by a number is the same as multiplication by its reciprocal. Because every division problem can be written as a multiplication problem and because we already know the rule for multiplication of two real numbers, we do not have to write a new rule for division of real numbers. We will simply replace our division problem with multiplication and use the rule we already have.

 EXAMPLES Write each division problem as an equivalent multiplication problem, and then multiply.

1. $\dfrac{6}{2} = 6\left(\dfrac{1}{2}\right) = 3$ **The product of two positives is positive**

2. $\dfrac{6}{-2} = 6\left(-\dfrac{1}{2}\right) = -3$

The product of a positive and a negative is a negative

3. $\dfrac{-6}{2} = -6\left(\dfrac{1}{2}\right) = -3$

4. $\dfrac{-6}{-2} = -6\left(-\dfrac{1}{2}\right) = 3$ **The product of two negatives is positive**

We must be careful when we are working with expressions such as $(-5)^2$ and -5^2 that we include the negative sign with the base only when parentheses indicate we are to do so.

Unless there are parentheses to indicate otherwise, we consider the base to be only the number directly below and to the left of the exponent. If we want to include a negative sign with the base, we must use parentheses.

To simplify a more complicated expression, we follow the same rule. For example,

$$7^2 - 3^2 = 49 - 9 \qquad \textbf{The bases are 7 and 3; the sign between the two terms is a subtraction sign}$$

For another example,

$$5^3 - 3^4 = 125 - 81 \qquad \textbf{We simplify exponents first, then subtract}$$

 EXAMPLES Simplify.

20. $\dfrac{5^2 - 3^2}{-5 + 3} = \dfrac{25 - 9}{-2}$ **Simplify numerator and denominator separately**

$= \dfrac{16}{-2}$

$= -8$

21. $\dfrac{(3+2)^2}{-3^2 - 2^2} = \dfrac{5^2}{-9 - 4}$ **Simplify numerator and denominator separately**

$= \dfrac{25}{-13}$

$= -\dfrac{25}{13}$

We can combine our knowledge of the properties of multiplication with our definition of division to simplify more expressions involving fractions. Here are two examples:

 EXAMPLES Simplify each expression.

22. $10\left(\dfrac{x}{2}\right) = 10\left(\dfrac{1}{2}x\right)$ **Dividing by 2 is the same as multiplying by $\frac{1}{2}$**

$= \left(10 \cdot \dfrac{1}{2}\right)x$ **Associative property**

$= 5x$ **Multiplication**

23. $a\left(\dfrac{3}{a} - 4\right) = a \cdot \dfrac{3}{a} - a \cdot 4$ **Distributive property**

$= 3 - 4a$ **Multiplication**

Division with the Number 0

For every division problem there is an associated multiplication problem involving the same numbers. For example, the following two problems say the same thing about the numbers 2, 3, and 6:

Division   *Multiplication*

$\dfrac{6}{3} = 2$ $6 = 2(3)$

We can use this relationship between division and multiplication to clarify division involving the number 0.

First, dividing 0 by a number other than 0 is allowed and always results in 0. To see this, consider dividing 0 by 5. We know the answer is 0 because of the relationship between multiplication and division. This is how we write it:

$$\frac{0}{5} = 0 \quad \text{because} \quad 0 = 0(5)$$

However, dividing a nonzero number by 0 is not allowed in the real numbers. Suppose we were attempting to divide 5 by 0. We don't know if there is an answer to this problem, but if there is, let's say the answer is a number that we can represent with the letter n. If 5 divided by 0 is a number n, then

$$\frac{5}{0} = n \quad \text{and} \quad 5 = n(0)$$

This is impossible, however, because no matter what number n is, when we multiply it by 0 the answer must be 0. It can never be 5. In algebra, we say expressions like $\frac{5}{0}$ are undefined because there is no answer to them; that is, division by 0 is not allowed in the real numbers.

The only other possibility for division involving the number 0 is 0 divided by 0. We will treat problems like $\frac{0}{0}$ as if they were undefined also.

LINKING OBJECTIVES AND EXAMPLES

Next to each objective we have listed the examples that are best described by that objective.

A	1–11
B	12–14
C	15–21

GETTING READY FOR CLASS

After reading through the preceding section, respond in your own words and in complete sentences.

1. Why do we define division in terms of multiplication?
2. What is the reciprocal of a number?
3. How do we divide fractions?
4. Why is division by 0 not allowed with real numbers?

Problem Set 1.7

Online support materials can be found at academic.cengage.com/login

Find the following quotients (divide).

▶ 1. $\dfrac{8}{-4}$

2. $\dfrac{10}{-5}$

▶ 3. $\dfrac{-48}{16}$

4. $\dfrac{-32}{4}$

5. $\dfrac{-7}{21}$

6. $\dfrac{-25}{100}$

▶ 7. $\dfrac{-39}{-13}$

8. $\dfrac{-18}{-6}$

9. $\dfrac{-6}{-42}$

10. $\dfrac{-4}{-28}$

11. $\dfrac{0}{-32}$

12. $\dfrac{0}{17}$

The following problems review all four operations with positive and negative numbers. Perform the indicated operations.

13. $-3 + 12$

14. $5 + (-10)$

15. $-3 - 12$

16. $5 - (-10)$

 = Videos available by instructor request

▶ = Online student support materials available at academic.cengage.com/login

17. $-3(12)$　　　　**18.** $5(-10)$

19. $-3 \div 12$　　　　**20.** $5 \div (-10)$

Divide and reduce all answers to lowest terms.

21. $\dfrac{4}{5} \div \dfrac{3}{4}$　　　　**22.** $\dfrac{6}{8} \div \dfrac{3}{4}$

23. $-\dfrac{5}{6} \div \left(-\dfrac{5}{8}\right)$　　**24.** $-\dfrac{7}{9} \div \left(-\dfrac{1}{6}\right)$

25. $\dfrac{10}{13} \div \left(-\dfrac{5}{4}\right)$　　**26.** $\dfrac{5}{12} \div \left(-\dfrac{10}{3}\right)$

27. $-\dfrac{5}{6} \div \dfrac{5}{6}$　　　　**28.** $-\dfrac{8}{9} \div \dfrac{8}{9}$

▸**29.** $-\dfrac{3}{4} \div \left(-\dfrac{3}{4}\right)$　　**30.** $-\dfrac{6}{7} \div \left(-\dfrac{6}{7}\right)$

The following problems involve more than one operation. Simplify as much as possible.

31. $\dfrac{3(-2)}{-10}$　　　　**32.** $\dfrac{4(-3)}{24}$

33. $\dfrac{-5(-5)}{-15}$　　　　**34.** $\dfrac{-7(-3)}{-35}$

35. $\dfrac{-8(-7)}{-28}$　　　　**36.** $\dfrac{-3(-9)}{-6}$

37. $\dfrac{27}{4 - 13}$　　　　**38.** $\dfrac{27}{13 - 4}$

39. $\dfrac{20 - 6}{5 - 5}$　　　　**40.** $\dfrac{10 - 12}{3 - 3}$

41. $\dfrac{-3 + 9}{2 \cdot 5 - 10}$　　**42.** $\dfrac{-4 + 8}{2 \cdot 4 - 8}$

▸**43.** $\dfrac{15(-5) - 25}{2(-10)}$　　**44.** $\dfrac{10(-3) - 20}{5(-2)}$

45. $\dfrac{27 - 2(-4)}{-3(5)}$　　**46.** $\dfrac{20 - 5(-3)}{10(-3)}$

47. $\dfrac{12 - 6(-2)}{12(-2)}$　　**48.** $\dfrac{3(-4) + 5(-6)}{10 - 6}$

49. $\dfrac{5^2 - 2^2}{-5 + 2}$　　　**50.** $\dfrac{7^2 - 4^2}{-7 + 4}$

51. $\dfrac{8^2 - 2^2}{8^2 + 2^2}$　　　**52.** $\dfrac{4^2 - 6^2}{4^2 + 6^2}$

53. $\dfrac{(5 + 3)^2}{-5^2 - 3^2}$　　**54.** $\dfrac{(7 + 2)^2}{-7^2 - 2^2}$

55. $\dfrac{(8 - 4)^2}{8^2 - 4^2}$　　**56.** $\dfrac{(6 - 2)^2}{6^2 - 2^2}$

▸**57.** $\dfrac{-4 \cdot 3^2 - 5 \cdot 2^2}{-8(7)}$　　**58.** $\dfrac{-2 \cdot 5^2 + 3 \cdot 2^3}{-3(13)}$

59. $\dfrac{3 \cdot 10^2 + 4 \cdot 10 + 5}{345}$　　**60.** $\dfrac{5 \cdot 10^2 + 6 \cdot 10 + 7}{567}$

61. $\dfrac{7 - [(2 - 3) - 4]}{-1 - 2 - 3}$　　**62.** $\dfrac{2 - [(3 - 5) - 8]}{-3 - 4 - 5}$

63. $\dfrac{6(-4) - 2(5 - 8)}{-6 - 3 - 5}$　　**64.** $\dfrac{3(-4) - 5(9 - 11)}{-9 - 2 - 3}$

65. $\dfrac{3(-5 - 3) + 4(7 - 9)}{5(-2) + 3(-4)}$

66. $\dfrac{-2(6 - 10) - 3(8 - 5)}{6(-3) - 6(-2)}$

67. $\dfrac{|3 - 9|}{3 - 9}$　　　　**68.** $\dfrac{|4 - 7|}{4 - 7}$

69. $\dfrac{2 + 0.15(10)}{10}$　　**70.** $\dfrac{5(5) + 250}{640(5)}$

▸**71.** $\dfrac{1 - 3}{3 - 1}$　　　　▸**72.** $\dfrac{25 - 16}{16 - 25}$

73. Simplify.

　　a. $\dfrac{5 - 2}{3 - 1}$　　　　**b.** $\dfrac{2 - 5}{1 - 3}$

74. Simplify.

　　a. $\dfrac{6 - 2}{3 - 5}$　　　　**b.** $\dfrac{2 - 6}{5 - 3}$

75. Simplify.

　　a. $\dfrac{-4 - 1}{5 - (-2)}$　　**b.** $\dfrac{1 - (-4)}{-2 - 5}$

76. Simplify.

　　a. $\dfrac{-6 - 1}{4 - (-5)}$　　**b.** $\dfrac{1 - (-6)}{-5 - 4}$

77. Simplify each expression.

　　a. $\dfrac{3 + 2.236}{2}$

　　b. $\dfrac{3 - 2.236}{2}$

　　c. $\dfrac{3 + 2.236}{2} + \dfrac{3 - 2.236}{2}$

78. Simplify each expression.

　　a. $\dfrac{1 + 1.732}{2}$

　　b. $\dfrac{1 - 1.732}{2}$

　　c. $\dfrac{1 + 1.732}{2} + \dfrac{1 - 1.732}{2}$

79. Simplify each expression.
　　a. $20 \div 4 \cdot 5$
　　b. $-20 \div 4 \cdot 5$
　　c. $20 \div (-4) \cdot 5$
　　d. $20 \div 4(-5)$
　　e. $-20 \div 4(-5)$

80. Simplify each expression.
   a. $32 \div 8 \cdot 4$
   b. $-32 \div 8 \cdot 4$
   c. $32 \div (-8) \cdot 4$
   d. $32 \div 8(-4)$
   e. $-32 \div 8(-4)$

81. Simplify each expression.
   a. $8 \div \frac{4}{5}$
   b. $8 \div \frac{4}{5} - 10$
   c. $8 \div \frac{4}{5}(-10)$
   d. $8 \div \left(-\frac{4}{5}\right) - 10$

82. Simplify each expression.
   a. $10 \div \frac{5}{6}$
   b. $10 \div \frac{5}{6} - 12$
   c. $10 \div \frac{5}{6}(-12)$
   d. $10 \div \left(-\frac{5}{6}\right) - 12$

Apply the distributive property.

83. $10\left(\frac{x}{2} + \frac{3}{5}\right)$ **84.** $6\left(\frac{x}{3} + \frac{5}{2}\right)$

85. $15\left(\frac{x}{5} + \frac{4}{3}\right)$ **86.** $6\left(\frac{x}{3} + \frac{1}{2}\right)$

87. $x\left(\frac{3}{x} + 1\right)$ **88.** $x\left(\frac{4}{x} + 3\right)$

89. $21\left(\frac{x}{7} - \frac{y}{3}\right)$ **90.** $36\left(\frac{x}{4} - \frac{y}{9}\right)$

91. $a\left(\frac{3}{a} - \frac{2}{a}\right)$ **92.** $a\left(\frac{7}{a} + \frac{1}{a}\right)$

93. $2y\left(\frac{1}{y} - \frac{1}{2}\right)$ **94.** $5y\left(\frac{3}{y} - \frac{4}{5}\right)$

Answer the following questions.

95. What is the quotient of -12 and -4?

96. The quotient of -4 and -12 is what number?

97. What number do we divide by -5 to get 2?

98. What number do we divide by -3 to get 4?

99. Twenty-seven divided by what number is -9?

100. Fifteen divided by what number is -3?

101. If the quotient of -20 and 4 is decreased by 3, what number results?

102. If -4 is added to the quotient of 24 and -8, what number results?

Applying the Concepts

103. **Investment** Suppose that you and 3 friends bought equal shares of an investment for a total of $15,000 and then sold it later for only $13,600. How much did each person lose?

104. **Investment** If 8 people invest $500 each in a stamp collection and after a year the collection is worth $3,800, how much did each person lose?

105. **Temperature Change** Suppose that the temperature outside is dropping at a constant rate. If the temperature is 75°F at noon and drops to 61°F by 4:00 in the afternoon, by how much did the temperature change each hour?

106. **Temperature Change** In a chemistry class, a thermometer is placed in a beaker of hot water. The initial temperature of the water is 165°F. After 10 minutes the water has cooled to 72°F. If the water temperature drops at a constant rate, by how much does the water temperature change each minute?

107. **Internet Mailing Lists** A company sells products on the Internet through an email list. They predict that they sell one $50 product for every 25 people on their mailing list.
   a. What is their projected revenue if their list contains 10,000 email addresses?
   b. What is their projected revenue if their list contains 25,000 email addresses?
   c. They can purchase a list of 5,000 email addresses for $5,000. Is this a wise purchase?

108. **Internet Mailing Lists** A new band has a following on the Internet. They sell their CDs through an email list. They predict that they sell one $15 CD for every 10 people on their mailing list.
   a. What is their projected revenue if their list contains 5,000 email addresses?
   b. What is their projected revenue if their list contains 20,000 email addresses?
   c. If they need to make $45,000, how many people do they need on their email list?

1.8 Subsets of the Real Numbers

OBJECTIVES

A Perform operations with sets.

B Associate numbers with subsets of the real numbers.

C Factor whole numbers into the product of prime factors.

D Reduce fractions to lowest terms.

In Section 1.2 we introduced the real numbers and defined them as the numbers associated with points on the real number line. At that time, we said the real numbers include whole numbers, fractions, and decimals, as well as other numbers that are not as familiar to us as these numbers. In this section we take a more detailed look at the kinds of numbers that make up the set of real numbers.

Let's begin with the definition of a set, the starting point for all branches of mathematics.

> **DEFINITION** A **set** is a collection of objects or things. The objects in the set are called **elements,** or **members,** of the set.

Sets are usually denoted by capital letters, and elements of sets are denoted by lowercase letters. We use braces, { }, to enclose the elements of a set.

To show that an element is contained in a set we use the symbol $\in$. That is,

$x \in A$ is read "x is an element (member) of set A"

For example, if A is the set $\{1, 2, 3\}$, then $2 \in A$. However, $5 \notin A$ means 5 is not an element of set A.

> **DEFINITION** Set A is a **subset** of set B, written $A \subset B$, if every element in A is also an element of B. That is:
>
> $A \subset B$ if and only if A is contained in B

EXAMPLES

1. The set of numbers used to count things is $\{1, 2, 3, \dots\}$. The dots mean the set continues indefinitely in the same manner. This is an example of an *infinite set.*

2. The set of all numbers represented by the dots on the faces of a regular die is $\{1, 2, 3, 4, 5, 6\}$. This set is a subset of the set in Example 1. It is an example of a *finite set* because it has a limited number of elements.

> **DEFINITION** The set with no members is called the **empty,** or **null, set.** It is denoted by the symbol $\varnothing$. The empty set is considered a subset of every set.

The diagrams shown here are called *Venn diagrams* after John Venn (1834–1923). They can be used to visualize operations with sets. The region inside the circle labeled *A* is set *A*; the region inside the circle labeled *B* is set *B*.

$A \cup B$ $A \cap B$

FIGURE 1 **The union of two sets**

FIGURE 2 **The intersection of two sets**

Operations with Sets

Two basic operations are used to combine sets: union and intersection.

> **DEFINITION** The **union** of two sets *A* and *B*, written $A \cup B$, is the set of all elements that are either in *A* or in *B*, or in both *A* and *B*. The key word here is *or*. For an element to be in $A \cup B$ it must be in *A* or *B*. In symbols, the definition looks like this:
>
> $$x \in A \cup B \qquad \text{if and only if} \qquad x \in A \text{ or } x \in B$$

> **DEFINITION** The **intersection** of two sets *A* and *B*, written $A \cap B$, is the set of elements in both *A* and *B*. The key word in this definition is the word *and*. For an element to be in $A \cap B$ it must be in both *A* and *B*, or
>
> $$x \in A \cap B \qquad \text{if and only if} \qquad x \in A \text{ and } x \in B$$

EXAMPLES   Let $A = \{1, 3, 5\}$, $B = \{0, 2, 4\}$, and $C = \{1, 2, 3, \dots\}$. Then

3. $A \cup B = \{0, 1, 2, 3, 4, 5\}$

4. $A \cap B = \varnothing$ (*A* and *B* have no elements in common.)

5. $A \cap C = \{1, 3, 5\} = A$

6. $B \cup C = \{0, 1, 2, 3, \dots\}$

Another notation we can use to describe sets is called *set-builder* notation. Here is how we write our definition for the union of two sets *A* and *B* using set-builder notation:

$$A \cup B = \{x \mid x \in A \text{ or } x \in B\}$$

The right side of this statement is read "the set of all *x* such that *x* is a member of *A* or *x* is a member of *B*." As you can see, the vertical line after the first *x* is read "such that."

EXAMPLE 7   If $A = \{1, 2, 3, 4, 5, 6\}$, find $C = \{x \mid x \in A \text{ and } x \geq 4\}$.

SOLUTION   We are looking for all the elements of *A* that are also greater than or equal to 4. They are 4, 5, and 6. Using set notation, we have

$$C = \{4, 5, 6\}$$

The numbers that make up the set of real numbers can be classified as *counting numbers, whole numbers, integers, rational numbers,* and *irrational numbers;* each is a subset of the real numbers. Here is a detailed description of the major subsets of the real numbers.

The *counting numbers* are the numbers with which we count. They are the numbers 1, 2, 3, and so on. The notation we use to specify a group of numbers like this is *set notation.* We use the symbols { and } to enclose the members of the set.

$$\text{Counting numbers} = \{1, 2, 3, \dots\}$$

EXAMPLE 8   Which of the numbers in the following set are not counting numbers?

$$\left\{-3, 0, \frac{1}{2}, 1, 1.5, 3\right\}$$

The numbers 2 and 3 are called *prime factors* of 12 because neither of them can be factored any further.

> **DEFINITION** If *a* and *b* represent integers, then *a* is said to be a **factor** (or divisor) of *b* if *a* divides *b* evenly; that is, if *a* divides *b* with no remainder.

Note

The number 15 is not a prime number because it has factors of 3 and 5; that is, $15 = 3 \cdot 5$. When a whole number larger than 1 is not prime, it is said to be *composite*.

> **DEFINITION** A **prime number** is any positive integer larger than 1 whose only positive factors (divisors) are itself and 1.

Here is a list of the first few prime numbers.

$$\text{Prime numbers} = \{2, 3, 5, 7, 11, 13, 17, 19, 23, 29, 31, 37, 41, \ldots\}$$

When a number is not prime, we can factor it into the product of prime numbers. To factor a number into the product of primes, we simply factor it until it cannot be factored further.

EXAMPLE 11   Factor the number 60 into the product of prime numbers.

SOLUTION   We begin by writing 60 as the product of any two positive integers whose product is 60, like 6 and 10:

$$60 = 6 \cdot 10$$

We then factor these numbers:

$$60 = 6 \cdot 10$$
$$= (2 \cdot 3) \cdot (2 \cdot 5)$$
$$= 2 \cdot 2 \cdot 3 \cdot 5$$
$$= 2^2 \cdot 3 \cdot 5$$

Note

It is customary to write the prime factors in order from smallest to largest.

EXAMPLE 12   Factor the number 630 into the product of primes.

SOLUTION   Let's begin by writing 630 as the product of 63 and 10:

Note

There are some "tricks" to finding the divisors of a number. For instance, if a number ends in 0 or 5, then it is divisible by 5. If a number ends in an even number (0, 2, 4, 6, or 8), then it is divisible by 2. A number is divisible by 3 if the sum of its digits is divisible by 3. For example, 921 is divisible by 3 because the sum of its digits is $9 + 2 + 1 = 12$, which is divisible by 3.

$$630 = 63 \cdot 10$$
$$= (7 \cdot 9) \cdot (2 \cdot 5)$$
$$= 7 \cdot 3 \cdot 3 \cdot 2 \cdot 5$$
$$= 2 \cdot 3^2 \cdot 5 \cdot 7$$

It makes no difference which two numbers we start with, as long as their product is 630. We always will get the same result because a number has only one set of prime factors.

$$630 = 18 \cdot 35$$
$$= 3 \cdot 6 \cdot 5 \cdot 7$$
$$= 3 \cdot 2 \cdot 3 \cdot 5 \cdot 7$$
$$= 2 \cdot 3^2 \cdot 5 \cdot 7$$

When we have factored a number into the product of its prime factors, we not only know what prime numbers divide the original number, but we also know all of the other numbers that divide it as well. For instance, if we were to factor 210 into its prime factors, we would have $210 = 2 \cdot 3 \cdot 5 \cdot 7$, which means that 2, 3, 5, and 7 divide 210, as well as any combination of products of 2, 3, 5, and 7; that is, because 3 and 7 divide 210, then so does their product 21. Because 3, 5, and 7 each divide 210, then so does their product 105:

$$
210 = 2 \cdot 3 \cdot 5 \cdot 7
$$

21 divides 210

105 divides 210

Although there are many ways in which factoring is used in arithmetic and algebra, one simple application is in reducing fractions to lowest terms.

Recall that we reduce fractions to lowest terms by dividing the numerator and denominator by the same number. We can use the prime factorization of numbers to help us reduce fractions with large numerators and denominators.

EXAMPLE 13   Reduce $\frac{210}{231}$ to lowest terms.

SOLUTION   First we factor 210 and 231 into the product of prime factors. Then we reduce to lowest terms by dividing the numerator and denominator by any :tors they have in common.

$$
\frac{210}{231} = \frac{2 \cdot 3 \cdot 5 \cdot 7}{3 \cdot 7 \cdot 11} \quad \textbf{Factor the numerator and denominator completely}
$$

$$
= \frac{2 \cdot \cancel{3} \cdot 5 \cdot \cancel{7}}{\cancel{3} \cdot \cancel{7} \cdot 11} \quad \textbf{Divide the numerator and denominator by } 3 \cdot 7
$$

$$
= \frac{2 \cdot 5}{11}
$$

$$
= \frac{10}{11}
$$

> **Note**
> The small lines we have drawn through the factors that are common to the numerator and denominator are used to indicate that we have divided the numerator and denominator by those factors.

When we are working with fractions or with division, some of the instructions we use are equivalent; they mean the same thing. For example, each of the problems below will yield the same result:

Reduce to lowest terms: $\dfrac{50}{-80}$.

Divide: $\dfrac{50}{-80}$.

Simplify: $\dfrac{50}{-80}$.

Whether you think of the problem as a division problem, a simplification, or reducing a fraction to lowest terms, the answer will be $-\frac{5}{8}$, or -0.625, if a decimal is more appropriate for the situation. Sometimes you will see the instruction *simplify*, and sometimes you will see the instruction *reduce*. In either case, you will work the problem in the same way.

LINKING OBJECTIVES AND EXAMPLES

Next to each **objective** we have listed the examples that are best described by that objective.

A	3–7
B	8–10
C	11, 12
D	13

GETTING READY FOR CLASS

After reading through the preceding section, respond in your own words and in complete sentences.

1. What is a whole number?
2. How are factoring and multiplication related?
3. Is every integer also a rational number? Explain.
4. What is a prime number?

Problem Set 1.8

Online support materials can be found at academic.cengage.com/login

For the following problems, let $A = \{0, 2, 4, 6\}$, $B = \{1, 2, 3, 4, 5\}$, and $C = \{1, 3, 5, 7\}$.

1. $A \cup B$

2. $A \cup C$

3. $A \cap B$

4. $A \cap C$

5. $B \cap C$

6. $B \cup C$

7. $A \cup (B \cap C)$

8. $C \cup (A \cap B)$

9. $\{x \mid x \in A \text{ and } x < 4\}$

10. $\{x \mid x \in B \text{ and } x > 3\}$

11. $\{x \mid x \in A \text{ and } x \notin B\}$

12. $\{x \mid x \in B \text{ and } x \notin C\}$

13. $\{x \mid x \in A \text{ or } x \in C\}$

14. $\{x \mid x \in A \text{ or } x \in B\}$

15. $\{x \mid x \in B \text{ and } x \neq 3\}$

16. $\{x \mid x \in C \text{ and } x \neq 5\}$

Given the numbers in the set $\{-3, -2.5, 0, 1, \frac{3}{2}, \sqrt{15}\}$:

▶ **17.** List all the whole numbers.

▶ **18.** List all the integers.

▶ **19.** List all the rational numbers.

▶ **20.** List all the irrational numbers.

▶ **21.** List all the real numbers.

Given the numbers in the set $\{-10, -8, -0.333\ldots, -2, 9, \frac{25}{3}, \pi\}$:

22. List all the whole numbers.

23. List all the integers.

24. List all the rational numbers.

25. List all the irrational numbers.

26. List all the real numbers.

Identify the following statements as either true or false.

27. Every whole number is also an integer.

28. The set of whole numbers is a subset of the set of integers.

29. A number can be both rational and irrational.

30. The set of rational numbers and the set of irrational numbers have some elements in common.

31. Some whole numbers are also negative integers.

32. Every rational number is also a real number.

33. All integers are also rational numbers.

34. The set of integers is a subset of the set of rational numbers.

■ = Videos available by instructor request

▶ = Online student support materials available at academic.cengage.com/login

Label each of the following numbers as *prime* or *composite*. If a number is composite, then factor it completely.

35. 48

36. 72

37. 37

38. 23

39. 1,023

40. 543

Factor the following into the product of primes. When the number has been factored completely, write its prime factors from smallest to largest.

41. 144

42. 288

43. 38

44. 63

45. 105

46. 210

47. 180

48. 900

49. 385

50. 1,925

51. 121

52. 546

53. 420

54. 598

55. 620

56. 2,310

Reduce each fraction to lowest terms by first factoring the numerator and denominator into the product of prime factors and then dividing out any factors they have in common.

57. $\dfrac{105}{165}$

58. $\dfrac{165}{385}$

59. $\dfrac{525}{735}$

60. $\dfrac{550}{735}$

61. $\dfrac{385}{455}$

62. $\dfrac{385}{735}$

63. $\dfrac{322}{345}$

64. $\dfrac{266}{285}$

65. $\dfrac{205}{369}$

66. $\dfrac{111}{185}$

67. $\dfrac{215}{344}$

68. $\dfrac{279}{310}$

The next two problems are intended to give you practice reading, and paying attention to, the instructions that accompany the problems you are working. You will see a number of problems like this throughout the book. Working these problems is an excellent way to get ready for a test or a quiz.

69. Work each problem according to the instructions given. (Note that each of these instructions could be replaced with the instruction *Simplify*.)
   a. Add: $50 + (-80)$
   b. Subtract: $50 - (-80)$
   c. Multiply: $50(-80)$
   d. Divide: $\dfrac{50}{-80}$

70. Work each problem according to the instructions given.
   a. Add: $-2.5 + 7.5$
   b. Subtract: $-2.5 - 7.5$
   c. Multiply: $-2.5(7.5)$
   d. Divide: $\dfrac{-2.5}{7.5}$

Simplify each expression without using a calculator.

71. $\dfrac{6.28}{9(3.14)}$

72. $\dfrac{12.56}{4(3.14)}$

73. $\dfrac{9.42}{2(3.14)}$

74. $\dfrac{12.56}{2(3.14)}$

75. $\dfrac{32}{0.5}$

76. $\dfrac{16}{0.5}$

77. $\dfrac{5,599}{11}$

78. $\dfrac{840}{80}$

79. Find the value of $\dfrac{2 + 0.15x}{x}$ for each of the values of x given below. Write your answers as decimals, to the nearest hundredth.
   a. $x = 10$
   b. $x = 15$
   c. $x = 20$

80. Find the value of $\dfrac{5x + 250}{640x}$ for each of the values of x given below. Write your answers as decimals, to the nearest thousandth.
   a. $x = 10$
   b. $x = 15$
   c. $x = 20$

81. Factor 6^3 into the product of prime factors by first factoring 6 and then raising each of its factors to the third power.

82. Factor 12^2 into the product of prime factors by first factoring 12 and then raising each of its factors to the second power.

83. Factor $9^4 \cdot 16^2$ into the product of prime factors by first factoring 9 and 16 completely.

84. Factor $10^2 \cdot 12^3$ into the product of prime factors by first factoring 10 and 12 completely.

85. Simplify the expression $3 \cdot 8 + 3 \cdot 7 + 3 \cdot 5$, and then factor the result into the product of primes. (Notice one of the factors of the answer is 3.)

86. Simplify the expression $5 \cdot 4 + 5 \cdot 9 + 5 \cdot 3$, and then factor the result into the product of primes.

Recall the Fibonacci sequence we introduced earlier in this chapter.

$$\text{Fibonacci sequence} = 1, 1, 2, 3, 5, 8, \dots$$

Any number in the Fibonacci sequence is a *Fibonacci number.*

87. The Fibonacci numbers are not a subset of which of the following sets: real numbers, rational numbers, irrational numbers, whole numbers?

88. Name three Fibonacci numbers that are prime numbers.

89. Name three Fibonacci numbers that are composite numbers.

90. Is the sequence of odd numbers a subset of the Fibonacci numbers?

1.9 Addition and Subtraction with Fractions

OBJECTIVES

A Add or subtract two or more fractions with the same denominator.

B Find the least common denominator for a set of fractions.

C Add or subtract fractions with different denominators.

D Extend a sequence of numbers containing fractions.

You may recall from previous math classes that to add two fractions with the same denominator, you simply add their numerators and put the result over the common denominator:

$$\frac{3}{4} + \frac{2}{4} = \frac{3+2}{4} = \frac{5}{4}$$

The reason we add numerators but do not add denominators is that we must follow the distributive property. To see this, you first have to recall that $\frac{3}{4}$ can be written as $3 \cdot \frac{1}{4}$, and $\frac{2}{4}$ can be written as $2 \cdot \frac{1}{4}$ (dividing by 4 is equivalent to multiplying by $\frac{1}{4}$). Here is the addition problem again, this time showing the use of the distributive property:

$$\frac{3}{4} + \frac{2}{4} = 3 \cdot \frac{1}{4} + 2 \cdot \frac{1}{4}$$

$$= (3 + 2) \cdot \frac{1}{4} \qquad \textbf{Distributive property}$$

$$= 5 \cdot \frac{1}{4}$$

$$= \frac{5}{4}$$

The problems below form a comprehensive review of the material in this chapter. They can be used to study for exams. If you would like to take a practice test on this chapter, you can use the odd-numbered problems. Give yourself an hour and work as many of the odd-numbered problems as possible. When you are finished, or when an hour has passed, check your answers with the answers in the back of the book. You can use the even-numbered problems for a second practice test.

The numbers in brackets refer to the sections of the text in which similar problems can be found.

Write the numerical expression that is equivalent to each phrase, and then simplify. [1.3, 1.4, 1.6, 1.7]

1. The sum of -7 and -10

2. Five added to the sum of -7 and 4

3. The sum of -3 and 12 increased by 5

4. The difference of 4 and 9

5. The difference of 9 and -3

6. The difference of -7 and -9

7. The product of -3 and -7 decreased by 6

8. Ten added to the product of 5 and -6

9. Twice the product of -8 and $3x$

10. The quotient of -25 and -5

Simplify. [1.2]

11. $|-1.8|$

12. $-|-10|$

For each number, give the opposite and the reciprocal. [1.2]

13. 6

14. $-\dfrac{12}{5}$

Multiply. [1.2, 1.6]

15. $\dfrac{1}{2}(-10)$

16. $\left(-\dfrac{4}{5}\right)\left(\dfrac{25}{16}\right)$

Add. [1.3]

17. $-9 + 12$

18. $-18 + (-20)$

19. $-2 + (-8) + [-9 + (-6)]$

20. $(-21) + 40 + (-23) + 5$

Subtract. [1.4]

21. $6 - 9$

22. $14 - (-8)$

23. $-12 - (-8)$

24. $4 - 9 - 15$

Find the products. [1.6]

25. $(-5)(6)$

26. $4(-3)$

27. $-2(3)(4)$

28. $(-1)(-3)(-1)(-4)$

Find the following quotients. [1.7]

29. $\dfrac{12}{-3}$

30. $-\dfrac{8}{9} \div \dfrac{4}{3}$

Simplify. [1.1, 1.6, 1.7]

31. $4 \cdot 5 + 3$

32. $9 \cdot 3 + 4 \cdot 5$

33. $2^3 - 4 \cdot 3^2 + 5^2$

34. $12 - 3(2 \cdot 5 + 7) + 4$

35. $20 + 8 \div 4 + 2 \cdot 5$

36. $2(3 - 5) - (2 - 8)$

37. $30 \div 3 \cdot 2$

38. $(-2)(3) - (4)(-3) - 9$

39. $3(4 - 7)^2 - 5(3 - 8)^2$

40. $(-5 - 2)(-3 - 7)$

41. $\dfrac{4(-3)}{-6}$

42. $\dfrac{3^2 + 5^2}{(3 - 5)^2}$

43. $\dfrac{15 - 10}{6 - 6}$

44. $\dfrac{2(-7) + (-11)(-4)}{7 - (-3)}$

State the property or properties that justify the following. [1.5]

45. $9(3y) = (9 \cdot 3)y$

46. $8(1) = 8$

47. $(4 + y) + 2 = (y + 4) + 2$

48. $5 + (-5) = 0$

49. $(4 + 2) + y = (4 + y) + 2$

50. $5(w - 6) = 5w - 30$

Use the associative property to rewrite each expression, and then simplify the result. [1.5]

51. $7 + (5 + x)$

52. $4(7a)$

53. $\dfrac{1}{9}(9x)$

54. $\dfrac{4}{5}\left(\dfrac{5}{4}y\right)$

Apply the distributive property to each of the following expressions. Simplify when possible. [1.5, 1.6]

55. $7(2x + 3)$

56. $3(2a - 4)$

57. $\dfrac{1}{2}(5x - 6)$

58. $-\dfrac{1}{2}(3x - 6)$

For the set $\{\sqrt{7}, -\frac{1}{3}, 0, 5, -4.5, \frac{2}{5}, \pi, -3\}$ list all the [1.8]

59. rational numbers

60. whole numbers

61. irrational numbers

62. integers

Factor into the product of primes. [1.8]

63. 90　　　　　　　　**64.** 840

Combine. [1.9]

65. $\dfrac{18}{35} + \dfrac{13}{42}$　　　　**66.** $\dfrac{x}{6} + \dfrac{7}{12}$

Find the next number in each sequence. [1.1, 1.2, 1.3, 1.6, 1.9]

67. $10, 7, 4, 1, \ldots$　　　　**68.** $10, -30, 90, -270, \ldots$

69. $1, 1, 2, 3, 5, \ldots$　　　　**70.** $4, 6, 8, 10, \ldots$

71. $1, \dfrac{1}{2}, 0, -\dfrac{1}{2}, \ldots$　　　　**72.** $1, -\dfrac{1}{2}, \dfrac{1}{4}, -\dfrac{1}{8}, \ldots$

Chapter 1 Projects
Basic Properties and Definitions

GROUP PROJECT Binary Numbers

Students and Instructors: The end of each chapter in this book will contain a section like this one containing two projects. The group project is intended to be done in class. The research projects are to be completed outside of class. They can be done in groups or individually. In my classes, I use the research projects for extra credit. I require all research projects to be done on a word processor and to be free of spelling errors.

Number of People 2 or 3

Time Needed 10 minutes

Equipment Paper and pencil

Background Our decimal number system is a base 10 number system. We have 10 digits—0, 1, 2, 3, 4, 5, 6, 7, 8, and 9—which we use to write all the numbers in our number system. The number 10 is the first number that is written with a combination of digits. Although our number system is very useful, there are other number systems that are more appropriate for some disciplines. For example, computers and computer programmers use both the binary number system, which is base 2, and the hexadecimal number system, which is base 16. The binary number system has only digits 0 and 1, which are used to write all the other numbers. Every number in our base 10 number system can be written in the base 2 number system as well.

Procedure To become familiar with the binary number system, we first learn to count in base 2. Imagine that the odometer on your car had only 0's and 1's. Here is what the odometer would look like for the first 6 miles the car was driven.

ODOMETER READING						MILEAGE
0	0	0	0	0	0	0
0	0	0	0	0	1	1
0	0	0	0	1	0	2
0	0	0	0	1	1	3
0	0	0	1	0	0	4
0	0	0	1	0	1	5
0	0	0	1	1	0	6

Continue the table at left to show the odometer reading for the first 32 miles the car is driven. At 32 miles, the odometer should read

1	0	0	0	0	0

Sophie Germain

Cheryl Slaughter

The photograph at the left shows the street sign in Paris named for the French mathematician Sophie Germain (1776–1831). Among her contributions to mathematics is her work with prime numbers. In this chapter we had an introductory look at some of the classifications for numbers, including the prime numbers. Within the prime numbers themselves, there are still further classifications. In fact, a Sophie Germain prime is a prime number P, for which both P and $2P + 1$ are primes. For example, the prime number 2 is the first Sophie Germain prime because both 2 and $2 \cdot 2 + 1 = 5$ are prime numbers. The next Germain prime is 3 because both 3 and $2 \cdot 3 + 1 = 7$ are primes.

Sophie Germain was born on April 1, 1776, in Paris, France. She taught herself mathematics by reading the books in her father's library at home. Today she is recognized most for her work in number theory, which includes her work with prime numbers. Research the life of Sophie Germain. Write a short essay that includes information on her work with prime numbers and how her results contributed to solving Fermat's Last Theorem almost 200 years later.

Linear Equations and Inequalities

2

Jeff Greenberg/The Image Works

Just before starting work on this edition of your text, I flew to Europe for vacation. From time to time the television screens on the plane displayed statistics about the flight. At one point during the flight the temperature outside the plane was −60°F. When I returned home, I did some research and found that the relationship between temperature T and altitude A can be described with the formula

$$T = -0.0035A + 70$$

when the temperature on the ground is 70°F. The table and the line graph also describe this relationship.

Air Temperature and Altitude	
Altitude (feet)	Temperature (°F)
0	70
10,000	35
20,000	0
30,000	−35
40,000	−70

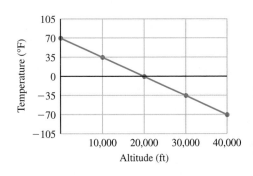

In this chapter we will start our work with formulas, and you will see how we use formulas to produce tables and line graphs like the ones above.

▶ Improve your grade and save time!
Go online to **academic.cengage.com/login** where you can
• Watch videos of instructors working through the in-text examples
• Follow step-by-step online tutorials of in-text examples and review questions
• Work practice problems
• Check your readiness for an exam by taking a pre-test and exploring the modules recommended in your Personalized Study plan
• Receive help from a live tutor online through vMentor™
Try it out! Log in with an access code or purchase access at **www.ichapters.com**.

If you have successfully completed Chapter 1, then you have made a good start at developing the study skills necessary to succeed in all math classes. Here is the list of study skills for this chapter.

1 Imitate Success

Your work should look like the work you see in this book and the work your instructor shows. The steps shown in solving problems in this book were written by someone who has been successful in mathematics. The same is true of your instructor. Your work should imitate the work of people who have been successful in mathematics.

2 List Difficult Problems

Begin to make lists of problems that give you the most difficulty. These are problems in which you are repeatedly making mistakes.

3 Begin to Develop Confidence with Word Problems

It seems that the major difference between those people who are good at working word problems and those who are not is confidence. The people with confidence know that no matter how long it takes them, they eventually will be able to solve the problem. Those without confidence begin by saying to themselves, "I'll never be able to work this problem." Are you like that? If you are, what you need to do is put your old ideas about you and word problems aside for a while and make a decision to be successful. Sometimes that's all it takes. Instead of telling yourself that you can't do word problems, that you don't like them, or that they're not good for anything anyway, decide to do whatever it takes to master them.

2.1 Simplifying Expressions

OBJECTIVES

A Simplify expressions by combining similar terms.

B Simplify expressions by applying the distributive property and then combining similar terms.

C Calculate the value of an expression for a given value of the variable.

If a cellular phone company charges $35 per month plus $0.25 for each minute, or fraction of a minute, that you use one of their cellular phones, then the amount of your monthly bill is given by the expression $35 + 0.25t$. To find the amount you will pay for using that phone 30 minutes in one month, you substitute 30 for t and simplify the resulting expression. This process is one of the topics we will study in this section.

As you will see in the next few sections, the first step in solving an equation is to simplify both sides as much as possible. In the first part of this section, we will practice simplifying expressions by combining what are called *similar* (or like) terms.

For our immediate purposes, a term is a number or a number and one or more variables multiplied together. For example, the number 5 is a term, as are the expressions $3x$, $-7y$, and $15xy$.

> **DEFINITION** Two or more terms with the same variable part are called **similar** (or **like**) terms.

The terms $3x$ and $4x$ are similar because their variable parts are identical. Likewise, the terms $18y$, $-10y$, and $6y$ are similar terms.

To simplify an algebraic expression, we simply reduce the number of terms in the expression. We accomplish this by applying the distributive property along with our knowledge of addition and subtraction of positive and negative real numbers. The following examples illustrate the procedure.

 EXAMPLES   Simplify by combining similar terms.

1. $3x + 4x = (3 + 4)x$ **Distributive property**
$\qquad\quad\ = 7x$ **Addition of 3 and 4**

2. $7a - 10a = (7 - 10)a$ **Distributive property**
$\qquad\qquad\ = -3a$ **Addition of 7 and −10**

3. $18y - 10y + 6y = (18 - 10 + 6)y$ **Distributive property**
$\qquad\qquad\qquad\ = 14y$ **Addition of 18, −10, and 6**

When the expression we intend to simplify is more complicated, we use the commutative and associative properties first.

 EXAMPLES   Simplify each expression.

4. $3x + 5 + 2x - 3 = 3x + 2x + 5 - 3$ **Commutative property**
$\qquad\qquad\qquad\quad = (3x + 2x) + (5 - 3)$ **Associative property**
$\qquad\qquad\qquad\quad = (3 + 2)x + (5 - 3)$ **Distributive property**
$\qquad\qquad\qquad\quad = 5x + 2$ **Addition**

5. $4a - 7 - 2a + 3 = (4a - 2a) + (-7 + 3)$ **Commutative and associative properties**

$\qquad\qquad\qquad\quad = (4 - 2)a + (-7 + 3)$ **Distributive property**
$\qquad\qquad\qquad\quad = 2a - 4$ **Addition**

6. $5x + 8 - x - 6 = (5x - x) + (8 - 6)$ **Commutative and associative properties**

$$= (5 - 1)x + (8 - 6)$$ **Distributive property**

$$= 4x + 2$$ **Addition**

Notice that in each case the result has fewer terms than the original expression. Because there are fewer terms, the resulting expression is said to be simpler than the original expression.

Simplifying Expressions Containing Parentheses

If an expression contains parentheses, it is often necessary to apply the distributive property to remove the parentheses before combining similar terms.

 EXAMPLE 7   Simplify the expression $5(2x - 8) - 3$.

SOLUTION   We begin by distributing the 5 across $2x - 8$. We then combine similar terms:

$$5(2x - 8) - 3 = 10x - 40 - 3$$ **Distributive property**

$$= 10x - 43$$

 EXAMPLE 8   Simplify $7 - 3(2y + 1)$.

SOLUTION   By the rule for order of operations, we must multiply before we add or subtract. For that reason, it would be incorrect to subtract 3 from 7 first. Instead, we multiply -3 and $2y + 1$ to remove the parentheses and then combine similar terms:

$$7 - 3(2y + 1) = 7 - 6y - 3$$ **Distributive property**

$$= -6y + 4$$

 EXAMPLE 9   Simplify $5(x - 2) - (3x + 4)$.

SOLUTION   We begin by applying the distributive property to remove the parentheses. The expression $-(3x + 4)$ can be thought of as $-1(3x + 4)$. Thinking of it in this way allows us to apply the distributive property:

$$-1(3x + 4) = -1(3x) + (-1)(4)$$

$$= -3x - 4$$

The complete solution looks like this:

$$5(x - 2) - (3x + 4) = 5x - 10 - 3x - 4$$ **Distributive property**

$$= 2x - 14$$ **Combine similar terms**

95. $(x + 3y) + 3(2x − y)$

96. $(2x − y) − 2(x + 3y)$

97. $3(2x + 3y) − 2(3x + 5y)$

98. $5(2x + 3y) − 3(3x + 5y)$

99. $-6\left(\dfrac{1}{2}x - \dfrac{1}{3}y\right) + 12\left(\dfrac{1}{4}x + \dfrac{2}{3}y\right)$

100. $6\left(\dfrac{1}{3}x + \dfrac{1}{2}y\right) - 4\left(x + \dfrac{3}{4}y\right)$

101. $0.08x + 0.09(x + 2{,}000)$

102. $0.06x + 0.04(x + 7{,}000)$

103. $0.10x + 0.12(x + 500)$

104. $0.08x + 0.06(x + 800)$

105. Find a so the expression $(5x + 4y) + a(2x − y)$ simplifies to an expression that does not contain y. Using that value of a, simplify the expression.

106. Find a so the expression $(5x + 4y) − a(x − 2y)$ simplifies to an expression that does not contain x. Using that value of a, simplify the expression.

Find the value of $b^2 − 4ac$ for the given values of a, b, and c. (You will see these problems later in the book.)

107. $a = 1, b = −5, c = −6$

108. $a = 1, b = −6, c = 7$

109. $a = 2, b = 4, c = −3$

110. $a = 3, b = 4, c = −2$

Applying the Concepts

111. Temperature and Altitude If the temperature on the ground is 70°F, then the temperature at A feet above the ground can be found from the expression $−0.0035A + 70$. Find the temperature at the following altitudes.
   a. 8,000 feet   **b.** 12,000 feet   **c.** 24,000 feet

Ed Curry/Corbis

112. Perimeter of a Rectangle The expression $2l + 2w$ gives the perimeter of a rectangle with length l and width w. Find the perimeter of the rectangles with the following lengths and widths.
   a. Length = 8 meters, width = 5 meters

5 meters

8 meters

   b. Length = 10 feet, width = 3 feet

3 feet

10 feet

113. Cellular Phone Rates A cellular phone company charges $35 per month plus $0.25 for each minute, or fraction of a minute, that you use one of their cellular phones. The expression $35 + 0.25t$ gives the amount of money you will pay for using one of their phones for t minutes a month. Find the monthly bill for using one of their phones.
   a. 10 minutes in a month
   b. 20 minutes in a month
   c. 30 minutes in a month

114. Cost of Bottled Water A water bottling company charges $7.00 per month for their water dispenser and $1.10 for each gallon of water delivered. If you have g gallons of water delivered in a month, then the expression $7 + 1.1g$ gives the amount of your bill for that month. Find the monthly bill for each of the following deliveries.

Cool, Refreshing Spring Water

for only **$7.00** per month and **$1.10** per gallon!

   a. 10 gallons
   b. 20 gallons
   c. 30 gallons

115. Taxes We all have to pay taxes. Suppose that 21% of your monthly pay is withheld for federal income taxes and another 8% is withheld for Social Security, state income tax, and other miscellaneous items. If G is your monthly pay before any money is deducted (your gross pay), then the amount of money that you take home each month is given by the expression $G - 0.21G - 0.08G$. Simplify this expression and then find your take-home pay if your gross pay is $1,250 per month.

116. Improving Your Quantitative Literacy The chart shows how the average cost per minute for using a cellular phone declined over a specific period of years. For each of the following years, use the chart to write an expression for the average cost to talk on a cell phone for x minutes, and then evaluate the expression to find the average cost to talk on a cell phone for an hour.

a. 1997
b. 2001
c. 2004

USA TODAY Snapshots®

Cost of cellular phone minutes continues to drop

Average cost per minute:

60¢
50¢
40¢
30¢
20¢   11¢
10¢
0¢
   96 97 98 99 00 01 02 03 04

Source: J.D. Power and Associates By Darryl Haralson and Sam Ward, USA TODAY

From *USA Today.* Copyright 2005. Reprinted with permission.

Maintaining Your Skills

From this point on, each problem set will contain a number of problems under the heading *Maintaining Your Skills.* These problems cover the most important skills you have learned in previous sections and chapters. Hopefully, by working these problems on a regular basis, you will keep yourself current on all the topics we have covered and possibly need less time to study for tests and quizzes.

117. $\dfrac{1}{8} - \dfrac{1}{6}$ **118.** $\dfrac{x}{8} - \dfrac{x}{6}$

119. $\dfrac{5}{9} - \dfrac{4}{3}$ **120.** $\dfrac{x}{9} - \dfrac{x}{3}$

121. $-\dfrac{7}{30} + \dfrac{5}{28}$ **122.** $-\dfrac{11}{105} + \dfrac{11}{30}$

Getting Ready for the Next Section

Problems under this heading, *Getting Ready for the Next Section,* are problems that you must be able to work in order to understand the material in the next section. The problems below are exactly the types of problems you will see in the explanations and examples in the next section.

Simplify.

123. $17 - 5$ **124.** $12 + (-2)$

125. $2 - 5$ **126.** $25 - 20$

127. $-2.4 + (-7.3)$ **128.** $8.1 + 2.7$

129. $-\dfrac{1}{2} + \left(-\dfrac{3}{4}\right)$ **130.** $-\dfrac{1}{6} + \left(-\dfrac{2}{3}\right)$

131. $4(2 \cdot 9 - 3) - 7 \cdot 9$

132. $5(3 \cdot 45 - 4) - 14 \cdot 45$

133. $4(2a - 3) - 7a$ **134.** $5(3a - 4) - 14a$

135. $-3 - \dfrac{1}{2}$ **136.** $-5 - \dfrac{1}{3}$

137. $\dfrac{4}{5} + \dfrac{1}{10} + \dfrac{3}{8}$ **138.** $\dfrac{3}{10} + \dfrac{7}{25} + \dfrac{3}{4}$

139. Find the value of $2x - 3$ when x is 5.

140. Find the value of $3x + 4$ when x is -2.

2.2 Addition Property of Equality

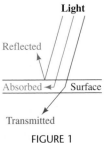

OBJECTIVES

A Check the solution to an equation by substitution.

B Use the addition property of equality to solve an equation.

When light comes into contact with any object, it is reflected, absorbed, and transmitted, as shown in Figure 1.

For a certain type of glass, 88% of the light hitting the glass is transmitted through to the other side, whereas 6% of the light is absorbed into the glass. To find the percent of light that is reflected by the glass, we can solve the equation

$$88 + R + 6 = 100$$

Light

Reflected

Absorbed — Surface

Transmitted

FIGURE 1

Solving equations of this type is what we study in this section. To solve an equation we must find all replacements for the variable that make the equation a true statement.

> **DEFINITION** The **solution set** for an equation is the set of all numbers that when used in place of the variable make the equation a true statement.

For example, the equation $x + 2 = 5$ has solution set $\{3\}$ because when x is 3 the equation becomes the true statement $3 + 2 = 5$, or $5 = 5$.

 EXAMPLE 1   Is 5 a solution to $2x - 3 = 7$?

SOLUTION   We substitute 5 for x in the equation, and then simplify to see if a true statement results. A true statement means we have a solution; a false statement indicates the number we are using is not a solution.

$$
\begin{aligned}
\text{When} \qquad & x = 5 \\
\text{the equation} \quad & 2x - 3 = 7 \\
\text{becomes} \quad & 2(5) - 3 \overset{?}{=} 7 \\
& 10 - 3 \overset{?}{=} 7 \\
& 7 = 7 \qquad \textbf{A true statement}
\end{aligned}
$$

> **Note**
> We can use a question mark over the equal signs to show that we don't know yet whether the two sides of the equation are equal.

Because $x = 5$ turns the equation into the true statement $7 = 7$, we know 5 is a solution to the equation.

EXAMPLE 2   Is -2 a solution to $8 = 3x + 4$?

SOLUTION   Substituting -2 for x in the equation, we have

$$
\begin{aligned}
8 &\overset{?}{=} 3(-2) + 4 \\
8 &\overset{?}{=} -6 + 4 \\
8 &= -2 \qquad \textbf{A false statement}
\end{aligned}
$$

Substituting -2 for x in the equation produces a false statement. Therefore, $x = -2$ is not a solution to the equation.

The important thing about an equation is its solution set. We therefore make the following definition to classify together all equations with the same solution set.

> **DEFINITION** Two or more equations with the same solution set are said to be **equivalent equations.**

Equivalent equations may look different but must have the same solution set.

 EXAMPLE 3

a. $x + 2 = 5$ and $x = 3$ are equivalent equations because both have solution set {3}.

b. $a - 4 = 3$, $a - 2 = 5$, and $a = 7$ are equivalent equations because they all have solution set {7}.

c. $y + 3 = 4$, $y - 8 = -7$, and $y = 1$ are equivalent equations because they all have solution set {1}.

If two numbers are equal and we increase (or decrease) both of them by the same amount, the resulting quantities are also equal. We can apply this concept to equations. Adding the same amount to both sides of an equation always produces an equivalent equation—one with the same solution set. This fact about equations is called the *addition property of equality* and can be stated more formally as follows.

Note

We will use this property many times in the future. Be sure you understand it completely by the time you finish this section.

> **Addition Property of Equality** For any three algebraic expressions A, B, and C,
>
> $$\text{if} \qquad A = B$$
> $$\text{then} \quad A + C = B + C$$
>
> *In words:* Adding the same quantity to both sides of an equation will not change the solution set.

This property is just as simple as it seems. We can add any amount to both sides of an equation and always be sure we have not changed the solution set.

Consider the equation $x + 6 = 5$. We want to solve this equation for the value of x that makes it a true statement. We want to end up with x on one side of the equal sign and a number on the other side. Because we want x by itself, we will add -6 to both sides:

$$x + 6 + (\mathbf{-6}) = 5 + (\mathbf{-6}) \qquad \textbf{Addition property of equality}$$
$$x + 0 = -1 \qquad \textbf{Addition}$$
$$x = -1$$

All three equations say the same thing about x. They all say that x is -1. All three equations are equivalent. The last one is just easier to read.

Here are some further examples of how the addition property of equality can be used to solve equations.

 EXAMPLE 4 Solve the equation $x - 5 = 12$ for x.

SOLUTION Because we want x alone on the left side, we choose to add $+5$ to both sides:

$$x - 5 + \mathbf{5} = 12 + \mathbf{5} \qquad \textbf{Addition property of equality}$$
$$x + 0 = 17$$
$$x = 17$$

To check our solution, we substitute 17 for x in the original equation:

When $\qquad\qquad x = 17$

the equation $\qquad x - 5 = 12$

becomes $\qquad\quad 17 - 5 \overset{?}{=} 12$

$\qquad\qquad\qquad\quad 12 = 12 \qquad$ **A true statement**

As you can see, our solution checks. The purpose for checking a solution to an equation is to catch any mistakes we may have made in the process of solving the equation.

EXAMPLE 5   Solve for a: $a + \frac{3}{4} = -\frac{1}{2}$.

SOLUTION Because we want a by itself on the left side of the equal sign, we add the opposite of $\frac{3}{4}$ to each side of the equation.

$$a + \frac{3}{4} + \left(-\frac{\mathbf{3}}{\mathbf{4}}\right) = -\frac{1}{2} + \left(-\frac{\mathbf{3}}{\mathbf{4}}\right) \qquad \text{\textbf{Addition property of equality}}$$

$$a + 0 = -\frac{1}{2} \cdot \frac{\mathbf{2}}{\mathbf{2}} + \left(-\frac{3}{4}\right) \qquad \text{\textbf{LCD on the right side is 4}}$$

$$a = -\frac{2}{4} + \left(-\frac{3}{4}\right) \qquad \text{$\frac{2}{4}$ \textbf{is equivalent to} $\frac{1}{2}$}$$

$$a = -\frac{5}{4} \qquad\qquad\qquad \text{\textbf{Add fractions}}$$

The solution is $a = -\frac{5}{4}$. To check our result, we replace a with $-\frac{5}{4}$ in the original equation. The left side then becomes $-\frac{5}{4} + \frac{3}{4}$, which reduces to $-\frac{1}{2}$, so our solution checks.

EXAMPLE 6   Solve for x: $7.3 + x = -2.4$.

SOLUTION Again, we want to isolate x, so we add the opposite of 7.3 to both sides:

$$7.3 + (-\mathbf{7.3}) + x = -2.4 + (-\mathbf{7.3}) \qquad \text{\textbf{Addition property of equality}}$$

$$0 + x = -9.7$$

$$x = -9.7$$

The addition property of equality also allows us to add variable expressions to each side of an equation.

EXAMPLE 7   Solve for x: $3x - 5 = 4x$.

SOLUTION Adding $-3x$ to each side of the equation gives us our solution.

$$3x - 5 = 4x$$

$$3x + (-\mathbf{3x}) - 5 = 4x + (-\mathbf{3x}) \qquad \text{\textbf{Distributive property}}$$

$$-5 = x$$

Sometimes it is necessary to simplify each side of an equation before using the addition property of equality. The reason we simplify both sides first is that we want as few terms as possible on each side of the equation before we use the addition property of equality. The following examples illustrate this procedure.

EXAMPLE 8   Solve $4(2a - 3) - 7a = 2 - 5$.

SOLUTION   We must begin by applying the distributive property to separate terms on the left side of the equation. Following that, we combine similar terms and then apply the addition property of equality.

$4(2a - 3) - 7a = 2 - 5$	**Original equation**
$8a - 12 - 7a = 2 - 5$	**Distributive property**
$a - 12 = -3$	**Simplify each side**
$a - 12 + \mathbf{12} = -3 + \mathbf{12}$	**Add 12 to each side**
$a = 9$	**Addition**

To check our solution, we replace a with 9 in the original equation.

$$4(2 \cdot 9 - 3) - 7 \cdot 9 \stackrel{?}{=} 2 - 5$$
$$4(15) - 63 \stackrel{?}{=} -3$$
$$60 - 63 \stackrel{?}{=} -3$$

$-3 = -3$	**A true statement**

> **Note**
>
> Again, we place a question mark over the equal sign because we don't know yet whether the expressions on the left and right side of the equal sign will be equal.

We can also add a term involving a variable to both sides of an equation.

EXAMPLE 9   Solve $3x - 5 = 2x + 7$.

SOLUTION   We can solve this equation in two steps. First, we add $-2x$ to both sides of the equation. When this has been done, x appears on the left side only. Second, we add 5 to both sides:

$3x + (\mathbf{-2x}) - 5 = 2x + (\mathbf{-2x}) + 7$	**Add $-2x$ to both sides**
$x - 5 = 7$	**Simplify each side**
$x - 5 + \mathbf{5} = 7 + \mathbf{5}$	**Add 5 to both sides**
$x = 12$	**Simplify each side**

> **Note**
>
> In my experience teaching algebra, I find that students make fewer mistakes if they think in terms of addition rather than subtraction. So, you are probably better off if you continue to use the addition property just the way we have used it in the examples in this section. But, if you are curious as to whether you can subtract the same number from both sides of an equation, the answer is yes.

> **A Note on Subtraction** Although the addition property of equality is stated for addition only, we can subtract the same number from both sides of an equation as well. Because subtraction is defined as addition of the opposite, subtracting the same quantity from both sides of an equation does not change the solution.
>
> | | |
> |---|---|
> | $x + 2 = 12$ | **Original equation** |
> | $x + 2 - \mathbf{2} = 12 - \mathbf{2}$ | **Subtract 2 from each side** |
> | $x = 10$ | **Subtraction** |

LINKING OBJECTIVES AND EXAMPLES

Next to each objective we have listed the examples that are best described by that objective.

A	1–3
B	4–9

GETTING READY FOR CLASS

After reading through the preceding section, respond in your own words and in complete sentences.

1. What is a solution to an equation?
2. What are equivalent equations?
3. Explain in words the addition property of equality.
4. How do you check a solution to an equation?

Problem Set 2.2

Online support materials can be found at academic.cengage.com/login

Find the solution for the following equations. Be sure to show when you have used the addition property of equality.

1. $x - 3 = 8$

2. $x - 2 = 7$

▶ 3. $x + 2 = 6$

4. $x + 5 = 4$

5. $a + \dfrac{1}{2} = -\dfrac{1}{4}$

6. $a + \dfrac{1}{3} = -\dfrac{5}{6}$

7. $x + 2.3 = -3.5$

8. $x + 7.9 = -3.4$

9. $y + 11 = -6$

10. $y - 3 = -1$

11. $x - \dfrac{5}{8} = -\dfrac{3}{4}$

12. $x - \dfrac{2}{5} = -\dfrac{1}{10}$

13. $m - 6 = 2m$

14. $3m - 10 = 4m$

15. $6.9 + x = 3.3$

16. $7.5 + x = 2.2$

17. $5a = 4a - 7$

18. $12a = -3 + 11a$

▶ 19. $-\dfrac{5}{9} = x - \dfrac{2}{5}$

20. $-\dfrac{7}{8} = x - \dfrac{4}{5}$

Simplify both sides of the following equations as much as possible, and then solve.

▶ 21. $4x + 2 - 3x = 4 + 1$

22. $5x + 2 - 4x = 7 - 3$

23. $8a - \dfrac{1}{2} - 7a = \dfrac{3}{4} + \dfrac{1}{8}$

24. $9a - \dfrac{4}{5} - 8a = \dfrac{3}{10} - \dfrac{1}{5}$

25. $-3 - 4x + 5x = 18$

26. $10 - 3x + 4x = 20$

27. $-11x + 2 + 10x + 2x = 9$

28. $-10x + 5 - 4x + 15x = 0$

29. $-2.5 + 4.8 = 8x - 1.2 - 7x$

30. $-4.8 + 6.3 = 7x - 2.7 - 6x$

31. $2y - 10 + 3y - 4y = 18 - 6$

32. $15 - 21 = 8x + 3x - 10x$

The following equations contain parentheses. Apply the distributive property to remove the parentheses, then simplify each side before using the addition property of equality.

33. $2(x + 3) - x = 4$

34. $5(x + 1) - 4x = 2$

35. $-3(x - 4) + 4x = 3 - 7$

36. $-2(x - 5) + 3x = 4 - 9$

37. $5(2a + 1) - 9a = 8 - 6$

38. $4(2a - 1) - 7a = 9 - 5$

39. $-(x + 3) + 2x - 1 = 6$

40. $-(x - 7) + 2x - 8 = 4$

▶ 41. $4y - 3(y - 6) + 2 = 8$

42. $7y - 6(y - 1) + 3 = 9$

43. $-3(2m - 9) + 7(m - 4) = 12 - 9$

44. $-5(m - 3) + 2(3m + 1) = 15 - 8$

= Videos available by instructor request

▶ = Online student support materials available at academic.cengage.com/login

Solve the following equations by the method used in Example 9 in this section. Check each solution in the original equation.

45. $4x = 3x + 2$ **46.** $6x = 5x - 4$

▶ **47.** $8a = 7a - 5$ **48.** $9a = 8a - 3$

49. $2x = 3x + 1$ **50.** $4x = 3x + 5$

51. $2y + 1 = 3y + 4$ **52.** $4y + 2 = 5y + 6$

53. $2m - 3 = m + 5$ **54.** $8m - 1 = 7m - 3$

55. $4x - 7 = 5x + 1$ **56.** $3x - 7 = 4x - 6$

57. $4x + \dfrac{4}{3} = 5x - \dfrac{2}{3}$ **58.** $2x + \dfrac{1}{4} = 3x - \dfrac{5}{4}$

59. $8a - 7.1 = 7a + 3.9$ **60.** $10a - 4.3 = 9a + 4.7$

61. Solve each equation.
 a. $2x = 3$
 b. $2 + x = 3$
 c. $2x + 3 = 0$
 d. $2x + 3 = -5$
 e. $2x + 3 = 7x - 5$

62. Solve each equation.
 a. $5t = 10$
 b. $5 + t = 10$
 c. $5t + 10 = 0$
 d. $5t + 10 = 12$
 e. $5t + 10 = 8t + 12$

Applying the Concepts

63. Light When light comes into contact with any object, it is reflected, absorbed, and transmitted, as shown in the following figure. If T represents the percent of light transmitted, R the percent of light reflected, and A the percent of light absorbed by a surface, then the equation $T + R + A = 100$ shows one way these quantities are related.

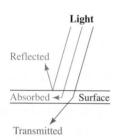

a. For glass, $T = 88$ and $A = 6$, meaning that 88% of the light hitting the glass is transmitted and 6% is absorbed. Substitute $T = 88$ and $A = 6$ into the equation $T + R + A = 100$ and solve for R to find the percent of light that is reflected.

b. For flat black paint, $A = 95$ and no light is transmitted, meaning that $T = 0$. What percent of light is reflected by flat black paint?

c. A pure white surface can reflect 98% of light, so $R = 98$. If no light is transmitted, what percent of light is absorbed by the pure white surface?

d. Typically, shiny gray metals reflect 70–80% of light. Suppose a thick sheet of aluminum absorbs 25% of light. What percent of light is reflected by this shiny gray metal? (Assume no light is transmitted.)

64. Improving Your Quantitative Literacy According to a survey done by *Seventeen* magazine in 2005, the average spending amount for girls going to the prom was $338.

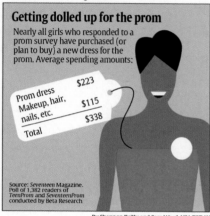

USA TODAY Snapshots®

Getting dolled up for the prom

Nearly all girls who responded to a prom survey have purchased (or plan to buy) a new dress for the prom. Average spending amounts:

Prom dress $223
Makeup, hair, nails, etc. $115
Total $338

Source: *Seventeen* Magazine. Poll of 1,382 readers of *TeenProm* and *SeventeenProm* conducted by Beta Research

By Shannon Reilly and Sam Ward, USA TODAY

From *USA Today.* Copyright 2005. Reprinted with permission.

a. Suppose Kendra spent $210 for a prom dress, then bought new shoes to go with the dress. If the total bill was $289, does the equation

$$210 + x = 289$$

describe this situation? If so, what does x represent?

b. Suppose Ava buys a prom dress on sale and then spends $120 on her makeup, nails, and hair. If the total bill is $219, and x represents a positive number, does the equation

$$x + 219 = 120$$

describe the situation?

65. Geometry The three angles shown in the triangle at the front of the tent in the following figure add up to 180°. Use this fact to write an equation containing x, and then solve the equation to find the number of degrees in the angle at the top of the triangle.

$$x \cdot \frac{y-1}{x} = \frac{3}{2} \cdot x \qquad \textbf{Multiply each side by } x$$

$$y - 1 = \frac{3}{2}x \qquad \textbf{Simplify each side}$$

$$y = \frac{3}{2}x + 1 \qquad \textbf{Add 1 to each side}$$

This is our solution. If we look back to the first step, we can justify our result on the left side of the equation this way: Dividing by x is equivalent to multiplying by its reciprocal $\frac{1}{x}$. Here is what it looks like when written out completely:

$$x \cdot \frac{y-1}{x} = x \cdot \frac{1}{x} \cdot (y - 1) = 1(y - 1) = y - 1$$

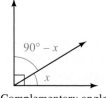

$90° - x$

x

Complementary angles

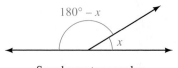

$180° - x$

x

Supplementary angles

FACTS FROM GEOMETRY

More on Complementary and Supplementary Angles

In Chapter 1 we defined complementary angles as angles that add to 90°; that is, if x and y are complementary angles, then

$$x + y = 90°$$

If we solve this formula for y, we obtain a formula equivalent to our original formula:

$$y = 90° - x$$

Because y is the complement of x, we can generalize by saying that the complement of angle x is the angle $90° - x$. By a similar reasoning process, we can say that the supplement of angle x is the angle $180° - x$. To summarize, if x is an angle, then

The complement of x is $90° - x$, and

The supplement of x is $180° - x$

If you go on to take a trigonometry class, you will see this formula again.

 EXAMPLE 6   Find the complement and the supplement of 25°.

SOLUTION   We can use the formulas above with $x = 25°$.

The complement of 25° is $90° - 25° = 65°$.

The supplement of 25° is $180° - 25° = 155°$.

Basic Percent Problems

The last examples in this section show how basic percent problems can be translated directly into equations. To understand these examples, you must recall that percent means "per hundred" that is, 75% is the same as $\frac{75}{100}$, 0.75, and, in reduced fraction form, $\frac{3}{4}$. Likewise, the decimal 0.25 is equivalent to 25%. To change a decimal to a percent, we move the decimal point two places to the right and write the % symbol. To change from a percent to a decimal, we drop the % symbol and move the decimal point two places to the left. The table that follows gives some of the most commonly used fractions and decimals and their equivalent percents.

Fraction	Decimal	Percent
$\frac{1}{2}$	0.5	50%
$\frac{1}{4}$	0.25	25%
$\frac{3}{4}$	0.75	75%
$\frac{1}{3}$	0.333 . . .	$33\frac{1}{3}$%
$\frac{2}{3}$	0.666 . . .	$66\frac{2}{3}$%
$\frac{1}{5}$	0.2	20%
$\frac{2}{5}$	0.4	40%
$\frac{3}{5}$	0.6	60%
$\frac{4}{5}$	0.8	80%

EXAMPLE 7   What number is 25% of 60?

SOLUTION   To solve a problem like this, we let x = the number in question (that is, the number we are looking for). Then, we translate the sentence directly into an equation by using an equal sign for the word "is" and multiplication for the word "of." Here is how it is done:

$$\underbrace{\text{What number}}_{x} \quad \underset{=}{\text{is}} \quad \underset{0.25}{25\%} \quad \underset{\cdot}{\text{of}} \quad \underset{60}{60?}$$

$$x = 15$$

Notice that we must write 25% as a decimal in order to do the arithmetic in the problem.

The number 15 is 25% of 60.

EXAMPLE 8   What percent of 24 is 6?

SOLUTION   Translating this sentence into an equation, as we did in Example 7, we have:

$$\underbrace{\text{What percent}}_{x} \quad \underset{\cdot}{\text{of}} \quad \underset{24}{24} \quad \underset{=}{\text{is}} \quad \underset{6}{6?}$$

$$\text{or} \qquad 24x = 6$$

Next, we multiply each side by $\frac{1}{24}$. (This is the same as dividing each side by 24.)

$$\frac{1}{24}(24x) = \frac{1}{24}(6)$$

$$x = \frac{6}{24}$$

$$= \frac{1}{4}$$

$$= 0.25, \text{ or } 25\%$$

The number 6 is 25% of 24.

EXAMPLE 9   45 is 75% of what number?

SOLUTION   Again, we translate the sentence directly:

45 is 75% of what number?

$$45 = 0.75 \cdot x$$

Next, we multiply each side by $\frac{1}{0.75}$ (which is the same as dividing each side by 0.75):

$$\frac{1}{0.75}(45) = \frac{1}{0.75}(0.75x)$$

$$\frac{45}{0.75} = x$$

$$60 = x$$

The number 45 is 75% of 60.

We can solve application problems involving percent by translating each problem into one of the three basic percent problems shown in Examples 7, 8, and 9.

EXAMPLE 10   The American Dietetic Association (ADA) recommends eating foods in which the calories from fat are less than 30% of the total calories. The nutrition labels from two kinds of granola bars are shown in Figure 1. For each bar, what percent of the total calories come from fat?

SOLUTION   The information needed to solve this problem is located toward the top of each label. Each serving of Bar I contains 210 calories, of which 70 calories come from fat. To find the percent of total calories that come from fat, we must answer this question:

70 is what percent of 210?

Nutrition Facts	
Serving Size 2 bars (47g)	
Servings Per Container 6	
Amount Per Serving	
Calories	210
Calories from Fat	70
	% Daily Value*
Total Fat 8g	12%
Saturated Fat 1g	5%
Cholesterol 0mg	0%
Sodium 150mg	6%
Total Carbohydrate 32g	11%
Dietary Fiber 2g	10%
Sugars 12g	
Protein 4g	

* Percent Daily Values are based on a 2,000 calorie diet. Your daily values may be higher or lower depending on your calorie needs.

Nutrition Facts	
Serving Size 1 bar (21g)	
Servings Per Container 8	
Amount Per Serving	
Calories	80
Calories from Fat	15
	% Daily Value*
Total Fat 1.5g	2%
Saturated Fat 0g	0%
Cholesterol 0mg	0%
Sodium 60mg	3%
Total Carbohydrate 16g	5%
Dietary Fiber 1g	4%
Sugars 5g	
Protein 2g	

* Percent Daily Values are based on a 2,000 calorie diet. Your daily values may be higher or lower depending on your calorie needs.

FIGURE 1

For Bar II, one serving contains 80 calories, of which 15 calories come from fat. To find the percent of total calories that come from fat, we must answer this question:

15 is what percent of 80?

Translating each equation into symbols, we have

70 is what percent of 210	15 is what percent of 80

$$70 = x \cdot 210$$

$$15 = x \cdot 80$$

$$x = \frac{70}{210}$$

$$x = \frac{15}{80}$$

$x = 0.33$ to the nearest hundredth

$x = 0.19$ to the nearest hundredth

$x = 33\%$

$x = 19\%$

Comparing the two bars, 33% of the calories in Bar I are fat calories, whereas 19% of the calories in Bar II are fat calories. According to the ADA, Bar II is the healthier choice.

LINKING OBJECTIVES AND EXAMPLES

Next to each **objective** we have listed the examples that are best described by that objective.

A	1, 2
B	3–5
C	6
D	7–10

GETTING READY FOR CLASS

After reading through the preceding section, respond in your own words and in complete sentences.

1. What is a formula?
2. How do you solve a formula for one of its variables?
3. What are complementary angles?
4. What does percent mean?

Problem Set 2.5

Online support materials can be found at academic.cengage.com/login

Use the formula $P = 2l + 2w$ to find the length l of a rectangular lot if

1. The width w is 50 feet and the perimeter P is 300 feet

2. The width w is 75 feet and the perimeter P is 300 feet

3. For the equation $2x + 3y = 6$,

   a. Find y when x is 0.

   b. Find x when y is 1.

   ▶ **c.** Find y when x is 3.

4. For the equation $2x - 5y = 20$,

   a. Find y when x is 0.

   b. Find x when y is 0.

   c. Find x when y is 2.

5. For the equation $y = -\frac{1}{3}x + 2$,

   a. Find y when x is 0.

   b. Find x when y is 3.

   c. Find y when x is 3.

6. For the equation $y = -\frac{2}{3}x + 1$,

   a. Find y when x is 0.

   b. Find x when y is -1.

   c. Find y when x is -3.

Use the equation $y = (x + 1)^2 - 3$ to find the value of y when

7. $x = -2$

8. $x = -1$

9. $x = 1$

10. $x = 2$

11. Use the formula $y = \frac{20}{x}$ to find y when

   a. $x = 10$ **b.** $x = 5$

12. Use the formula $y = 2x^2$ to find y when

   a. $x = 5$ **b.** $x = -6$

= Videos available by instructor request

▶ = Online student support materials available at academic.cengage.com/login

13. Use the formula $y = Kx$ to find K when

    a. $y = 15$ and $x = 3$ **b.** $y = 72$ and $x = 4$

14. Use the formula $y = Kx^2$ to find K when

    a. $y = 32$ and $x = 4$ **b.** $y = 45$ and $x = 3$

15. If $y = \dfrac{K}{x}$, find K if

    a. x is 5 and y is 4.

    b. x is 5 and y is 15.

16. If $I = \dfrac{K}{d^2}$, find K if

    a. $I = 200$ and $d = 10$.

    b. $I = 200$ and $d = 5$.

Solve each of the following for the indicated variable.

▶ **17.** $A = lw$ for l **18.** $d = rt$ for r

19. $V = lwh$ for h **20.** $PV = nRT$ for P

▶ **21.** $P = a + b + c$ for a

22. $P = a + b + c$ for b

23. $x - 3y = -1$ for x

24. $x + 3y = 2$ for x

25. $-3x + y = 6$ for y

26. $2x + y = -17$ for y

27. $2x + 3y = 6$ for y

28. $4x + 5y = 20$ for y

29. $P = 2l + 2w$ for w

30. $P = 2l + 2w$ for l

31. $h = vt + 16t^2$ for v

32. $h = vt - 16t^2$ for v

33. $A = \pi r^2 + 2\pi rh$ for h

34. $A = 2\pi r^2 + 2\pi rh$ for h

▶ **35.** Solve for y.

    a. $y - 3 = -2(x + 4)$

    b. $y - 5 = 4(x - 3)$

36. Solve for y.

    a. $y + 1 = -\dfrac{2}{3}(x - 3)$

    b. $y - 3 = -\dfrac{2}{3}(x + 3)$

37. Solve for y.

    a. $y - 1 = \dfrac{3}{4}(x - 1)$

    b. $y + 2 = \dfrac{3}{4}(x - 4)$

38. Solve for y.

    a. $y + 3 = \dfrac{3}{2}(x - 2)$

    b. $y + 4 = \dfrac{4}{3}(x - 3)$

39. Solve for y.

    a. $\dfrac{y - 1}{x} = \dfrac{3}{5}$

    b. $\dfrac{y - 2}{x} = \dfrac{1}{2}$

    c. $\dfrac{y - 3}{x} = 4$

40. Solve for y.

    a. $\dfrac{y + 1}{x} = -\dfrac{3}{5}$

    b. $\dfrac{y + 2}{x} = -\dfrac{1}{2}$

    c. $\dfrac{y + 3}{x} = -4$

Solve each formula for y.

41. $\dfrac{x}{7} - \dfrac{y}{3} = 1$ **42.** $\dfrac{x}{4} - \dfrac{y}{9} = 1$

43. $-\dfrac{1}{4}x + \dfrac{1}{8}y = 1$

44. $-\dfrac{1}{9}x + \dfrac{1}{3}y = 1$

The next two problems are intended to give you practice reading, and paying attention to, the instructions that accompany the problems you are working. As we have mentioned previously, working these problems is an excellent way to get ready for a test or a quiz.

45. Work each problem according to the instructions given.

    a. Solve: $4x + 5 = 20$

    b. Find the value of $4x + 5$ when x is 3.

    c. Solve for y: $4x + 5y = 20$

    d. Solve for x: $4x + 5y = 20$

46. Work each problem according to the instructions given.

    a. Solve: $-2x + 1 = 4$

    b. Find the value of $-2x + 1$ when x is 8.

    c. Solve for y: $-2x + y = 20$

    d. Solve for x: $-2x + y = 20$

Find the complement and the supplement of each angle.

47. $30°$ **48.** $60°$

49. $45°$ **50.** $15°$

Translate each of the following into an equation, and then solve that equation.

51. What number is 25% of 40?

52. What number is 75% of 40?

53. What number is 12% of 2,000?

54. What number is 9% of 3,000?

55. What percent of 28 is 7?

56. What percent of 28 is 21?

57. What percent of 40 is 14?

58. What percent of 20 is 14?

59. 32 is 50% of what number?

60. 16 is 50% of what number?

61. 240 is 12% of what number?

62. 360 is 12% of what number?

Applying the Concepts

More About Temperatures As we mentioned in Chapter 1, in the U.S. system, temperature is measured on the Fahrenheit scale. In the metric system, temperature is measured on the Celsius scale. On the Celsius scale, water boils at 100 degrees and freezes at 0 degrees. To denote a temperature of 100 degrees on the Celsius scale, we write

      100°C, which is read "100 degrees Celsius"

Table 1 is intended to give you an intuitive idea of the relationship between the two temperature scales. Table 2 gives the formulas, in both symbols and words, that are used to convert between the two scales.

63. Let F = 212 in the formula $C = \frac{5}{9}(F - 32)$, and solve for C. Does the value of C agree with the information in Table 1?

TABLE 1

Situation	Temperature	
	Fahrenheit	Celsius
Water freezes	32°F	0°C
Room temperature	68°F	20°C
Normal body temperature	98.6°F	37°C
Water boils	212°F	100°C
Bake cookies	365°F	185°C

TABLE 2

To Convert from	Formula in Symbols	Formula in Words
Fahrenheit to Celsius	$C = \frac{5}{9}(F - 32)$	Subtract 32, multiply by 5, then divide by 9.
Celsius to Fahrenheit	$F = \frac{9}{5}C + 32$	Multiply by $\frac{9}{5}$, then add 32.

64. Let C = 100 in the formula $F = \frac{9}{5}C + 32$, and solve for F. Does the value of F agree with the information in Table 1?

65. Let F = 68 in the formula $C = \frac{5}{9}(F - 32)$, and solve for C. Does the value of C agree with the information in Table 1?

66. Let C = 37 in the formula $F = \frac{9}{5}C + 32$, and solve for F. Does the value of F agree with the information in Table 1?

67. Solve the formula $F = \frac{9}{5}C + 32$ for C.

68. Solve the formula $C = \frac{5}{9}(F - 32)$ for F.

Nutrition Labels The nutrition label in Figure 2 is from a quart of vanilla ice cream. The label in Figure 3 is from a pint of vanilla frozen yogurt. Use the information on these labels for problems 69–72. Round your answers to the nearest tenth of a percent.

Nutrition Facts
Serving Size 1/2 cup (65g)
Servings 8

Amount/Serving		
Calories 150	Calories from Fat 90	
		% Daily Value*
Total Fat 10g		16%
Saturated Fat 6g		32%
Cholesterol 35mg		12%
Sodium 30mg		1%
Total Carbohydrate 14g		5%
Dietary Fiber 0g		0%
Sugars 11g		
Protein 2g		
Vitamin A 6%	•	Vitamin C 0%
Calcium 6%	•	Iron 0%
* Percent Daily Values are based on a 2,000 calorie diet.		

FIGURE 2 **Vanilla ice cream**

Nutrition Facts

Serving Size 1/2 cup (98g)
Servings Per Container 4

Amount Per Serving	
Calories 160	Calories from Fat 25

	% Daily Value*
Total Fat 2.5g	4%
Saturated Fat 1.5g	7%
Cholesterol 45mg	15%
Sodium 55mg	2%
Total Carbohydrate 26g	9%
Dietary Fiber 0g	0%
Sugars 19g	
Protein 8g	

Vitamin A 0%	•	Vitamin C 0%
Calcium 25%	•	Iron 0%

* Percent Daily Values are based on a 2,000 calorie diet.

FIGURE 3   **Vanilla frozen yogurt**

69. What percent of the calories in one serving of the vanilla ice cream are fat calories?

70. What percent of the calories in one serving of the frozen yogurt are fat calories?

71. One serving of frozen yogurt is 98 grams, of which 26 grams are carbohydrates. What percent of one serving are carbohydrates?

72. One serving of vanilla ice cream is 65 grams. What percent of one serving is sugar?

Circumference The circumference of a circle is given by the formula $C = 2\pi r$. Find r if

73. The circumference C is 44 meters and π is $\frac{22}{7}$

74. The circumference C is 176 meters and π is $\frac{22}{7}$

75. The circumference is 9.42 inches and π is 3.14

76. The circumference is 12.56 inches and π is 3.14

Volume The volume of a cylinder is given by the formula $V = \pi r^2 h$. Find the height h if

77. The volume V is 42 cubic feet, the radius is $\frac{7}{22}$ feet, and π is $\frac{22}{7}$

78. The volume V is 84 cubic inches, the radius is $\frac{7}{11}$ inches, and π is $\frac{22}{7}$

79. The volume is 6.28 cubic centimeters, the radius is 3 centimeters, and π is 3.14

80. The volume is 12.56 cubic centimeters, the radius is 2 centimeters, and π is 3.14

Maintaining Your Skills

The problems that follow review some of the more important skills you have learned in previous sections and chapters. You can consider the time you spend working these problems as time spent studying for exams.

81. a. $27 - (-68)$
   b. $27 + (-68)$
   c. $-27 - 68$
   d. $-27 + 68$

82. a. $55 - (-29)$
   b. $55 + (-29)$
   c. $-55 - 29$
   d. $-55 + 29$

83. a. $-32 - (-41)$
   b. $-32 + (-41)$
   c. $-32 + 41$
   d. $-32 - 41$

84 a. $-56 - (-35)$
   b. $-56 + (-35)$
   c. $-56 + 35$
   d. $-56 - 35$

Getting Ready for the Next Section

To understand all of the explanations and examples in the next section you must be able to work the problems below.

Write an equivalent expression in English. Include the words *sum* and *difference* when possible.

85. $4 + 1 = 5$

86. $7 + 3 = 10$

87. $6 - 2 = 4$

88. $8 - 1 = 7$

89. $x - 5 = -12$

90. $2x + 3 = 7$

91. $x + 3 = 4(x - 3)$

92. $2(2x - 5) = 2x - 34$

For each of the following expressions, write an equivalent equation.

93. Twice the sum of 6 and 3 is 18.

94. Four added to the product of 3 and -1 is 1.

95. The sum of twice 5 and 3 is 13.

96. Twice the difference of 8 and 2 is 12.

97. The sum of a number and five is thirteen.

98. The difference of ten and a number is negative eight.

99. Five times the sum of a number and seven is thirty.

100. Five times the difference of twice a number and six is negative twenty.

2.6 Applications

OBJECTIVES

A Apply the Blueprint for Problem Solving to a variety of application problems.

As you begin reading through the examples in this section, you may find yourself asking why some of these problems seem so contrived. The title of the section is "Applications," but many of the problems here don't seem to have much to do with "real life." You are right about that. Example 3 is what we refer to as an "age problem." But imagine a conversation in which you ask someone how old her children are and she replies, "Bill is 6 years older than Tom. Three years ago the sum of their ages was 21. You figure it out." Although many of the "application" problems in this section are contrived, they are also good for practicing the strategy we will use to solve all application problems.

To begin this section, we list the steps used in solving application problems. We call this strategy the *Blueprint for Problem Solving*. It is an outline that will overlay the solution process we use on all application problems.

BLUEPRINT FOR PROBLEM SOLVING

STEP 1: *Read* the problem, and then mentally *list* the items that are known and the items that are unknown.

STEP 2: *Assign a variable* to one of the unknown items. (In most cases this will amount to letting x = the item that is asked for in the problem.) Then *translate* the other *information* in the problem to expressions involving the variable.

STEP 3: *Reread* the problem, and then *write an equation,* using the items and variables listed in steps 1 and 2, that describes the situation.

STEP 4: *Solve the equation* found in step 3.

STEP 5: *Write* your *answer* using a complete sentence.

STEP 6: *Reread* the problem, and *check* your solution with the original words in the problem.

English	Algebra
The sum of a and b	$a + b$
The difference of a and b	$a - b$
The product of a and b	$a \cdot b$
The quotient of a and b	$\dfrac{a}{b}$
of	$\cdot$ (multiply)
is	= (equals)
A number	x
4 more than x	$x + 4$
4 times x	$4x$
4 less than x	$x - 4$

There are a number of substeps within each of the steps in our blueprint. For instance, with steps 1 and 2 it is always a good idea to draw a diagram or picture if it helps visualize the relationship between the items in the problem. In other cases a table helps organize the information. As you gain more experience using the blueprint to solve application problems, you will find additional techniques that expand the blueprint.

To help with problems of the type shown next in Example 1, in the margin are some common English words and phrases and their mathematical translations.

SOLUTION

Step 1: **Read and list.**

> *Known items:*   The type of coins, the total value of the coins, and that there are 8 more dimes than nickels.
>
> *Unknown items:*   The number of nickels and the number of dimes

Step 2: **Assign a variable, and translate information.**

> If we let $x =$ the number of nickels, then $x + 8 =$ the number of dimes. Because the value of each nickel is 5 cents, the amount of money in nickels is $5x$. Similarly, because each dime is worth 10 cents, the amount of money in dimes is $10(x + 8)$. The table summarizes the information we have so far.

	Nickels	Dimes
Number	x	$x + 8$
Value (in cents)	$5x$	$10(x + 8)$

Step 3: **Reread, and write an equation.**

> Because the total value of all the coins is 245 cents, the equation that describes this situation is

Amount of money in nickels		Amount of money in dimes		Total amount of money
$5x$	$+$	$10(x + 8)$	$=$	245

Step 4: **Solve the equation.**

> To solve the equation, we apply the distributive property first.

$$5x + 10x + 80 = 245 \qquad \textbf{Distributive property}$$

$$15x + 80 = 245 \qquad \textbf{Add } 5x \textbf{ and } 10x$$

$$15x = 165 \qquad \textbf{Add } -80 \textbf{ to each side}$$

$$x = 11 \qquad \textbf{Divide each side by 15}$$

Step 5: **Write the answer.**

> The number of nickels is $x = 11$.
> The number of dimes is $x + 8 = 11 + 8 = 19$.

Step 6: **Reread, and check.**

> To check our results

$$
\begin{aligned}
11 \text{ nickels are worth } 5(11) &= 55 \text{ cents} \\
19 \text{ dimes are worth } 10(19) &= 190 \text{ cents} \\
\hline
\text{The total value is } 245 \text{ cents} &= \$2.45
\end{aligned}
$$

When you begin working the problems in the problem set that follows, there are a couple of things to remember. The first is that you may have to read the problems a number of times before you begin to see how to solve them. The second thing to remember is that word problems are not always solved correctly the first time you try them. Sometimes it takes a couple of attempts and some wrong answers before you can set up and solve these problems correctly.

GETTING READY FOR CLASS

After reading through the preceding section, respond in your own words and in complete sentences.

1. What is the first step in the Blueprint for Problem Solving?
2. What is the last thing you do when solving an application problem?
3. What good does it do you to solve application problems even when they don't have much to do with real life?
4. Write an application problem whose solution depends on solving the equation $2x + 3 = 7$.

LINKING OBJECTIVES AND EXAMPLES

Next to each **objective** we have listed the examples that are best described by that objective.

A   1–5

Problem Set 2.6

Online support materials can be found at academic.cengage.com/login

Solve the following word problems. Follow the steps given in the Blueprint for Problem Solving.

Number Problems

1. The sum of a number and five is thirteen. Find the number.

2. The difference of ten and a number is negative eight. Find the number.

3. The sum of twice a number and four is fourteen. Find the number.

4. The difference of four times a number and eight is sixteen. Find the number.

5. Five times the sum of a number and seven is thirty. Find the number.

6. Five times the difference of twice a number and six is negative twenty. Find the number.

▶ **7.** One number is two more than another. Their sum is eight. Find both numbers.

8. One number is three less than another. Their sum is fifteen. Find the numbers.

9. One number is four less than three times another. If their sum is increased by five, the result is twenty-five. Find the numbers.

10. One number is five more than twice another. If their sum is decreased by ten, the result is twenty-two. Find the numbers.

Age Problems

11. Shelly is 3 years older than Michele. Four years ago the sum of their ages was 67. Find the age of each person now.

	Four Years Ago	Now
Shelly	$x - 1$	$x + 3$
Michele	$x - 4$	x

12. Cary is 9 years older than Dan. In 7 years the sum of their ages will be 93. Find the age of each man now.

	Now	In Seven Years
Cary	$x + 9$	
Dan	x	$x + 7$

13. Cody is twice as old as Evan. Three years ago the sum of their ages was 27. Find the age of each boy now.

	Three Years Ago	Now
Cody		
Evan	$x - 3$	x

14. Justin is 2 years older than Ethan. In 9 years the sum of their ages will be 30. Find the age of each boy now.

	Now	In Nine Years
Justin		
Ethan	x	

15. Fred is 4 years older than Barney. Five years ago the sum of their ages was 48. How old are they now?

	Five Years Ago	Now
Fred		
Barney		x

16. Tim is 5 years older than JoAnn. Six years from now the sum of their ages will be 79. How old are they now?

	Now	Six Years From Now
Tim		
JoAnn	x	

17. Jack is twice as old as Lacy. In 3 years the sum of their ages will be 54. How old are they now?

18. John is 4 times as old as Martha. Five years ago the sum of their ages was 50. How old are they now?

19. Pat is 20 years older than his son Patrick. In 2 years Pat will be twice as old as Patrick. How old are they now?

20. Diane is 23 years older than her daughter Amy. In 6 years Diane will be twice as old as Amy. How old are they now?

Perimeter Problems

21. The perimeter of a square is 36 inches. Find the length of one side.

22. The perimeter of a square is 44 centimeters. Find the length of one side.

23. The perimeter of a square is 60 feet. Find the length of one side.

24. The perimeter of a square is 84 meters. Find the length of one side.

25. One side of a triangle is three times the shortest side. The third side is 7 feet more than the shortest side. The perimeter is 62 feet. Find all three sides.

26. One side of a triangle is half the longest side. The third side is 10 meters less than the longest side. The perimeter is 45 meters. Find all three sides.

27. One side of a triangle is half the longest side. The third side is 12 feet less than the longest side. The perimeter is 53 feet. Find all three sides.

28. One side of a triangle is 6 meters more than twice the shortest side. The third side is 9 meters more than the shortest side. The perimeter is 75 meters. Find all three sides.

29. The length of a rectangle is 5 inches more than the width. The perimeter is 34 inches. Find the length and width.

$x + 5$

30. The width of a rectangle is 3 feet less than the length. The perimeter is 10 feet. Find the length and width.

31. The length of a rectangle is 7 inches more than twice the width. The perimeter is 68 inches. Find the length and width.

32. The length of a rectangle is 4 inches more than three times the width. The perimeter is 72 inches. Find the length and width.

33. The length of a rectangle is 6 feet more than three times the width. The perimeter is 36 feet. Find the length and width.

34. The length of a rectangle is 3 feet less than twice the width. The perimeter is 54 feet. Find the length and width.

Coin Problems

35. Marissa has $4.40 in quarters and dimes. If she has 5 more quarters than dimes, how many of each coin does she have?

	Dimes	Quarters
Number	x	$x + 5$
Value (cents)	$10(x)$	$25(x + 5)$

36. Kendra has $2.75 in dimes and nickels. If she has twice as many dimes as nickels, how many of each coin does she have?

	Nickels	Dimes
Number	x	$2x$
Value (cents)	$5(x)$	

37. Tanner has $4.35 in nickels and quarters. If he has 15 more nickels than quarters, how many of each coin does he have?

	Nickels	Quarters
Number	$x + 15$	x
Value (cents)		

38. Connor has $9.00 in dimes and quarters. If he has twice as many quarters as dimes, how many of each coin does he have?

	Dimes	Quarters
Number	x	$2x$
Value (cents)		

▶ **39.** Sue has $2.10 in dimes and nickels. If she has 9 more dimes than nickels, how many of each coin does she have?

40. Mike has $1.55 in dimes and nickels. If he has 7 more nickels than dimes, how many of each coin does he have?

41. Katie has a collection of nickels, dimes, and quarters with a total value of $4.35. There are 3 more dimes than nickels and 5 more quarters than nickels. How many of each coin is in her collection?

	Nickels	Dimes	Quarters
Number	x		
Value			

42. Mary Jo has $3.90 worth of nickels, dimes, and quarters. The number of nickels is 3 more than the number of dimes. The number of quarters is 7 more than the number of dimes. How many of each coin is in her collection?

	Nickels	Dimes	Quarters
Number			
Value			

43. Cory has a collection of nickels, dimes, and quarters with a total value of $2.55. There are 6 more dimes than nickels and twice as many quarters as nickels. How many of each coin is in her collection?

	Nickels	Dimes	Quarters
Number	x		
Value			

44. Kelly has a collection of nickels, dimes, and quarters with a total value of $7.40. There are four more nickels than dimes and twice as many quarters as nickels. How many of each coin is in her collection?

	Nickels	Dimes	Quarters
Number			
Value			

Maintaining Your Skills

Write an equivalent statement in English.

45. $4 < 10$

46. $4 \leq 10$

47. $9 \geq -5$

48. $x - 2 > 4$

Place the symbol $<$ or the symbol $>$ between the quantities in each expression.

49. 12   20

50. -12 20

51. -8 -6

52. -10 -20

Simplify.

53. $|8 - 3| - |5 - 2|$

54. $|9 - 2| - |10 - 8|$

55. $15 - |9 - 3(7 - 5)|$

56. $10 - |7 - 2(5 - 3)|$

Getting Ready for the Next Section

To understand all of the explanations and examples in the next section you must be able to work the problems below.

Simplify the following expressions.

57. $x + 2x + 2x$

58. $x + 2x + 3x$

59. $x + 0.075x$

60. $x + 0.065x$

61. $0.09(x + 2,000)$

62. $0.06(x + 1,500)$

Solve each of the following equations.

63. $0.05x + 0.06(x - 1,500) = 570$

64. $0.08x + 0.09(x + 2,000) = 690$

65. $x + 2x + 3x = 180$

66. $2x + 3x + 5x = 180$

2.7 More Applications

OBJECTIVES

A Apply the Blueprint for Problem Solving to a variety of application problems.

Now that you have worked through a number of application problems using our blueprint, you probably have noticed that step 3, in which we write an equation that describes the situation, is the key step. Anyone with experience solving application problems will tell you that there will be times when your first attempt at step 3 results in the wrong equation. Remember, mistakes are part of the process of learning to do things correctly. Many times the correct equation will become obvious after you have written an equation that is partially wrong. In any case it is better to write an equation that is partially wrong and be actively involved with the problem than to write nothing at all. Application problems, like other problems in algebra, are not always solved correctly the first time.

Consecutive Integers

Our first example involves consecutive integers. When we ask for consecutive integers, we mean integers that are next to each other on the number line, like 5 and 6, or 13 and 14, or -4 and -3. In the dictionary, consecutive is defined as following one another in uninterrupted order. If we ask for consecutive *odd* integers, then we mean odd integers that follow one another on the number line. For example, 3 and 5, 11 and 13, and -9 and -7 are consecutive odd integers. As you can see, to get from one odd integer to the next consecutive odd integer we add 2.

If we are asked to find two consecutive integers and we let x equal the first integer, the next one must be $x + 1$, because consecutive integers always differ by 1. Likewise, if we are asked to find two consecutive odd or even integers, and

FACTS FROM GEOMETRY

Labeling Triangles and the Sum of the Angles in a Triangle
One way to label the important parts of a triangle is to label the vertices with capital letters and the sides with small letters, as shown in Figure 1.

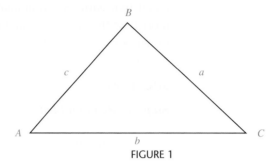

FIGURE 1

In Figure 1, notice that side a is opposite vertex A, side b is opposite vertex B, and side c is opposite vertex C. Also, because each vertex is the vertex of one of the angles of the triangle, we refer to the three interior angles as A, B, and C.

In any triangle, the sum of the interior angles is $180°$. For the triangle shown in Figure 1, the relationship is written

$$A + B + C = 180°$$

EXAMPLE 3   The angles in a triangle are such that one angle is twice the smallest angle, whereas the third angle is three times as large as the smallest angle. Find the measure of all three angles.

SOLUTION

Step 1: **Read and list.**
 Known items: The sum of all three angles is $180°$, one angle is twice the smallest angle, the largest angle is three times the smallest angle.
 Unknown items: The measure of each angle

Step 2: **Assign a variable, and translate information.**
 Let x be the smallest angle, then $2x$ will be the measure of another angle and $3x$ will be the measure of the largest angle.

Step 3: **Reread, and write an equation.**
 When working with geometric objects, drawing a generic diagram sometimes will help us visualize what it is that we are asked to find. In Figure 2, we draw a triangle with angles A, B, and C.

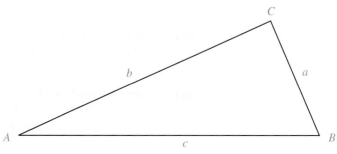

FIGURE 2

We can let the value of $A = x$, the value of $B = 2x$, and the value of $C = 3x$. We know that the sum of angles A, B, and C will be 180°, so our equation becomes

$$x + 2x + 3x = 180°$$

Step 4: *Solve the equation.*

$$x + 2x + 3x = 180°$$
$$6x = 180°$$
$$x = 30°$$

Step 5: *Write the answer.*

The smallest angle A measures 30°
Angle B measures $2x$, or $2(30°) = 60°$
Angle C measures $3x$, or $3(30°) = 90°$

Step 6: *Reread, and check.*

The angles must add to 180°:

$$A + B + C = 180°$$

$$30° + 60° + 90° \stackrel{?}{=} 180°$$

$$180° = 180° \qquad \textbf{Our answers check} \qquad$$

LINKING OBJECTIVES AND EXAMPLES

Next to each **objective** we have listed the examples that are best described by that objective.

A 1–3

GETTING READY FOR CLASS

After reading through the preceding section, respond in your own words and in complete sentences.

1. How do we label triangles?
2. What rule is always true about the three angles in a triangle?
3. Write an application problem whose solution depends on solving the equation x + 0.075x = 500.
4. Write an application problem whose solution depends on solving the equation 0.05x + 0.06(x + 200) = 67.

Problem Set 2.7

Online support materials can be found at academic.cengage.com/login

Consecutive Integer Problems

1. The sum of two consecutive integers is 11. Find the numbers.

2. The sum of two consecutive integers is 15. Find the numbers.

3. The sum of two consecutive integers is −9. Find the numbers.

4. The sum of two consecutive integers is −21. Find the numbers.

5. The sum of two consecutive odd integers is 28. Find the numbers.

6. The sum of two consecutive odd integers is 44. Find the numbers.

7. The sum of two consecutive even integers is 106. Find the numbers.

8. The sum of two consecutive even integers is 66. Find the numbers.

9. The sum of two consecutive even integers is −30. Find the numbers.

10. The sum of two consecutive odd integers is −76. Find the numbers.

11. The sum of three consecutive odd integers is 57. Find the numbers.

12. The sum of three consecutive odd integers is −51. Find the numbers.

13. The sum of three consecutive even integers is 132. Find the numbers.

14. The sum of three consecutive even integers is −108. Find the numbers.

Interest Problems

15. Suppose you invest money in two accounts. One of the accounts pays 8% annual interest, whereas the other pays 9% annual interest. If you have $2,000 more invested at 9% than you have invested at 8%, how much do you have invested in each account if the total amount of interest you earn in a year is $860? (Begin by completing the following table.)

	Dollars Invested at 8%	Dollars Invested at 9%
Number of	x	
Interest on		

16. Suppose you invest a certain amount of money in an account that pays 11% interest annually, and $4,000 more than that in an account that pays 12% annually. How much money do you have in each account if the total interest for a year is $940?

	Dollars Invested at 11%	Dollars Invested at 12%
Number of	x	
Interest on		

▶ 17. Tyler has two savings accounts that his grandparents opened for him. The two accounts pay 10% and 12% in annual interest; there is $500 more in the account that pays 12% than there is in the other account. If the total interest for a year is $214, how much money does he have in each account?

18. Travis has a savings account that his parents opened for him. It pays 6% annual interest. His uncle also opened an account for him, but it pays 8% annual interest. If there is $800 more in the account that pays 6%, and the total interest from both accounts is $104, how much money is in each of the accounts?

19. A stockbroker has money in three accounts. The interest rates on the three accounts are 8%, 9%, and 10%. If she has twice as much money invested at 9% as she has invested at 8%, three times as much at 10% as she has at 8%, and the total interest for the year is $280, how much is invested at each rate? (*Hint:* Let x = the amount invested at 8%.)

20. An accountant has money in three accounts that pay 9%, 10%, and 11% in annual interest. He has twice as much invested at 9% as he does at 10% and three times as much invested at 11% as he does at 10%. If the total interest from the three accounts is $610 for the year, how much is invested at each rate? (*Hint:* Let x = the amount invested at 10%.)

Triangle Problems

21. Two angles in a triangle are equal and their sum is equal to the third angle in the triangle. What are the measures of each of the three interior angles?

22. One angle in a triangle measures twice the smallest angle, whereas the largest angle is six times the smallest angle. Find the measures of all three angles.

23. The smallest angle in a triangle is $\frac{1}{5}$ as large as the largest angle. The third angle is twice the smallest angle. Find the three angles.

24. One angle in a triangle is half the largest angle but three times the smallest. Find all three angles.

25. A right triangle has one 37° angle. Find the other two angles.

26. In a right triangle, one of the acute angles is twice as large as the other acute angle. Find the measure of the two acute angles.

27. One angle of a triangle measures 20° more than the smallest, while a third angle is twice the smallest. Find the measure of each angle.

28. One angle of a triangle measures 50° more than the smallest, while a third angle is three times the smallest. Find the measure of each angle.

Miscellaneous Problems

29. **Ticket Prices** Miguel is selling tickets to a barbecue. Adult tickets cost $6.00 and children's tickets cost $4.00. He sells six more children's tickets than adult tickets. The total amount of money he collects is $184. How many adult tickets and how many children's tickets did he sell?

	Adult	**Child**
Number	x	
Income		

30. **Working Two Jobs** Maggie has a job working in an office for $10 an hour and another job driving a tractor for $12 an hour. One week she works in the office twice as long as she drives the tractor. Her total income for that week is $416. How many hours did she spend at each job?

Job	**Office**	**Tractor**
Hours Worked		x
Wages Earned		

31. **Phone Bill** The cost of a long-distance phone call is $0.41 for the first minute and $0.32 for each additional minute. If the total charge for a long-distance call is $5.21, how many minutes was the call?

32. **Phone Bill** Danny, who is 1 year old, is playing with the telephone when he accidentally presses one of the buttons his mother has programmed to dial her friend Sue's number. Sue answers the phone and realizes Danny is on the other end. She talks to Danny, trying to get him to hang up. The cost for a call is $0.23 for the first minute and $0.14 for every minute after that. If the total charge for the call is $3.73, how long did it take Sue to convince Danny to hang up the phone?

33. **Hourly Wages** JoAnn works in the publicity office at the state university. She is paid $12 an hour for the first 35 hours she works each week and $18 an hour for every hour after that. If she makes $492 one week, how many hours did she work?

34. **Hourly Wages** Diane has a part-time job that pays her $6.50 an hour. During one week she works 26 hours and is paid $178.10. She realizes when she sees her check that she has been given a raise. How much per hour is that raise?

35. **Office Numbers** Professors Wong and Gil have offices in the mathematics building at Miami Dade College. Their office numbers are consecutive odd integers with a sum of 14,660. What are the office numbers of these two professors?

36. **Cell Phone Numbers** Diana and Tom buy two cell phones. The phone numbers assigned to each are consecutive integers with a sum of 11,109,295. If the smaller number is Diana's, what are their phone numbers?

37. **Age** Marissa and Kendra are 2 years apart in age. Their ages are two consecutive even integers. Kendra is the younger of the two. If Marissa's age is added to twice Kendra's age, the result is 26. How old is each girl?

38. **Age** Justin's and Ethan's ages form two consecutive odd integers. What is the difference of their ages?

39. **Arrival Time** Jeff and Carla Cole are driving separately from San Luis Obispo, California, to the north shore of Lake Tahoe, a distance of 425 miles. Jeff leaves San Luis Obispo at 11:00 AM and averages 55 miles per hour on the drive, Carla leaves later, at 1:00 PM but averages 65 miles per hour. Which person arrives in Lake Tahoe first?

40. **Piano Lessons** Tyler is taking piano lessons. Because he doesn't practice as often as his parents would like him to, he has to pay for part of the lessons himself. His parents pay him $0.50 to do the laun-

dry and $1.25 to mow the lawn. In one month, he does the laundry 6 more times than he mows the lawn. If his parents pay him $13.50 that month, how many times did he mow the lawn?

At one time, the Texas Junior College Teachers Association annual conference was held in Austin. At that time a taxi ride in Austin was $1.25 for the first $\frac{1}{5}$ of a mile and $0.25 for each additional $\frac{1}{5}$ of a mile. Use this information for Problems 41 and 42.

41. Cost of a Taxi Ride If the distance from one of the convention hotels to the airport is 7.5 miles, how much will it cost to take at taxi from that hotel to the airport?

42. Cost of a Taxi Ride Suppose the distance from one of the hotels to one of the western dance clubs in Austin is 12.4 miles. If the fare meter in the taxi gives the charge for that trip as $16.50, is the meter working correctly?

43. Geometry The length and width of a rectangle are consecutive even integers. The perimeter is 44 meters. Find the length and width.

44. Geometry The length and width of a rectangle are consecutive odd integers. The perimeter is 128 meters. Find the length and width.

45. Geometry The angles of a triangle are three consecutive integers. Find the measure of each angle.

46. Geometry The angles of a triangle are three consecutive even integers. Find the measure of each angle.

Ike and Nancy Lara give western dance lessons at the Elks Lodge on Sunday nights. The lessons cost $3.00 for members of the lodge and $5.00 for nonmembers. Half of the money collected for the lesson is paid to Ike and Nancy. The Elks Lodge keeps the other half. One Sunday night Ike counts 36 people in the dance lesson. Use this information to work Problems 47 through 50.

47. Dance Lessons What is the least amount of money Ike and Nancy will make?

48. Dance Lessons What is the largest amount of money Ike and Nancy will make?

49. Dance Lessons At the end of the evening, the Elks Lodge gives Ike and Nancy a check for $80 to cover half of the receipts. Can this amount be correct?

50. Dance Lessons Besides the number of people in the dance lesson, what additional information does Ike need to know to always be sure he is being paid the correct amount?

Maintaining Your Skills

The problems that follow review some of the more important skills you have learned in previous sections and chapters. You can consider the time you spend working these problems as time spent studying for exams.

Simplify the expression $36x - 12$ for each of the following values of x.

51. $\frac{1}{4}$ **52.** $\frac{1}{6}$

53. $\frac{1}{9}$ **54.** $\frac{3}{2}$

55. $\frac{1}{3}$ **56.** $\frac{5}{12}$

57. $\frac{5}{9}$ **58.** $\frac{2}{3}$

Find the value of each expression when $x = -4$.

59. $3(x - 4)$ **60.** $-3(x - 4)$

61. $-5x + 8$ **62.** $5x + 8$

63. $\dfrac{x - 14}{36}$ **64.** $\dfrac{x - 12}{36}$

65. $\dfrac{16}{x} + 3x$ **66.** $\dfrac{16}{x} - 3x$

67. $7x - \dfrac{12}{x}$ **68.** $7x + \dfrac{12}{x}$

69. $8\left(\dfrac{x}{2} + 5\right)$ **70.** $-8\left(\dfrac{x}{2} + 5\right)$

Getting Ready for the Next Section

To understand all of the explanations and examples in the next section you must be able to work the problems below.

Solve the following equations.

71. a. $x - 3 = 6$
    b. $x + 3 = 6$
    c. $-x - 3 = 6$
    d. $-x + 3 = 6$

72. a. $x - 7 = 16$

   b. $x + 7 = 16$

   c. $-x - 7 = 16$

   d. $-x + 7 = 16$

73. a. $\dfrac{x}{4} = -2$

   b. $-\dfrac{x}{4} = -2$

   c. $\dfrac{x}{4} = 2$

   d. $-\dfrac{x}{4} = 2$

74. a. $3a = 15$

   b. $3a = -15$

   c. $-3a = 15$

   d. $-3a = -15$

75. $2.5x - 3.48 = 4.9x + 2.07$

76. $2(1 - 3x) + 4 = 4x - 14$

77. $3(x - 4) = -2$

78. Solve $2x - 3y = 6$ for y.

2.8   Linear Inequalities

OBJECTIVES

A   Use the addition property for inequalities to solve an inequality.

B   Use the multiplication property for inequalities to solve an inequality.

C   Use both the addition and multiplication properties to solve an inequality.

D   Graph the solution set for an inequality.

E   Translate and solve application problems involving inequalities.

Linear inequalities are solved by a method similar to the one used in solving linear equations. The only real differences between the methods are in the multiplication property for inequalities and in graphing the solution set.

An inequality differs from an equation only with respect to the comparison symbol between the two quantities being compared. In place of the equal sign, we use $<$ (less than), $\leq$ (less than or equal to), $>$ (greater than), or $\geq$ (greater than or equal to). The addition property for inequalities is almost identical to the addition property for equality.

> **Addition Property for Inequalities**   For any three algebraic expressions A, B, and C,
>
> $$\text{if} \qquad A < B$$
> $$\text{then} \qquad A + C < B + C$$
>
> *In words:*   Adding the same quantity to both sides of an inequality will not change the solution set.

It makes no difference which inequality symbol we use to state the property. Adding the same amount to both sides always produces an inequality equivalent to the original inequality. Also, because subtraction can be thought of as addition of the opposite, this property holds for subtraction as well as addition.

 EXAMPLE 1   Solve the inequality $x + 5 < 7$.

SOLUTION   To isolate x, we add -5 to both sides of the inequality:

$$x + 5 < 7$$
$$x + 5 + (\mathbf{-5}) < 7 + (\mathbf{-5}) \qquad \textbf{Addition property for inequalities}$$
$$x < 2$$

We can go one step further here and graph the solution set. The solution set is all real numbers less than 2. To graph this set, we simply draw a straight line and label the center 0 (zero) for reference. Then we label the 2 on the right side of zero and extend an arrow beginning at 2 and pointing to the left. We use an open circle at 2 because it is not included in the solution set. Here is the graph.

EXAMPLE 2   Solve $x - 6 \leq -3$.

SOLUTION   Adding 6 to each side will isolate x on the left side:

$$x - 6 \leq -3$$
$$x - 6 + \mathbf{6} \leq -3 + \mathbf{6} \qquad \textbf{Add 6 to both sides}$$
$$x \leq 3$$

The graph of the solution set is

Notice that the dot at the 3 is darkened because 3 is included in the solution set. We always will use open circles on the graphs of solution sets with $<$ or $>$ and closed (darkened) circles on the graphs of solution sets with $\leq$ or $\geq$.

To see the idea behind the multiplication property for inequalities, we will consider three true inequality statements and explore what happens when we multiply both sides by a positive number and then what happens when we multiply by a negative number.

Consider the following three true statements:

$$3 < 5 \qquad -3 < 5 \qquad -5 < -3$$

Now multiply both sides by the positive number 4:

$$4(3) < 4(5) \qquad 4(-3) < 4(5) \qquad 4(-5) < 4(-3)$$
$$12 < 20 \qquad\quad -12 < 20 \qquad\quad -20 < -12$$

In each case, the inequality symbol in the result points in the same direction it did in the original inequality. We say the "sense" of the inequality doesn't change when we multiply both sides by a positive quantity.

Notice what happens when we go through the same process but multiply both sides by -4 instead of 4:

$$3 < 5 \qquad\qquad -3 < 5 \qquad\qquad -5 < -3$$

$$-4(3) > -4(5) \qquad -4(-3) > -4(5) \qquad -4(-5) > -4(-3)$$
$$-12 > -20 \qquad\qquad 12 > -20 \qquad\qquad 20 > 12$$

In each case, we have to change the direction in which the inequality symbol points to keep each statement true. Multiplying both sides of an inequality by a negative quantity *always* reverses the sense of the inequality. Our results are summarized in the multiplication property for inequalities.

Note

This discussion is intended to show why the multiplication property for inequalities is written the way it is. You may want to look ahead to the property itself and then come back to this discussion if you are having trouble making sense out of it.

Note

Because division is defined in terms of multiplication, this property is also true for division. We can divide both sides of an inequality by any nonzero number we choose. If that number happens to be negative, we must also reverse the direction of the inequality symbol.

Multiplication Property for Inequalities For any three algebraic expressions A, B, and C,

$$\text{if} \quad A < B$$
$$\text{then} \quad AC < BC \quad \text{when } C \text{ is positive}$$
$$\text{and} \quad AC > BC \quad \text{when } C \text{ is negative}$$

In words: Multiplying both sides of an inequality by a positive number does not change the solution set. When multiplying both sides of an inequality by a negative number, it is necessary to reverse the inequality symbol to produce an equivalent inequality.

We can multiply both sides of an inequality by any nonzero number we choose. If that number happens to be negative, we must also reverse the sense of the inequality.

 EXAMPLE 3 Solve $3a < 15$ and graph the solution.

SOLUTION We begin by multiplying each side by $\frac{1}{3}$. Because $\frac{1}{3}$ is a positive number, we do not reverse the direction of the inequality symbol:

$$3a < 15$$
$$\frac{1}{3}(3a) < \frac{1}{3}(15) \qquad \textbf{Multiply each side by } \tfrac{1}{3}$$
$$a < 5$$

 EXAMPLE 4 Solve $-3a \leq 18$, and graph the solution.

SOLUTION We begin by multiplying both sides by $-\frac{1}{3}$. Because $-\frac{1}{3}$ is a negative number, we must reverse the direction of the inequality symbol at the same time that we multiply by $-\frac{1}{3}$.

$$-3a \leq 18$$
$$-\frac{1}{3}(-3a) \geq -\frac{1}{3}(18) \qquad \begin{array}{l}\textbf{Multiply both sides by } -\tfrac{1}{3} \textbf{ and reverse}\\ \textbf{the direction of the inequality symbol}\end{array}$$
$$a \geq -6$$

EXAMPLE 5 Solve $-\dfrac{x}{4} > 2$ and graph the solution.

SOLUTION To isolate x, we multiply each side by -4. Because -4 is a negative number, we also must reverse the direction of the inequality symbol:

$$-\frac{x}{4} > 2$$
$$-4\left(-\frac{x}{4}\right) < -4(2) \qquad \begin{array}{l}\textbf{Multiply each side by } -4, \textbf{ and reverse}\\ \textbf{the direction of the inequality symbol}\end{array}$$
$$x < -8$$

To solve more complicated inequalities, we use the following steps.

> **Strategy for Solving Linear Inequalities in One Variable**
>
> **Step 1a:** Use the distributive property to separate terms, if necessary.
>
> **1b:** If fractions are present, consider multiplying both sides by the LCD to eliminate the fractions. If decimals are present, consider multiplying both sides by a power of 10 to clear the inequality of decimals.
>
> **1c:** Combine similar terms on each side of the inequality.
>
> **Step 2:** Use the addition property for inequalities to get all variable terms on one side of the inequality and all constant terms on the other side.
>
> **Step 3:** Use the multiplication property for inequalities to get x by itself on one side of the inequality.
>
> **Step 4:** Graph the solution set.

 EXAMPLE 6 Solve $2.5x - 3.48 < -4.9x + 2.07$.

SOLUTION We have two methods we can use to solve this inequality. We can simply apply our properties to the inequality the way it is currently written and work with the decimal numbers, or we can eliminate the decimals to begin with and solve the resulting inequality.

Method 1 Working with the decimals.

$$2.5x - 3.48 < -4.9x + 2.07 \qquad \text{Original inequality}$$

$$2.5x + \mathbf{4.9x} - 3.48 < -4.9x + \mathbf{4.9x} + 2.07 \qquad \text{Add } 4.9x \text{ to each side}$$

$$7.4x - 3.48 < 2.07$$

$$7.4x - 3.48 + \mathbf{3.48} < 2.07 + \mathbf{3.48} \qquad \text{Add } 3.48 \text{ to each side}$$

$$7.4x < 5.55$$

$$\frac{7.4x}{\mathbf{7.4}} < \frac{5.55}{\mathbf{7.4}} \qquad \text{Divide each side by } 7.4$$

$$x < 0.75$$

Method 2 Eliminating the decimals in the beginning.

Because the greatest number of places to the right of the decimal point in any of the numbers is 2, we can multiply each side of the inequality by 100 and we will be left with an equivalent inequality that contains only whole numbers.

$$2.5x - 3.48 < -4.9x + 2.07 \qquad \text{Original inequality}$$

$$\mathbf{100}(2.5x - 3.48) < \mathbf{100}(-4.9x + 2.07) \qquad \text{Multiply each side by 100}$$

$$\mathbf{100}(2.5x) - \mathbf{100}(3.48) < \mathbf{100}(-4.9x) + \mathbf{100}(2.07) \qquad \text{Distributive property}$$

$$250x - 348 < -490x + 207 \qquad \text{Multiplication}$$

$$740x - 348 < 207 \qquad \text{Add } 490x \text{ to each side}$$

$$740x < 555 \qquad \textbf{Add 348 to each side}$$

$$\frac{740x}{\textbf{740}} < \frac{555}{\textbf{740}} \qquad \textbf{Divide each side by 740}$$

$$x < 0.75$$

The solution by either method is $x < 0.75$. Here is the graph:

 EXAMPLE 7   Solve $3(x - 4) \geq -2$.

SOLUTION

$$3x - 12 \geq -2 \qquad \textbf{Distributive property}$$

$$3x - 12 + \textbf{12} \geq -2 + \textbf{12} \qquad \textbf{Add 12 to both sides}$$

$$3x \geq 10$$

$$\frac{\textbf{1}}{\textbf{3}}(3x) \geq \frac{\textbf{1}}{\textbf{3}}(10) \qquad \textbf{Multiply both sides by } \tfrac{1}{3}$$

$$x \geq \frac{10}{3}$$

 EXAMPLE 8   Solve and graph $2(1 - 3x) + 4 < 4x - 14$.

SOLUTION

$$2 - 6x + 4 < 4x - 14 \qquad \textbf{Distributive property}$$

$$-6x + 6 < 4x - 14 \qquad \textbf{Simplify}$$

$$-6x + 6 + (\textbf{-6}) < 4x - 14 + (\textbf{-6}) \qquad \textbf{Add } -6 \textbf{ to both sides}$$

$$-6x < 4x - 20$$

$$-6x + (\textbf{-4x}) < 4x + (\textbf{-4x}) - 20 \qquad \textbf{Add } -4x \textbf{ to both sides}$$

$$-10x < -20$$

$$\left(-\frac{\textbf{1}}{\textbf{10}}\right)(-10x) \geq \left(-\frac{\textbf{1}}{\textbf{10}}\right)(-20) \qquad \begin{array}{l}\textbf{Multiply by } -\tfrac{1}{10}\textbf{, reverse} \\ \textbf{the sense of the inequality}\end{array}$$

$$x > 2$$

 EXAMPLE 9   Solve $2x - 3y < 6$ for y.

SOLUTION   We can solve this inequality for y by first adding $-2x$ to each side and then multiplying each side by $-\frac{1}{3}$. When we multiply by $-\frac{1}{3}$ we must reverse the direction of the inequality symbol. Because this is an inequality in two variables, we will not graph the solution.

$$2x - 3y < 6 \qquad \text{Original inequality}$$

$$2x + (\mathbf{-2x}) - 3y < (\mathbf{-2x}) + 6 \qquad \text{Add } \mathbf{-2x} \text{ to each side}$$

$$-3y < -2x + 6$$

$$-\frac{1}{3}(-3y) > -\frac{1}{3}(-2x + 6) \qquad \text{Multiply each side by } -\frac{1}{3}$$

$$y > \frac{2}{3}x - 2 \qquad \text{Distributive property} \qquad \text{▨}$$

When working application problems that involve inequalities, the phrases "at least" and "at most" translate as follows:

In Words	In Symbols
x is at least 30	$x \geq 30$
x is at most 20	$x \leq 20$

Applying the Concepts

Our next example is similar to an example done earlier in this chapter. This time it involves an inequality instead of an equation.

We can modify our Blueprint for Problem Solving to solve application problems whose solutions depend on writing and then solving inequalities.

 EXAMPLE 10   The sum of two consecutive odd integers is at most 28. What are the possibilities for the first of the two integers?

SOLUTION When we use the phrase "their sum is at most 28," we mean that their sum is less than or equal to 28.

Step 1: ***Read and list.***
    Known items: Two consecutive odd integers. Their sum is less than or equal to 28.
    Unknown items: The numbers in question.

Step 2: ***Assign a variable, and translate information.***
    If we let $x =$ the first of the two consecutive odd integers, then $x + 2$ is the next consecutive one.

Step 3: ***Reread, and write an inequality.***
    Their sum is at most 28.

$$x + (x + 2) \leq 28$$

Step 4: ***Solve the inequality.***

$$2x + 2 \leq 28 \qquad \text{Simplify the left side}$$

$$2x \leq 26 \qquad \text{Add } -2 \text{ to each side}$$

$$x \leq 13 \qquad \text{Multiply each side by } \frac{1}{2}$$

Step 5: Write the answer.

The first of the two integers must be an odd integer that is less than or equal to 13. The second of the two integers will be two more than whatever the first one is.

Step 6: Reread, and check.

Suppose the first integer is 13. The next consecutive odd integer is 15. The sum of 15 and 13 is 28. If the first odd integer is less than 13, the sum of it and the next consecutive odd integer will be less than 28.

LINKING OBJECTIVES AND EXAMPLES

Next to each **objective** we have listed the examples that are best described by that objective.

A	1, 2
B	3–5
C	6–9
D	1–8
E	10

GETTING READY FOR CLASS

After reading through the preceding section, respond in your own words and in complete sentences.

1. State the addition property for inequalities.
2. How is the multiplication property for inequalities different from the multiplication property of equality?
3. When do we reverse the direction of an inequality symbol?
4. Under what conditions do we not change the direction of the inequality symbol when we multiply both sides of an inequality by a number?

Problem Set 2.8

Online support materials can be found at academic.cengage.com/login

Solve the following inequalities using the addition property of inequalities. Graph each solution set.

1. $x - 5 < 7$

2. $x + 3 < -5$

3. $a - 4 \leq 8$

4. $a + 3 \leq 10$

5. $x - 4.3 > 8.7$

6. $x - 2.6 > 10.4$

7. $y + 6 \geq 10$

8. $y + 3 \geq 12$

9. $2 < x - 7$

10. $3 < x + 8$

Solve the following inequalities using the multiplication property of inequalities. If you multiply both sides by a negative number, be sure to reverse the direction of the inequality symbol. Graph the solution set.

▶ 11. $3x < 6$

12. $2x < 14$

13. $5a \leq 25$

14. $4a \leq 16$

15. $\dfrac{x}{3} > 5$

16. $\dfrac{x}{7} > 1$

17. $-2x > 6$

18. $-3x \geq 9$

19. $-3x \geq -18$

20. $-8x \geq -24$

21. $-\dfrac{x}{5} \leq 10$

22. $-\dfrac{x}{9} \geq -1$

23. $-\dfrac{2}{3}y > 4$

24. $-\dfrac{3}{4}y > 6$

Solve the following inequalities. Graph the solution set in each case.

▶ 25. $2x - 3 < 9$

26. $3x - 4 < 17$

27. $-\dfrac{1}{5}y - \dfrac{1}{3} \leq \dfrac{2}{3}$

28. $-\dfrac{1}{6}y - \dfrac{1}{2} \leq \dfrac{2}{3}$

29. $-7.2x + 1.8 > -19.8$

☐ = Videos available by instructor request

▶ = Online student support materials available at academic.cengage.com/login

30. $-7.8x - 1.3 > 22.1$

31. $\frac{2}{3}x - 5 \le 7$ **32.** $\frac{3}{4}x - 8 \le 1$

33. $-\frac{2}{5}a - 3 > 5$ **34.** $-\frac{4}{5}a - 2 > 10$

35. $5 - \frac{3}{5}y > -10$ **36.** $4 - \frac{5}{6}y > -11$

37. $0.3(a + 1) \le 1.2$ **38.** $0.4(a - 2) \le 0.4$

▶ **39.** $2(5 - 2x) \le -20$ **40.** $7(8 - 2x) > 28$

41. $3x - 5 > 8x$ **42.** $8x - 4 > 6x$

43. $\frac{1}{3}y - \frac{1}{2} \le \frac{5}{6}y + \frac{1}{2}$

44. $\frac{7}{6}y + \frac{4}{3} \le \frac{11}{6}y - \frac{7}{6}$

45. $-2.8x + 8.4 < -14x - 2.8$

46. $-7.2x - 2.4 < -2.4x + 12$

47. $3(m - 2) - 4 \ge 7m + 14$

48. $2(3m - 1) + 5 \ge 8m - 7$

49. $3 - 4(x - 2) \le -5x + 6$

50. $8 - 6(x - 3) \le -4x + 12$

Solve each of the following inequalities for y.

51. $3x + 2y < 6$ **52.** $-3x + 2y < 6$

53. $2x - 5y > 10$ **54.** $-2x - 5y > 5$

55. $-3x + 7y \le 21$ **56.** $-7x + 3y \le 21$

57. $2x - 4y \ge -4$ **58.** $4x - 2y \ge -8$

The next two problems are intended to give you practice reading, and paying attention to, the instructions that accompany the problems you are working.

59. Work each problem according to the instructions given.
  a. Evaluate when $x = 0$: $-5x + 3$
  b. Solve: $-5x + 3 = -7$
  c. Is 0 a solution to $-5x + 3 < -7$
  d. Solve: $-5x + 3 < -7$

60. Work each problem according to the instructions given.
  a. Evaluate when $x = 0$: $-2x - 5$
  b. Solve: $-2x - 5 = 1$
  c. Is 0 a solution to $-2x - 5 > 1$
  d. Solve: $-2x - 5 > 1$

For each graph below, write an inequality whose solution is the graph.

61.

62.

63.

64.

Applying the Concepts

65. Consecutive Integers The sum of two consecutive integers is at least 583. What are the possibilities for the first of the two integers?

66. Consecutive Integers The sum of two consecutive integers is at most 583. What are the possibilities for the first of the two integers?

67. Number Problems The sum of twice a number and six is less than ten. Find all solutions.

68. Number Problems Twice the difference of a number and three is greater than or equal to the number increased by five. Find all solutions.

69. Number Problems The product of a number and four is greater than the number minus eight. Find the solution set.

70. Number Problems The quotient of a number and five is less than the sum of seven and two. Find the solution set.

71. Geometry Problems The length of a rectangle is 3 times the width. If the perimeter is to be at least 48 meters, what are the possible values for the width? (If the perimeter is at least 48 meters, then it is greater than or equal to 48 meters.)

72. Geometry Problems The length of a rectangle is 3 more than twice the width. If the perimeter is to be at least 51 meters, what are the possible values for the width? (If the perimeter is at least 51 meters, then it is greater than or equal to 51 meters.)

73. **Geometry Problems** The numerical values of the three sides of a triangle are given by three consecutive even integers. If the perimeter is greater than 24 inches, what are the possibilities for the shortest side?

74. **Geometry Problems** The numerical values of the three sides of a triangle are given by three consecutive odd integers. If the perimeter is greater than 27 inches, what are the possibilities for the shortest side?

75. **Car Heaters** If you have ever gotten in a cold car early in the morning you know that the heater does not work until the engine warms up. This is because the heater relies on the heat coming off the engine. Write an equation using an inequality sign to express when the heater will work if the heater works only after the engine is 100°F.

76. **Exercise** When Kate exercises, she either swims or runs. She wants to spend a minimum of 8 hours a week exercising, and she wants to swim 3 times the amount she runs. What is the minimum amount of time she must spend doing each exercise?

77. **Profit and Loss** Movie theaters pay a certain price for the movies that you and I see. Suppose a theater pays $1,500 for each showing of a popular movie. If they charge $7.50 for each ticket they sell, then they will lose money if ticket sales are less than $1,500. However, they will make a profit if ticket sales are greater than $1,500. What is the range of tickets they can sell and still lose money? What is the range of tickets they can sell and make a profit?

78. **Stock Sales** Suppose you purchase x shares of a stock at $12 per share. After 6 months you decide to sell all your shares at $20 per share. Your broker charges you $15 for the trade. If your profit is at least $3,985, how many shares did you purchase in the first place?

Maintaining Your Skills

The problems that follow review some of the more important skills you have learned in previous sections and chapters. You can consider the time you spend working these problems as time spent studying for exams.

Apply the distributive property, then simplify.

79. $\frac{1}{6}(12x + 6)$ 80. $\frac{3}{5}(15x - 10)$

81. $\frac{2}{3}(-3x - 6)$ 82. $\frac{3}{4}(-4x - 12)$

83. $3\left(\frac{5}{6}a + \frac{4}{9}\right)$ 84. $2\left(\frac{3}{4}a - \frac{5}{6}\right)$

85. $-3\left(\frac{2}{3}a + \frac{5}{6}\right)$

86. $-4\left(\frac{5}{6}a + \frac{4}{9}\right)$

Apply the distributive property, then find the LCD and simplify.

87. $\frac{1}{2}x + \frac{1}{6}x$ 88. $\frac{1}{2}x - \frac{3}{4}x$

89. $\frac{2}{3}x - \frac{5}{6}x$ 90. $\frac{1}{3}x + \frac{3}{5}x$

91. $\frac{3}{4}x + \frac{1}{6}x$ 92. $\frac{3}{2}x - \frac{2}{3}x$

93. $\frac{2}{5}x + \frac{5}{8}x$ 94. $\frac{3}{5}x - \frac{3}{8}x$

Getting Ready for the Next Section

Solve each inequality. Do not graph.

95. $2x - 1 \geq 3$

96. $3x + 1 \geq 7$

97. $-2x > -8$

98. $-3x > -12$

99. $-3 \leq 4x + 1$

100. $4x + 1 \leq 9$

Chapter 2 SUMMARY

EXAMPLES

1. The terms $2x$, $5x$, and $-7x$ are all similar because their variable parts are the same.

Similar Terms [2.1]

A *term* is a number or a number and one or more variables multiplied together. *Similar terms* are terms with the same variable part.

Simplifying Expressions [2.1]

2. Simplify $3x + 4x$.
$$3x + 4x = (3 + 4)x$$
$$= 7x$$

In this chapter we simplified expressions that contained variables by using the distributive property to combine similar terms.

Solution Set [2.2]

3. The solution set for the equation $x + 2 = 5$ is {3} because when x is 3 the equation is $3 + 2 = 5$ or $5 = 5$.

The *solution set* for an equation (or inequality) is all the numbers that, when used in place of the variable, make the equation (or inequality) a true statement.

Equivalent Equations [2.2]

4. The equation $a - 4 = 3$ and $a - 2 = 5$ are equivalent because both have solution set {7}.

Two equations are called *equivalent* if they have the same solution set.

Addition Property of Equality [2.2]

5. Solve $x - 5 = 12$.
$$x - 5 + \mathbf{5} = 12 + \mathbf{5}$$
$$x + 0 = 17$$
$$x = 17$$

When the same quantity is added to both sides of an equation, the solution set for the equation is unchanged. Adding the same amount to both sides of an equation produces an equivalent equation.

Multiplication Property of Equality [2.3]

6. Solve $3x = 18$.
$$\frac{1}{3}(3x) = \frac{1}{3}(18)$$
$$x = 6$$

If both sides of an equation are multiplied by the same nonzero number, the solution set is unchanged. Multiplying both sides of an equation by a nonzero quantity produces an equivalent equation.

Strategy for Solving Linear Equations in One Variable [2.4]

7. Solve $2(x + 3) = 10$.
$$2x + 6 = 10$$
$$2x + 6 + (\mathbf{-6}) = 10 + (\mathbf{-6})$$
$$2x = 4$$
$$\frac{1}{2}(2x) = \frac{1}{2}(4)$$
$$x = 2$$

Step 1a: Use the distributive property to separate terms, if necessary.

1b: If fractions are present, consider multiplying both sides by the LCD to eliminate the fractions. If decimals are present, consider multiplying both sides by a power of 10 to clear the equation of decimals.

1c: Combine similar terms on each side of the equation.

Step 2: Use the addition property of equality to get all variable terms on one side of the equation and all constant terms on the other side. A variable term is a term that contains the variable (for example, 5x). A constant term is a term that does not contain the variable (the number 3, for example.)

Step 3: Use the multiplication property of equality to get x (that is, 1x) by itself on one side of the equation.

Step 4: Check your solution in the original equation to be sure that you have not made a mistake in the solution process.

Formulas [2.5]

8. Solving $P = 2l + 2w$ for l, we have

$$P - 2w = 2l$$
$$\frac{P - 2w}{2} = l$$

A formula is an equation with more than one variable. To solve a formula for one of its variables, we use the addition and multiplication properties of equality to move everything except the variable in question to one side of the equal sign so the variable in question is alone on the other side.

Blueprint for Problem Solving [2.6, 2.7]

Step 1: **Read** the problem, and then mentally **list** the items that are known and the items that are unknown.

Step 2: **Assign a variable** to one of the unknown items. (In most cases this will amount to letting $x =$ the item that is asked for in the problem.) Then **translate** the other **information** in the problem to expressions involving the variable.

Step 3: **Reread** the problem, and then **write an equation,** using the items and variables listed in steps 1 and 2, that describes the situation.

Step 4: **Solve the equation** found in step 3.

Step 5: **Write** your **answer** using a complete sentence.

Step 6: **Reread** the problem, and **check** your solution with the original words in the problem.

Addition Property for Inequalities [2.8]

9. Solve $x + 5 < 7$.
$$x + 5 + (\mathbf{-5}) < 7 + (\mathbf{-5})$$
$$x < 2$$

Adding the same quantity to both sides of an inequality produces an equivalent inequality, one with the same solution set.

Multiplication Property for Inequalities [2.8]

10. Solve $-3a \leq 18$.
$$-\frac{1}{\mathbf{3}}(-3a) \geq -\frac{1}{\mathbf{3}}(18)$$
$$a \geq -6$$

Multiplying both sides of an inequality by a positive number never changes the solution set. If both sides are multiplied by a negative number, the sense of the inequality must be reversed to produce an equivalent inequality.

Strategy for Solving Linear Inequalities in One Variable [2.8]

11. Solve $3(x - 4) \geq -2$.
$$3x - 12 \geq -2$$
$$3x - 12 + \mathbf{12} \geq -2 + \mathbf{12}$$
$$3x \geq 10$$
$$\frac{\mathbf{1}}{\mathbf{3}}(3x) \geq \frac{\mathbf{1}}{\mathbf{3}}(10)$$
$$x \geq \frac{10}{3}$$

Step 1a: Use the distributive property to separate terms, if necessary.

1b: If fractions are present, consider multiplying both sides by the LCD to eliminate the fractions. If decimals are present, consider multiplying both sides by a power of 10 to clear the inequality of decimals.

1c: Combine similar terms on each side of the inequality.

Step 2: Use the addition property for inequalities to get all variable terms on one side of the inequality and all constant terms on the other side.

Step 3: Use the multiplication property for inequalities to get x by itself on one side of the inequality.

Step 4: Graph the solution set.

 COMMON MISTAKES

1. Trying to subtract away coefficients (the number in front of variables) when solving equations. For example:
$$4x = 12$$
$$4x - \mathbf{4} = 12 - \mathbf{4}$$
$$x = 8 \leftarrow \text{Mistake}$$

   It is not incorrect to add (-4) to both sides; it's just that $4x - 4$ is not equal to x. Both sides should be multiplied by $\frac{1}{4}$ to solve for x.

2. Forgetting to reverse the direction of the inequality symbol when multiplying both sides of an inequality by a negative number. For instance:
$$-3x < 12$$
$$-\frac{1}{3}(-3x) < -\frac{1}{3}(12) \leftarrow \text{Mistake}$$
$$x < -4$$

   It is not incorrect to multiply both sides by $-\frac{1}{3}$. But if we do, we must also reverse the sense of the inequality.

The problems below form a comprehensive review of the material in this chapter. They can be used to study for exams. If you would like to take a practice test on this chapter, you can use the odd-numbered problems. Give yourself an hour and work as many of the odd-numbered problems as possible. When you are finished, or when an hour has passed, check your answers with the answers in the back of the book. You can use the even-numbered problems for a second practice test.

The numbers in brackets refer to the sections of the text in which similar problems can be found.

Simplify each expression as much as possible. [2.1]

1. $5x - 8x$

2. $6x - 3 - 8x$

3. $-a + 2 + 5a - 9$

4. $5(2a - 1) - 4(3a - 2)$

5. $6 - 2(3y + 1) - 4$

6. $4 - 2(3x - 1) - 5$

Find the value of each expression when x is 3. [2.1]

7. $7x - 2$

8. $-4x - 5 + 2x$

9. $-x - 2x - 3x$

Find the value of each expression when x is -2. [2.1]

10. $5x - 3$

11. $-3x + 2$

12. $7 - x - 3$

Solve each equation. [2.2, 2.3]

13. $x + 2 = -6$

14. $x - \dfrac{1}{2} = \dfrac{4}{7}$

15. $10 - 3y + 4y = 12$

16. $-3 - 4 = -y - 2 + 2y$

17. $2x = -10$

18. $3x = 0$

19. $\dfrac{x}{3} = 4$

20. $-\dfrac{x}{4} = 2$

21. $3a - 2 = 5a$

22. $\dfrac{7}{10}a = \dfrac{1}{5}a + \dfrac{1}{2}$

23. $3x + 2 = 5x - 8$

24. $6x - 3 = x + 7$

25. $0.7x - 0.1 = 0.5x - 0.1$

26. $0.2x - 0.3 = 0.8x - 0.3$

Solve each equation. Be sure to simplify each side first. [2.4]

27. $2(x - 5) = 10$

28. $12 = 2(5x - 4)$

29. $\dfrac{1}{2}(3t - 2) + \dfrac{1}{2} = \dfrac{5}{2}$

30. $\dfrac{3}{5}(5x - 10) = \dfrac{2}{3}(9x + 3)$

31. $2(3x + 7) = 4(5x - 1) + 18$

32. $7 - 3(y + 4) = 10$

Use the formula $4x - 5y = 20$ to find y if [2.5]

33. x is 5

34. x is 0

35. x is -5

36. x is 10

Solve each of the following formulas for the indicated variable. [2.5]

37. $2x - 5y = 10$ for y

38. $5x - 2y = 10$ for y

39. $V = \pi r^2 h$ for h

40. $P = 2l + 2w$ for w

41. What number is 86% of 240? [2.5]

42. What percent of 2,000 is 180? [2.5]

Solve each of the following word problems. In each case, be sure to show the equation that describes the situation. [2.6, 2.7]

43. Number Problem The sum of twice a number and 6 is 28. Find the number.

44. Geometry The length of a rectangle is 5 times as long as the width. If the perimeter is 60 meters, find the length and the width.

45. Investing A man invests a certain amount of money in an account that pays 9% annual interest. He invests $300 more than that in an account that pays 10% annual interest. If his total interest after a year is $125, how much does he have invested in each account?

46. Coin Problem A collection of 15 coins is worth $1.00. If the coins are dimes and nickels, how many of each coin are there?

Solve each inequality. [2.8]

47. $-2x < 4$

48. $-5x > -10$

49. $-\dfrac{a}{2} \le -3$

50. $-\dfrac{a}{3} > 5$

Solve each inequality, and graph the solution. [2.8]

51. $-4x + 5 > 37$

52. $2x + 10 < 5x - 11$

53. $2(3t + 1) + 6 \ge 5(2t + 4)$

GROUP PROJECT   Tables and Graphs

Number of People   2-3

Time Needed   5–10 minutes

Equipment   Pencil and graph paper

Background   Building tables is a method of visualizing information. We can build a table from a situation (as below) or from an equation. In this project, we will first build a table and then write an equation from the information in the table.

Procedure   A parking meter, which accepts only dimes and quarters, is emptied at the end of each day. The amount of money in the meter at the end of one particular day is $3.15.

1. Complete the following table so that all possible combinations of dimes and quarters, along with the total number of coins, is shown. Remember, although the number of coins will vary, the value of the dimes and quarters must total $3.15.

2. From the information in the table, answer the following questions.

   a. What is the maximum possible number of coins taken from the meter?

   b. What is the minimum possible number of coins taken from the meter?

   c. When is the number of dimes equal to the number of quarters?

3. Let x = the number of dimes and y = the number of quarters. Write an equation in two variables such that the value of the dimes added to the value of the quarters is $3.15.

Number of Dimes	Quarters	Total Coins	Value
29	1	30	$3.15
24			$3.15
			$3.15
			$3.15
			$3.15
			$3.15

Stand and Deliver

The Kobal Collection

The 1988 film *Stand and Deliver* starring Edward James Olmos and Lou Diamond Phillips is based on a true story. Olmos, in his portrayal of high-school math teacher Jaime Escalante, earned an Academy Award nomination for best actor.

Watch the movie *Stand and Deliver*. After briefly describing the movie, explain how Escalante's students became successful in math. Make a list of specific things you observe that the students had to do to become successful. Indicate which items on this list you think will also help you become successful.

Graphing and Systems of Equations

3

Klaus Hackenberg/zefa/Corbis

Whenen light comes into contact with a surface that does not transmit light, then all the light that contacts the surface is either reflected off the surface or absorbed into the surface. If we let R represent the percentage of light reflected and A represent the percentage of light absorbed, then the relationship between these two variables can be written as

$$R + A = 100$$

which is a linear equation in two variables. The following table and graph show the same relationship as that described by the equation. The table is a numerical description; the graph is a visual description.

Reflected and Absorbed Light	
Percent Reflected	Percent Absorbed
0	100
20	80
40	60
60	40
80	20
100	0

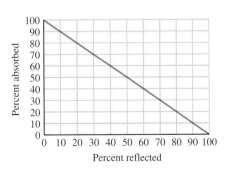

In this chapter we learn how to build tables and draw graphs from linear equations in two variables.

▶ Improve your grade and save time!
Go online to **academic.cengage.com/login** where you can
• Watch videos of instructors working through the in-text examples
• Follow step-by-step online tutorials of in-text examples and review questions
• Work practice problems
• Check your readiness for an exam by taking a pre-test and exploring the modules recommended in your Personalized Study plan
• Receive help from a live tutor online through vMentor™
Try it out! Log in with an access code or purchase access at **www.ichapters.com**.

Try to arrange your daily study habits so that you have very little studying to do the night before your next exam. The next two goals will help you achieve this.

1 Review with the Exam in Mind

Each day you should review material that will be covered on the next exam. Your review should consist of working problems. Preferably, the problems you work should be problems from your list of difficult problems.

2 Pay Attention to Instructions

Each of the following is a valid instruction with respect to the equation $y = 3x - 2$, and the result of applying the instructions will be different in each case:

Find x when y is 10.	(Section 2.5)
Solve for x.	(Section 2.5)
Graph the equation.	(Section 3.3)
Find the intercepts.	(Section 3.4)

There are many things to do with the equation $y = 3x - 2$. If you train yourself to pay attention to the instructions that accompany a problem as you work through the assigned problems, you will not find yourself confused about what to do with a problem when you see it on a test.

Paired Data and Graphing Ordered Pairs

OBJECTIVES

A Create a bar chart, scatter diagram, or line graph from a table of data.

B Graph ordered pairs on a rectangular coordinate system.

In Chapter 1 we showed the relationship between the table of values for the speed of a race car and the corresponding bar chart. Table 1 and Figure 1 from the introduction of Chapter 1 are reproduced here for reference. In Figure 1, the horizontal line that shows the elapsed time in seconds is called the *horizontal axis,* and the vertical line that shows the speed in miles per hour is called the *vertical axis.*

The data in Table 1 are called *paired data* because the information is organized so that each number in the first column is paired with a specific number in the second column. Each pair of numbers is associated with one of the solid bars in Figure 1. For example, the third bar in the bar chart is associated with the pair of numbers 3 seconds and 162.8 miles per hour. The first number, 3 seconds, is associated with the horizontal axis, and the second number, 162.8 miles per hour, is associated with the vertical axis.

TABLE 1

Speed of a Race Car

Time in Seconds	Speed in Miles per Hour
0	0
1	72.7
2	129.9
3	162.8
4	192.2
5	212.4
6	228.1

Scatter Diagrams and Line Graphs

The information in Table 1 can be visualized with a *scatter diagram* and *line graph* as well. Figure 2 is a scatter diagram of the information in Table 1. We use dots instead of the bars shown in Figure 1 to show the speed of the race car at each second during the race. Figure 3 is called a *line graph.* It is constructed by taking the dots in Figure 2 and connecting each one to the next with a straight line. Notice that we have labeled the axes in these two figures a little differently than we did with the bar chart by making the axes intersect at the number 0.

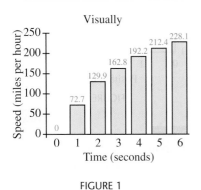

FIGURE 1

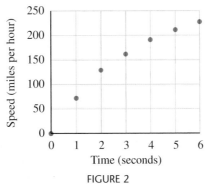

FIGURE 2

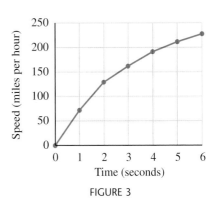

FIGURE 3

The number sequences we have worked with in the past can also be written as paired data by associating each number in the sequence with its position in the sequence. For instance, in the sequence of odd numbers

$$1, 3, 5, 7, 9, \ldots$$

the number 7 is the fourth number in the sequence. Its position is 4, and its value is 7. Here is the sequence of odd numbers written so that the position of each term is noted:

Position   1, 2, 3, 4, 5, . . .

Value   1, 3, 5, 7, 9, . . .

Getting Ready for the Next Section

69. Let $2x + 3y = 6$.
    a. Find x if $y = 4$.   **b.** Find x if $y = -2$.
    c. Find y if $x = 3$.   **d.** Find y if $x = 9$.

70. Let $2x - 5y = 20$.
    a. Find x if $y = 0$.   **b.** Find x if $y = -6$.
    c. Find y if $x = 0$.   **d.** Find y if $x = 5$.

71. Let $y = 2x - 1$.
    a. Find x if $y = 7$.   **b.** Find x if $y = 3$.
    c. Find y if $x = 0$.   **d.** Find y if $x = 5$.

72. Let $y = 3x - 2$.
    a. Find x if $y = 4$.   **b.** Find x if $y = 7$.
    c. Find y if $x = 0$.   **d.** Find y if $x = -3$.

3.2 Solutions to Linear Equations in Two Variables

OBJECTIVES

A Find solutions to linear equations in two variables.

B Decide if an ordered pair is a solution to a linear equation in two variables.

In this section we will begin to investigate equations in two variables. As you will see, equations in two variables have pairs of numbers for solutions. Because we know how to use paired data to construct tables, histograms, and other charts, we can take our work with paired data further by using equations in two variables to construct tables of paired data. Let's begin this section by reviewing the relationship between equations in one variable and their solutions.

If we solve the equation $3x - 2 = 10$, the solution is $x = 4$. If we graph this solution, we simply draw the real number line and place a dot at the point whose coordinate is 4. The relationship between linear equations in one variable, their solutions, and the graphs of those solutions look like this:

Equation	Solution	Graph of Solution Set
$3x - 2 = 10$	$x = 4$	
$x + 5 = 7$	$x = 2$	
$2x = -6$	$x = -3$	

When the equation has one variable, the solution is a single number whose graph is a point on a line.

Now, consider the equation $2x + y = 3$. The first thing we notice is that there are two variables instead of one. Therefore, a solution to the equation $2x + y = 3$ will be not a single number but a pair of numbers, one for x and one for y, that makes the equation a true statement. One pair of numbers that works is $x = 2$, $y = -1$ because when we substitute them for x and y in the equation, we get a true statement.

$$2(2) + (-1) \overset{?}{=} 3$$

$$4 - 1 = 3$$

$$3 = 3 \quad \textbf{A true statement}$$

The pair of numbers $x = 2, y = -1$ is written as $(2, -1)$. As you know from Section 3.1, $(2, -1)$ is called an *ordered pair* because it is a pair of numbers written in a specific order. The first number is always associated with the variable x,

Note

If this discussion seems a little long and confusing, you may want to look over some of the examples first and then come back and read this. Remember, it isn't always easy to read material in mathematics. What is important is that you understand what you are doing when you work problems. The reading is intended to assist you in understanding what you are doing. It is important to read everything in the book, but you don't always have to read it in the order it is written.

and the second number is always associated with the variable y. We call the first number in the ordered pair the *x-coordinate* (or x component) and the second number the *y-coordinate* (or y component) of the ordered pair.

Let's look back to the equation $2x + y = 3$. The ordered pair $(2, -1)$ is not the only solution. Another solution is $(0, 3)$ because when we substitute 0 for x and 3 for y we get

$$2(0) + 3 \stackrel{?}{=} 3$$

$$0 + 3 = 3$$

$$3 = 3 \qquad \textbf{A true statement}$$

Still another solution is the ordered pair $(5, -7)$ because

$$2(5) + (-7) \stackrel{?}{=} 3$$

$$10 - 7 = 3$$

$$3 = 3 \qquad \textbf{A true statement}$$

As a matter of fact, for any number we want to use for x, there is another number we can use for y that will make the equation a true statement. There is an infinite number of ordered pairs that satisfy (are solutions to) the equation $2x + y = 3$; we have listed just a few of them.

 EXAMPLE 1   Given the equation $2x + 3y = 6$, complete the following ordered pairs so they will be solutions to the equation: $(0, \ ), (\ , 1), (3, \ )$.

SOLUTION   To complete the ordered pair $(0, \ )$, we substitute 0 for x in the equation and then solve for y:

$$2(0) + 3y = 6$$

$$3y = 6$$

$$y = 2$$

The ordered pair is $(0, 2)$.

To complete the ordered pair $(\ , 1)$, we substitute 1 for y in the equation and solve for x:

$$2x + 3(1) = 6$$

$$2x + 3 = 6$$

$$2x = 3$$

$$x = \frac{3}{2}$$

The ordered pair is $(\frac{3}{2}, 1)$.

To complete the ordered pair $(3, \ )$, we substitute 3 for x in the equation and solve for y:

$$2(3) + 3y = 6$$

$$6 + 3y = 6$$

$$3y = 0$$

$$y = 0$$

The ordered pair is $(3, 0)$.

Notice in each case that once we have used a number in place of one of the variables, the equation becomes a linear equation in one variable. We then use the method explained in Chapter 2 to solve for that variable.

 EXAMPLE 2   Complete the following table for the equation $2x - 5y = 20$.

x	y
0	
	2
	0
−5	

SOLUTION   Filling in the table is equivalent to completing the following ordered pairs: $(0, \ )$, $(\ , 2)$, $(\ , 0)$, $(-5, \ )$. So we proceed as in Example 1.

When $x = 0$, we have

$$2(0) - 5y = 20$$
$$0 - 5y = 20$$
$$-5y = 20$$
$$y = -4$$

When $y = 2$, we have

$$2x - 5(2) = 20$$
$$2x - 10 = 20$$
$$2x = 30$$
$$x = 15$$

When $y = 0$, we have

$$2x - 5(0) = 20$$
$$2x - 0 = 20$$
$$2x = 20$$
$$x = 10$$

When $x = -5$, we have

$$2(-5) - 5y = 20$$
$$-10 - 5y = 20$$
$$-5y = 30$$
$$y = -6$$

The completed table looks like this:

x	y
0	−4
15	2
10	0
−5	−6

which is equivalent to the ordered pairs $(0, -4)$, $(15, 2)$, $(10, 0)$, and $(-5, -6)$.

 EXAMPLE 3   Complete the following table for the equation $y = 2x - 1$.

x	y
0	
5	
	7
	3

SOLUTION   When $x = 0$, we have   When $x = 5$, we have

$$y = 2(0) - 1$$ $$y = 2(5) - 1$$

$$y = 0 - 1$$ $$y = 10 - 1$$

$$y = -1$$ $$y = 9$$

When $y = 7$, we have   When $y = 3$, we have

$$7 = 2x - 1$$ $$3 = 2x - 1$$

$$8 = 2x$$ $$4 = 2x$$

$$4 = x$$ $$2 = x$$

The completed table is

x	y
0	-1
5	9
4	7
2	3

which means the ordered pairs $(0, -1)$, $(5, 9)$, $(4, 7)$, and $(2, 3)$ are among the solutions to the equation $y = 2x - 1$. ▨

 EXAMPLE 4   Which of the ordered pairs $(2, 3)$, $(1, 5)$, and $(-2, -4)$ are solutions to the equation $y = 3x + 2$?

SOLUTION   If an ordered pair is a solution to the equation, then it must satisfy the equation; that is, when the coordinates are used in place of the variables in the equation, the equation becomes a true statement.

Try $(2, 3)$ in $y = 3x + 2$:

$$3 \stackrel{?}{=} 3(2) + 2$$

$$3 = 6 + 2$$

$$3 = 8 \qquad \textbf{A false statement}$$

Try $(1, 5)$ in $y = 3x + 2$:

$$5 \stackrel{?}{=} 3(1) + 2$$

$$5 = 3 + 2$$

$$5 = 5 \qquad \textbf{A true statement}$$

Try $(-2, -4)$ in $y = 3x + 2$:

$$-4 \stackrel{?}{=} 3(-2) + 2$$

$$-4 = -6 + 2$$

$$-4 = -4 \qquad \textbf{A true statement}$$

The ordered pairs $(1, 5)$ and $(-2, -4)$ are solutions to the equation $y = 3x + 2$, and $(2, 3)$ is not. ▨

LINKING OBJECTIVES AND EXAMPLES

Next to each **objective** we have listed the examples that are best described by that objective.

A	1–3
B	4

GETTING READY FOR CLASS

After reading through the preceding section, respond in your own words and in complete sentences.

1. How can you tell if an ordered pair is a solution to an equation?
2. How would you find a solution to $y = 3x - 5$?
3. Why is $(3, 2)$ not a solution to $y = 3x - 5$?
4. How many solutions are there to an equation that contains two variables?

Problem Set 3.2

Online support materials can be found at academic.cengage.com/login

For each equation, complete the given ordered pairs.

▶ **1.** $2x + y = 6$ $(0, \ ), (\ , 0), (\ , -6)$

2. $3x - y = 5$ $(0, \ ), (1, \ ), (\ , 5)$

3. $3x + 4y = 12$ $(0, \ ), (\ , 0), (-4, \ )$

4. $5x - 5y = 20$ $(0, \ ), (\ , -2), (1, \ )$

5. $y = 4x - 3$ $(1, \ ), (\ , 0), (5, \ )$

6. $y = 3x - 5$ $(\ , 13), (0, \ ), (-2, \ )$

7. $y = 7x - 1$ $(2, \ ), (\ , 6), (0, \ )$

8. $y = 8x + 2$ $(3, \ ), (\ , 0), (\ , -6)$

▶ **9.** $x = -5$ $(\ , 4), (\ , -3), (\ , 0)$

10. $y = 2$ $(5, \ ), (-8, \ ), \left(\dfrac{1}{2}, \ \right)$

For each of the following equations, complete the given table.

▶ **11.** $y = 3x$

x	y
1	3
-3	
	12
	18

12. $y = -2x$

x	y
-4	
0	
	10
	12

13. $y = 4x$

x	y
0	
	-2
-3	
	12

14. $y = -5x$

x	y
3	
	0
-2	
	-20

15. $x + y = 5$

x	y
2	
3	
	0
	-4

16. $x - y = 8$

x	y
0	
4	
	-3
	-2

17. $2x - y = 4$

x	y
	0
	2
1	
-3	

18. $3x - y = 9$

x	y
	0
	-9
5	
-4	

19. $y = 6x - 1$

x	y
0	
	-7
-3	
	8

20. $y = 5x + 7$

x	y
0	
-2	
-4	
	-8

= Videos available by instructor request

▶ = Online student support materials available at academic.cengage.com/login

For the following equations, tell which of the given ordered pairs are solutions.

21. $2x - 5y = 10$ $(2, 3), (0, -2), \left(\dfrac{5}{2}, 1\right)$

22. $3x + 7y = 21$ $(0, 3), (7, 0), (1, 2)$

23. $y = 7x - 2$ $(1, 5), (0, -2), (-2, -16)$

24. $y = 8x - 3$ $(0, 3), (5, 16), (1, 5)$

▶ **25.** $y = 6x$ $(1, 6), (-2, -12), (0, 0)$

26. $y = -4x$ $(0, 0), (2, 4), (-3, 12)$

27. $x + y = 0$ $(1, 1), (2, -2), (3, 3)$

28. $x - y = 1$ $(0, 1), (0, -1), (1, 2)$

29. $x = 3$ $(3, 0), (3, -3), (5, 3)$

30. $y = -4$ $(3, -4), (-4, 4), (0, -4)$

Applying the Concepts

31. Perimeter If the perimeter of a rectangle is 30 inches, then the relationship between the length l and the width w is given by the equation

$$2l + 2w = 30$$

What is the length when the width is 3 inches?

32. Perimeter The relationship between the perimeter P of a square and the length of its side s is given by the formula $P = 4s$. If each side of a square is 5 inches, what is the perimeter? If the perimeter of a square is 28 inches, how long is a side?

33. Janai earns $12 per hour working as a math tutor. We can express the amount she earns each week, y, for working x hours with the equation $y = 12x$. Indicate with a *yes* or *no* which of the following could be one of Janai's paychecks. If you answer no, explain your answer.
   a. $60 for working 5 hours

   b. $100 for working nine hours

   c. $80 for working 7 hours

   d. $168 for working 14 hours

34. Erin earns $15 per hour working as a graphic designer. We can express the amount she earns each week, y, for working x hours with the equation $y = 15x$. Indicate with a *yes* or *no* which of the following could be one of Erin's paychecks. If you answer no, explain your answer.
   a. $75 for working 5 hours

   b. $125 for working 9 hours.

   c. $90 for working 6 hours

   d. $500 for working 35 hours.

35. The equation $V = -45{,}000t + 600{,}000$ can be used to find the value, V, of a small crane at the end of t years.
   a. What is the value of the crane at the end of 5 years?

   b. When is the crane worth $330,000?

   c. Is it true that the crane will be worth $150,000 after 9 years?

   d. How much did the crane cost?

36. The equation $P = -400t + 2{,}500$, can be used to find the price, P, of a notebook computer at the end of t years.
   a. What is the value of the notebook computer at the end of 4 years?

   b. When is the notebook computer worth $1,700?

   c. Is it true that the notebook computer will be worth $100 after 5 years?

   d. How much did the notebook computer cost?

Maintaining Your Skills

37. $\dfrac{11(-5) - 17}{2(-6)}$

38. $\dfrac{12(-4) + 15}{3(-11)}$

39. $\dfrac{13(-6) + 18}{4(-5)}$

40. $\dfrac{9^2 - 6^2}{-9 - 6}$

41. $\dfrac{7^2 - 5^2}{(7 - 5)^2}$

42. $\dfrac{7^2 - 2^2}{-7 - 2}$

43. $\dfrac{-3 \cdot 4^2 - 3 \cdot 2^4}{-3(8)}$

44. $\dfrac{-4(8 - 13) - 2(6 - 11)}{-5(3) + 5}$

Getting Ready for the Next Section

45. Find y when x is 4 in the formula $3x + 2y = 6$.

46. Find y when x is 0 in the formula $3x + 2y = 6$.

47. Find y when x is 0 in $y = -\dfrac{1}{3}x + 2$.

48. Find y when x is 3 in $y = -\dfrac{1}{3}x + 2$.

49. Find y when x is 2 in $y = \dfrac{3}{2}x - 3$.

50. Find y when x is 4 in $y = \dfrac{3}{2}x - 3$.

51. Solve $5x + y = 4$ for y.

52. Solve $-3x + y = 5$ for y.

53. Solve $3x - 2y = 6$ for y.

54. Solve $2x - 3y = 6$ for y.

3.3 Graphing Linear Equations in Two Variables

OBJECTIVES

A Graph a linear equation in two variables.

B Graph horizontal lines, vertical lines, and lines through the origin.

In this section we will use the rectangular coordinate system introduced in Section 3.1 to obtain a visual picture of *all* solutions to a linear equation in two variables. The process we use to obtain a visual picture of all solutions to an equation is called *graphing*. The picture itself is called the *graph* of the equation. Graphing equations is an important part of algebra. It is a powerful tool that we can use to help apply algebra to the world around us.

 EXAMPLE 1   Graph the solution set for $x + y = 5$.

SOLUTION   We know from the previous section that an infinite number of ordered pairs are solutions to the equation $x + y = 5$. We can't possibly list them all. What we can do is list a few of them and see if there is any pattern to their graphs.

Some ordered pairs that are solutions to $x + y = 5$ are $(0, 5)$, $(2, 3)$, $(3, 2)$, $(5, 0)$. The graph of each is shown in Figure 1.

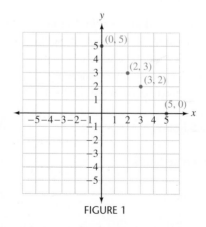

FIGURE 1

Now, by passing a straight line through these points we can graph the solution set for the equation $x + y = 5$. Linear equations in two variables always have graphs that are straight lines. The graph of the solution set for $x + y = 5$ is shown in Figure 2.

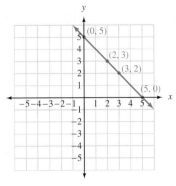

FIGURE 2

Every ordered pair that satisfies $x + y = 5$ has its graph on the line, and any point on the line has coordinates that satisfy the equation. So, there is a one-to-one correspondence between points on the line and solutions to the equation.

Our ability to graph an equation as we have done in Example 1 is due to the invention of the rectangular coordinate system. The French philosopher René Descartes (1596–1650) is the person usually credited with the invention of the rectangular coordinate system. As a philosopher, Descartes is responsible for the statement "I think, therefore I am." Until Descartes invented his coordinate system in 1637, algebra and geometry were treated as separate subjects. The rectangular coordinate system allows us to connect algebra and geometry by associating geometric shapes with algebraic equations.

Here is the precise definition for a linear equation in two variables.

> **DEFINITION** Any equation that can be put in the form $ax + by = c$, where a, b, and c are real numbers and a and b are not both 0, is called a **linear equation in two variables.** The graph of any equation of this form is a straight line (that is why these equations are called "linear"). The form $ax + by = c$ is called **standard form.**

To graph a linear equation in two variables, we simply graph its solution set; that is, we draw a line through all the points whose coordinates satisfy the equation. Here are the steps to follow.

Note

The meaning of the *convenient numbers* referred to in step 1 will become clear as you read the next two examples.

> **To Graph a Linear Equation in Two Variables**
> **Step 1:** Find any three ordered pairs that satisfy the equation. This can be done by using a convenient number for one variable and solving for the other variable.
>
> **Step 2:** Graph the three ordered pairs found in step 1. Actually, we need only two points to graph a straight line. The third point serves as a check. If all three points do not line up, there is a mistake in our work.
>
> **Step 3:** Draw a straight line through the three points graphed in step 2.

EXAMPLE 2   Graph the equation $y = 3x - 1$.

SOLUTION   Because $y = 3x - 1$ can be put in the form $ax + by = c$, it is a linear equation in two variables. Hence, the graph of its solution set is a straight line. We can find some specific solutions by substituting numbers for x and then solving for the corresponding values of y. We are free to choose any numbers for x, so let's use 0, 2, and -1.

Let $x = 0$: $y = 3(0) - 1$
$y = 0 - 1$
$y = -1$

The ordered pair $(0, -1)$ is one solution.

Let $x = 2$: $y = 3(2) - 1$
$y = 6 - 1$
$y = 5$

The ordered pair $(2, 5)$ is a second solution.

Let $x = -1$: $y = 3(-1) - 1$
$y = -3 - 1$
$y = -4$

The ordered pair $(-1, -4)$ is a third solution.

In table form

x	y
0	-1
2	5
-1	-4

> **Note**
>
> It may seem that we have simply picked the numbers 0, 2, and -1 out of the air and used them for x. In fact we have done just that. Could we have used numbers other than these? The answer is yes, we can substitute any number for x; there will always be a value of y to go with it.

Next, we graph the ordered pairs $(0, -1)$, $(2, 5)$, $(-1, -4)$ and draw a straight line through them.

The line we have drawn in Figure 3 is the graph of $y = 3x - 1$.

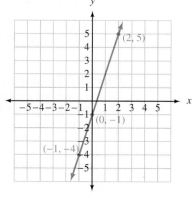

FIGURE 3

Example 2 again illustrates the connection between algebra and geometry that we mentioned previously. Descartes's rectangular coordinate system allows us to associate the equation $y = 3x - 1$ (an algebraic concept) with a specific straight line (a geometric concept). The study of the relationship between equations in algebra and their associated geometric figures is called *analytic geometry.* The rectangular coordinate system often is referred to as the *Cartesian coordinate system* in honor of Descartes.

EXAMPLE 3   Graph the equation $y = -\frac{1}{3}x + 2$.

SOLUTION   We need to find three ordered pairs that satisfy the equation. To do so, we can let x equal any numbers we choose and find corresponding values of

y. But, every value of x we substitute into the equation is going to be multiplied by $-\frac{1}{3}$. Let's use numbers for x that are divisible by 3, like -3, 0, and 3. That way, when we multiply them by $-\frac{1}{3}$, the result will be an integer.

$$\text{Let } x = -3: \quad y = -\frac{1}{3}(-3) + 2$$

$$y = 1 + 2$$

$$y = 3$$

| In table form |

x	y
-3	3
0	2
3	1

The ordered pair $(-3, 3)$ is one solution.

$$\text{Let } x = 0: \quad y = -\frac{1}{3}(0) + 2$$

$$y = 0 + 2$$

$$y = 2$$

The ordered pair $(0, 2)$ is a second solution.

$$\text{Let } x = 3: \quad y = -\frac{1}{3}(3) + 2$$

$$y = -1 + 2$$

$$y = 1$$

The ordered pair $(3, 1)$ is a third solution.

> ### Note
>
> In Example 3 the values of x we used, -3, 0, and 3, are referred to as convenient values of x because they are easier to work with than some other numbers. For instance, if we let $x = 2$ in our original equation, we would have to add $-\frac{2}{3}$ and 2 to find the corresponding value of y. Not only would the arithmetic be more difficult but also the ordered pair we obtained would have a fraction for its y-coordinate, making it more difficult to graph accurately.

Graphing the ordered pairs $(-3, 3)$, $(0, 2)$, and $(3, 1)$ and drawing a straight line through their graphs, we have the graph of the equation $y = -\frac{1}{3}x + 2$, as shown in Figure 4.

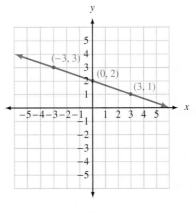

FIGURE 4

 EXAMPLE 4   Graph the solution set for $3x - 2y = 6$.

SOLUTION   It will be easier to find convenient values of x to use in the equation if we first solve the equation for y. To do so, we add $-3x$ to each side, and then we multiply each side by $-\frac{1}{2}$.

$$3x - 2y = 6 \qquad \qquad \text{Original equation}$$

$$-2y = -3x + 6 \qquad \text{Add } -3x \text{ to each side}$$

$$-\frac{1}{2}(-2y) = -\frac{1}{2}(-3x + 6) \qquad \textbf{Multiply each side by } -\frac{1}{2}$$

$$y = \frac{3}{2}x - 3 \qquad \textbf{Simplify each side}$$

Now, because each value of x will be multiplied by $\frac{3}{2}$, it will be to our advantage to choose values of x that are divisible by 2. That way, we will obtain values of y that do not contain fractions. This time, let's use 0, 2, and 4 for x.

When $x = 0$: $\quad y = \frac{3}{2}(0) - 3$

$$y = 0 - 3$$

$$y = -3 \qquad \textbf{(0, −3) is one solution}$$

When $x = 2$: $\quad y = \frac{3}{2}(2) - 3$

$$y = 3 - 3$$

$$y = 0 \qquad \textbf{(2, 0) is a second solution}$$

When $x = 4$: $\quad y = \frac{3}{2}(4) - 3$

$$y = 6 - 3$$

$$y = 3 \qquad \textbf{(4, 3) is a third solution}$$

Graphing the ordered pairs $(0, -3)$, $(2, 0)$, and $(4, 3)$ and drawing a line through them, we have the graph shown in Figure 5.

Note

After reading through Example 4, many students ask why we didn't use −2 for x when we were finding ordered pairs that were solutions to the original equation. The answer is, we could have. If we were to let $x = -2$, the corresponding value of y would have been −6. As you can see by looking at the graph in Figure 5, the ordered pair $(-2, -6)$ is on the graph.

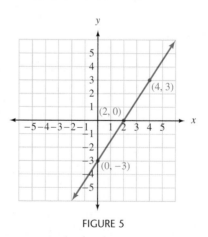

FIGURE 5

 EXAMPLE 5   Graph each of the following lines.

a. $y = \frac{1}{2}x$ b. $x = 3$ c. $y = -2$

SOLUTION

a. The line $y = \frac{1}{2}x$ passes through the origin because $(0, 0)$ satisfies the equation. To sketch the graph we need at least one more point on the line. When x is 2, we obtain the point $(2, 1)$, and when x is -4, we obtain the point $(-4, -2)$. The graph of $y = \frac{1}{2}x$ is shown in Figure 6a.

b. The line $x = 3$ is the set of all points whose x-coordinate is 3. The variable y does not appear in the equation, so the y-coordinate can be any number. Note that we can write our equation as a linear equation in two variables by writing it as $x + 0y = 3$. Because the product of 0 and y will always be 0, y can be any number. The graph of $x = 3$ is the vertical line shown in Figure 6b.

c. The line $y = -2$ is the set of all points whose y-coordinate is -2. The variable x does not appear in the equation, so the x-coordinate can be any number. Again, we can write our equation as a linear equation in two variables by writing it as $0x + y = -2$. Because the product of 0 and x will always be 0, x can be any number. The graph of $y = -2$ is the horizontal line shown in Figure 6c.

FIGURE 6A   FIGURE 6B   FIGURE 6C

FACTS FROM GEOMETRY

Special Equations and Their Graphs

For the equations below, m, a, and b are real numbers.

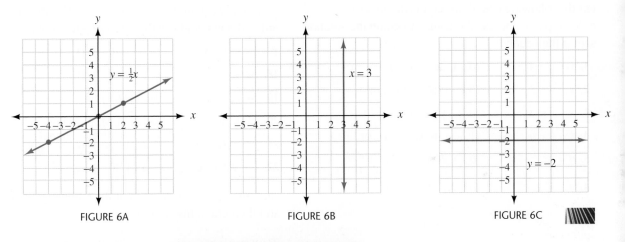

Through the Origin   Vertical Line   Horizontal Line

$y = mx$ $x = a$ $y = b$

FIGURE 7A Any equation of the form $y = mx$ has a graph that passes through the origin.

FIGURE 7B Any equation of the form $x = a$ has a vertical line for its graph.

FIGURE 7C Any equation of the form $y = b$ has a horizontal line for its graph.

LINKING OBJECTIVES AND EXAMPLES

Next to each **objective** we have listed the examples that are best described by that objective.

A 1–4

B 5

GETTING READY FOR CLASS

After reading through the preceding section, respond in your own words and in complete sentences.

1. Explain how you would go about graphing the line $x + y = 5$.

2. When graphing straight lines, why is it a good idea to find three points, when every straight line is determined by only two points?

3. What kind of equations have vertical lines for graphs?

4. What kind of equations have horizontal lines for graphs?

Problem Set 3.3

Online support materials can be found at academic.cengage.com/login

For the following equations, complete the given ordered pairs, and use the results to graph the solution set for the equation.

1. $x + y = 4$ $(0,), (2,), (, 0)$

2. $x - y = 3$ $(0,), (2,), (, 0)$

▶ **3.** $x + y = 3$ $(0,), (2,), (, -1)$

4. $x - y = 4$ $(1,), (-1,), (, 0)$

5. $y = 2x$ $(0,), (-2,), (2,)$

6. $y = \dfrac{1}{2}x$ $(0,), (-2,), (2,)$

7. $y = \dfrac{1}{3}x$ $(-3,), (0,), (3,)$

8. $y = 3x$ $(-2,), (0,), (2,)$

9. $y = 2x + 1$ $(0,), (-1,), (1,)$

10. $y = -2x + 1$ $(0,), (-1,), (1,)$

11. $y = 4$ $(0,), (-1,), (2,)$

12. $x = 3$ $(, -2), (, 0), (, 5)$

13. $y = \dfrac{1}{2}x + 3$ $(-2,), (0,), (2,)$

14. $y = \dfrac{1}{2}x - 3$ $(-2,), (0,), (2,)$

▶ **15.** $y = -\dfrac{2}{3}x + 1$ $(-3,), (0,), (3,)$

16. $y = -\dfrac{2}{3}x - 1$ $(-3,), (0,), (3,)$

Solve each equation for y. Then, complete the given ordered pairs, and use them to draw the graph.

17. $2x + y = 3$ $(-1,), (0,), (1,)$

18. $3x + y = 2$ $(-1,), (0,), (1,)$

19. $3x + 2y = 6$ $(0,), (2,), (4,)$

20. $2x + 3y = 6$ $(0,), (3,), (6,)$

21. $-x + 2y = 6$ $(-2,), (0,), (2,)$

22. $-x + 3y = 6$ $(-3,), (0,), (3,)$

Find three solutions to each of the following equations, and then graph the solution set.

23. $y = -\dfrac{1}{2}x$ **24.** $y = -2x$

25. $y = 3x - 1$ **26.** $y = -3x - 1$

▶ **27.** $-2x + y = 1$ **28.** $-3x + y = 1$

29. $3x + 4y = 8$ **30.** $3x - 4y = 8$

▶ **31.** $x = -2$ **32.** $y = 3$

▶ **33.** $y = 2$ **34.** $x = -3$

Graph each equation.

35. $y = \dfrac{3}{4}x + 1$ **36.** $y = \dfrac{2}{3}x + 1$

37. $y = \dfrac{1}{3}x + \dfrac{2}{3}$ **38.** $y = \dfrac{1}{2}x + \dfrac{1}{2}$

39. $y = \dfrac{2}{3}x + \dfrac{2}{3}$ **40.** $y = -\dfrac{3}{4}x + \dfrac{3}{2}$

= Videos available by instructor request

▶ = Online student support materials available at academic.cengage.com/login

For each equation in each table below, indicate whether the graph is horizontal (H), or vertical (V), or whether it passes through the origin (O).

41.

Equation	H, V, and/or O
$x = 3$	
$y = 3$	
$y = 3x$	
$y = 0$	

42.

Equation	H, V, and/or O
$x = \dfrac{1}{2}$	
$y = \dfrac{1}{2}$	
$y = \dfrac{1}{2}x$	
$x = 0$	

43.

Equation	H, V, and/or O
$x = -\dfrac{3}{5}$	
$y = -\dfrac{3}{5}$	
$y = -\dfrac{3}{5}x$	
$x = 0$	

44.

Equation	H, V, and/or O
$x = -4$	
$y = -4$	
$y = -4x$	
$y = 0$	

The next two problems are intended to give you practice reading, and paying attention to, the instructions that accompany the problems you are working. Working these problems is an excellent way to get ready for a test or a quiz.

45. Work each problem according to the instructions given.
   a. Solve: $2x + 5 = 10$
   b. Find x when y is 0: $2x + 5y = 10$
   c. Find y when x is 0: $2x + 5y = 10$
   d. Graph: $2x + 5y = 10$
   e. Solve for y: $2x + 5y = 10$

46. Work each problem according to the instructions given.
   a. Solve: $x - 2 = 6$
   b. Find x when y is 0: $x - 2y = 6$
   c. Find y when x is 0: $x - 2y = 6$
   d. Graph: $x - 2y = 6$
   e. Solve for y: $x - 2y = 6$

Maintaining Your Skills

Apply the distributive property.

47. $\dfrac{1}{2}(4x + 10)$

48. $\dfrac{1}{2}(6x - 12)$

49. $\dfrac{2}{3}(3x - 9)$

50. $\dfrac{1}{3}(2x + 12)$

51. $\dfrac{3}{4}(4x + 10)$

52. $\dfrac{3}{4}(8x - 6)$

53. $\dfrac{3}{5}(10x + 15)$

54. $\dfrac{2}{5}(5x - 10)$

55. $5\left(\dfrac{2}{5}x + 10\right)$

56. $3\left(\dfrac{2}{3}x + 5\right)$

57. $4\left(\dfrac{3}{2}x - 7\right)$

58. $4\left(\dfrac{3}{4}x + 5\right)$

59. $\dfrac{3}{4}(2x + 12y)$

60. $\dfrac{3}{4}(8x - 16y)$

61. $\dfrac{1}{2}(5x - 10y) + 6$

62. $\dfrac{1}{3}(5x - 15y) - 5$

Getting Ready for the Next Section

63. Let $3x - 2y = 6$.
   a. Find x when $y = 0$.
   b. Find y when $x = 0$.

64. Let $2x - 5y = 10$.
   a. Find x when $y = 0$.
   b. Find y when $x = 0$.

65. Let $-x + 2y = 4$.
   a. Find x when $y = 0$.
   b. Find y when $x = 0$.

66. Let $3x - y = 6$.
   a. Find x when $y = 0$.
   b. Find y when $x = 0$.

67. Let $y = -\dfrac{1}{3}x + 2$.

    a. Find x when $y = 0$.

    b. Find y when $x = 0$.

68. Let $y = \dfrac{3}{2}x - 3$.

    a. Find x when $y = 0$.

    b. Find y when $x = 0$.

3.4 More on Graphing: Intercepts

OBJECTIVES

A Find the intercepts of a line from the equation of the line.

B Use intercepts to graph a line.

In this section we continue our work with graphing lines by finding the points where a line crosses the axes of our coordinate system. To do so, we use the fact that any point on the x-axis has a y-coordinate of 0 and any point on the y-axis has an x-coordinate of 0. We begin with the following definition.

> **DEFINITION** The **x-intercept** of a straight line is the x-coordinate of the point where the graph crosses the x-axis. The **y-intercept** is defined similarly. It is the y-coordinate of the point where the graph crosses the y-axis.

If the x-intercept is a, then the point $(a, 0)$ lies on the graph. (This is true because any point on the x-axis has a y-coordinate of 0.)

If the y-intercept is b, then the point $(0, b)$ lies on the graph. (This is true because any point on the y-axis has an x-coordinate of 0.)

Graphically, the relationship is shown in Figure 1.

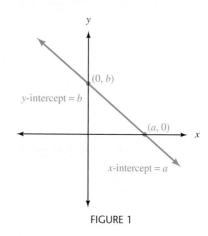

FIGURE 1

 EXAMPLE 1 Find the x- and y-intercepts for $3x - 2y = 6$, and then use them to draw the graph.

SOLUTION To find where the graph crosses the x-axis, we let $y = 0$. (The y-coordinate of any point on the x-axis is 0.)

x-intercept:

$$
\begin{aligned}
\text{When} \qquad\qquad y &= 0 \\
\text{the equation} \qquad 3x - 2y &= 6 \\
\text{becomes} \qquad 3x - 2(0) &= 6 \\
3x - 0 &= 6 \\
x &= 2 \qquad \textbf{Multiply each side by } \tfrac{1}{3}
\end{aligned}
$$

The graph crosses the x-axis at $(2, 0)$, which means the x-intercept is 2.

y-intercept:

When	$x = 0$
the equation	$3x - 2y = 6$
becomes	$3(0) - 2y = 6$
	$0 - 2y = 6$
	$-2y = 6$
	$y = -3$ **Multiply each side by $-\frac{1}{2}$**

The graph crosses the *y*-axis at $(0, -3)$, which means the *y*-intercept is -3.

Plotting the *x*- and *y*-intercepts and then drawing a line through them, we have the graph of $3x - 2y = 6$, as shown in Figure 2.

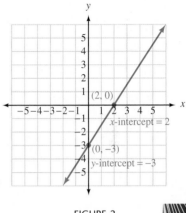

FIGURE 2

EXAMPLE 2   Graph $-x + 2y = 4$ by finding the intercepts and using them to draw the graph.

SOLUTION   Again, we find the *x*-intercept by letting $y = 0$ in the equation and solving for *x*. Similarly, we find the *y*-intercept by letting $x = 0$ and solving for *y*.

x-intercept:

When	$y = 0$
the equation	$-x + 2y = 4$
becomes	$-x + 2(0) = 4$
	$-x + 0 = 4$
	$-x = 4$
	$x = -4$ **Multiply each side by -1**

The *x*-intercept is -4, indicating that the point $(-4, 0)$, is on the graph of $-x + 2y = 4$.

y-intercept:

When	$x = 0$
the equation	$-x + 2y = 4$
becomes	$-0 + 2y = 4$
	$2y = 4$
	$y = 2$ **Multiply each side by $\frac{1}{2}$**

The *y*-intercept is 2, indicating that the point (0, 2) is on the graph of $-x + 2y = 4$.

Plotting the intercepts and drawing a line through them, we have the graph of $-x + 2y = 4$, as shown in Figure 3.

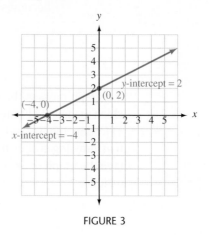

FIGURE 3

Graphing a line by finding the intercepts, as we have done in Examples 1 and 2, is an easy method of graphing if the equation has the form $ax + by = c$ and both the numbers *a* and *b* divide the number *c* evenly.

In our next example we use the intercepts to graph a line in which *y* is given in terms of *x*.

 EXAMPLE 3   Use the intercepts for $y = -\frac{1}{3}x + 2$ to draw its graph.

SOLUTION   We graphed this line previously in Example 3 of Section 3.3 by substituting three different values of *x* into the equation and solving for *y*. This time we will graph the line by finding the intercepts.

***x*-intercept:**

$$\text{When} \qquad y = 0$$

$$\text{the equation} \qquad y = -\frac{1}{3}x + 2$$

$$\text{becomes} \qquad 0 = -\frac{1}{3}x + 2$$

$$-2 = -\frac{1}{3}x \qquad \text{Add } -2 \text{ to each side}$$

$$6 = x \qquad \text{Multiply each side by } -3$$

The *x*-intercept is 6, which means the graph passes through the point (6, 0).

***y*-intercept:**

$$\text{When} \qquad x = 0$$

$$\text{the equation} \qquad y = -\frac{1}{3}x + 2$$

$$\text{becomes} \qquad y = -\frac{1}{3}(0) + 2$$

$$y = 2$$

The y-intercept is 2, which means the graph passes through the point (0, 2).

The graph of $y = -\frac{1}{3}x + 2$ is shown in Figure 4. Compare this graph, and the method used to obtain it, with Example 3 in Section 3.3.

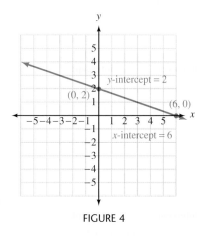

FIGURE 4

LINKING OBJECTIVES AND EXAMPLES

Next to each **objective** we have listed the examples that are best described by that objective.

A 1–3

B 1–3

GETTING READY FOR CLASS

After reading through the preceding section, respond in your own words and in complete sentences.

1. What is the x-intercept for a graph?

2. What is the y-intercept for a graph?

3. How do we find the y-intercept for a line from the equation?

4. How do we graph a line using its intercepts?

Problem Set 3.4

Online support materials can be found at academic.cengage.com/login

Find the x- and y-intercepts for the following equations. Then use the intercepts to graph each equation.

1. $2x + y = 4$

2. $2x + y = 2$

3. $-x + y = 3$

4. $-x + y = 4$

5. $-x + 2y = 2$

6. $-x + 2y = 4$

7. $5x + 2y = 10$

8. $2x + 5y = 10$

9. $-4x + 5y = 20$

10. $-5x + 4y = 20$

11. $3x - 4y = -4$

12. $-2x + 3y = 3$

13. $x - 3y = 2$

14. $x - 2y = 1$

15. $2x - 3y = -2$

16. $3x + 4y = 6$

17. $y = 2x - 6$

18. $y = 2x + 6$

19. $y = 2x - 1$

20. $y = -2x - 1$

21. $y = \frac{1}{2}x + 3$

22. $y = \frac{1}{2}x - 3$

23. $y = -\frac{1}{3}x - 2$

24. $y = -\frac{1}{3}x + 2$

For each of the following lines, the x-intercept and the y-intercept are both 0, which means the graph of each will go through the origin, (0, 0). Graph each line by finding a point on each, other than the origin, and then drawing a line through that point and the origin.

25. $y = -2x$

26. $y = \frac{1}{2}x$

EXAMPLE 2 Solve the following system by graphing.

$$x + 2y = 8$$
$$2x - 3y = 2$$

SOLUTION Graphing each equation on the same coordinate system, we have the lines shown in Figure 2.

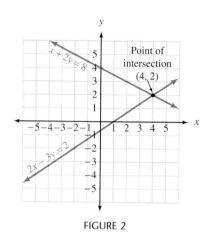

FIGURE 2

From Figure 2, we can see the solution for our system is (4, 2). We check this solution as follows.

When	$x = 4$	When		$x = 4$
and	$y = 2$	and		$y = 2$
the equation	$x + 2y = 8$	the equation		$2x - 3y = 2$
becomes	$4 + 2(2) \overset{?}{=} 8$	becomes		$2(4) - 3(2) \overset{?}{=} 2$
	$4 + 4 = 8$			$8 - 6 = 2$
	$8 = 8$			$2 = 2$

The point (4, 2) satisfies both equations and, therefore, must be the solution to our system.

EXAMPLE 3 Solve this system by graphing.

$$y = 2x - 3$$
$$x = 3$$

SOLUTION Graphing both equations on the same set of axes, we have Figure 3.

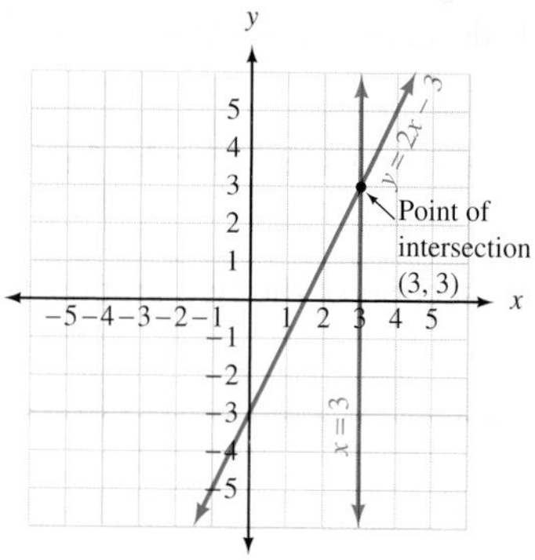

FIGURE 3

The solution to the system is the point (3, 3).

30. Wo
giv
a.
b.
c.
d.
e.

31. As
diff
if th
The
the

32. The
Solʋ

33. A se
of e
equ
follo
grap

34. The
Solʋ

Apply

35. Job (
posi
$8.0

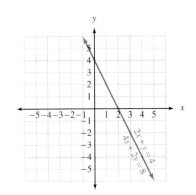

FIGURE 5

  EXAMPLE 4   Solve by graphing.

$$y = x - 2$$
$$y = x + 1$$

SOLUTION Graphing both equations produces the lines shown in Figure 4. We can see in Figure 4 that the lines are parallel and therefore do not intersect. Our system has no ordered pair as a solution because there is no ordered pair that satisfies both equations. We say the solution set is the empty set and write ∅.

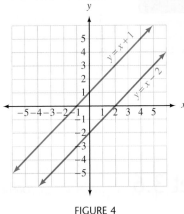

FIGURE 4

Example 4 is one example of two special cases associated with linear systems. The other special case happens when the two graphs coincide. Here is an example.

> **Note**
> We sometimes use special vocabulary to describe the special cases shown in Examples 4 and 5. When a system of equations has no solution because the lines are parallel (as in Example 4), we say the system is *inconsistent*. When the lines coincide (as in Example 5), we say the system is *dependent*.

  EXAMPLE 5   Graph the system.

$$2x + y = 4$$
$$4x + 2y = 8$$

SOLUTION Both graphs are shown in Figure 5. The two graphs coincide. The reason becomes apparent when we multiply both sides of the first equation by 2:

$$2x + y = 4$$

$$\mathbf{2}(2x + y) = \mathbf{2}(4) \qquad \textbf{Multiply both sides by 2}$$

$$4x + 2y = 8$$

The equations have the same solution set. Any ordered pair that is a solution to one is a solution to the system. The system has an infinite number of solutions. (Any point on the line is a solution to the system.)

Probl

Solve the fo
graphing.

▶ **1.** $x + y =$
$x - y =$

3. $x + y =$
$-x + y =$

5. $x + y =$
$-x + y =$

▶ **7.** $3x - 2y$
$x - y$

9. $6x - 2y$
$3x + y =$

11. $4x + y =$
$3x - y =$

13. $x + 2y =$
$2x - y =$

15. $3x - 5y$
$-2x + y =$

17. $y = 2x +$
$y = -2x -$

 EXAMPLE 3   Solve the following system by solving the first equation for x and then using the substitution method:

$$x - 3y = -1$$

$$2x - 3y = 4$$

SOLUTION   We solve the first equation for x by adding $3y$ to both sides to get

$$x = 3y - 1$$

Using this value of x in the second equation, we have

$$2(3y - 1) - 3y = 4$$

$$6y - 2 - 3y = 4$$

$$3y - 2 = 4$$

$$3y = 6$$

$$y = 2$$

Next, we find x.

When $y = 2$

the equation $x = 3y - 1$

becomes $x = 3(2) - 1$

$$x = 6 - 1$$

$$x = 5$$

The solution to our system is $(5, 2)$.

Here are the steps to use in solving a system of equations by the substitution method.

> **Strategy for Solving a System of Equations by the Substitution Method**
>
> **Step 1:** Solve either one of the equations for x or y. (This step is not necessary if one of the equations is already in the correct form, as in Examples 1 and 2.)
>
> **Step 2:** Substitute the expression for the variable obtained in step 1 into the other equation and solve it.
>
> **Step 3:** Substitute the solution from step 2 into any equation in the system that contains both variables and solve it.
>
> **Step 4:** Check your results, if necessary.

 EXAMPLE 4   Solve by substitution.

$$-2x + 4y = 14$$

$$-3x + y = 6$$

SOLUTION   We can solve either equation for either variable. If we look at the system closely, it becomes apparent that solving the second equation for y is the easiest way to go. If we add $3x$ to both sides of the second equation, we have

$$y = 3x + 6$$

Substituting the expression $3x + 6$ back into the first equation in place of y yields the following result.

$$-2x + 4(3x + 6) = 14$$

$$-2x + 12x + 24 = 14$$

$$10x + 24 = 14$$

$$10x = -10$$

$$x = -1$$

Substituting $x = -1$ into the equation $y = 3x + 6$ leaves us with

$$y = 3(-1) + 6$$

$$y = -3 + 6$$

$$y = 3$$

The solution to our system is $(-1, 3)$.

EXAMPLE 5   Solve by substitution.

$$4x + 2y = 8$$

$$y = -2x + 4$$

SOLUTION   Substituting the expression $-2x + 4$ for y from the second equation into the first equation, we have

$$4x + 2(-2x + 4) = 8$$

$$4x - 4x + 8 = 8$$

$$8 = 8 \quad \textbf{A true statement}$$

Both variables have been eliminated, and we are left with a true statement. Recall from the last section that a true statement in this situation tells us the lines coincide; that is, the equations $4x + 2y = 8$ and $y = -2x + 4$ have exactly the same graph. Any point on that graph has coordinates that satisfy both equations and is a solution to the system.

 EXAMPLE 6   The following table shows two contract rates charged by GTE Wireless for cellular phone use. At how many minutes will the two rates cost the same amount?

	Flat Rate	Plus	Per Minute Charge
Plan 1	$15		$1.50
Plan 2	$24.95		$0.75

SOLUTION If we let y = the monthly charge for x minutes of phone use, then the equations for each plan are

$$\text{Plan 1:} \quad y = 1.5x + 15$$

$$\text{Plan 2:} \quad y = 0.75x + 24.95$$

We can solve this system by substitution by replacing the variable y in Plan 2 with the expression $1.5x + 15$ from Plan 1. If we do so, we have

$$1.5x + 15 = 0.75x + 24.95$$

$$0.75x + 15 = 24.95$$

$$0.75x = 9.95$$

$$x = 13.27 \qquad \text{\textbf{to the nearest hundredth}}$$

The monthly bill is based on the number of minutes you use the phone, with any fraction of a minute moving you up to the next minute. If you talk for a total of 13 minutes, you are billed for 13 minutes. If you talk for 13 minutes, 10 seconds, you are billed for 14 minutes. The number of minutes on your bill always will be a whole number. So, to calculate the cost for talking 13.27 minutes, we would replace x with 14 and find y. Let's compare the two plans at $x = 13$ minutes and at $x = 14$ minutes.

$$\text{Plan 1:} \quad y = 1.5x + 15$$

$$\text{When} \quad x = 13, y = \$34.50$$

$$\text{When} \quad x = 14, y = \$36.00$$

$$\text{Plan 2:} \quad y = 0.75x + 24.95$$

$$\text{When} \quad x = 13, y = \$34.70$$

$$\text{When} \quad x = 14, y = \$35.45$$

The two plans will never give the same cost for talking x minutes. If you talk 13 or less minutes, Plan 1 will cost less. If you talk for more than 13 minutes, you will be billed for 14 minutes, and Plan 2 will cost less than Plan 1.

GETTING READY FOR CLASS

After reading through the preceding section, respond in your own words and in complete sentences.

1. What is the first step in solving a system of linear equations by substitution?
2. When would substitution be more efficient than the elimination method in solving two linear equations?
3. What does it mean when we solve a system of linear equations by the substitution method and we end up with the statement 8 = 8?
4. How would you begin solving the following system using the substitution method?

$$x + y = 2$$
$$y = 2x - 1$$

LINKING OBJECTIVES AND EXAMPLES

Next to each **objective** we have listed the examples that are best described by that objective.

A   1–6

Solve the following systems by substitution. Substitute the expression in the second equation into the first equation and solve.

▶ **1.** $x + y = 11$
$\quad y = 2x - 1$

2. $x - y = -3$
$\quad y = 3x + 5$

3. $x + y = 20$
$\quad y = 5x + 2$

4. $3x - y = -1$
$\quad x = 2y - 7$

5. $-2x + y = -1$
$\quad y = -4x + 8$

6. $4x - y = 5$
$\quad y = -4x + 1$

7. $3x - 2y = -2$
$\quad x = -y + 6$

8. $2x - 3y = 17$
$\quad x = -y + 6$

9. $5x - 4y = -16$
$\quad y = 4$

10. $6x + 2y = 18$
$\quad x = 3$

11. $5x + 4y = 7$
$\quad y = -3x$

12. $10x + 2y = -6$
$\quad y = -5x$

Solve the following systems by solving one of the equations for x or y and then using the substitution method.

13. $x + 3y = 4$
$\quad x - 2y = -1$

14. $x - y = 5$
$\quad x + 2y = -1$

▶ **15.** $2x + y = 1$
$\quad x - 5y = 17$

16. $2x - 2y = 2$
$\quad x - 3y = -7$

17. $3x + 5y = -3$
$\quad x - 5y = -5$

18. $2x - 4y = -4$
$\quad x + 2y = 8$

19. $5x + 3y = 0$
$\quad x - 3y = -18$

20. $x - 3y = -5$
$\quad x - 2y = 0$

21. $-3x - 9y = 7$
$\quad x + 3y = 12$

22. $2x + 6y = -18$
$\quad x + 3y = -9$

Solve the following systems using the substitution method.

23. $5x - 8y = 7$
$\quad y = 2x - 5$

24. $3x + 4y = 10$
$\quad y = 8x - 15$

▶ **25.** $7x - 6y = -1$
$\quad x = 2y - 1$

26. $4x + 2y = 3$
$\quad x = 4y - 3$

27. $-3x + 2y = 6$
$\quad y = 3x$

28. $-2x - y = -3$
$\quad y = -3x$

29. $5x - 6y = -4$
$\quad x = y$

30. $2x - 4y = 0$
$\quad y = x$

31. $3x + 3y = 9$
$\quad y = 2x - 12$

32. $7x + 6y = -9$
$\quad y = -2x + 1$

33. $7x - 11y = 16$
$\quad y = 10$

34. $9x - 7y = -14$
$\quad x = 7$

35. $-4x + 4y = -8$
$\quad y = x - 2$

36. $-4x + 2y = -10$
$\quad y = 2x - 5$

Solve each system.

▶ **37.** $2x + 5y = 36$
$\quad y = 12 - x$

▶ **38.** $3x + 6y = 120$
$\quad y = 25 - x$

▶ **39.** $5x + 2y = 54$
$\quad y = 18 - x$

▶ **40.** $10x + 5y = 320$
$\quad y = 40 - x$

▶ **41.** $2x + 2y = 96$
$\quad y = 2x$

▶ **42.** $x + y = 22$
$\quad y = x + 9$

Solve each system by substitution. You can eliminate the decimals if you like, but you don't have to. The solution will be the same in either case.

43. $0.05x + 0.10y = 1.70$
$\quad y = 22 - x$

44. $0.20x + 0.50y = 3.60$
$\quad y = 12 - x$

The next two problems are intended to give you practice reading, and paying attention to, the instructions that accompany the problems you are working. Working these problems is an excellent way to get ready for a test or quiz.

45. Work each problem according to the instructions given.
  a. Solve: $4y - 5 = 20$
  b. Solve for y: $4x - 5y = 20$
  c. Solve for x: $x - y = 5$
  d. Solve the system: $4x - 5y = 20$
            $x - y = 5$

46. Work each problem according to the instructions given.
  a. Solve: $2x - 1 = 4$
  b. Solve for y: $2x - y = 4$
  c. Solve for x: $x + 3y = 9$
  d. Solve the system: $2x - y = 4$
            $x + 3y = 9$

Applying the Concepts

47. Gas Mileage Daniel is trying to decide whether to buy a car or a truck. The truck he is considering will cost him $150 a month in loan payments, and it gets 20 miles per gallon in gas mileage. The car will cost $180 a month in loan payments, but it gets 35 miles per gallon in gas mileage. Daniel estimates that he will pay $1.40 per gallon for gas. This means that the monthly cost to drive the truck x miles will be $y = \frac{1.40}{20}x + 150$. The total monthly cost to drive the car x miles will be $y = \frac{1.40}{35}x + 180$. The following figure shows the graph of each equation.

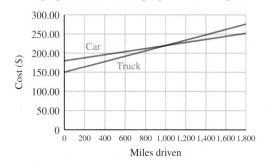

a. At how many miles do the car and the truck cost the same to operate?

b. If Daniel drives more than 1,200 miles, which will be cheaper?

c. If Daniel drives fewer than 800 miles, which will be cheaper?

d. Why do the graphs appear in the first quadrant only?

48. Video Production Pat runs a small company that duplicates videotapes. The daily cost and daily revenue for a company duplicating videos are shown in the following figure. The daily cost for duplicating x videos is $y = \frac{6}{5}x + 20$; the daily revenue (the amount of money he brings in each day) for duplicating x videos is $y = 1.7x$. The graphs of the two lines are shown in the following figure.

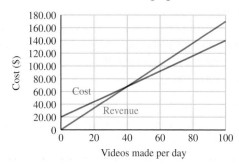

a. Pat will "break even" when his cost and his revenue are equal. How many videos does he need to duplicate to break even?

b. Pat will incur a loss when his revenue is less than his cost. If he duplicates 30 videos in one day, will he incur a loss?

c. Pat will make a profit when his revenue is larger than his costs. For what values of x will Pat make a profit?

d. Why does the graph appear in the first quadrant only?

Maintaining Your Skills

49. $6(3 + 4) + 5$

50. $[(1 + 2)(2 + 3)] + (4 \div 2)$

51. $1^2 + 2^2 + 3^2$ **52.** $(1 + 2 + 3)^2$

53. $5(6 + 3 \cdot 2) + 4 + 3 \cdot 2$

54. $(1 + 2)^3 + [(2 \cdot 3) + (4 \cdot 5)]$

55. $(1^3 + 2^3) + [(2 \cdot 3) + (4 \cdot 5)]$

56. $[2(3 + 4 + 5)] \div 3$

57. $(2 \cdot 3 + 4 + 5) \div 3$

58. $10^4 + 10^3 + 10^2 + 10^1$

59. $6 \cdot 10^3 + 5 \cdot 10^2 + 4 \cdot 10^1$

60. $5 \cdot 10^3 + 2 \cdot 10^2 + 8 \cdot 10^1$

61. $1 \cdot 10^3 + 7 \cdot 10^2 + 6 \cdot 10^1 + 0$

62. $4(2 - 1) + 5(3 - 2)$

63. $4 \cdot 2 - 1 + 5 \cdot 3 - 2$ **64.** $2^3 + 3^2 \cdot 4 - 5$

65. $(2^3 + 3^2) \cdot 4 - 5$ **66.** $4^2 - 2^4 + (2 \cdot 2)^2$

67. $2(2^2 + 3^2) + 3(3^2)$ **68.** $2 \cdot 2^2 + 3^2 + 3 \cdot 3^2$

Getting Ready for the Next Section

69. One number is eight more than five times another, their sum is 26. Find the numbers.

70. One number is three less than four times another; their sum is 27. Find the numbers.

71. The difference of two positive numbers is nine. The larger number is six less than twice the smaller number. Find the numbers.

72. The difference of two positive numbers is 17. The larger number is one more than twice the smaller number. Find the numbers.

73. The length of a rectangle is 5 inches more than three times the width. The perimeter is 58 inches. Find the length and width.

74. The length of a rectangle is 3 inches less than twice the width. The perimeter is 36 inches. Find the length and width.

75. John has $1.70 in nickels and dimes in his pocket. He has four more nickels than he does dimes. How many of each does he have?

76. Jamie has $2.65 in dimes and quarters in her pocket. She has two more dimes than she does quarters. How many of each does she have?

Solve the systems by any method.

77. $y = 5x + 2$
    $x + y = 20$

78. $y = 2x + 3$
    $x + y = 9$

79. $x + y = 15,000$
    $0.06x + 0.07y = 980$

80. $x + y = 22$
    $0.05x + 0.1y = 1.70$

3.9   Applications

OBJECTIVES

A Apply the Blueprint for Problem Solving to a variety of application problems involving systems of equations.

I often have heard students remark about the word problems in elementary algebra: "What does this have to do with real life?" Most of the word problems we will encounter don't have much to do with "real life." We are actually just practicing. Ultimately, all problems requiring the use of algebra are word problems; that is, they are stated in words first, then translated to symbols. The problem then is solved by some system of mathematics, like algebra. Most real applications involve calculus or higher levels of mathematics. So, if the problems we solve are upsetting or frustrating to you, then you probably are taking them too seriously.

The word problems in this section have two unknown quantities. We will write two equations in two variables (each of which represents one of the unknown quantities), which of course is a system of equations. We then solve the system by one of the methods developed in the previous sections of this chapter. Here are the steps to follow in solving these word problems.

Blueprint for Problem Solving Using a System of Equations

Step 1: **Read** the problem, and then mentally **list** the items that are known and the items that are unknown.

Step 2: **Assign variables** to each of the unknown items; that is, let $x =$ one of the unknown items and $y =$ the other unknown item. Then **translate** the other **information** in the problem to expressions involving the two variables.

Step 3: **Reread** the problem, and then **write a system of equations,** using the items and variables listed in steps 1 and 2, that describes the situation.

Step 4: **Solve the system** found in step 3.

Step 5: **Write** your **answers** using complete sentences.

Step 6: **Reread** the problem, and **check** your solution with the original words in the problem.

Remember, the more problems you work, the more problems you will be able to work. If you have trouble getting started on the problem set, come back to the examples and work through them yourself. The examples are similar to the problems found in the problem set.

Number Problem

EXAMPLE 1   One number is 2 more than 5 times another number. Their sum is 20. Find the two numbers.

SOLUTION   Applying the steps in our blueprint, we have

Step 1: We know that the two numbers have a sum of 20 and that one of them is 2 more than 5 times the other. We don't know what the numbers themselves are.

Step 2: Let x represent one of the numbers and y represent the other. "One number is 2 more than 5 times another" translates to

$$y = 5x + 2$$

"Their sum is 20" translates to

$$x + y = 20$$

Step 3: The system that describes the situation must be

$$x + y = 20$$
$$y = 5x + 2$$

Step 4: We can solve this system by substituting the expression $5x + 2$ in the second equation for y in the first equation:

$$x + (5x + 2) = 20$$
$$6x + 2 = 20$$
$$6x = 18$$
$$x = 3$$

Using $x = 3$ in either of the first two equations and then solving for y, we get $y = 17$.

Step 5: So 17 and 3 are the numbers we are looking for.

Step 6: The number 17 is 2 more than 5 times 3, and the sum of 17 and 3 is 20.

> *Note*
> We are using the substitution method here because the system we are solving is one in which the substitution method is the more convenient method.

Interest Problem

EXAMPLE 2   Mr. Hicks had $15,000 to invest. He invested part at 6% and the rest at 7%. If he earns $980 in interest, how much did he invest at each rate?

SOLUTION   Remember, step 1 is done mentally.

Step 1: We do not know the specific amounts invested in the two accounts. We do know that their sum is $15,000 and that the interest rates on the two accounts are 6% and 7%.

EXAMPLES Simplify the following expressions.

7. $\left(\dfrac{x}{2}\right)^3 = \dfrac{x^3}{2^3} = \dfrac{x^3}{8}$

8. $\left(\dfrac{5}{y}\right)^2 = \dfrac{5^2}{y^2} = \dfrac{25}{y^2}$

9. $\left(\dfrac{2}{3}\right)^4 = \dfrac{2^4}{3^4} = \dfrac{16}{81}$

Zero and One as Exponents

We have two special exponents left to deal with before our rules for exponents are complete: 0 and 1. To obtain an expression for x^1, we will solve a problem two different ways:

$$\left.\begin{array}{l} \dfrac{x^3}{x^2} = \dfrac{x \cdot x \cdot x}{x \cdot x} = x \\[2em] \dfrac{x^3}{x^2} = x^{3-2} = x^1 \end{array}\right\} \quad \textbf{Hence } x^1 = x$$

Stated generally, this rule says that $a^1 = a$. This seems reasonable and we will use it since it is consistent with our property of division using the same base.

We use the same procedure to obtain an expression for x^0:

$$\left.\begin{array}{l} \dfrac{5^2}{5^2} = \dfrac{25}{25} = 1 \\[2em] \dfrac{5^2}{5^2} = 5^{2-2} = 5^0 \end{array}\right\} \quad \textbf{Hence } 5^0 = 1$$

It seems, therefore, that the best definition of x^0 is 1 for all x except $x = 0$. In the case of $x = 0$, we have 0^0, which we will not define. This definition will probably seem awkward at first. Most people would like to define x^0 as 0 when they first encounter it. Remember, the zero in this expression is an exponent, so x^0 does not mean to multiply by zero. Thus, we can make the general statement that $a^0 = 1$ for all real numbers except $a = 0$.

Here are some examples involving the exponents 0 and 1.

 EXAMPLES Simplify the following expressions:

10. $8^0 = 1$

11. $8^1 = 8$

12. $4^0 + 4^1 = 1 + 4 = 5$

13. $(2x^2y)^0 = 1$

Here is a summary of the definitions and properties of exponents we have developed so far. For each definition or property in the list, a and b are real numbers, and r and s are integers.

Definitions	Properties
$a^{-r} = \dfrac{1}{a^r} = \left(\dfrac{1}{a}\right)^r \quad a \neq 0$	1. $a^r \cdot a^s = a^{r+s}$
$a^1 = a$	2. $(a^r)^s = a^{rs}$
$a^0 = 1 \quad a \neq 0$	3. $(ab)^r = a^r b^r$
	4. $\dfrac{a^r}{a^s} = a^{r-s} \quad a \neq 0$
	5. $\left(\dfrac{a}{b}\right)^r = \dfrac{a^r}{b^r} \quad b \neq 0$

Here are some additional examples. These examples use a combination of the preceding properties and definitions.

EXAMPLES Simplify each expression. Write all answers with positive exponents only:

14. $\dfrac{(5x^3)^2}{x^4} = \dfrac{25x^6}{x^4}$ **Properties 2 and 3**

$\phantom{\dfrac{(5x^3)^2}{x^4}} = 25x^2$ **Property 4**

15. $\dfrac{x^{-8}}{(x^2)^3} = \dfrac{x^{-8}}{x^6}$ **Property 2**

$\phantom{\dfrac{x^{-8}}{(x^2)^3}} = x^{-8-6}$ **Property 4**

$\phantom{\dfrac{x^{-8}}{(x^2)^3}} = x^{-14}$ **Subtraction**

$\phantom{\dfrac{x^{-8}}{(x^2)^3}} = \dfrac{1}{x^{14}}$ **Definition of negative exponents**

16. $\left(\dfrac{y^5}{y^3}\right)^2 = \dfrac{(y^5)^2}{(y^3)^2}$ **Property 5**

$\phantom{\left(\dfrac{y^5}{y^3}\right)^2} = \dfrac{y^{10}}{y^6}$ **Property 2**

$\phantom{\left(\dfrac{y^5}{y^3}\right)^2} = y^4$ **Property 4**

Notice in Example 16 that we could have simplified inside the parentheses first and then raised the result to the second power:

$$\left(\dfrac{y^5}{y^3}\right)^2 = (y^2)^2 = y^4$$

17. $(3x^5)^{-2} = \dfrac{1}{(3x^5)^2}$ **Definition of negative exponents**

$\phantom{(3x^5)^{-2}} = \dfrac{1}{9x^{10}}$ **Properties 2 and 3**

18. $x^{-8} \cdot x^5 = x^{-8+5}$ **Property 1**

$\phantom{x^{-8} \cdot x^5} = x^{-3}$ **Addition**

$\phantom{x^{-8} \cdot x^5} = \dfrac{1}{x^3}$ **Definition of negative exponents**

19. $\dfrac{(a^3)^2 a^{-4}}{(a^{-4})^3} = \dfrac{a^6 a^{-4}}{a^{-12}}$ **Property 2**

$\phantom{\dfrac{(a^3)^2 a^{-4}}{(a^{-4})^3}} = \dfrac{a^2}{a^{-12}}$ **Property 1**

$\phantom{\dfrac{(a^3)^2 a^{-4}}{(a^{-4})^3}} = a^{14}$ **Property 4**

In the next two examples we use division to compare the area and volume of geometric figures.

 EXAMPLE 20 Suppose you have two squares, one of which is larger than the other. If the length of a side of the larger square is 3 times as long as the length of a side of the smaller square, how many of the smaller squares will it take to cover up the larger square?

SOLUTION If we let x represent the length of a side of the smaller square, then the length of a side of the larger square is $3x$. The area of each square, along with a diagram of the situation, is given in Figure 1.

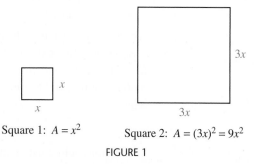

Square 1: $A = x^2$ ⠀⠀⠀ Square 2: $A = (3x)^2 = 9x^2$

FIGURE 1

To find out how many smaller squares it will take to cover up the larger square, we divide the area of the larger square by the area of the smaller square.

$$\frac{\text{Area of square 2}}{\text{Area of square 1}} = \frac{9x^2}{x^2} = 9$$

It will take 9 of the smaller squares to cover the larger square.

 EXAMPLE 21 Suppose you have two boxes, each of which is a cube. If the length of a side in the second box is 3 times as long as the length of a side of the first box, how many of the smaller boxes will fit inside the larger box?

SOLUTION If we let x represent the length of a side of the smaller box, then the length of a side of the larger box is $3x$. The volume of each box, along with a diagram of the situation, is given in Figure 2.

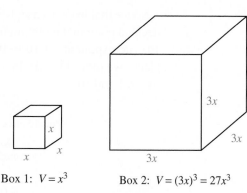

Box 1: $V = x^3$ ⠀⠀⠀ Box 2: $V = (3x)^3 = 27x^3$

FIGURE 2

To find out how many smaller boxes will fit inside the larger box, we divide the volume of the larger box by the volume of the smaller box.

$$\frac{\text{Volume of box 2}}{\text{Volume of box 1}} = \frac{27x^3}{x^3} = 27$$

We can fit 27 of the smaller boxes inside the larger box.

More on Scientific Notation

Now that we have completed our list of definitions and properties of exponents, we can expand the work we did previously with scientific notation.

Recall that a number is in scientific notation when it is written in the form

$$n \times 10^r$$

where $1 \le n < 10$ and r is an integer.

Since negative exponents give us reciprocals, we can use negative exponents to write very small numbers in scientific notation. For example, the number 0.00057, when written in scientific notation, is equivalent to 5.7×10^{-4}. Here's why:

$$5.7 \times 10^{-4} = 5.7 \times \frac{1}{10^4} = 5.7 \times \frac{1}{10,000} = \frac{5.7}{10,000} = 0.00057$$

The table below lists some other numbers in both scientific notation and expanded form.

 EXAMPLE 22

Number Written the Long Way		Number Written Again in Scientific Notation
376,000	=	3.76×10^5
49,500	=	4.95×10^4
3,200	=	3.2×10^3
591	=	5.91×10^2
46	=	4.6×10^1
8	=	8×10^0
0.47	=	4.7×10^{-1}
0.093	=	9.3×10^{-2}
0.00688	=	6.88×10^{-3}
0.0002	=	2×10^{-4}
0.000098	=	9.8×10^{-5}

Notice that in each case, when the number is written in scientific notation, the decimal point in the first number is placed so that the number is between 1 and 10. The exponent on 10 in the second number keeps track of the number of places we moved the decimal point in the original number to get a number between 1 and 10:

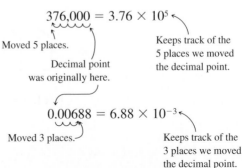

$$376,000 = 3.76 \times 10^5$$

Moved 5 places.

Decimal point was originally here.

Keeps track of the 5 places we moved the decimal point.

$$0.00688 = 6.88 \times 10^{-3}$$

Moved 3 places.

Keeps track of the 3 places we moved the decimal point.

LINKING OBJECTIVES AND EXAMPLES

Next to each **objective** we have listed the examples that are best described by that objective.

A	1–3
B	4–6
C	7–9
D	10–13
E	14–19
F	22

GETTING READY FOR CLASS

After reading through the preceding section, respond in your own words and in complete sentences.

1. How do you divide two expressions containing exponents when they each have the same base?
2. Explain the difference between 3^2 and 3^{-2}.
3. If a positive base is raised to a negative exponent, can the result be a negative number?
4. Explain what happens when we use 0 as an exponent.

Problem Set 4.2

Online support materials can be found at academic.cengage.com/login

Write each of the following with positive exponents, and then simplify, when possible.

1. 3^{-2} **2.** 3^{-3}

3. 6^{-2} **4.** 2^{-6}

5. 8^{-2} **6.** 3^{-4}

7. 5^{-3} **8.** 9^{-2}

9. $2x^{-3}$ **10.** $5x^{-1}$

11. $(2x)^{-3}$ **12.** $(5x)^{-1}$

13. $(5y)^{-2}$ **14.** $5y^{-2}$

15. 10^{-2} **16.** 10^{-3}

17. Complete the following table.

Number x	Square x^2	Power of 2 2^x
−3		
−2		
−1		
0		
1		
2		
3		

18. Complete the following table.

Number x	Cube x^3	Power of 3 3^x
−3		
−2		
−1		
0		
1		
2		
3		

Use Property 4 to simplify each of the following expressions. Write all answers that contain exponents with positive exponents only.

19. $\dfrac{5^1}{5^3}$ **20.** $\dfrac{7^6}{7^8}$

21. $\dfrac{x^{10}}{x^4}$ **22.** $\dfrac{x^4}{x^{10}}$

      = Videos available by instructor request

▶ = Online student support materials available at academic.cengage.com/login

23. $\dfrac{4^3}{4^0}$ **24.** $\dfrac{4^0}{4^3}$

25. $\dfrac{(2x)^7}{(2x)^4}$ **26.** $\dfrac{(2x)^4}{(2x)^7}$

27. $\dfrac{6^{11}}{6}$ **28.** $\dfrac{8^7}{8}$

29. $\dfrac{6}{6^{11}}$ **30.** $\dfrac{8}{8^7}$

31. $\dfrac{2^{-5}}{2^3}$ **32.** $\dfrac{2^{-5}}{2^{-3}}$

▶ **33.** $\dfrac{2^5}{2^{-3}}$ **34.** $\dfrac{2^{-3}}{2^{-5}}$

35. $\dfrac{(3x)^{-5}}{(3x)^{-8}}$ **36.** $\dfrac{(2x)^{-10}}{(2x)^{-15}}$

Simplify the following expressions. Any answers that contain exponents should contain positive exponents only.

37. $(3xy)^4$ **38.** $(4xy)^3$

39. 10^0 **40.** 10^1

41. $(2a^2b)^1$ **42.** $(2a^2b)^0$

43. $(7y^3)^{-2}$ **44.** $(5y^4)^{-2}$

45. $x^{-3}x^{-5}$ **46.** $x^{-6} \cdot x^8$

47. $y^7 \cdot y^{-10}$ **48.** $y^{-4} \cdot y^{-6}$

49. $\dfrac{(x^2)^3}{x^4}$ **50.** $\dfrac{(x^5)^3}{x^{10}}$

51. $\dfrac{(a^4)^3}{(a^3)^2}$ **52.** $\dfrac{(a^5)^3}{(a^5)^2}$

53. $\dfrac{y^7}{(y^2)^8}$ **54.** $\dfrac{y^2}{(y^3)^4}$

55. $\left(\dfrac{y^7}{y^2}\right)^8$ **56.** $\left(\dfrac{y^2}{y^3}\right)^4$

57. $\dfrac{(x^{-2})^3}{x^{-5}}$ **58.** $\dfrac{(x^2)^{-3}}{x^{-5}}$

59. $\left(\dfrac{x^{-2}}{x^{-5}}\right)^3$ **60.** $\left(\dfrac{x^2}{x^{-5}}\right)^{-3}$

▶ **61.** $\dfrac{(a^3)^2(a^4)^5}{(a^5)^2}$ **62.** $\dfrac{(a^4)^8(a^2)^5}{(a^3)^4}$

63. $\dfrac{(a^{-2})^3(a^4)^2}{(a^{-3})^{-2}}$ **64.** $\dfrac{(a^{-5})^{-3}(a^7)^{-1}}{(a^{-3})^5}$

65. Let $x = 2$ in each of the following expressions and simplify.

  a. $\dfrac{x^7}{x^2}$ **b.** x^5

  c. $\dfrac{x^2}{x^7}$ **d.** x^{-5}

66. Let $x = -1$ in each of the following expressions and simplify.

  a. $\dfrac{x^{12}}{x^9}$ **b.** x^3

  c. $\dfrac{x^{11}}{x^9}$ **d.** x^2

67. Write each expression as a perfect square.

  a. $\dfrac{1}{25} = \left(-\right)^2$ **b.** $\dfrac{1}{64} = \left(-\right)^2$

  c. $\dfrac{1}{x^2} = \left(-\right)^2$ **d.** $\dfrac{1}{x^4} = \left(-\right)^2$

68. Write each expression as a perfect cube.

  a. $\dfrac{1}{125} = \left(-\right)^3$ **b.** $\dfrac{1}{27} = \left(-\right)^3$

  c. $\dfrac{x^6}{125} = \left(-\right)^3$ **d.** $\dfrac{x^3}{27} = \left(-\right)^3$

69. Complete the following table, and then construct a line graph of the information in the table.

Number x	Power of 2 2^x
-3	
-2	
-1	
0	
1	
2	
3	

Notice that the procedure we used in both of these examples is very similar to multiplication and division of monomials, for which we multiplied or divided coefficients and added or subtracted exponents.

Addition and Subtraction of Monomials

Addition and subtraction of monomials will be almost identical since subtraction is defined as addition of the opposite. With multiplication and division of monomials, the key was rearranging the numbers and variables using the commutative and associative properties. With addition, the key is application of the distributive property. We sometimes use the phrase *combine monomials* to describe addition and subtraction of monomials.

> **DEFINITION** Two terms (monomials) with the same variable part (same variables raised to the same powers) are called **similar** (or *like*) **terms.**

You can add only similar terms. This is because the distributive property (which is the key to addition of monomials) cannot be applied to terms that are not similar.

 EXAMPLES Combine the following monomials.

8. $-3x^2 + 15x^2 = (-3 + 15)x^2$ **Distributive property**

$= 12x^2$ **Add coefficients**

9. $9x^2y - 20x^2y = (9 - 20)x^2y$ **Distributive property**

$= -11x^2y$ **Add coefficients**

10. $5x^2 + 8y^2$ **In this case we cannot apply the distributive property, so we cannot add the monomials**

The next examples show how we simplify expressions containing monomials when more than one operation is involved.

 EXAMPLE 11 Simplify $\dfrac{(6x^4y)(3x^7y^5)}{9x^5y^2}$.

SOLUTION We begin by multiplying the two monomials in the numerator:

$$\frac{(6x^4y)(3x^7y^5)}{9x^5y^2} = \frac{18x^{11}y^6}{9x^5y^2} \quad \textbf{Simplify numerator}$$

$$= 2x^6y^4 \quad \textbf{Divide}$$

EXAMPLE 12 Simplify $\dfrac{(6.8 \times 10^5)(3.9 \times 10^{-7})}{7.8 \times 10^{-4}}$.

SOLUTION We group the numbers between 1 and 10 separately from the powers of 10:

$$\frac{(6.8)(3.9)}{7.8} \times \frac{(10^5)(10^{-7})}{10^{-4}} = 3.4 \times 10^{5+(-7)-(-4)}$$

$$= 3.4 \times 10^2$$

EXAMPLE 13 Simplify $\dfrac{14x^5}{2x^2} + \dfrac{15x^8}{3x^5}$.

SOLUTION Simplifying each expression separately and then combining similar terms gives

$$\dfrac{14x^5}{2x^2} + \dfrac{15x^8}{3x^5} = 7x^3 + 5x^3 \quad \textbf{Divide}$$

$$= 12x^3 \quad \textbf{Add}$$

Our work with exponents and division allows us to multiply fractions and other expressions involving exponents. For example,

$$x^5 \cdot \dfrac{6}{x^2} = \dfrac{x^5}{1} \cdot \dfrac{6}{x^2} = \dfrac{6x^5}{x^2} = 6x^3$$

It is not necessary to show the intermediate steps in a problem like this. We are showing them here just so you can see why we subtract exponents to get the x^3 in the answer.

 EXAMPLES Apply the distributive property, then simplify, if possible.

14. $x^2\left(1 - \dfrac{6}{x}\right) = x^2 \cdot 1 - x^2 \cdot \dfrac{6}{x} = x^2 - \dfrac{6x^2}{x} = x^2 - 6x$

15. $ab\left(\dfrac{1}{b} - \dfrac{1}{a}\right) = ab \cdot \dfrac{1}{b} - ab \cdot \dfrac{1}{a} = \dfrac{ab}{b} - \dfrac{ab}{a} = a - b$

 EXAMPLE 16 A rectangular solid is twice as long as it is wide and one-half as high as it is wide. Write an expression for the volume.

SOLUTION We begin by making a diagram of the object (Figure 1) with the dimensions labeled as given in the problem.

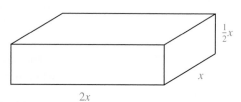

FIGURE 1

The volume is the product of the three dimensions:

$$V = 2x \cdot x \cdot \dfrac{1}{2}x = x^3$$

The box has the same volume as a cube with side x, as shown in Figure 2.

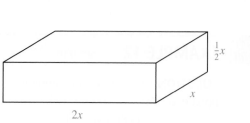

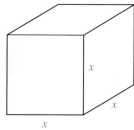

Equal Volumes

FIGURE 2

LINKING OBJECTIVES AND EXAMPLES

Next to each **objective** we have listed the examples that are best described by that objective.

A	1, 2, 11
B	3–5, 11
C	6, 7, 12
D	8–10

GETTING READY FOR CLASS

After reading through the preceding section, respond in your own words and in complete sentences.

1. What is a monomial?
2. Describe how you would multiply $3x^2$ and $5x^2$.
3. Describe how you would add $3x^2$ and $5x^2$.
4. Describe how you would multiply two numbers written in scientific notation.

Problem Set 4.3

Online support materials can be found at academic.cengage.com/login

Multiply.

1. $(3x^4)(4x^3)$

2. $(6x^5)(-2x^2)$

3. $(-2y^4)(8y^7)$

4. $(5y^{10})(2y^5)$

5. $(8x)(4x)$

6. $(7x)(5x)$

7. $(10a^3)(10a)(2a^2)$

8. $(5a^4)(10a)(10a^4)$

9. $(6ab^2)(-4a^2b)$

10. $(-5a^3b)(4ab^4)$

▸ **11.** $(4x^2y)(3x^3y^3)(2xy^4)$

12. $(5x^6)(-10xy^4)(-2x^2y^6)$

Divide. Write all answers with positive exponents only.

13. $\dfrac{15x^3}{5x^2}$

14. $\dfrac{25x^5}{5x^4}$

15. $\dfrac{18y^9}{3y^{12}}$

16. $\dfrac{24y^4}{-8y^7}$

17. $\dfrac{32a^3}{64a^4}$

18. $\dfrac{25a^5}{75a^6}$

▸ **19.** $\dfrac{21a^2b^3}{-7ab^5}$

20. $\dfrac{32a^5b^6}{8ab^5}$

21. $\dfrac{3x^3y^2z}{27xy^2z^3}$

22. $\dfrac{5x^5y^4z}{30x^3yz^2}$

23. Fill in the table.

a	b	ab	$\dfrac{a}{b}$	$\dfrac{b}{a}$
10	$5x$			
$20x^3$	$6x^2$			
$25x^5$	$5x^4$			
$3x^{-2}$	$3x^2$			
$-2y^4$	$8y^7$			

24. Fill in the table.

a	b	ab	$\dfrac{a}{b}$	$\dfrac{b}{a}$
$10y$	$2y^2$			
$10y^2$	$2y$			
$5y^3$	15			
5	$15y^3$			
$4y^{-3}$	$4y^3$			

■ = Videos available by instructor request

▸ = Online student support materials available at academic.cengage.com/login

Find each product. Write all answers in scientific notation.

25. $(3 \times 10^3)(2 \times 10^5)$

26. $(4 \times 10^8)(1 \times 10^6)$

27. $(3.5 \times 10^4)(5 \times 10^{-6})$

28. $(7.1 \times 10^5)(2 \times 10^{-8})$

29. $(5.5 \times 10^{-3})(2.2 \times 10^{-4})$

30. $(3.4 \times 10^{-2})(4.5 \times 10^{-6})$

Find each quotient. Write all answers in scientific notation.

31. $\dfrac{8.4 \times 10^5}{2 \times 10^2}$

32. $\dfrac{9.6 \times 10^{20}}{3 \times 10^6}$

33. $\dfrac{6 \times 10^8}{2 \times 10^{-2}}$

34. $\dfrac{8 \times 10^{12}}{4 \times 10^{-3}}$

35. $\dfrac{2.5 \times 10^{-6}}{5 \times 10^{-4}}$

36. $\dfrac{4.5 \times 10^{-8}}{9 \times 10^{-4}}$

Combine by adding or subtracting as indicated.

37. $3x^2 + 5x^2$

38. $4x^3 + 8x^3$

39. $8x^5 - 19x^5$

40. $75x^6 - 50x^6$

41. $2a + a - 3a$

42. $5a + a - 6a$

43. $10x^3 - 8x^3 + 2x^3$

44. $7x^5 + 8x^5 - 12x^5$

45. $20ab^2 - 19ab^2 + 30ab^2$

46. $18a^3b^2 - 20a^3b^2 + 10a^3b^2$

47. Fill in the table.

a	b	ab	$a + b$
$5x$	$3x$		
$4x^2$	$2x^2$		
$3x^3$	$6x^3$		
$2x^4$	$-3x^4$		
x^5	$7x^5$		

48. Fill in the table.

a	b	ab	$a - b$
$2y$	$3y$		
$-2y$	$3y$		
$4y^2$	$5y^2$		
y^3	$-3y^3$		
$5y^4$	$7y^4$		

Simplify. Write all answers with positive exponents only.

49. $\dfrac{(3x^2)(8x^5)}{6x^4}$

50. $\dfrac{(7x^3)(6x^8)}{14x^5}$

51. $\dfrac{(9a^2b)(2a^3b^4)}{18a^5b^7}$

52. $\dfrac{(21a^5b)(2a^8b^4)}{14ab}$

53. $\dfrac{(4x^3y^2)(9x^4y^{10})}{(3x^5y)(2x^6y)}$

54. $\dfrac{(5x^4y^4)(10x^3y^3)}{(25xy^5)(2xy^7)}$

Simplify each expression, and write all answers in scientific notation.

55. $\dfrac{(6 \times 10^8)(3 \times 10^5)}{9 \times 10^7}$

56. $\dfrac{(8 \times 10^4)(5 \times 10^{10})}{2 \times 10^7}$

57. $\dfrac{(5 \times 10^3)(4 \times 10^{-5})}{2 \times 10^{-2}}$

58. $\dfrac{(7 \times 10^6)(4 \times 10^{-4})}{1.4 \times 10^{-3}}$

59. $\dfrac{(2.8 \times 10^{-7})(3.6 \times 10^4)}{2.4 \times 10^3}$

60. $\dfrac{(5.4 \times 10^2)(3.5 \times 10^{-9})}{4.5 \times 10^6}$

Simplify.

61. $\dfrac{18x^4}{3x} + \dfrac{21x^7}{7x^4}$

62. $\dfrac{24x^{10}}{6x^4} + \dfrac{32x^7}{8x}$

63. $\dfrac{45a^6}{9a^4} - \dfrac{50a^8}{2a^6}$

64. $\dfrac{16a^9}{4a} - \dfrac{28a^{12}}{4a^4}$

65. $\dfrac{6x^7y^4}{3x^2y^2} + \dfrac{8x^5y^8}{2y^6}$

66. $\dfrac{40x^{10}y^{10}}{8x^2y^5} + \dfrac{10x^8y^8}{5y^3}$

Apply the distributive property.

67. $xy\left(x + \dfrac{1}{y}\right)$

68. $xy\left(y + \dfrac{1}{x}\right)$

69. $xy\left(\dfrac{1}{y} + \dfrac{1}{x}\right)$

70. $xy\left(\dfrac{1}{x} - \dfrac{1}{y}\right)$

71. $x^2\left(1 - \dfrac{4}{x^2}\right)$

72. $x^2\left(1 - \dfrac{9}{x^2}\right)$

73. $x^2\left(1 - \dfrac{1}{x} - \dfrac{6}{x^2}\right)$

74. $x^2\left(1 - \dfrac{5}{x} + \dfrac{6}{x^2}\right)$

75. $x^2\left(1 - \dfrac{5}{x}\right)$

76. $x^2\left(1 - \dfrac{3}{x}\right)$

77. $x^2\left(1 - \dfrac{8}{x}\right)$ **78.** $x^2\left(1 - \dfrac{6}{x}\right)$

79. Divide each monomial by $5a^2$.
   a. $10a^2$
   b. $-15a^2b$
   c. $25a^2b^2$

80. Divide each monomial by $36x^2$.
   a. $6x^2a$
   b. $12x^2a$
   c. $-6x^2a$

81. Divide each monomial by $8x^2y$.
   a. $24x^3y^2$
   b. $16x^2y^2$
   c. $-4x^2y^3$

82. Divide each monomial by $7x^2y$.
   a. $21x^3y^2$
   b. $14x^2y^2$
   c. $-7x^2y^3$

Maintaining Your Skills

Find the value of each expression when $x = -2$.

83. $4x$ **84.** $-3x$

85. $-2x + 5$ **86.** $-4x - 1$

87. $x^2 + 5x + 6$ **88.** $x^2 - 5x + 6$

For each of the following equations complete the given ordered pairs so each is a solution to the equation, and then use the ordered pairs to graph the equation.

89. $y = 2x + 2$ $(-2, \ )$, $(0, \ )$, $(2, \ )$

90. $y = 2x - 3$ $(-1, \ )$, $(0, \ )$, $(2, \ )$

91. $y = \dfrac{1}{3}x + 1$ $(-3, \ )$, $(0, \ )$, $(3, \ )$

92. $y = \dfrac{1}{2}x - 2$ $(-2, \ )$, $(0, \ )$, $(2, \ )$

Getting Ready for the Next Section

Simplify.

93. $3 - 8$ **94.** $-5 + 7$

95. $-1 + 7$ **96.** $1 - 8$

97. $3(5)^2 + 1$ **98.** $3(-2)^2 - 5(-2) + 4$

99. $2x^2 + 4x^2$ **100.** $3x^2 - x^2$

101. $-5x + 7x$ **102.** $x - 2x$

103. $-(2x + 9)$

104. $-(4x^2 - 2x - 6)$

105. Find the value of $2x + 3$ when $x = 4$.

106. Find the value of $(3x)^2$ when $x = 3$.

4.4   Addition and Subtraction of Polynomials

OBJECTIVES

A   Add and subtract polynomials.

B   Find the value of a polynomial for a given value of the variable.

In this section we will extend what we learned in Section 4.3 to expressions called polynomials. We begin this section with the definition of a polynomial.

> **DEFINITION** A **polynomial** is a finite sum of monomials (terms).

Examples The following are polynomials:

$$3x^2 + 2x + 1 \qquad 15x^2y + 21xy^2 - y^2 \qquad 3a - 2b + 4c - 5d$$

Polynomials can be further classified by the number of terms they contain. A polynomial with two terms is called a binomial. If it has three terms, it is a trinomial. As stated before, a monomial has only one term.

> **DEFINITION** The **degree** of a polynomial in one variable is the highest power to which the variable is raised.

Examples

$3x^5 + 2x^3 + 1$	**A trinomial of degree 5**
$2x + 1$	**A binomial of degree 1**
$3x^2 + 2x + 1$	**A trinomial of degree 2**
$3x^5$	**A monomial of degree 5**
-9	**A monomial of degree 0**

There are no new rules for adding one or more polynomials. We rely only on our previous knowledge. Here are some examples.

 EXAMPLE 1   Add $(2x^2 - 5x + 3) + (4x^2 + 7x - 8)$.

SOLUTION   We use the commutative and associative properties to group similar terms together and then apply the distributive property to add:

$$(2x^2 - 5x + 3) + (4x^2 + 7x - 8)$$

$$= (2x^2 + 4x^2) + (-5x + 7x) + (3 - 8) \qquad \text{\textbf{Commutative and associative properties}}$$

$$= (2 + 4)x^2 + (-5 + 7)x + (3 - 8) \qquad \text{\textbf{Distributive property}}$$

$$= 6x^2 + 2x - 5 \qquad \text{\textbf{Addition}}$$

The results here indicate that to add two polynomials, we add coefficients of similar terms.

 EXAMPLE 2   Add $x^2 + 3x + 2x + 6$.

SOLUTION   The only similar terms here are the two middle terms. We combine them as usual to get

$$x^2 + 3x + 2x + 6 = x^2 + 5x + 6$$

You will recall from Chapter 1 the definition of subtraction: $a - b = a + (-b)$. To subtract one expression from another, we simply add its opposite. The letters a and b in the definition can each represent polynomials. The opposite of a polynomial is the polynomial with opposite terms. When you subtract one polynomial from another you subtract each of its terms.

 EXAMPLE 3   Subtract $(3x^2 + x + 4) - (x^2 + 2x + 3)$.

SOLUTION   To subtract $x^2 + 2x + 3$, we change the sign of each of its terms and add. If you are having trouble remembering why we do this, remember that we can think of $-(x^2 + 2x + 3)$ as $-1(x^2 + 2x + 3)$. If we distribute the -1 across $x^2 + 2x + 3$, we get $-x^2 - 2x - 3$:

$$(3x^2 + x + 4) - (x^2 + 2x + 3)$$

$$= 3x^2 + x + 4 - x^2 - 2x - 3 \qquad \text{\textbf{Take the opposite of each term in the second polynomial}}$$

$$= (3x^2 - x^2) + (x - 2x) + (4 - 3)$$

$$= 2x^2 - x + 1$$

EXAMPLE 4   Subtract $-4x^2 + 5x - 7$ from $x^2 - x - 1$.

SOLUTION   The polynomial $x^2 - x - 1$ comes first, then the subtraction sign, and finally the polynomial $-4x^2 + 5x - 7$ in parentheses.

$$(x^2 - x - 1) - (-4x^2 + 5x - 7)$$

$$= x^2 - x - 1 + 4x^2 - 5x + 7 \qquad \text{Take the opposite of each term in the second polynomial}$$

$$= (x^2 + 4x^2) + (-x - 5x) + (-1 + 7)$$

$$= 5x^2 - 6x + 6$$

The last topic we want to consider in this section is finding the value of a polynomial for a given value of the variable.

To find the value of the polynomial $3x^2 + 1$ when x is 5, we replace x with 5 and simplify the result:

When $x = 5$

the polynomial $3x^2 + 1$

becomes $3(5)^2 + 1 = 3(25) + 1$

$$= 75 + 1 = 76$$

There are two important points to remember when adding or subtracting polynomials. First, to add or subtract two polynomials, you always add or subtract *coefficients* of similar terms. Second, the exponents never increase in value when you are adding or subtracting similar terms.

 EXAMPLE 5   Find the value of $3x^2 - 5x + 4$ when $x = -2$.

SOLUTION

When $x = -2$

the polynomial $3x^2 - 5x + 4$

becomes $3(-2)^2 - 5(-2) + 4 = 3(4) + 10 + 4$

$$= 12 + 10 + 4 = 26$$

LINKING OBJECTIVES AND EXAMPLES

Next to each **objective** we have listed the examples that are best described by that objective.

A   1–4

B   5

GETTING READY FOR CLASS

After reading through the preceding section, respond in your own words and in complete sentences.

1. What are similar terms?
2. What is the degree of a polynomial?
3. Describe how you would subtract one polynomial from another.
4. How would you find the value of $3x^2 - 5x + 4$ when x is -2?

Identify each of the following polynomials as a trinomial, binomial, or monomial, and give the degree in each case.

1. $2x^3 - 3x^2 + 1$

▶**2.** $4x^2 - 4x + 1$

3. $5 + 8a - 9a^3$

4. $6 + 12x^3 + x^4$

5. $2x - 1$

6. $4 + 7x$

7. $45x^2 - 1$

8. $3a^3 + 8$

9. $7a^2$

▶**10.** $90x$

11. -4

12. 56

Perform the following additions and subtractions.

13. $(2x^2 + 3x + 4) + (3x^2 + 2x + 5)$

▶**14.** $(x^2 + 5x + 6) + (x^2 + 3x + 4)$

15. $(3a^2 - 4a + 1) + (2a^2 - 5a + 6)$

16. $(5a^2 - 2a + 7) + (4a^2 - 3a + 2)$

17. $x^2 + 4x + 2x + 8$

18. $x^2 + 5x - 3x - 15$

19. $6x^2 - 3x - 10x + 5$

20. $10x^2 + 30x - 2x - 6$

21. $x^2 - 3x + 3x - 9$

22. $x^2 - 5x + 5x - 25$

23. $3y^2 - 5y - 6y + 10$

24. $y^2 - 18y + 2y - 12$

25. $(6x^3 - 4x^2 + 2x) + (9x^2 - 6x + 3)$

26. $(5x^3 + 2x^2 + 3x) + (2x^2 + 5x + 1)$

27. $\left(\dfrac{2}{3}x^2 - \dfrac{1}{5}x - \dfrac{3}{4}\right) + \left(\dfrac{4}{3}x^2 - \dfrac{4}{5}x + \dfrac{7}{4}\right)$

28. $\left(\dfrac{3}{8}x^3 - \dfrac{5}{7}x^2 - \dfrac{2}{5}\right) + \left(\dfrac{5}{8}x^3 - \dfrac{2}{7}x^2 + \dfrac{7}{5}\right)$

29. $(a^2 - a - 1) - (-a^2 + a + 1)$

▶**30.** $(5a^2 - a - 6) - (-3a^2 - 2a + 4)$

31. $\left(\dfrac{5}{9}x^3 + \dfrac{1}{3}x^2 - 2x + 1\right) - \left(\dfrac{2}{3}x^3 + x^2 + \dfrac{1}{2}x - \dfrac{3}{4}\right)$

32. $\left(4x^3 - \dfrac{2}{5}x^2 + \dfrac{3}{8}x - 1\right) - \left(\dfrac{9}{2}x^3 + \dfrac{1}{4}x^2 - x + \dfrac{5}{6}\right)$

33. $(4y^2 - 3y + 2) + (5y^2 + 12y - 4) - (13y^2 - 6y + 20)$

34. $(2y^2 - 7y - 8) - (6y^2 + 6y - 8) + (4y^2 - 2y + 3)$

Simplify.

▶**35.** $(x^2 - 5x) - (x^2 - 3x)$

▶**36.** $(-2x + 8) - (-2x + 6)$

▶**37.** $(6x^2 - 11x) - (6x^2 - 15x)$

▶**38.** $(10x^2 - 3x) - (10x^2 - 50x)$

▶**39.** $(x^3 + 3x^2 + 9x) - (3x^2 + 9x + 27)$

▶**40.** $(x^3 + 2x^2 + 4x) - (2x^2 + 4x - 8)$

▶**41.** $(x^3 + 4x^2 + 4x) + (2x^2 + 8x + 8)$

▶**42.** $(x^3 + 2x^2 + x) + (x^2 + 2x + 1)$

▶**43.** $(x^2 - 4) - (x^2 - 4x + 4)$

▶**44.** $(4x^2 - 9) - (4x^2 - 12x + 9)$

45. Subtract $10x^2 + 23x - 50$ from $11x^2 - 10x + 13$.

46. Subtract $2x^2 - 3x + 5$ from $4x^2 - 5x + 10$.

▶**47.** Subtract $3y^2 + 7y - 15$ from $11y^2 + 11y + 11$.

48. Subtract $15y^2 - 8y - 2$ from $3y^2 - 3y + 2$.

49. Add $50x^2 - 100x - 150$ to $25x^2 - 50x + 75$.

50. Add $7x^2 - 8x + 10$ to $-8x^2 + 2x - 12$.

51. Subtract $2x + 1$ from the sum of $3x - 2$ and $11x + 5$.

52. Subtract $3x - 5$ from the sum of $5x + 2$ and $9x - 1$.

53. Find the value of the polynomial $x^2 - 2x + 1$ when x is 3.

54. Find the value of the polynomial $(x - 1)^2$ when x is 3.

55. Find the value of $100p^2 - 1,300p + 4,000$ when
　　a. $p = 5$
　　b. $p = 8$

56. Find the value of $100p^2 - 800p + 1,200$ when
　　a. $p = 2$
　　b. $p = 6$

　= Videos available by instructor request

▶ = Online student support materials available at academic.cengage.com/login

57. Find the value of $600 + 1,000x - 100x^2$ when
   a. $x = 8$
   b. $x = -2$

58. Find the value of $500 + 800x - 100x^2$ when
   a. $x = 6$
   b. $x = -1$

Applying the Concepts

59. Packaging A crystal ball with a diameter of 6 inches is being packaged for shipment. If the crystal ball is placed inside a circular cylinder with radius 3 inches and height 6 inches, how much volume will need to be filled with padding? (The volume of a sphere with radius r is $\frac{4}{3}\pi r^3$, and the volume of a right circular cylinder with radius r and height h is $\pi r^2 h$.)

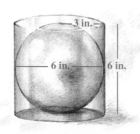

60. Packaging Suppose the circular cylinder of Problem 45 has a radius of 4 inches and a height of 7 inches. How much volume will need to be filled with padding?

Maintaining Your Skills

61. $3x(-5x)$

62. $-3x(-7x)$

63. $2x(3x^2)$

64. $x^2(3x)$

65. $3x^2(2x^2)$

66. $4x^2(2x^2)$

Getting Ready for the Next Section

Simplify.

67. $(-5)(-1)$

68. $3(-4)$

69. $(-1)(6)$

70. $(-7)8$

71. $(5x)(-4x)$

72. $(3x)(2x)$

73. $3x(-7)$

74. $3x(-1)$

75. $5x + (-3x)$

76. $-3x - 10x$

Multiply.

77. $3(2x - 6)$

78. $-4x(x + 5)$

4.5 Multiplication with Polynomials

OBJECTIVES

A Multiply a monomial with a polynomial.

B Multiply two binomials.

C Multiply two polynomials.

We begin our discussion of multiplication of polynomials by finding the product of a monomial and a trinomial.

EXAMPLE 1   Multiply $3x^2(2x^2 + 4x + 5)$.

SOLUTION   Applying the distributive property gives us

$$3x^2(2x^2 + 4x + 5) = 3x^2(2x^2) + 3x^2(4x) + 3x^2(5) \qquad \textbf{Distributive property}$$

$$= 6x^4 + 12x^3 + 15x^2 \qquad\qquad \textbf{Multiplication}$$

The distributive property is the key to multiplication of polynomials. We can use it to find the product of any two polynomials. There are some shortcuts we can use in certain situations, however. Let's look at an example that involves the product of two binomials.

 **EXAMPLE 2**   Multiply $(3x - 5)(2x - 1)$.

SOLUTION

$$(3x - 5)(2x - 1) = 3x(2x - 1) - 5(2x - 1)$$
$$= 3x(2x) + 3x(-1) + (-5)(2x) + (-5)(-1)$$
$$= 6x^2 - 3x - 10x + 5$$
$$= 6x^2 - 13x + 5$$

If we look closely at the second and third lines of work in this example, we can see that the terms in the answer come from all possible products of terms in the first binomial with terms in the second binomial. This result is generalized as follows.

> **Rule** To multiply any two polynomials, multiply each term in the first with each term in the second.

There are two ways we can put this rule to work.

FOIL Method

If we look at the original problem in Example 2 and then at the answer, we see that the first term in the answer came from multiplying the first terms in each binomial:

$$3x \cdot 2x = 6x^2 \quad \text{FIRST}$$

The middle term in the answer came from adding the products of the two outside terms and the two inside terms in each binomial:

$$3x(-1) = -3x \qquad \text{OUTSIDE}$$
$$-5(2x) = \underline{-10x} \qquad \text{INSIDE}$$
$$-13x$$

The last term in the answer came from multiplying the two last terms:

$$-5(-1) = 5 \quad \text{LAST}$$

To summarize the FOIL method, we will multiply another two binomials.

  EXAMPLE 3   Multiply $(2x + 3)(5x - 4)$.

SOLUTION

$$(2x + 3)(5x - 4) = \underbrace{2x(5x)} + \underbrace{2x(-4)} + \underbrace{3(5x)} + \underbrace{3(-4)}$$

$$\text{FIRST} \qquad \text{OUTSIDE} \qquad \text{INSIDE} \qquad \text{LAST}$$

$$= 10x^2 - 8x + 15x - 12$$
$$= 10x^2 + 7x - 12$$

With practice $-8x + 15x = 7x$ can be done mentally.

COLUMN Method

The FOIL method can be applied only when multiplying two binomials. To find products of polynomials with more than two terms, we use what is called the COLUMN method.

The COLUMN method of multiplying two polynomials is very similar to long multiplication with whole numbers. It is just another way of finding all possible products of terms in one polynomial with terms in another polynomial.

EXAMPLE 4   Multiply $(2x + 3)(3x^2 - 2x + 1)$.

SOLUTION

$$
\begin{array}{r}
3x^2 - 2x + 1 \\
2x + 3 \\
\hline
6x^3 - 4x^2 + 2x \qquad \leftarrow 2x(3x^2 - 2x + 1) \\
9x^2 - 6x + 3 \leftarrow \; 3(3x^2 - 2x + 1) \\
\hline
6x^3 + 5x^2 - 4x + 3 \leftarrow \textbf{Add similar terms}
\end{array}
$$

It will be to your advantage to become very fast and accurate at multiplying polynomials. You should be comfortable using either method. The following examples illustrate the three types of multiplication.

EXAMPLES   Multiply:

5. $4a^2(2a^2 - 3a + 5) = 4a^2(2a^2) + 4a^2(-3a) + 4a^2(5)$
$$= 8a^4 - 12a^3 + 20a^2$$

6. $(x - 2)(y + 3) = x(y) + x(3) + (-2)(y) + (-2)(3)$
$$\qquad\qquad\qquad\quad \text{F} \qquad \text{O} \qquad \text{I} \qquad \text{L}$$
$$= xy + 3x - 2y - 6$$

7. $(x + y)(a - b) = x(a) + x(-b) + y(a) + y(-b)$
$$\qquad\qquad\qquad\quad \text{F} \qquad \text{O} \qquad \text{I} \qquad \text{L}$$
$$= xa - xb + ya - yb$$

8. $(5x - 1)(2x + 6) = 5x(2x) + 5x(6) + (-1)(2x) + (-1)(6)$
$$\qquad\qquad\qquad\qquad \text{F} \qquad \text{O} \qquad \text{I} \qquad \text{L}$$
$$= 10x^2 + 30x + (-2x) + (-6)$$
$$= 10x^2 + 28x - 6$$

EXAMPLE 9   The length of a rectangle is 3 more than twice the width. Write an expression for the area of the rectangle.

SOLUTION   We begin by drawing a rectangle and labeling the width with x. Since the length is 3 more than twice the width, we label the length with $2x + 3$.

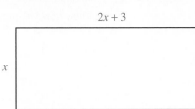

Since the area A of a rectangle is the product of the length and width, we write our formula for the area of this rectangle as

$$A = x(2x + 3)$$
$$A = 2x^2 + 3x \qquad \textbf{Multiply}$$

Revenue

Suppose that a store sells x items at p dollars per item. The total amount of money obtained by selling the items is called the *revenue*. It can be found by multiplying the number of items sold, x, by the price per item, p. For example, if 100 items are sold for $6 each, the revenue is $100(6) = \$600$. Similarly, if 500 items are sold for $8 each, the total revenue is $500(8) = \$4{,}000$. If we denote the revenue with the letter R, then the formula that relates R, x, and p is

Revenue = (number of items sold)(price of each item)

In symbols: $R = xp$.

EXAMPLE 10   A store selling diskettes for home computers knows from past experience that it can sell x diskettes each day at a price of p dollars per diskette, according to the equation $x = 800 - 100p$. Write a formula for the daily revenue that involves only the variables R and p.

SOLUTION   From our previous discussion we know that the revenue R is given by the formula

$$R = xp$$

But, since $x = 800 - 100p$, we can substitute $800 - 100p$ for x in the revenue equation to obtain

$$R = (800 - 100p)p$$
$$R = 800p - 100p^2$$

This last formula gives the revenue, R, in terms of the price, p.

GETTING READY FOR CLASS

After reading through the preceding section, respond in your own words and in complete sentences.

1. How do we multiply two polynomials?
2. Describe how the distributive property is used to multiply a monomial and a polynomial.
3. Describe how you would use the FOIL method to multiply two binomials.
4. Show how the product of two binomials can be a trinomial.

31. $\dfrac{a^2b^2 - ab^2}{-ab^2}$

32. $\dfrac{a^2b^2c + ab^2c^2}{abc}$

33. $\dfrac{x^3 - 3x^2y + xy^2}{x}$

34. $\dfrac{x^2 - 3xy^2 + xy^3}{x}$

35. $\dfrac{10a^2 - 15a^2b + 25a^2b^2}{5a^2}$

36. $\dfrac{11a^2b^2 - 33ab}{-11ab}$

37. $\dfrac{26x^2y^2 - 13xy}{-13xy}$

38. $\dfrac{6x^2y^2 - 3xy}{6xy}$

39. $\dfrac{4x^2y^2 - 2xy}{4xy}$

40. $\dfrac{6x^2a + 12x^2b - 6x^2c}{36x^2}$

41. $\dfrac{5a^2x - 10ax^2 + 15a^2x^2}{20a^2x^2}$

42. $\dfrac{12ax - 9bx + 18cx}{6x^2}$

▶ **43.** $\dfrac{16x^5 + 8x^2 + 12x}{12x^3}$

44. $\dfrac{27x^2 - 9x^3 - 18x^4}{-18x^3}$

Divide. Assume all variables represent positive numbers.

45. $\dfrac{9a^{5m} - 27a^{3m}}{3a^{2m}}$

46. $\dfrac{26a^{3m} - 39a^{5m}}{13a^{3m}}$

47. $\dfrac{10x^{5m} - 25x^{3m} + 35x^m}{5x^m}$

48. $\dfrac{18x^{2m} + 24x^{4m} - 30x^{6m}}{6x^{2m}}$

Simplify each numerator, and then divide.

49. $\dfrac{2x^3(3x + 2) - 3x^2(2x - 4)}{2x^2}$

50. $\dfrac{5x^2(6x - 3) + 6x^3(3x - 1)}{3x}$

51. $\dfrac{(x + 2)^2 - (x - 2)^2}{2x}$

52. $\dfrac{(x - 3)^2 - (x + 3)^2}{3x}$

53. $\dfrac{(x + 5)^2 + (x + 5)(x - 5)}{2x}$

54. $\dfrac{(x - 4)^2 + (x + 4)(x - 4)}{2x}$

55. Find the value of each expression when x is 2.

   a. $2x + 3$

   b. $\dfrac{10x + 15}{5}$

   c. $10x + 3$

56. Find the value of each expression when x is 5.

   a. $3x + 2$

   b. $\dfrac{6x^2 + 4x}{2x}$

   c. $6x^2 + 2$

57. Evaluate each expression for $x = 10$.

   a. $\dfrac{3x + 8}{2}$

   b. $3x + 4$

   c. $\dfrac{3}{2}x + 4$

58. Evaluate each expression for $x = 10$.

   a. $\dfrac{5x - 6}{2}$

   b. $5x - 3$

   c. $\dfrac{5}{2}x - 3$

59. Find the value of each expression when x is 2.

   a. $2x^2 - 3x$

   b. $\dfrac{10x^3 - 15x^2}{5x}$

60. Find the value of each expression when a is 3.

   a. $-4a + 2$

   b. $\dfrac{8a^2 - 4a}{-2a}$

Maintaining Your Skills

Solve each system of equations by the elimination method.

61. $x + y = 6$
    $x - y = 8$

62. $2x + y = 5$
    $-x + y = -4$

63. $2x - 3y = -5$
    $x + y = 5$

64. $2x - 4y = 10$
    $3x - 2y = -1$

Solve each system by the substitution method.

65. $x + y = 2$
$\quad\quad y = 2x - 1$

66. $2x - 3y = 4$
$\quad\quad x = 3y - 1$

67. $4x + 2y = 8$
$\quad\quad y = -2x + 4$

68. $4x + 2y = 8$
$\quad\quad y = -2x + 5$

Getting Ready for the Next Section

Divide.

69. $27\overline{)3,962}$

70. $13\overline{)18,780}$

71. $\dfrac{2x^2 + 5x}{x}$

72. $\dfrac{7x^2 + 9x^3 + 3x^7}{x^3}$

Multiply.

73. $(x - 3)x$

74. $(x - 3)(-2)$

75. $2x^2(x - 5)$

76. $10x(x - 5)$

Subtract.

77. $(x^2 - 5x) - (x^2 - 3x)$

78. $(2x^3 + 0x^2) - (2x^3 - 10x^2)$

79. $(-2x + 8) - (-2x + 6)$

80. $(4x - 14) - (4x - 10)$

4.8 Dividing a Polynomial by a Polynomial

OBJECTIVES

A Divide a polynomial by a polynomial.

Since long division for polynomials is very similar to long division with whole numbers, we will begin by reviewing a division problem with whole numbers. You may realize when looking at Example 1 that you don't have a very good idea why you proceed as you do with long division. What you do know is that the process always works. We are going to approach the explanations in this section in much the same manner; that is, we won't always be sure why the steps we will use are important, only that they always produce the correct result.

EXAMPLE 1 Divide $27\overline{)3,962}$.

SOLUTION

$$
\begin{array}{r}
1 \quad\quad \leftarrow \textbf{Estimate 27 into 39} \\
27\overline{)3,962} \\
2\,7 \quad\quad \leftarrow \textbf{Multiply } 1 \times 27 = 27 \\
\overline{1\,2} \quad\quad \leftarrow \textbf{Subtract } 39 - 27 = 12
\end{array}
$$

$$
\begin{array}{r}
1 \quad\quad \\
27\overline{)3,962} \\
2\,7{\downarrow} \quad\quad \\
\overline{1\,26} \quad \leftarrow \textbf{Bring down the 6}
\end{array}
$$

These are the four basic steps in long division. Estimate, multiply, subtract, and bring down the next term. To finish the problem, we simply perform the same four steps again:

$$
\begin{array}{r}
14 \quad \leftarrow \textbf{4 is the estimate} \\
27\overline{)3{,}962} \\
2\ 7\downarrow \\
\hline
1\ 26 \\
\underline{1\ 08} \quad \leftarrow \textbf{Multiply to get 108} \\
182 \quad \leftarrow \textbf{Subtract to get 18, then bring down the 2}
\end{array}
$$

One more time.

$$
\begin{array}{r}
146 \quad \leftarrow \textbf{6 is the estimate} \\
27\overline{)3{,}962} \\
2\ 7 \\
\hline
1\ 26 \\
1\ 08 \\
\hline
182 \\
\underline{162} \quad \leftarrow \textbf{Multiply to get 162} \\
20 \quad \leftarrow \textbf{Subtract to get 20}
\end{array}
$$

Since there is nothing left to bring down, we have our answer.

$$
\frac{3{,}962}{27} = 146 + \frac{20}{27} \qquad \text{or} \qquad 146\frac{20}{27}
$$

Here is how it works with polynomials.

 EXAMPLE 2   Divide $\dfrac{x^2 - 5x + 8}{x - 3}$.

SOLUTION

$$
\begin{array}{r}
x \quad\quad\quad \leftarrow \textbf{Estimate } x^2 \div x = x \\
x - 3\overline{)\ x^2 - 5x + 8} \\
\underset{-}{\cancel{x^2}} \underset{+}{\cancel{-}} 3x \quad\quad \leftarrow \textbf{Multiply } x(x - 3) = x^2 - 3x \\
\hline
-2x \quad\quad \leftarrow \textbf{Subtract } (x^2 - 5x) - (x^2 - 3x) = -2x
\end{array}
$$

$$
\begin{array}{r}
x \quad\quad\quad\quad \\
x - 3\overline{)\ x^2 - 5x + 8} \\
\underset{-}{\cancel{x^2}} \underset{+}{\cancel{-}} 3x \quad\quad\quad \\
\hline
-2x + 8 \quad \leftarrow \textbf{Bring down the 8}
\end{array}
$$

Notice that to subtract one polynomial from another, we add its opposite. That is why we change the signs on $x^2 - 3x$ and add what we get to $x^2 - 5x$. (To subtract the second polynomial, simply change the signs and add.)

We perform the same four steps again:

$$
\begin{array}{r}
x - 2 \quad \leftarrow \textbf{-2 is the estimate } (-2x \div x = -2) \\
x - 3\overline{)\ x^2 - 5x + 8} \\
\underset{-}{\cancel{x^2}} \underset{+}{\cancel{-}} 3x \quad\ \downarrow \\
\hline
-2x + 8 \\
\underset{+}{\cancel{-}} 2x \underset{-}{\cancel{+}} 6 \quad \leftarrow \textbf{Multiply } -2(x - 3) = -2x + 6 \\
\hline
2 \quad \leftarrow \textbf{Subtract } (-2x + 8) - (-2x + 6) = 2
\end{array}
$$

Since there is nothing left to bring down, we have our answer:

$$\frac{x^2 - 5x + 8}{x - 3} = x - 2 + \frac{2}{x - 3}$$

To check our answer, we multiply $(x - 3)(x - 2)$ to get $x^2 - 5x + 6$. Then, adding on the remainder, 2, we have $x^2 - 5x + 8$.

 EXAMPLE 3   Divide $\dfrac{6x^2 - 11x - 14}{2x - 5}$.

SOLUTION

$$
\begin{array}{r}
3x + 2 \\
2x - 5 \overline{)\; 6x^2 - 11x - 14\;} \\
6x^2 - 15x \\
\hline
+\; 4x - 14 \\
4x - 10 \\
\hline
-\; 4
\end{array}
$$

$$\frac{6x^2 - 11x - 14}{2x - 5} = 3x + 2 + \frac{-4}{2x - 5}$$

One last step is sometimes necessary. The two polynomials in a division problem must both be in descending powers of the variable and cannot skip any powers from the highest power down to the constant term.

EXAMPLE 4   Divide $\dfrac{2x^3 - 3x + 2}{x - 5}$.

SOLUTION   The problem will be much less confusing if we write $2x^3 - 3x + 2$ as $2x^3 + 0x^2 - 3x + 2$. Adding $0x^2$ does not change our original problem.

$$
\begin{array}{r}
2x^2 \\
x - 5 \overline{)\; 2x^3 + 0x^2 - 3x + 2\;} \\
2x^3 - 10x^2 \\
\hline
+\; 10x^2 - 3x
\end{array}
$$

← **Estimate** $2x^3 \div x = 2x^2$

← **Multiply** $2x^2(x - 5) = 2x^3 - 10x^2$

← **Subtract:**
$(2x^3 + 0x^2) - (2x^3 - 10x^2) = 10x^2$
Bring down the next term

Adding the term $0x^2$ gives us a column in which to write $10x^2$. (Remember, you can add and subtract only similar terms.)

Here is the completed problem:

$$
\begin{array}{r}
2x^2 + 10x + 47 \\
x - 5 \overline{)\; 2x^3 + 0x^2 - 3x + 2\;} \\
2x^3 - 10x^2 \\
\hline
+\; 10x^2 - 3x \\
10x^2 - 50x \\
\hline
+\; 47x + 2 \\
47x - 235 \\
\hline
237
\end{array}
$$

Our answer is $\dfrac{2x^3 - 3x + 2}{x - 5} = 2x^2 + 10x + 47 + \dfrac{237}{x - 5}$.

As you can see, long division with polynomials is a mechanical process. Once you have done it correctly a couple of times, it becomes very easy to produce the correct answer.

GETTING READY FOR CLASS

After reading through the preceding section, respond in your own words and in complete sentences.

1. What are the four steps used in long division with whole numbers?
2. How is division of two polynomials similar to long division with whole numbers?
3. What are the four steps used in long division with polynomials?
4. How do we use 0 when dividing the polynomial $2x^3 - 3x + 2$ by $x - 5$?

LINKING OBJECTIVES AND EXAMPLES

Next to each **objective** we have listed the examples that are best described by that objective.

A 2–4

Problem Set 4.8

Online support materials can be found at academic.cengage.com/login

Divide.

1. $\dfrac{x^2 - 5x + 6}{x - 3}$

2. $\dfrac{x^2 - 5x + 6}{x - 2}$

3. $\dfrac{a^2 + 9a + 20}{a + 5}$

4. $\dfrac{a^2 + 9a + 20}{a + 4}$

5. $\dfrac{x^2 - 6x + 9}{x - 3}$

6. $\dfrac{x^2 + 10x + 25}{x + 5}$

7. $\dfrac{2x^2 + 5x - 3}{2x - 1}$

8. $\dfrac{4x^2 + 4x - 3}{2x - 1}$

9. $\dfrac{2a^2 - 9a - 5}{2a + 1}$

10. $\dfrac{4a^2 - 8a - 5}{2a + 1}$

11. $\dfrac{x^2 + 5x + 8}{x + 3}$

12. $\dfrac{x^2 + 5x + 4}{x + 3}$

13. $\dfrac{a^2 + 3a + 2}{a + 5}$

14. $\dfrac{a^2 + 4a + 3}{a + 5}$

15. $\dfrac{x^2 + 2x + 1}{x - 2}$

16. $\dfrac{x^2 + 6x + 9}{x - 3}$

17. $\dfrac{x^2 + 5x - 6}{x + 1}$

18. $\dfrac{x^2 - x - 6}{x + 1}$

19. $\dfrac{a^2 + 3a + 1}{a + 2}$

20. $\dfrac{a^2 - a + 3}{a + 1}$

21. $\dfrac{2x^2 - 2x + 5}{2x + 4}$

22. $\dfrac{15x^2 + 19x - 4}{3x + 8}$

▶ 23. $\dfrac{6a^2 + 5a + 1}{2a + 3}$

24. $\dfrac{4a^2 + 4a + 3}{2a + 1}$

25. $\dfrac{6a^3 - 13a^2 - 4a + 15}{3a - 5}$

26. $\dfrac{2a^3 - a^2 + 3a + 2}{2a + 1}$

Fill in the missing terms in the numerator, and then use long division to find the quotients (see Example 4).

27. $\dfrac{x^3 + 4x + 5}{x + 1}$

28. $\dfrac{x^3 + 4x^2 - 8}{x + 2}$

29. $\dfrac{x^3 - 1}{x - 1}$

30. $\dfrac{x^3 + 1}{x + 1}$

31. $\dfrac{x^3 - 8}{x - 2}$

32. $\dfrac{x^3 + 27}{x + 3}$

= Videos available by instructor request

▶ = Online student support materials available at academic.cengage.com/login

33. Find the value of each expression when x is 3.

   a. $x^2 + 2x + 4$

   b. $\dfrac{x^3 - 8}{x - 2}$

   c. $x^2 - 4$

34. Find the value of each expression when x is 2.

   a. $x^2 - 3x + 9$

   b. $\dfrac{x^3 + 27}{x + 3}$

   c. $x^2 + 9$

35. Find the value of each expression when x is 4.

   a. $x + 3$

   b. $\dfrac{x^2 + 9}{x + 3}$

   c. $x - 3 + \dfrac{18}{x + 3}$

36. Find the value of each expression when x is 2.

   a. $x + 1$

   b. $\dfrac{x^2 + 1}{x + 1}$

   c. $x - 1 + \dfrac{2}{x + 1}$

Long Division Use the information in the table to find the monthly payment for auto insurance for the cities below. Round to the nearest cent.

USA TODAY Snapshots®

Cities with the most expensive auto insurance annual premiums

Detroit   **$5,162**
Philadelphia   **$4,142**
Newark, N.J.   **$3,482**
Los Angeles   **$3,225**
New York City   **$3,127**

Note: Based on drivers over a minimum age with clean driving records and including comprehensive, collision, bodily injury, property damage and uninsured motorist coverage.

Source: Runzheimer International

By Darryl Haralson and Marcy E. Mullins, USA TODAY

37. Detroit

38. Philadelphia

39. Newark, N.J.

40. Los Angeles

Maintaining Your Skills

Use systems of equations to solve the following word problems.

41. **Number Problem** The sum of two numbers is 25. One of the numbers is 4 times the other. Find the numbers.

42. **Number Problem** The sum of two numbers is 24. One of the numbers is 3 more than twice the other. Find the numbers.

43. **Investing** Suppose you have a total of $1,200 invested in two accounts. One of the accounts pays 8% annual interest and the other pays 9% annual interest. If your total interest for the year is $100, how much money did you invest in each of the accounts?

44. **Investing** If you invest twice as much money in an account that pays 12% annual interest as you do in an account that pays 11% annual interest, how much do you have in each account if your total interest for a year is $210?

45. **Money Problem** If you have a total of $160 in $5 bills and $10 bills, how many of each type of bill do you have if you have 4 more $10 bills than $5 bills?

46. **Coin Problem** Suppose you have 20 coins worth a total of $2.80. If the coins are all nickels and quarters, how many of each type do you have?

47. **Mixture Problem** How many gallons of 20% antifreeze solution and 60% antifreeze solution must be mixed to get 16 gallons of 35% antifreeze solution?

48. **Mixture Problem** A chemist wants to obtain 80 liters of a solution that is 12% hydrochloric acid. How many liters of 10% hydrochloric acid solution and 20% hydrochloric acid solution should he mix to do so?

5.1 The Greatest Common Factor and Factoring by Grouping

OBJECTIVES

A Factor the greatest common factor from a polynomial.

B Factor by grouping.

In Chapter 1 we used the following diagram to illustrate the relationship between multiplication and factoring.

$$\text{Factors} \rightarrow 3 \cdot 5 = 15 \leftarrow \text{Product}$$

A similar relationship holds for multiplication of polynomials. Reading the following diagram from left to right, we say the product of the binomials $x + 2$ and $x + 3$ is the trinomial $x^2 + 5x + 6$. However, if we read in the other direction, we can say that $x^2 + 5x + 6$ factors into the product of $x + 2$ and $x + 3$.

$$\text{Factors} \rightarrow (x + 2)(x + 3) = x^2 + 5x + 6 \leftarrow \text{Product}$$

In this chapter we develop a systematic method of factoring polynomials.

In this section we will apply the distributive property to polynomials to factor from them what is called the greatest common factor.

> **DEFINITION** The **greatest common factor** for a polynomial is the largest monomial that divides (is a factor of) each term of the polynomial.

We use the term *largest monomial* to mean the monomial with the greatest coefficient and highest power of the variable.

 EXAMPLE 1   Find the greatest common factor for the polynomial:

$$3x^5 + 12x^2$$

SOLUTION   The terms of the polynomial are $3x^5$ and $12x^2$. The largest number that divides the coefficients is 3, and the highest power of x that is a factor of x^5 and x^2 is x^2. Therefore, the greatest common factor for $3x^5 + 12x^2$ is $3x^2$; that is, $3x^2$ is the largest monomial that divides each term of $3x^5 + 12x^2$.

 EXAMPLE 2   Find the greatest common factor for:

$$8a^3b^2 + 16a^2b^3 + 20a^3b^3$$

SOLUTION   The largest number that divides each of the coefficients is 4. The highest power of the variable that is a factor of a^3b^2, a^2b^3, and a^3b^3 is a^2b^2. The greatest common factor for $8a^3b^2 + 16a^2b^3 + 20a^3b^3$ is $4a^2b^2$. It is the largest monomial that is a factor of each term.

Once we have recognized the greatest common factor of a polynomial, we can apply the distributive property and factor it out of each term. We rewrite the polynomial as the product of its greatest common factor with the polynomial that remains after the greatest common factor has been factored from each term in the original polynomial.

EXAMPLE 3   Factor the greatest common factor from $3x - 15$.

SOLUTION   The greatest common factor for the terms $3x$ and 15 is 3. We can rewrite both $3x$ and 15 so that the greatest common factor 3 is showing in each term. It is important to realize that $3x$ means $3 \cdot x$. The 3 and the x are not "stuck" together:

$$3x - 15 = 3 \cdot x - 3 \cdot 5$$

Now, applying the distributive property, we have:

$$3 \cdot x - 3 \cdot 5 = 3(x - 5)$$

To check a factoring problem like this, we can multiply 3 and $x - 5$ to get $3x - 15$, which is what we started with. Factoring is simply a procedure by which we change sums and differences into products. In this case we changed the difference $3x - 15$ into the product $3(x - 5)$. Note, however, that we have not changed the meaning or value of the expression. The expression we end up with is equivalent to the expression we started with.

 EXAMPLE 4   Factor the greatest common factor from:

$$5x^3 - 15x^2$$

SOLUTION   The greatest common factor is $5x^2$. We rewrite the polynomial as:

$$5x^3 - 15x^2 = 5x^2 \cdot x - 5x^2 \cdot 3$$

Then we apply the distributive property to get:

$$5x^2 \cdot x - 5x^2 \cdot 3 = 5x^2(x - 3)$$

To check our work, we simply multiply $5x^2$ and $(x - 3)$ to get $5x^3 - 15x^2$, which is our original polynomial.

EXAMPLE 5   Factor the greatest common factor from:

$$16x^5 - 20x^4 - 8x^3$$

SOLUTION   The greatest common factor is $4x^3$. We rewrite the polynomial so we can see the greatest common factor $4x^3$ in each term; then we apply the distributive property to factor it out.

$$16x^5 - 20x^4 - 8x^3 = 4x^3 \cdot 4x^2 - 4x^3 \cdot 5x - 4x^3 \cdot 2$$

$$= 4x^3(4x^2 - 5x - 2)$$

EXAMPLE 6   Factor the greatest common factor from:

$$6x^3y - 18x^2y^2 - 12xy^3$$

SOLUTION   The greatest common factor is $6xy$. We rewrite the polynomial in terms of $6xy$ and then apply the distributive property as follows:

$$6x^3y - 18x^2y^2 - 12xy^3 = 6xy \cdot x^2 - 6xy \cdot 3xy - 6xy \cdot 2y^2$$

$$= 6xy(x^2 - 3xy - 2y^2)$$

EXAMPLE 7   Factor the greatest common factor from:

$$3a^2b - 6a^3b^2 + 9a^3b^3$$

SOLUTION   The greatest common factor is $3a^2b$:

$$3a^2b - 6a^3b^2 + 9a^3b^3 = 3a^2b(1) - 3a^2b(2ab) + 3a^2b(3ab^2)$$
$$= 3a^2b(1 - 2ab + 3ab^2)$$

Factoring by Grouping

To develop our next method of factoring, called *factoring by grouping,* we start by examining the polynomial $xc + yc$. The greatest common factor for the two terms is c. Factoring c from each term we have:

$$xc + yc = c(x + y)$$

Note

But suppose that c itself was a more complicated expression, such as $a + b$, so that the expression we were trying to factor was $x(a + b) + y(a + b)$, instead of $xc + yc$. The expression $x(a + b) + y(a + b)$ has $(a + b)$ common to each term. Factoring this common factor from each term looks like this:

$$x(a + b) + y(a + b) = (a + b)(x + y)$$

To see how all of this applies to factoring polynomials, consider the polynomial

$$xy + 3x + 2y + 6$$

There is no greatest common factor other than the number 1. However, if we group the terms together two at a time, we can factor an x from the first two terms and a 2 from the last two terms:

$$xy + 3x + 2y + 6 = x(y + 3) + 2(y + 3)$$

The expression on the right can be thought of as having two terms: $x(y + 3)$ and $2(y + 3)$. Each of these expressions contains the common factor $y + 3$, which can be factored out using the distributive property:

$$x(y + 3) + 2(y + 3) = (y + 3)(x + 2)$$

This last expression is in factored form. The process we used to obtain it is called factoring by grouping. Here are some additional examples.

The phrase *greatest common factor* doesn't fit what we are doing here as well as it fit the examples in the beginning of this section. Certainly our original definition of the greatest common factor can be applied to the expression $xc + yc$ to obtain a greatest common factor of c. However, for the expression

$$x(a + b) + y(a + b)$$

the common factor $(a + b)$ is not itself a monomial, and so it doesn't fit our definition of greatest common factor. However, the idea of factoring out as much as possible from each term still applies to this expression. The expression $(a + b)$ is common to both terms, and so we factor it from each term using the distributive property.

 EXAMPLE 8   Factor $ax + bx + ay + by$.

SOLUTION   We begin by factoring x from the first two terms and y from the last two terms:

$$ax + bx + ay + by = x(a + b) + y(a + b)$$
$$= (a + b)(x + y)$$

To convince yourself that this is factored correctly, multiply the two factors $(a + b)$ and $(x + y)$.

EXAMPLE 9   Factor by grouping: $3ax - 2a + 15x - 10$.

SOLUTION   First, we factor a from the first two terms and 5 from the last two terms. Then, we factor $3x - 2$ from the remaining two expressions:

$$3ax - 2a + 15x - 10 = a(3x - 2) + 5(3x - 2)$$

$$= (3x - 2)(a + 5)$$

Again, multiplying $(3x - 2)$ and $(a + 5)$ will convince you that these are the correct factors.

 EXAMPLE 10   Factor $2x^2 + 5ax - 2xy - 5ay$.

SOLUTION   From the first two terms we factor x. From the second two terms we must factor $-y$ so that the binomial that remains after we do so matches the binomial produced by the first two terms:

$$2x^2 + 5ax - 2xy - 5ay = x(2x + 5a) - y(2x + 5a)$$

$$= (2x + 5a)(x - y)$$

Another way to accomplish the same result is to use the commutative property to interchange the middle two terms, and then factor by grouping:

$$2x^2 + 5ax - 2xy - 5ay = 2x^2 - 2xy + 5ax - 5ay \qquad \textbf{Commutative}$$
$$\textbf{property}$$

$$= 2x(x - y) + 5a(x - y)$$

$$= (x - y)(2x + 5a)$$

This is the same result we obtained previously.

 EXAMPLE 11   Factor $6x^2 - 3x - 4x + 2$ by grouping.

SOLUTION   The first two terms have $3x$ in common, and the last two terms have either a 2 or a -2 in common. Suppose we factor $3x$ from the first two terms and 2 from the last two terms. We get:

$$6x^2 - 3x - 4x + 2 = 3x(2x - 1) + 2(-2x + 1)$$

We can't go any further because there is no common factor that will allow us to factor further. However, if we factor -2, instead of 2, from the last two terms, our problem is solved:

$$6x^2 - 3x - 4x + 2 = 3x(2x - 1) - 2(2x - 1)$$

$$= (2x - 1)(3x - 2)$$

In this case, factoring -2 from the last two terms gives us an expression that can be factored further.

You should convince yourself that these factors are correct by finding their product.

LINKING OBJECTIVES AND EXAMPLES

Next to each **objective** we have listed the examples that are best described by that objective.

A   1, 2, 5

B   3, 4

GETTING READY FOR CLASS

After reading through the preceding section, respond in your own words and in complete sentences.

1. When the leading coefficient of a trinomial is 1, what is the relationship between the other two coefficients and the factors of the trinomial?

2. When factoring polynomials, what should you look for first?

3. How can you check to see that you have factored a trinomial correctly?

4. Describe how you would find the factors of $x^2 + 8x + 12$.

Problem Set 5.2

Online support materials can be found at academic.cengage.com/login

Factor the following trinomials.

1. $x^2 + 7x + 12$

2. $x^2 + 7x + 10$

3. $x^2 + 3x + 2$

4. $x^2 + 7x + 6$

▶ **5.** $a^2 + 10a + 21$

6. $a^2 - 7a + 12$

7. $x^2 - 7x + 10$

8. $x^2 - 3x + 2$

9. $y^2 - 10y + 21$

10. $y^2 - 7y + 6$

11. $x^2 - x - 12$

12. $x^2 - 4x - 5$

13. $y^2 + y - 12$

14. $y^2 + 3y - 18$

15. $x^2 + 5x - 14$

16. $x^2 - 5x - 24$

17. $r^2 - 8r - 9$

18. $r^2 - r - 2$

19. $x^2 - x - 30$

20. $x^2 + 8x + 12$

21. $a^2 + 15a + 56$

22. $a^2 - 9a + 20$

23. $y^2 - y - 42$

24. $y^2 + y - 42$

25. $x^2 + 13x + 42$

26. $x^2 - 13x + 42$

Factor the following problems completely. First, factor out the greatest common factor, and then factor the remaining trinomial.

27. $2x^2 + 6x + 4$

28. $3x^2 - 6x - 9$

▶ **29.** $3a^2 - 3a - 60$

30. $2a^2 - 18a + 28$

31. $100x^2 - 500x + 600$

32. $100x^2 - 900x + 2,000$

▶ **33.** $100p^2 - 1,300p + 4,000$

34. $100p^2 - 1,200p + 3,200$

35. $x^4 - x^3 - 12x^2$

36. $x^4 - 11x^3 + 24x^2$

▤ = Videos available by instructor request

▶ = Online student support materials available at academic.cengage.com/login

37. $2r^3 + 4r^2 - 30r$

38. $5r^3 + 45r^2 + 100r$

39. $2y^4 - 6y^3 - 8y^2$

40. $3r^3 - 3r^2 - 6r$

41. $x^5 + 4x^4 + 4x^3$

42. $x^5 + 13x^4 + 42x^3$

43. $3y^4 - 12y^3 - 15y^2$

44. $5y^4 - 10y^3 + 5y^2$

45. $4x^4 - 52x^3 + 144x^2$

46. $3x^3 - 3x^2 - 18x$

Factor the following trinomials.

47. $x^2 + 5xy + 6y^2$

48. $x^2 - 5xy + 6y^2$

49. $x^2 - 9xy + 20y^2$

50. $x^2 + 9xy + 20y^2$

51. $a^2 + 2ab - 8b^2$

52. $a^2 - 2ab - 8b^2$

53. $a^2 - 10ab + 25b^2$

54. $a^2 + 6ab + 9b^2$

55. $a^2 + 10ab + 25b^2$

56. $a^2 - 6ab + 9b^2$

57. $x^2 + 2xa - 48a^2$

58. $x^2 - 3xa - 10a^2$

59. $x^2 - 5xb - 36b^2$

60. $x^2 - 13xb + 36b^2$

Factor completely.

61. $x^4 - 5x^2 + 6$

62. $x^6 - 2x^3 - 15$

63. $x^2 - 80x - 2,000$

64. $x^2 - 190x - 2,000$

65. $x^2 - x + \dfrac{1}{4}$

66. $x^2 - \dfrac{2}{3}x + \dfrac{1}{9}$

67. $x^2 + 0.6x + 0.08$

68. $x^2 + 0.8x + 0.15$

69. If one of the factors of $x^2 + 24x + 128$ is $x + 8$, what is the other factor?

70. If one factor of $x^2 + 260x + 2,500$ is $x + 10$, what is the other factor?

71. What polynomial, when factored, gives $(4x + 3)(x - 1)$?

72. What polynomial factors to $(4x - 3)(x + 1)$?

Maintaining Your Skills

Simplify each expression. Write using only positive exponents.

73. $\left(-\dfrac{2}{5}\right)^2$

74. $\left(-\dfrac{3}{8}\right)^2$

75. $(3a^3)^2(2a^2)^3$

76. $(-4x^4)^2(2x^5)^4$

77. $\dfrac{(4x)^{-7}}{(4x)^{-5}}$

78. $\dfrac{(2x)^{-3}}{(2x)^{-5}}$

79. $\dfrac{12a^5b^3}{72a^2b^5}$

80. $\dfrac{25x^5y^3}{50x^2y^7}$

81. $\dfrac{15x^{-5}y^3}{45x^2y^5}$

82. $\dfrac{25a^2b^7}{75a^5b^3}$

83. $(-7x^3y)(3xy^4)$

84. $(9a^6b^4)(6a^4b^3)$

85. $(-5a^3b^{-1})(4a^{-2}b^4)$

86. $(-3a^2b^{-4})(6a^5b^{-2})$

87. $(9a^2b^3)(-3a^3b^5)$

88. $(-7a^5b^8)(6a^7b^4)$

Getting Ready for the Next Section

Multiply using the FOIL method.

89. $(6a + 1)(a + 2)$

90. $(6a - 1)(a - 2)$

91. $(3a + 2)(2a + 1)$

92. $(3a - 2)(2a - 1)$

93. $(6a + 2)(a + 1)$

94. $(3a + 1)(2a + 2)$

More Trinomials to Factor

OBJECTIVES

A Factor a trinomial whose leading coefficient is other than 1.

B Factor a polynomial by first factoring out the greatest common factor and then factoring the polynomial that remains.

We now will consider trinomials whose greatest common factor is 1 and whose leading coefficient (the coefficient of the squared term) is a number other than 1. We present two methods for factoring trinomials of this type. The first method involves listing possible factors until the correct pair of factors is found. This requires a certain amount of trial and error. The second method is based on the factoring by grouping process that we covered previously. Either method can be used to factor trinomials whose leading coefficient is a number other than 1.

Method 1: Factoring $ax^2 + bx + c$ by trial and error

Suppose we want to factor the trinomial $2x^2 - 5x - 3$. We know the factors (if they exist) will be a pair of binomials. The product of their first terms is $2x^2$ and the product of their last term is -3. Let us list all the possible factors along with the trinomial that would result if we were to multiply them together. Remember, the middle term comes from the product of the inside terms plus the product of the outside terms.

Binomial Factors	First Term	Middle Term	Last Term
$(2x - 3)(x + 1)$	$2x^2$	$-x$	-3
$(2x + 3)(x - 1)$	$2x^2$	$+x$	-3
$(2x - 1)(x + 3)$	$2x^2$	$+5x$	-3
$(2x + 1)(x - 3)$	$2x^2$	$-5x$	-3

We can see from the last line that the factors of $2x^2 - 5x - 3$ are $(2x + 1)$ and $(x - 3)$. There is no straightforward way, as there was in the previous section, to find the factors, other than by trial and error or by simply listing all the possibilities. We look for possible factors that, when multiplied, will give the correct first and last terms, and then we see if we can adjust them to give the correct middle term.

EXAMPLE 1 Factor $6a^2 + 7a + 2$.

SOLUTION We list all the possible pairs of factors that, when multiplied together, give a trinomial whose first term is $6a^2$ and whose last term is $+2$.

Binomial Factors	First Term	Middle Term	Last Term
$(6a + 1)(a + 2)$	$6a^2$	$+13a$	$+2$
$(6a - 1)(a - 2)$	$6a^2$	$-13a$	$+2$
$(3a + 2)(2a + 1)$	$6a^2$	$+7a$	$+2$
$(3a - 2)(2a - 1)$	$6a^2$	$-7a$	$+2$

Note
Remember, we can always check our results by multiplying the factors we have and comparing that product with our original polynomial.

The factors of $6a^2 + 7a + 2$ are $(3a + 2)$ and $(2a + 1)$.

Check: $(3a + 2)(2a + 1) = 6a^2 + 7a + 2$

Notice that in the preceding list we did not include the factors $(6a + 2)$ and $(a + 1)$. We do not need to try these since the first factor has a 2 common to each term and so could be factored again, giving $2(3a + 1)(a + 1)$. Since our original trinomial, $6a^2 + 7a + 2$, did *not* have a greatest common factor of 2, neither of its factors will.

 EXAMPLE 2   Factor $4x^2 - x - 3$.

SOLUTION   We list all the possible factors that, when multiplied, give a trinomial whose first term is $4x^2$ and whose last term is -3.

Binomial Factors	First Term	Middle Term	Last Term
$(4x + 1)(x - 3)$	$4x^2$	$-11x$	-3
$(4x - 1)(x + 3)$	$4x^2$	$+11x$	-3
$(4x + 3)(x - 1)$	$4x^2$	$-x$	-3
$(4x - 3)(x + 1)$	$4x^2$	$+x$	-3
$(2x + 1)(2x - 3)$	$4x^2$	$-4x$	-3
$(2x - 1)(2x + 3)$	$4x^2$	$+4x$	-3

The third line shows that the factors are $(4x + 3)$ and $(x - 1)$.

$$\text{Check:} \quad (4x + 3)(x - 1) = 4x^2 - x - 3$$

You will find that the more practice you have at factoring this type of trinomial, the faster you will get the correct factors. You will pick up some shortcuts along the way, or you may come across a system of eliminating some factors as possibilities. Whatever works best for you is the method you should use. Factoring is a very important tool, and you must be good at it.

 EXAMPLE 3   Factor $12y^3 + 10y^2 - 12y$.

SOLUTION   We begin by factoring out the greatest common factor, $2y$:

$$12y^3 + 10y^2 - 12y = 2y(6y^2 + 5y - 6)$$

Note

Once again, the first step in any factoring problem is to factor out the greatest common factor if it is other than 1.

We now list all possible factors of a trinomial with the first term $6y^2$ and last term -6, along with the associated middle terms.

Possible Factors	Middle Term When Multiplied
$(3y + 2)(2y - 3)$	$-5y$
$(3y - 2)(2y + 3)$	$+5y$
$(6y + 1)(y - 6)$	$-35y$
$(6y - 1)(y + 6)$	$+35y$

The second line gives the correct factors. The complete problem is:

$$12y^3 + 10y^2 - 12y = 2y(6y^2 + 5y - 6)$$
$$= 2y(3y - 2)(2y + 3)$$

EXAMPLE 4   Factor $30x^2y - 5xy^2 - 10y^3$.

SOLUTION   The greatest common factor is $5y$:

$$30x^2y - 5xy^2 - 10y^3 = 5y(6x^2 - xy - 2y^2)$$
$$= 5y(2x + y)(3x - 2y)$$

Maintaining Your Skills

Perform the following additions and subtractions.

69. $(6x^3 - 4x^2 + 2x) + (9x^2 - 6x + 3)$

70. $(6x^3 - 4x^2 + 2x) - (9x^2 - 6x + 3)$

71. $(-7x^4 + 4x^3 - 6x) + (8x^4 + 7x^3 - 9)$

72. $(-7x^4 + 4x^3 - 6x) - (8x^4 + 7x^3 - 9)$

73. $(2x^5 + 3x^3 + 4x) + (5x^3 - 6x - 7)$

74. $(2x^5 + 3x^3 + 4x) - (5x^3 - 6x - 7)$

75. $(-8x^5 - 5x^4 + 7) + (7x^4 + 2x^2 + 5)$

76. $(-8x^5 - 5x^4 + 7) - (7x^4 + 2x^2 + 5)$

77. $\dfrac{24x^3y^7}{6x^{-2}y^4} + \dfrac{27x^{-2}y^{10}}{9x^{-7}y^7}$

78. $\dfrac{15x^8y^4}{5x^2y^2} - \dfrac{4x^7y^5}{2xy^3}$

79. $\dfrac{18a^5b^9}{3a^3b^6} - \dfrac{48a^{-3}b^{-1}}{16a^{-5}b^{-4}}$

80. $\dfrac{54a^{-3}b^5}{6a^{-7}b^{-2}} - \dfrac{32a^6b^5}{8a^2b^{-2}}$

Getting Ready for the Next Section

Multiply each of the following.

81. $(x + 3)(x - 3)$

82. $(x - 4)(x + 4)$

83. $(x + 5)(x - 5)$

84. $(x - 6)(x + 6)$

85. $(x + 7)(x - 7)$

86. $(x - 8)(x + 8)$

87. $(x + 9)(x - 9)$

88. $(x - 10)(x + 10)$

89. $(2x - 3y)(2x + 3y)$

90. $(5x - 6y)(5x + 6y)$

91. $(x^2 + 4)(x + 2)(x - 2)$

92. $(x^2 + 9)(x + 3)(x - 3)$

93. $(x + 3)^2$

94. $(x - 4)^2$

95. $(x + 5)^2$

96. $(x - 6)^2$

97. $(x + 7)^2$

98. $(x - 8)^2$

99. $(x + 9)^2$

100. $(x - 10)^2$

101. $(2x + 3)^2$

102. $(3x - y)^2$

103. $(4x - 2y)^2$

104. $(5x - 6y)^2$

5.4 The Difference of Two Squares

OBJECTIVES

A Factor the difference of two squares.

B Factor a perfect square trinomial.

C Factor a polynomial by first factoring out the greatest common factor and then factoring the polynomial that remains.

In Chapter 5 we listed the following three special products:

$$(a + b)^2 = (a + b)(a + b) = a^2 + 2ab + b^2$$
$$(a - b)^2 = (a - b)(a - b) = a^2 - 2ab + b^2$$
$$(a + b)(a - b) = a^2 - b^2$$

Since factoring is the reverse of multiplication, we can also consider the three special products as three special factorings:

$$a^2 + 2ab + b^2 = (a + b)^2$$
$$a^2 - 2ab + b^2 = (a - b)^2$$
$$a^2 - b^2 = (a + b)(a - b)$$

Any trinomial of the form $a^2 + 2ab + b^2$ or $a^2 - 2ab + b^2$ can be factored by the methods of Section 6.3. The last line is the factoring to obtain the difference of two squares. The difference of two squares always factors in this way. Again, these are patterns you must be able to recognize on sight.

EXAMPLE 1   Factor $16x^2 - 25$.

SOLUTION   We can see that the first term is a perfect square, and the last term is also. This fact becomes even more obvious if we rewrite the problem as:

$$16x^2 - 25 = (4x)^2 - (5)^2$$

The first term is the square of the quantity $4x$, and the last term is the square of 5. The completed problem looks like this:

$$16x^2 - 25 = (4x)^2 - (5)^2$$
$$= (4x + 5)(4x - 5)$$

To check our results, we multiply:

$$(4x + 5)(4x - 5) = 16x^2 + 20x - 20x - 25$$
$$= 16x^2 - 25$$

EXAMPLE 2   Factor $36a^2 - 1$.

SOLUTION   We rewrite the two terms to show they are perfect squares and then factor. Remember, 1 is its own square, $1^2 = 1$.

$$36a^2 - 1 = (6a)^2 - (1)^2$$
$$= (6a + 1)(6a - 1)$$

To check our results, we multiply:

$$(6a + 1)(6a - 1) = 36a^2 + 6a - 6a - 1$$
$$= 36a^2 - 1$$

EXAMPLE 3   Factor $x^4 - y^4$.

SOLUTION x^4 is the perfect square $(x^2)^2$, and y^4 is $(y^2)^2$:

$$x^4 - y^4 = (x^2)^2 - (y^2)^2$$
$$= (x^2 - y^2)(x^2 + y^2)$$

Note

If you think the sum of two squares $x^2 + y^2$ factors, you should try it. Write down the factors you think it has, and then multiply them using the FOIL method. You won't get $x^2 + y^2$.

The factor $(x^2 - y^2)$ is itself the difference of two squares and therefore can be factored again. The factor $(x^2 + y^2)$ is the *sum* of two squares and cannot be factored again. The complete solution is this:

$$x^4 - y^4 = (x^2)^2 - (y^2)^2$$
$$= (x^2 - y^2)(x^2 + y^2)$$
$$= (x + y)(x - y)(x^2 + y^2)$$

EXAMPLE 4   Factor $25x^2 - 60x + 36$.

SOLUTION   Although this trinomial can be factored by the method we used in Section 6.3, we notice that the first and last terms are the perfect squares $(5x)^2$ and $(6)^2$. Before going through the method for factoring trinomials by listing all possible factors, we can check to see if $25x^2 - 60x + 36$ factors to $(5x - 6)^2$. We need only multiply to check:

$$(5x - 6)^2 = (5x - 6)(5x - 6)$$
$$= 25x^2 - 30x - 30x + 36$$
$$= 25x^2 - 60x + 36$$

The trinomial $25x^2 - 60x + 36$ factors to $(5x - 6)(5x - 6) = (5x - 6)^2$.

 EXAMPLE 5   Factor $5x^2 + 30x + 45$.

SOLUTION   We begin by factoring out the greatest common factor, which is 5. Then we notice that the trinomial that remains is a perfect square trinomial:

$$5x^2 + 30x + 45 = 5(x^2 + 6x + 9)$$
$$= 5(x + 3)^2$$

 EXAMPLE 6   Factor $(x - 3)^2 - 25$.

SOLUTION   This example has the form $a^2 - b^2$, where a is $x - 3$ and b is 5. We factor it according to the formula for the difference of two squares:

$$(x - 3)^2 - 25 = (x - 3)^2 - 5^2 \qquad \textbf{Write 25 as } 5^2$$
$$= [(x - 3) - 5][(x - 3) + 5] \qquad \textbf{Factor}$$
$$= (x - 8)(x + 2) \qquad \textbf{Simplify}$$

Notice in this example we could have expanded $(x - 3)^2$, subtracted 25, and then factored to obtain the same result:

$$(x - 3)^2 - 25 = x^2 - 6x + 9 - 25 \qquad \textbf{Expand } (x - 3)^2$$
$$= x^2 - 6x - 16 \qquad \textbf{Simplify}$$
$$= (x - 8)(x + 2) \qquad \textbf{Factor}$$

GETTING READY FOR CLASS

After reading through the preceding section, respond in your own words and in complete sentences.

1. Describe how you factor the difference of two squares.
2. What is a perfect square trinomial?
3. How do you know when you've factored completely?
4. Describe how you would factor $25x^2 - 60x + 36$.

Factor the following.

▶ **1.** $x^2 - 9$

2. $x^2 - 25$

3. $a^2 - 36$

4. $a^2 - 64$

5. $x^2 - 49$

6. $x^2 - 121$

7. $4a^2 - 16$

8. $4a^2 + 16$

9. $9x^2 + 25$

10. $16x^2 - 36$

11. $25x^2 - 169$

12. $x^2 - y^2$

▶ **13.** $9a^2 - 16b^2$

14. $49a^2 - 25b^2$

15. $9 - m^2$

16. $16 - m^2$

17. $25 - 4x^2$

18. $36 - 49y^2$

19. $2x^2 - 18$

20. $3x^2 - 27$

21. $32a^2 - 128$

22. $3a^3 - 48a$

23. $8x^2y - 18y$

24. $50a^2b - 72b$

25. $a^4 - b^4$

26. $a^4 - 16$

27. $16m^4 - 81$

28. $81 - m^4$

29. $3x^3y - 75xy^3$

30. $2xy^3 - 8x^3y$

Factor the following.

31. $x^2 - 2x + 1$

32. $x^2 - 6x + 9$

33. $x^2 + 2x + 1$

34. $x^2 + 6x + 9$

▶ **35.** $a^2 - 10a + 25$

36. $a^2 + 10a + 25$

37. $y^2 + 4y + 4$

38. $y^2 - 8y + 16$

39. $x^2 - 4x + 4$

40. $x^2 + 8x + 16$

41. $m^2 - 12m + 36$

42. $m^2 + 12m + 36$

▶ **43.** $4a^2 + 12a + 9$

44. $9a^2 - 12a + 4$

45. $49x^2 - 14x + 1$

46. $64x^2 - 16x + 1$

47. $9y^2 - 30y + 25$

48. $25y^2 + 30y + 9$

49. $x^2 + 10xy + 25y^2$

50. $25x^2 + 10xy + y^2$

51. $9a^2 + 6ab + b^2$

52. $9a^2 - 6ab + b^2$

Factor.

▶ **53.** $y^2 - 3y + \dfrac{9}{4}$

▶ **54.** $y^2 + 3y + \dfrac{9}{4}$

▶ **55.** $a^2 + a + \dfrac{1}{4}$

▶ **56.** $a^2 - 5a + \dfrac{25}{4}$

▶ **57.** $x^2 - 7x + \dfrac{49}{4}$

▶ **58.** $x^2 + 9x + \dfrac{81}{4}$

▶ **59.** $x^2 - \dfrac{3}{4}x + \dfrac{9}{64}$

▶ **60.** $x^2 - \dfrac{3}{2}x + \dfrac{9}{16}$

▶ **61.** $t^2 - \dfrac{2}{5}t + \dfrac{1}{25}$

▶ **62.** $t^2 + \dfrac{6}{5}t + \dfrac{9}{25}$

Factor the following by first factoring out the greatest common factor.

63. $3a^2 + 18a + 27$

64. $4a^2 - 16a + 16$

65. $2x^2 + 20xy + 50y^2$

66. $3x^2 + 30xy + 75y^2$

67. $5x^3 + 30x^2y + 45xy^2$

68. $12x^2y - 36xy^2 + 27y^3$

Factor by grouping the first three terms together.

69. $x^2 + 6x + 9 - y^2$

70. $x^2 + 10x + 25 - y^2$

71. $x^2 + 2xy + y^2 - 9$

72. $a^2 + 2ab + b^2 - 25$

73. Find a value for b so that the polynomial $x^2 + bx + 49$ factors to $(x + 7)^2$.

74. Find a value of b so that the polynomial $x^2 + bx + 81$ factors to $(x + 9)^2$.

75. Find the value of c for which the polynomial $x^2 + 10x + c$ factors to $(x + 5)^2$.

76. Find the value of a for which the polynomial $ax^2 + 12x + 9$ factors to $(2x + 3)^2$.

= Videos available by instructor request

▶ = Online student support materials available at academic.cengage.com/login

LINKING OBJECTIVES AND EXAMPLES

Next to each **objective** we have listed the examples that are best described by that objective.

A   1–6

GETTING READY FOR CLASS

After reading through the preceding section, respond in your own words and in complete sentences.

1. When is an equation in standard form?
2. What is the first step in solving an equation by factoring?
3. Describe the zero-factor property in your own words.
4. Describe how you would solve the equation $2x^2 - 5x = 12$.

Problem Set 5.7

Online support materials can be found at academic.cengage.com/login

The following equations are already in factored form. Use the special zero factor property to set the factors to 0 and solve.

1. $(x + 2)(x - 1) = 0$

2. $(x + 3)(x + 2) = 0$

3. $(a - 4)(a - 5) = 0$

4. $(a + 6)(a - 1) = 0$

▶ 5. $x(x + 1)(x - 3) = 0$

6. $x(2x + 1)(x - 5) = 0$

7. $(3x + 2)(2x + 3) = 0$

8. $(4x - 5)(x - 6) = 0$

9. $m(3m + 4)(3m - 4) = 0$

10. $m(2m - 5)(3m - 1) = 0$

11. $2y(3y + 1)(5y + 3) = 0$

12. $3y(2y - 3)(3y - 4) = 0$

Solve the following equations

13. $x^2 + 3x + 2 = 0$

14. $x^2 - x - 6 = 0$

▶ 15. $x^2 - 9x + 20 = 0$

16. $x^2 + 2x - 3 = 0$

17. $a^2 - 2a - 24 = 0$

18. $a^2 - 11a + 30 = 0$

19. $100x^2 - 500x + 600 = 0$

20. $100x^2 - 300x + 200 = 0$

21. $x^2 = -6x - 9$

22. $x^2 = 10x - 25$

23. $a^2 - 16 = 0$

24. $a^2 - 36 = 0$

25. $2x^2 + 5x - 12 = 0$

26. $3x^2 + 14x - 5 = 0$

27. $9x^2 + 12x + 4 = 0$

28. $12x^2 - 24x + 9 = 0$

▶ 29. $a^2 + 25 = 10a$

30. $a^2 + 16 = 8a$

31. $2x^2 = 3x + 20$

32. $6x^2 = x + 2$

33. $3m^2 = 20 - 7m$

34. $2m^2 = -18 + 15m$

35. $4x^2 - 49 = 0$

36. $16x^2 - 25 = 0$

▶ 37. $x^2 + 6x = 0$

38. $x^2 - 8x = 0$

39. $x^2 - 3x = 0$

40. $x^2 + 5x = 0$

41. $2x^2 = 8x$

42. $2x^2 = 10x$

43. $3x^2 = 15x$

44. $5x^2 = 15x$

45. $1,400 = 400 + 700x - 100x^2$

46. $2,700 = 700 + 900x - 100x^2$

47. $6x^2 = -5x + 4$

48. $9x^2 = 12x - 4$

49. $x(2x - 3) = 20$

50. $x(3x - 5) = 12$

51. $t(t + 2) = 80$

52. $t(t + 2) = 99$

53. $4,000 = (1,300 - 100p)p$

54. $3,200 = (1,200 - 100p)p$

55. $x(14 - x) = 48$

56. $x(12 - x) = 32$

57. $(x + 5)^2 = 2x + 9$

58. $(x + 7)^2 = 2x + 13$

59. $(y - 6)^2 = y - 4$

60. $(y + 4)^2 = y + 6$

61. $10^2 = (x + 2)^2 + x^2$

62. $15^2 = (x + 3)^2 + x^2$

63. $2x^3 + 11x^2 + 12x = 0$

64. $3x^3 + 17x^2 + 10x = 0$

▶ 65. $4y^3 - 2y^2 - 30y = 0$

☐ = Videos available by instructor request

▶ = Online student support materials available at academic.cengage.com/login

66. $9y^3 + 6y^2 - 24y = 0$

67. $8x^3 + 16x^2 = 10x$

68. $24x^3 - 22x^2 = -4x$

69. $20a^3 = -18a^2 + 18a$

70. $12a^3 = -2a^2 + 10a$

▶ **71.** $16t^2 - 32t + 12 = 0$

▶ **72.** $16t^2 - 64t + 48 = 0$

Simplify each side as much as possible, then solve the equation.

▶ **73.** $(a - 5)(a + 4) = -2a$

▶ **74.** $(a + 2)(a - 3) = -2a$

▶ **75.** $3x(x + 1) - 2x(x - 5) = -42$

▶ **76.** $4x(x - 2) - 3x(x - 4) = -3$

▶ **77.** $2x(x + 3) = x(x + 2) - 3$

▶ **78.** $3x(x - 3) = 2x(x - 4) + 6$

▶ **79.** $a(a - 3) + 6 = 2a$

▶ **80.** $a(a - 4) + 8 = 2a$

▶ **81.** $15(x + 20) + 15x = 2x(x + 20)$

▶ **82.** $15(x + 8) + 15x = 2x(x + 8)$

▶ **83.** $15 = a(a + 2)$

▶ **84.** $6 = a(a - 5)$

Use factoring by grouping to solve the following equations.

85. $x^3 + 3x^2 - 4x - 12 = 0$

86. $x^3 + 5x^2 - 9x - 45 = 0$

87. $x^3 + x^2 - 16x - 16 = 0$

88. $4x^3 + 12x^2 - 9x - 27 = 0$

89. Find a quadratic equation that has two solutions: $x = 3$ and $x = 5$. Write your answer in standard form.

90. Find a quadratic equation that has two solutions: $x = 9$ and $x = 1$. Write your answer in standard form.

91. Find a quadratic equation that has the two given solutions.

    a. $x = 3$ and $x = 2$.

    b. $x = 1$ and $x = 6$.

    c. $x = 3$ and $x = -2$.

92. Find a quadratic equation that has the two given solutions.

    a. $x = 4$ and $x = 5$.

    b. $x = 2$ and $x = 10$.

    c. $x = -4$ and $x = 5$.

Maintaining Your Skills

Use the properties of exponents to simplify each expression.

93. 2^{-3}

94. 5^{-2}

95. $\dfrac{x^5}{x^{-3}}$

96. $\dfrac{x^{-2}}{x^{-5}}$

97. $\dfrac{(x^2)^3}{(x^{-3})^4}$

98. $\dfrac{(x^2)^{-4}(x^{-2})^3}{(x^{-3})^{-5}}$

99. Write the number 0.0056 in scientific notation.

100. Write the number 2.34×10^{-4} in expanded form.

101. Write the number 5,670,000,000 in scientific notation.

102. Write the number 0.00000567 in scientific notation.

Getting Ready for the Next Section

Write each sentence as an algebraic equation.

103. The product of two consecutive integers is 72.

104. The product of two consecutive even integers is 80.

105. The product of two consecutive odd integers is 99.

106. The product of two consecutive odd integers is 63.

107. The product of two consecutive even integers is 10 less than 5 times their sum.

108. The product of two consecutive odd integers is 1 less than 4 times their sum.

The following word problems are taken from the book *Academic Algebra,* written by William J. Milne and published by the American Book Company in 1901. Solve each problem.

109. **Cost of a Bicycle and a Suit** A bicycle and a suit cost $90. How much did each cost, if the bicycle cost 5 times as much as the suit?

110. **Cost of a Cow and a Calf** A man bought a cow and a calf for $36, paying 8 times as much for the cow as for the calf. What was the cost of each?

111. **Cost of a House and a Lot** A house and a lot cost $3,000. If the house cost 4 times as much as the lot, what was the cost of each?

112. **Daily Wages** A plumber and two helpers together earned $7.50 per day. How much did each earn per day, if the plumber earned 4 times as much as each helper?

5.8 Applications

In this section we will look at some application problems, the solutions to which require solving a quadratic equation. We will also introduce the Pythagorean theorem, one of the oldest theorems in the history of mathematics. The person whose name we associate with the theorem, Pythagoras (of Samos), was a Greek philosopher and mathematician who lived from about 560 B.C. to 480 B.C. According to the British philosopher Bertrand Russell, Pythagoras was "intellectually one of the most important men that ever lived."

Also in this section, the solutions to the examples show only the essential steps from our Blueprint for Problem Solving. Recall that step 1 is done mentally; we read the problem and mentally list the items that are known and the items that are unknown. This is an essential part of problem solving. However, now that you have had experience with application problems, you are doing step 1 automatically.

Number Problems

 EXAMPLE 1   The product of two consecutive odd integers is 63. Find the integers.

SOLUTION   Let x = the first odd integer; then $x + 2$ = the second odd integer. An equation that describes the situation is:

$$x(x + 2) = 63 \quad \text{**Their product is 63**}$$

We solve the equation:

$$x(x + 2) = 63$$
$$x^2 + 2x = 63$$
$$x^2 + 2x - 63 = 0$$
$$(x - 7)(x + 9) = 0$$
$$x - 7 = 0 \quad \text{or} \quad x + 9 = 0$$
$$x = 7 \quad \text{or} \quad x = -9$$

If the first odd integer is 7, the next odd integer is $7 + 2 = 9$. If the first odd integer is -9, the next consecutive odd integer is $-9 + 2 = -7$. We have two pairs of consecutive odd integers that are solutions. They are 7, 9 and -9, -7.

We check to see that their products are 63:

$$7(9) = 63$$

$$-7(-9) = 63$$

Suppose we know that the sum of two numbers is 50. We want to find a way to represent each number using only one variable. If we let x represent one of the two numbers, how can we represent the other? Let's suppose for a moment that x turns out to be 30. Then the other number will be 20, because their sum is 50; that is, if two numbers add up to 50 and one of them is 30, then the other must be $50 - 30 = 20$. Generalizing this to any number x, we see that if two numbers have a sum of 50 and one of the numbers is x, then the other must be $50 - x$. The table that follows shows some additional examples.

If Two Numbers Have a Sum of	And One of Them Is	Then the Other Must Be
50	x	$50 - x$
100	x	$100 - x$
10	y	$10 - y$
12	n	$12 - n$

Now, let's look at an example that uses this idea.

EXAMPLE 2
The sum of two numbers is 13. Their product is 40. Find the numbers.

SOLUTION If we let x represent one of the numbers, then $13 - x$ must be the other number because their sum is 13. Since their product is 40, we can write:

$$x(13 - x) = 40 \qquad \text{The product of the two numbers is 40}$$

$$13x - x^2 = 40 \qquad \text{Multiply the left side}$$

$$x^2 - 13x = -40 \qquad \text{Multiply both sides by } -1 \text{ and reverse the order of the terms on the left side}$$

$$x^2 - 13x + 40 = 0 \qquad \text{Add 40 to each side}$$

$$(x - 8)(x - 5) = 0 \qquad \text{Factor the left side}$$

$$x - 8 = 0 \quad \text{or} \quad x - 5 = 0$$

$$x = 8 \qquad\qquad x = 5$$

The two solutions are 8 and 5. If x is 8, then the other number is $13 - x = 13 - 8 = 5$. Likewise, if x is 5, the other number is $13 - x = 13 - 5 = 8$. Therefore, the two numbers we are looking for are 8 and 5. Their sum is 13 and their product is 40.

Geometry Problems

Many word problems dealing with area can best be described algebraically by quadratic equations.

 EXAMPLE 3   The length of a rectangle is 3 more than twice the width. The area is 44 square inches. Find the dimensions (find the length and width).

SOLUTION As shown in Figure 1, let x = the width of the rectangle. Then $2x + 3$ = the length of the rectangle because the length is three more than twice the width.

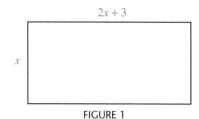

FIGURE 1

Since the area is 44 square inches, an equation that describes the situation is

$$x(2x + 3) = 44 \quad \textbf{Length} \cdot \textbf{width} = \textbf{area}$$

We now solve the equation:

$$x(2x + 3) = 44$$
$$2x^2 + 3x = 44$$
$$2x^2 + 3x - 44 = 0$$
$$(2x + 11)(x - 4) = 0$$
$$2x + 11 = 0 \quad \text{or} \quad x - 4 = 0$$
$$x = -\frac{11}{2} \quad \text{or} \quad x = 4$$

The solution $x = -\frac{11}{2}$ cannot be used since length and width are always given in positive units. The width is 4. The length is 3 more than twice the width or $2(4) + 3 = 11$.

Width = 4 inches

Length = 11 inches

The solutions check in the original problem since $4(11) = 44$.

 EXAMPLE 4   The numerical value of the area of a square is twice its perimeter. What is the length of its side?

SOLUTION As shown in Figure 2, let x = the length of its side. Then x^2 = the area of the square and $4x$ = the perimeter of the square:

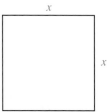

FIGURE 2

An equation that describes the situation is

$$x^2 = 2(4x) \qquad \textbf{The area is 2 times the perimeter}$$

$$x^2 = 8x$$

$$x^2 - 8x = 0$$

$$x(x - 8) = 0$$

$$x(x - 8) = 0$$

$$x = 0 \qquad \text{or} \qquad x - 8 = 0$$

$$x = 8$$

Since $x = 0$ does not make sense in our original problem, we use $x = 8$. If the side has length 8, then the perimeter is $4(8) = 32$ and the area is $8^2 = 64$. Since 64 is twice 32, our solution is correct.

FACTS FROM GEOMETRY

The Pythagorean Theorem
Next, we will work some problems involving the Pythagorean theorem, which we mentioned in the introduction to this section. It may interest you to know that Pythagoras formed a secret society around the year 540 B.C. Known as the Pythagoreans, members kept no written record of their work; everything was handed down by spoken word. They influenced not only mathematics, but religion, science, medicine, and music as well. Among other things, they discovered the correlation between musical notes and the reciprocals of counting numbers, $\frac{1}{2}, \frac{1}{3}, \frac{1}{4}$, and so on. In their daily lives, they followed strict dietary and moral rules to achieve a higher rank in future lives.

Pythagorean Theorem In any right triangle (Figure 3), the square of the longer side (called the hypotenuse) is equal to the sum of the squares of the other two sides (called legs).

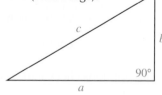

$$c^2 = a^2 + b^2$$

FIGURE 3

EXAMPLE 5
The three sides of a right triangle are three consecutive integers. Find the lengths of the three sides.

SOLUTION   Let x = the first integer (shortest side)

then $x + 1$ = the next consecutive integer

and $x + 2$ = the third consecutive integer (longest side)

A diagram of the triangle is shown in Figure 4.

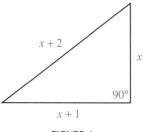

FIGURE 4

The Pythagorean theorem tells us that the square of the longest side $(x + 2)^2$ is equal to the sum of the squares of the two shorter sides, $(x + 1)^2 + x^2$. Here is the equation:

$$(x + 2)^2 = (x + 1)^2 + x^2$$

$x^2 + 4x + 4 = x^2 + 2x + 1 + x^2$	**Expand squares**
$x^2 - 2x - 3 = 0$	**Standard form**
$(x - 3)(x + 1) = 0$	**Factor**
$x - 3 = 0 \quad$ or $\quad x + 1 = 0$	**Set factors to 0**
$x = 3 \quad$ or $\quad x = -1$	

Since a triangle cannot have a side with a negative number for its length, we must not use -1 for a solution to our original problem; therefore, the shortest side is 3. The other two sides are the next two consecutive integers, 4 and 5.

EXAMPLE 6 The hypotenuse of a right triangle is 5 inches, and the lengths of the two legs (the other two sides) are given by two consecutive integers. Find the lengths of the two legs.

SOLUTION If we let $x =$ the length of the shortest side, then the other side must be $x + 1$. A diagram of the triangle is shown in Figure 5.

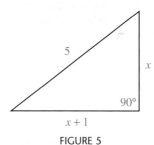

FIGURE 5

The Pythagorean theorem tells us that the square of the longest side, 5^2, is equal to the sum of the squares of the two shorter sides, $x^2 + (x + 1)^2$. Here is the equation:

$5^2 = x^2 + (x + 1)^2$	**Pythagorean theorem**
$25 = x^2 + x^2 + 2x + 1$	**Expand 5^2 and $(x + 1)^2$**
$25 = 2x^2 + 2x + 1$	**Simplify the right side**
$0 = 2x^2 + 2x - 24$	**Add -25 to each side**
$0 = 2(x^2 + x - 12)$	**Begin factoring**
$0 = 2(x + 4)(x - 3)$	**Factor completely**
$x + 4 = 0 \quad$ or $\quad x - 3 = 0$	**Set variable factors to 0**
$x = -4 \quad$ or $\quad x = 3$	

Since a triangle cannot have a side with a negative number for its length, we cannot use -4; therefore, the shortest side must be 3 inches. The next side is

$x + 1 = 3 + 1 = 4$ inches. Since the hypotenuse is 5, we can check our solutions with the Pythagorean theorem as shown in Figure 6.

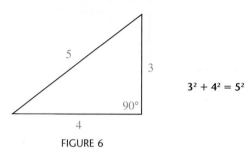

$3^2 + 4^2 = 5^2$

FIGURE 6

 EXAMPLE 7 A company can manufacture x hundred items for a total cost of $C = 300 + 500x - 100x^2$. How many items were manufactured if the total cost is $900?

SOLUTION We are looking for x when C is 900. We begin by substituting 900 for C in the cost equation. Then we solve for x:

When $\qquad C = 900$

the equation $\qquad C = 300 + 500x - 100x^2$

becomes $\qquad 900 = 300 + 500x - 100x^2$

We can write this equation in standard form by adding -300, $-500x$, and $100x^2$ to each side. The result looks like this:

$$100x^2 - 500x + 600 = 0$$
$$100(x^2 - 5x + 6) = 0 \qquad \textbf{Begin factoring}$$
$$100(x - 2)(x - 3) = 0 \qquad \textbf{Factor completely}$$
$$x - 2 = 0 \quad \text{or} \quad x - 3 = 0 \qquad \textbf{Set variable factors to 0}$$
$$x = 2 \quad \text{or} \quad x = 3$$

Our solutions are 2 and 3, which means that the company can manufacture 2 hundred items or 3 hundred items for a total cost of $900.

Note

If you are planning on taking finite mathematics, statistics, or business calculus in the future, Examples 7 and 8 will give you a head start on some of the problems you will see in those classes.

 EXAMPLE 8 A manufacturer of small portable radios knows that the number of radios she can sell each week is related to the price of the radios by the equation $x = 1{,}300 - 100p$ (x is the number of radios and p is the price per radio). What price should she charge for the radios to have a weekly revenue of $4,000?

SOLUTION First, we must find the revenue equation. The equation for total revenue is $R = xp$, where x is the number of units sold and p is the price per unit. Since we want R in terms of p, we substitute $1{,}300 - 100p$ for x in the equation $R = xp$:

If $\qquad R = xp$

and $\qquad x = 1{,}300 - 100p$

then $\qquad R = (1{,}300 - 100p)p$

Step 5: As a final check, see if any of the factors you have written can be factored further. If you have overlooked a common factor, you can catch it here.

Strategy for Solving a Quadratic Equation [5.7]

6. Solve $x^2 - 6x = -8$.
$$x^2 - 6x + 8 = 0$$
$$(x - 4)(x - 2) = 0$$
$$x - 4 = 0 \quad \text{or} \quad x - 2 = 0$$
$$x = 4 \quad \text{or} \quad x = 2$$
Both solutions check.

Step 1: Write the equation in standard form:
$$ax^2 + bx + c = 0$$

Step 2: Factor completely.

Step 3: Set each variable factor equal to 0.

Step 4: Solve the equations found in step 3.

Step 5: Check solutions, if necessary.

The Pythagorean Theorem [5.8]

7. The hypotenuse of a right triangle is 5 inches, and the lengths of the two legs (the other two sides) are given by two consecutive integers. Find the lengths of the two legs.

In any right triangle, the square of the longest side (called the hypotenuse) is equal to the sum of the squares of the other two sides (called legs).

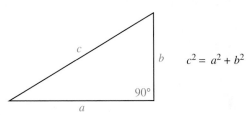

$$c^2 = a^2 + b^2$$

If we let x = the length of the shortest side, then the other side must be $x + 1$. The Pythagorean theorem tells us that
$$5^2 = x^2 + (x + 1)^2$$
$$25 = x^2 + x^2 + 2x + 1$$
$$25 = 2x^2 + 2x + 1$$
$$0 = 2x^2 + 2x - 24$$
$$0 = 2(x^2 + x - 12)$$
$$0 = 2(x + 4)(x - 3)$$
$$x + 4 = 0 \quad \text{or} \quad x - 3 = 0$$
$$x = -4 \quad \text{or} \quad x = 3$$
Since a triangle cannot have a side with a negative number for its length, we cannot use -4. One leg is $x = 3$ and the other leg is $x + 1 = 3 + 1 = 4$.

! COMMON MISTAKES

It is a mistake to apply the zero-factor property to numbers other than zero. For example, consider the equation $(x - 3)(x + 4) = 18$. A fairly common mistake is to attempt to solve it with the following steps:
$$(x - 3)(x + 4) = 18$$
$$x - 3 = 18 \quad \text{or} \quad x + 4 = 18 \leftarrow \textbf{Mistake}$$
$$x = 21 \quad \text{or} \quad x = 14$$

These are obviously not solutions, as a quick check will verify:

Check: $x = 21$ Check: $x = 14$

$(21 - 3)(21 + 4) \overset{?}{=} 18$ $(14 - 3)(14 + 4) \overset{?}{=} 18$

$18 \cdot 25 = 18$ $11 \cdot 18 = 18$

$450 = 18 \xleftarrow{\text{false statements}} 198 = 18$

The mistake is in setting each factor equal to 18. It is not necessarily true that when the product of two numbers is 18, either one of them is itself 18. The correct solution looks like this:

$$(x - 3)(x + 4) = 18$$
$$x^2 + x - 12 = 18$$
$$x^2 + x - 30 = 0$$
$$(x + 6)(x - 5) = 0$$

$$x + 6 = 0 \quad \text{or} \quad x - 5 = 0$$
$$x = -6 \quad \text{or} \quad x = 5$$

To avoid this mistake, remember that before you factor a quadratic equation, you must write it in standard form. It is in standard form only when 0 is on one side and decreasing powers of the variable are on the other.

The problems below form a comprehensive review of the material in this chapter. They can be used to study for exams. If you would like to take a practice test on this chapter, you can use the odd-numbered problems. Give yourself an hour and work as many of the odd-numbered problems as possible. When you are finished, or when an hour has passed, check your answers with the answers in the back of the book. You can use the even-numbered problems for a second practice test.

The numbers in brackets refer to the sections of the text in which similar problems can be found.

Factor the following by factoring out the greatest common factor. [5.1]

1. $10x - 20$

2. $4x^3 - 9x^2$

3. $5x - 5y$

4. $7x^3 + 2x$

5. $8x + 4$

6. $2x^2 + 14x + 6$

7. $24y^2 - 40y + 48$

8. $30xy^3 - 45x^3y^2$

9. $49a^3 - 14b^3$

10. $6ab^2 + 18a^3b^3 - 24a^2b$

Factor by grouping. [5.1]

11. $xy + bx + ay + ab$

12. $xy + 4x - 5y - 20$

13. $2xy + 10x - 3y - 15$

14. $5x^2 - 4ax - 10bx + 8ab$

Factor the following trinomials. [5.2]

15. $y^2 + 9y + 14$

16. $w^2 + 15w + 50$

17. $a^2 - 14a + 48$

18. $r^2 - 18r + 72$

19. $y^2 + 20y + 99$

20. $y^2 + 8y + 12$

Factor the following trinomials. [5.3]

21. $2x^2 + 13x + 15$

22. $4y^2 - 12y + 5$

23. $5y^2 + 11y + 6$

24. $20a^2 - 27a + 9$

25. $6r^2 + 5rt - 6t^2$

26. $10x^2 - 29x - 21$

Factor the following if possible. [5.4]

27. $n^2 - 81$

28. $4y^2 - 9$

29. $x^2 + 49$

30. $36y^2 - 121x^2$

31. $64a^2 - 121b^2$

32. $64 - 9m^2$

Factor the following. [5.4]

33. $y^2 + 20y + 100$

34. $m^2 - 16m + 64$

35. $64t^2 + 16t + 1$

36. $16n^2 - 24n + 9$

37. $4r^2 - 12rt + 9t^2$

38. $9m^2 + 30mn + 25n^2$

Factor the following. [5.2]

39. $2x^2 + 20x + 48$

40. $a^3 - 10a^2 + 21a$

41. $3m^3 - 18m^2 - 21m$

42. $5y^4 + 10y^3 - 40y^2$

Factor the following trinomials. [5.3]

43. $8x^2 + 16x + 6$

44. $3a^3 - 14a^2 - 5a$

45. $20m^3 - 34m^2 + 6m$

46. $30x^2y - 55xy^2 + 15y^3$

Factor the following. [5.4]

47. $4x^2 + 40x + 100$

48. $4x^3 + 12x^2 + 9x$

49. $5x^2 - 45$

50. $12x^3 - 27xy^2$

Factor the following. [5.5]

51. $8a^3 - b^3$

52. $27x^3 + 8y^3$

53. $125x^3 - 64y^3$

Factor the following polynomials completely. [5.6]

54. $6a^3b + 33a^2b^2 + 15ab^3$

55. $x^5 - x^3$

56. $4y^6 + 9y^4$

57. $12x^5 + 20x^4y - 8x^3y^2$

58. $30a^4b + 35a^3b^2 - 15a^2b^3$

59. $18a^3b^2 + 3a^2b^3 - 6ab^4$

Solve. [5.7]

60. $(x - 5)(x + 2) = 0$

61. $3(2y + 5)(2y - 5) = 0$

62. $m^2 + 3m = 10$

63. $a^2 - 49 = 0$

64. $m^2 - 9m = 0$

65. $6y^2 = -13y - 6$

66. $9x^4 + 9x^3 = 10x^2$

Solve the following word problems. [5.8]

67. Number Problem The product of two consecutive even integers is 120. Find the two integers.

68. Number Problem The product of two consecutive integers is 110. Find the two integers.

69. Number Problem The product of two consecutive odd integers is 1 less than 3 times their sum. Find the integers.

70. Number Problem The sum of two numbers is 20. Their product is 75. Find the numbers.

71. Number Problem One number is 1 less than twice another. Their product is 66. Find the numbers.

72. Geometry The height of a triangle is 8 times the base. The area is 16 square inches. Find the base.

GROUP PROJECT   Visual Factoring

Number of People 2 or 3

Time Needed 10–15 minutes

Equipment Pencil, graph paper, and scissors

Background When a geometric figure is divided into smaller figures, the area of the original figure and the area of any rearrangement of the smaller figures must be the same. We can use this fact to help visualize some factoring problems.

Procedure Use the diagram below to work the following problems.

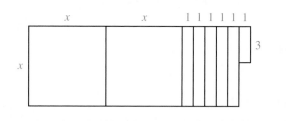

1. Write a polynomial involving x that gives the area of the diagram.

2. Factor the polynomial found in Part 1.

3. Copy the figure onto graph paper, then cut along the lines so that you end up with 2 squares and 6 rectangles.

4. Rearrange the pieces from Part 3 to show that the factorization you did in Part 2 is correct.

Factoring and Internet Security

The security of the information on computers is directly related to factoring of whole numbers. The key lies in the fact that multiplying whole numbers is a straightforward, simple task, whereas factoring can be very time-consuming. For example, multiplying the numbers 1,234 and 3,433 to obtain 4,236,322 takes very little time, even if done by hand. But given the number 4,236,322, finding its factors, even with a calculator or computer, is more than you want to try. The discipline that studies how to make and break codes is cryptography. The current Web browsers, such as Internet Explorer and Netscape, use a system called RSA public-key cryptosystem invented by Adi Shamir of Israel's Weizmann Institute of Science. In 1999 Shamir announced that he had found a method of factoring large numbers quickly that will put the current Internet security system at risk.

Research the connection between computer security and factoring, the RSA cryptosystem, and the current state of security on the Internet, and then write an essay summarizing your results.

Rational Expressions

Rubberball/Getty Images

First Bank of San Luis Obispo charges $2.00 per month and $0.15 per check for a regular checking account. If we write x checks in one month, the total monthly cost of the checking account will be $C = 2.00 + 0.15x$. From this formula we see that the more checks we write in a month, the more we pay for the account. But, it is also true that the more checks we write in a month, the lower the cost per check. To find the average cost per check, we divide the total cost by the number of checks written:

$$\text{Average cost} = A = \frac{C}{X} = \frac{2.00 + 0.15x}{X}$$

We can use this formula to create Table 1 and Figure 1, giving us a visual interpretation of the relationship between the number of checks written and the average cost per check.

TABLE 1
Average Cost

Number of Checks	Average Cost Per Check
1	2.15
2	1.15
5	0.55
10	0.35
15	0.28
20	0.25

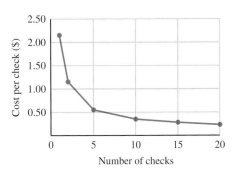

As you can see, if we write one check per month, the cost per check is relatively high, $2.15. However, if we write 20 checks per month, each check costs us only $0.25. Using average cost per check is a good way to compare different checking accounts. The expression $\dfrac{2.00 + 0.15x}{X}$ in the average cost formula is a rational expression. When you have finished this chapter you will have a good working knowledge of rational expressions.

This is the last chapter in which we will mention study skills. You know by now what works best for you and what you have to do to achieve your goals for this course. From now on, it is simply a matter of sticking with the things that work for you, and avoiding the things that do not work. It seems simple, but as with anything that takes effort, it is up to you to see that you maintain the skills that get you where you want to be in the course.

If you intend to take more classes in mathematics, and you want to ensure your success in those classes, then you can work toward this goal: *Become a student who can learn mathematics on your own.* Most people who have degrees in mathematics were students who could learn mathematics on their own. This doesn't mean that you have to learn it all on your own; it simply means that if you have to, you can learn it on your own. Attaining this goal gives you independence and puts you in control of your success in any math class you take.

OBJECTIVES

A Find the restrictions on the variable in a rational expression.

B Reduce a rational expression to lowest terms.

C Work problems involving ratios.

In Chapter 1 we defined the set of rational numbers to be the set of all numbers that could be put in the form $\frac{a}{b}$, where a and b are integers ($b \neq 0$):

$$\text{Rational numbers} = \left\{ \frac{a}{b} \,\middle|\, a \text{ and } b \text{ are integers, } b \neq 0 \right\}$$

A *rational expression* is any expression that can be put in the form $\frac{P}{Q}$, where P and Q are polynomials and $Q \neq 0$:

$$\text{Rational expressions} = \left\{ \frac{P}{Q} \,\middle|\, P \text{ and } Q \text{ are polynomials, } Q \neq 0 \right\}$$

Each of the following is an example of a rational expression:

$$\frac{2x + 3}{x} \qquad \frac{x^2 - 6x + 9}{x^2 - 4} \qquad \frac{5}{x^2 + 6} \qquad \frac{2x^2 + 3x + 4}{2}$$

For the rational expression

$$\frac{x^2 - 6x + 9}{x^2 - 4}$$

the polynomial on top, $x^2 - 6x + 9$, is called the numerator, and the polynomial on the bottom, $x^2 - 4$, is called the denominator. The same is true of the other rational expressions.

We must be careful that we do not use a value of the variable that will give us a denominator of zero. Remember, division by zero is not defined.

EXAMPLES

State the restrictions on the variable in the following rational expressions:

1. $\dfrac{x + 2}{x - 3}$

SOLUTION The variable x can be any real number except $x = 3$ since, when $x = 3$, the denominator is $3 - 3 = 0$. We state this restriction by writing $x \neq 3$.

2. $\dfrac{5}{x^2 - x - 6}$

SOLUTION If we factor the denominator, we have $x^2 - x - 6 = (x - 3)(x + 2)$. If either of the factors is zero, the whole denominator is zero. Our restrictions are $x \neq 3$ and $x \neq -2$ since either one makes $x^2 - x - 6 = 0$.

We will not always list each restriction on a rational expression, but we should be aware of them and keep in mind that no rational expression can have a denominator of zero.

The two fundamental properties of rational expressions are listed next. We will use these two properties many times in this chapter.

Properties of Rational Expressions

Property 1
Multiplying the numerator and denominator of a rational expression by the same nonzero quantity will not change the value of the rational expression.

Property 2
Dividing the numerator and denominator of a rational expression by the same nonzero quantity will not change the value of the rational expression.

We can use Property 2 to reduce rational expressions to lowest terms. Since this process is almost identical to the process of reducing fractions to lowest terms, let's recall how the fraction $\frac{6}{15}$ is reduced to lowest terms:

$$\frac{6}{15} = \frac{2 \cdot 3}{5 \cdot 3} \qquad \text{\textbf{Factor numerator and denominator}}$$

$$= \frac{2 \cdot \cancel{3}}{5 \cdot \cancel{3}} \qquad \text{\textbf{Divide out the common factor, 3}}$$

$$= \frac{2}{5} \qquad \text{\textbf{Reduce to lowest terms}}$$

The same procedure applies to reducing rational expressions to lowest terms. The process is summarized in the following rule.

Rule To reduce a rational expression to lowest terms, first factor the numerator and denominator completely and then divide both the numerator and denominator by any factors they have in common.

 EXAMPLE 3   Reduce $\dfrac{x^2 - 9}{x^2 + 5x + 6}$ to lowest terms.

SOLUTION   We begin by factoring:

$$\frac{x^2 - 9}{x^2 + 5x + 6} = \frac{(x - 3)(x + 3)}{(x + 2)(x + 3)}$$

Notice that both polynomials contain the factor $(x + 3)$. If we divide the numerator by $(x + 3)$, we are left with $(x - 3)$. If we divide the denominator by $(x + 3)$, we are left with $(x + 2)$. The complete solution looks like this:

$$\frac{x^2 - 9}{x^2 + 5x + 6} = \frac{(x - 3)\cancel{(x + 3)}}{(x + 2)\cancel{(x + 3)}} \qquad \text{\textbf{Factor the numerator and denominator completely}}$$

$$= \frac{x - 3}{x + 2} \qquad \text{\textbf{Divide out the common factor, }} x + 3$$

It is convenient to draw a line through the factors as we divide them out. It is especially helpful when the problems become longer.

EXAMPLE 4   Reduce to lowest terms $\dfrac{10a + 20}{5a^2 - 20}$.

SOLUTION   We begin by factoring out the greatest common factor from the numerator and denominator:

$$\frac{10a + 20}{5a^2 - 20} = \frac{10(a + 2)}{5(a^2 - 4)}$$ **Factor the greatest common factor from the numerator and denominator**

$$= \frac{10\cancel{(a + 2)}}{5\cancel{(a + 2)}(a - 2)}$$ **Factor the denominator as the difference of two squares**

$$= \frac{2}{a - 2}$$ **Divide out the common factors 5 and $a + 2$**

EXAMPLE 5   Reduce $\dfrac{2x^3 + 2x^2 - 24x}{x^3 + 2x^2 - 8x}$ to lowest terms.

SOLUTION   We begin by factoring the numerator and denominator completely. Then we divide out all factors common to the numerator and denominator. Here is what it looks like:

$$\frac{2x^3 + 2x^2 - 24x}{x^3 + 2x^2 - 8x} = \frac{2x(x^2 + x - 12)}{x(x^2 + 2x - 8)}$$ **Factor out the greatest common factor first**

$$= \frac{2\cancel{x}(x - 3)\cancel{(x + 4)}}{\cancel{x}(x - 2)\cancel{(x + 4)}}$$ **Factor the remaining trinomials**

$$= \frac{2(x - 3)}{x - 2}$$ **Divide out the factors common to the numerator and denominator**

EXAMPLE 6   Reduce $\dfrac{x - 5}{x^2 - 25}$ to lowest terms.

SOLUTION

$$\frac{x - 5}{x^2 - 25} = \frac{\cancel{x - 5}}{\cancel{(x - 5)}(x + 5)}$$ **Factor numerator and denominator completely**

$$= \frac{1}{x + 5}$$ **Divide out the common factor, $x - 5$**

EXAMPLE 7   Reduce $\dfrac{x^3 + y^3}{x^2 - y^2}$ to lowest terms.

SOLUTION   We begin by factoring the numerator and denominator completely. (Remember, we can only reduce to lowest terms when the numerator and denominator are in factored form. Trying to reduce before factoring will only lead to mistakes.)

$$\frac{x^3 + y^3}{x^2 - y^2} = \frac{\cancel{(x + y)}(x^2 - xy + y^2)}{\cancel{(x + y)}(x - y)}$$ **Factor**

$$= \frac{x^2 - xy + y^2}{x - y}$$ **Divide out the common factor**

Ratios

For the rest of this section we will concern ourselves with ratios, a topic closely related to reducing fractions and rational expressions to lowest terms. Let's start with a definition.

> **DEFINITION** If a and b are any two numbers, $b \neq 0$, then the **ratio** of a and b is
>
> $$\frac{a}{b}$$

As you can see, ratios are another name for fractions or rational numbers. They are a way of comparing quantities. Since we also can think of $\frac{a}{b}$ as the quotient of a and b, ratios are also quotients. The following table gives some ratios in words and as fractions.

Note
With ratios it is common to leave the 1 in the denominator.

Ratio	As a Fraction	In Lowest Terms
25 to 75	$\frac{25}{75}$	$\frac{1}{3}$
8 to 2	$\frac{8}{2}$	$\frac{4}{1}$
20 to 16	$\frac{20}{16}$	$\frac{5}{4}$

 EXAMPLE 8 A solution of hydrochloric acid (HCl) and water contains 49 milliliters of water and 21 milliliters of HCl. Find the ratio of HCl to water and of HCl to the total volume of the solution.

SOLUTION The ratio of HCl to water is 21 to 49, or

$$\frac{21}{49} = \frac{3}{7}$$

The amount of total solution volume is $49 + 21 = 70$ milliliters. Therefore, the ratio of HCl to total solution is 21 to 70, or

$$\frac{21}{70} = \frac{3}{10}$$

Rate Equation

Many of the problems in this chapter will use what is called the *rate equation*. You use this equation on an intuitive level when you are estimating how long it will take you to drive long distances. For example, if you drive at 50 miles per hour for 2 hours, you will travel 100 miles. Here is the rate equation:

$$\text{Distance} = \text{rate} \cdot \text{time}$$
$$d = r \cdot t$$

The rate equation has two equivalent forms, the most common of which is obtained by solving for r. Here it is:

$$r = \frac{d}{t}$$

The rate r in the rate equation is the ratio of distance to time and also is referred to as *average speed*. The units for rate are miles per hour, feet per second, kilometers per hour, and so on.

EXAMPLE 9
The Forest chair lift at the Northstar ski resort in Lake Tahoe is 5,603 feet long. If a ride on this chair lift takes 11 minutes, what is the average speed of the lift in feet per minute?

SOLUTION To find the speed of the lift, we find the ratio of distance covered to time. (Our answer is rounded to the nearest whole number.)

$$\text{Rate} = \frac{\text{distance}}{\text{time}} = \frac{5{,}603 \text{ feet}}{11 \text{ minutes}} = \frac{5{,}603}{11} \text{ feet/minute} = 509 \text{ feet/minute}$$

Note how we separate the numerical part of the problem from the units. In the next section, we will convert this rate to miles per hour.

EXAMPLE 10
A Ferris wheel was built in St. Louis in 1986. It is named *Colossus*. The circumference of the wheel is 518 feet. It has 40 cars, each of which holds 6 passengers. A trip around the wheel takes 40 seconds. Find the average speed of a rider on *Colossus*.

Solution To find the average speed, we divide the distance traveled, which in this case is the circumference, by the time it takes to travel once around the wheel.

$$r = \frac{d}{t} = \frac{518 \text{ feet}}{40 \text{ seconds}} = 13.0 \text{ (rounded)}$$

The average speed of a rider on the *Colossus* is 13.0 feet per second.

In the next section, you will convert the ratio into an equivalent ratio that gives the speed of the rider in miles per hour.

LINKING OBJECTIVES AND EXAMPLES

Next to each **objective** we have listed the examples that are best described by that objective.

A	1, 2
B	3–7
C	8–10

GETTING READY FOR CLASS

After reading through the preceding section, respond in your own words and in complete sentences.

1. How do you reduce a rational expression to lowest terms?
2. What are the properties we use to manipulate rational expressions?
3. For what values of the variable is a rational expression undefined?
4. What is a ratio?

1. Simplify each expression.

a. $\dfrac{5 + 1}{25 - 1}$

b. $\dfrac{x + 1}{x^2 - 1}$

c. $\dfrac{x^2 - x}{x^2 - 1}$

d. $\dfrac{x^3 - 1}{x^2 - 1}$

e. $\dfrac{x^3 - 1}{x^3 - x^2}$

2. Simplify each expression.

a. $\dfrac{25 - 30 + 9}{25 - 9}$

b. $\dfrac{x^2 - 6x + 9}{x^2 - 9}$

c. $\dfrac{x^2 - 10x + 9}{x^2 - 9x}$

d. $\dfrac{x^2 + 3x + ax + 3a}{x^2 - 9}$

e. $\dfrac{x^3 + 27}{x^3 - 9x}$

Reduce the following rational expressions to lowest terms, if possible. Also, specify any restrictions on the variable in Problems 1 through 10.

3. $\dfrac{a - 3}{a^2 - 9}$

4. $\dfrac{a + 4}{a^2 - 16}$

5. $\dfrac{x + 5}{x^2 - 25}$

6. $\dfrac{x - 2}{x^2 - 4}$

7. $\dfrac{2x^2 - 8}{4}$

8. $\dfrac{5x - 10}{x - 2}$

9. $\dfrac{2x - 10}{3x - 6}$

10. $\dfrac{4x - 8}{x - 2}$

▶ **11.** $\dfrac{10a + 20}{5a + 10}$

12. $\dfrac{11a + 33}{6a + 18}$

13. $\dfrac{5x^2 - 5}{4x + 4}$

14. $\dfrac{7x^2 - 28}{2x + 4}$

15. $\dfrac{x - 3}{x^2 - 6x + 9}$

16. $\dfrac{x^2 - 10x + 25}{x - 5}$

17. $\dfrac{3x + 15}{3x^2 + 24x + 45}$

18. $\dfrac{5x + 15}{5x^2 + 40x + 75}$

19. $\dfrac{a^2 - 3a}{a^3 - 8a^2 + 15a}$

20. $\dfrac{a^2 + 3a}{a^3 - 2a^2 - 15a}$

21. $\dfrac{3x - 2}{9x^2 - 4}$

22. $\dfrac{2x - 3}{4x^2 - 9}$

23. $\dfrac{x^2 + 8x + 15}{x^2 + 5x + 6}$

24. $\dfrac{x^2 - 8x + 15}{x^2 - x - 6}$

25. $\dfrac{2m^3 - 2m^2 - 12m}{m^2 - 5m + 6}$

26. $\dfrac{2m^3 + 4m^2 - 6m}{m^2 - m - 12}$

▶ **27.** $\dfrac{x^3 + 3x^2 - 4x}{x^3 - 16x}$

28. $\dfrac{3a^2 - 8a + 4}{9a^3 - 4a}$

29. $\dfrac{4x^3 - 10x^2 + 6x}{2x^3 + x^2 - 3x}$

30. $\dfrac{3a^3 - 8a^2 + 5a}{4a^3 - 5a^2 + 1a}$

▶ **31.** $\dfrac{4x^2 - 12x + 9}{4x^2 - 9}$

32. $\dfrac{5x^2 + 18x - 8}{5x^2 + 13x - 6}$

33. $\dfrac{x + 3}{x^4 - 81}$

34. $\dfrac{x^2 + 9}{x^4 - 81}$

35. $\dfrac{3x^2 + x - 10}{x^4 - 16}$

36. $\dfrac{5x^2 - 26x + 24}{x^4 - 64}$

37. $\dfrac{42x^3 - 20x^2 - 48x}{6x^2 - 5x - 4}$

38. $\dfrac{36x^3 + 132x^2 - 135x}{6x^2 + 25x - 9}$

39. $\dfrac{x^3 - y^3}{x^2 - y^2}$

40. $\dfrac{x^3 + y^3}{x^2 - y^2}$

41. $\dfrac{x^3 + 8}{x^2 - 4}$

42. $\dfrac{x^3 - 125}{x^2 - 25}$

43. $\dfrac{x^3 + 8}{x^2 + x - 2}$

44. $\dfrac{x^2 - 2x - 3}{x^3 - 27}$

To reduce each of the following rational expressions to lowest terms, you will have to use factoring by grouping. Be sure to factor each numerator and denominator completely before dividing out any common factors. (Remember, factoring by grouping takes two steps.)

45. $\dfrac{xy + 3x + 2y + 6}{xy + 3x + 5y + 15}$

46. $\dfrac{xy + 7x + 4y + 28}{xy + 3x + 4y + 12}$

= Videos available by instructor request

▶ = Online student support materials available at academic.cengage.com/login

Multiply or divide as indicated. Be sure to reduce all answers to lowest terms. (The numerator and denominator of the answer should not have any factors in common.)

1. $\dfrac{x + y}{3} \cdot \dfrac{6}{x + y}$

2. $\dfrac{x - 1}{x + 1} \cdot \dfrac{5}{x - 1}$

▶ **3.** $\dfrac{2x + 10}{x^2} \cdot \dfrac{x^3}{4x + 20}$

4. $\dfrac{3x^4}{3x - 6} \cdot \dfrac{x - 2}{x^2}$

5. $\dfrac{9}{2a - 8} \div \dfrac{3}{a - 4}$

6. $\dfrac{8}{a^2 - 25} \div \dfrac{16}{a + 5}$

7. $\dfrac{x + 1}{x^2 - 9} \div \dfrac{2x + 2}{x + 3}$

8. $\dfrac{11}{x - 2} \div \dfrac{22}{2x^2 - 8}$

9. $\dfrac{a^2 + 5a}{7a} \cdot \dfrac{4a^2}{a^2 + 4a}$

10. $\dfrac{4a^2 + 4a}{a^2 - 25} \cdot \dfrac{a^2 - 5a}{8a}$

▶ **11.** $\dfrac{y^2 - 5y + 6}{2y + 4} \div \dfrac{2y - 6}{y + 2}$

12. $\dfrac{y^2 - 7y}{3y^2 - 48} \div \dfrac{y^2 - 9}{y^2 - 7y + 12}$

13. $\dfrac{2x - 8}{x^2 - 4} \cdot \dfrac{x^2 + 6x + 8}{x - 4}$

14. $\dfrac{x^2 + 5x + 1}{7x - 7} \cdot \dfrac{x - 1}{x^2 + 5x + 1}$

15. $\dfrac{x - 1}{x^2 - x - 6} \cdot \dfrac{x^2 + 5x + 6}{x^2 - 1}$

16. $\dfrac{x^2 - 3x - 10}{x^2 - 4x + 3} \cdot \dfrac{x^2 - 5x + 6}{x^2 - 3x - 10}$

17. $\dfrac{a^2 + 10a + 25}{a + 5} \div \dfrac{a^2 - 25}{a - 5}$

18. $\dfrac{a^2 + a - 2}{a^2 + 5a + 6} \div \dfrac{a - 1}{a}$

19. $\dfrac{y^3 - 5y^2}{y^4 + 3y^3 + 2y^2} \div \dfrac{y^2 - 5y + 6}{y^2 - 2y - 3}$

20. $\dfrac{y^2 - 5y}{y^2 + 7y + 12} \div \dfrac{y^3 - 7y^2 + 10y}{y^2 + 9y + 18}$

21. $\dfrac{2x^2 + 17x + 21}{x^2 + 2x - 35} \cdot \dfrac{x^3 - 125}{2x^2 - 7x - 15}$

22. $\dfrac{x^2 + x - 42}{4x^2 + 31x + 21} \cdot \dfrac{4x^2 - 5x - 6}{x^3 - 8}$

23. $\dfrac{2x^2 + 10x + 12}{4x^2 + 24x + 32} \cdot \dfrac{2x^2 + 18x + 40}{x^2 + 8x + 15}$

24. $\dfrac{3x^2 - 3}{6x^2 + 18x + 12} \cdot \dfrac{2x^2 - 8}{x^2 - 3x + 2}$

25. $\dfrac{2a^2 + 7a + 3}{a^2 - 16} \div \dfrac{4a^2 + 8a + 3}{2a^2 - 5a - 12}$

26. $\dfrac{3a^2 + 7a - 20}{a^2 + 3a - 4} \div \dfrac{3a^2 - 2a - 5}{a^2 - 2a + 1}$

▶ **27.** $\dfrac{4y^2 - 12y + 9}{y^2 - 36} \div \dfrac{2y^2 - 5y + 3}{y^2 + 5y - 6}$

28. $\dfrac{5y^2 - 6y + 1}{y^2 - 1} \div \dfrac{16y^2 - 9}{4y^2 + 7y + 3}$

29. $\dfrac{x^2 - 1}{6x^2 + 42x + 60} \cdot \dfrac{7x^2 + 17x + 6}{x^3 + 1} \cdot \dfrac{6x + 30}{7x^2 - 11x - 6}$

30. $\dfrac{4x^2 - 1}{3x - 15} \cdot \dfrac{4x^2 - 17x - 15}{4x^2 - 9x - 9} \cdot \dfrac{3x - 9}{8x^3 - 1}$

31. $\dfrac{18x^3 + 21x^2 - 60x}{21x^2 - 25x - 4} \cdot \dfrac{28x^2 - 17x - 3}{16x^3 + 28x^2 - 30x}$

32. $\dfrac{56x^3 + 54x^2 - 20x}{8x^2 - 2x - 15} \cdot \dfrac{6x^2 + 5x - 21}{63x^3 + 129x^2 - 42x}$

The next two problems are intended to give you practice reading, and paying attention to, the instructions that accompany the problems you are working. Working these problems is an excellent way to get ready for a test or quiz.

33. Work each problem according to the instructions given.

   a. Simplify: $\dfrac{9 - 1}{27 - 1}$

   b. Reduce: $\dfrac{x^2 - 1}{x^3 - 1}$

   c. Multiply: $\dfrac{x^2 - 1}{x^3 - 1} \cdot \dfrac{x - 1}{x + 1}$

   d. Divide: $\dfrac{x^2 - 1}{x^3 - 1} \div \dfrac{x - 1}{x^2 + x + 1}$

34. Work each problem according to the instructions given.

   a. Simplify: $\dfrac{16 - 9}{16 + 24 + 9}$

   b. Reduce: $\dfrac{4x^2 - 9}{4x^2 + 12x + 9}$

   c. Multiply: $\dfrac{4x^2 - 9}{4x^2 + 12x + 9} \cdot \dfrac{2x + 3}{2x - 3}$

   d. Divide: $\dfrac{4x^2 - 9}{4x^2 + 12x + 9} \div \dfrac{2x + 3}{2x - 3}$

Multiply the following expressions using the method shown in Examples 5 and 6 in this section.

▶ **35.** $(x^2 - 9)\left(\dfrac{2}{x + 3}\right)$

⬛ = Videos available by instructor request

▶ = Online student support materials available at academic.cengage.com/login

6.2 Multiplication and Division of Rational Expressions

36. $(x^2 - 9)\left(\dfrac{-3}{x - 3}\right)$

37. $(x^2 - x - 6)\left(\dfrac{x + 1}{x - 3}\right)$

38. $(x^2 - 2x - 8)\left(\dfrac{x + 3}{x - 4}\right)$

39. $(x^2 - 4x - 5)\left(\dfrac{-2x}{x + 1}\right)$

40. $(x^2 - 6x + 8)\left(\dfrac{4x}{x - 2}\right)$

Each of the following problems involves some factoring by grouping. Remember, before you can divide out factors common to the numerators and denominators of a product, you must factor completely.

41. $\dfrac{x^2 - 9}{x^2 - 3x} \cdot \dfrac{2x + 10}{xy + 5x + 3y + 15}$

42. $\dfrac{x^2 - 16}{x^2 - 4x} \cdot \dfrac{3x + 18}{xy + 6x + 4y + 24}$

43. $\dfrac{2x^2 + 4x}{x^2 - y^2} \cdot \dfrac{x^2 + 3x + xy + 3y}{x^2 + 5x + 6}$

44. $\dfrac{x^2 - 25}{3x^2 + 3xy} \cdot \dfrac{x^2 + 4x + xy + 4y}{x^2 + 9x + 20}$

45. $\dfrac{x^3 - 3x^2 + 4x - 12}{x^4 - 16} \cdot \dfrac{3x^2 + 5x - 2}{3x^2 - 10x + 3}$

46. $\dfrac{x^3 - 5x^2 + 9x - 45}{x^4 - 81} \cdot \dfrac{5x^2 + 18x + 9}{5x^2 - 22x - 15}$

Simplify each expression. Work inside parentheses first, and then divide out common factors.

47. $\left(1 - \dfrac{1}{2}\right)\left(1 - \dfrac{1}{3}\right)\left(1 - \dfrac{1}{4}\right)\left(1 - \dfrac{1}{5}\right)$

48. $\left(1 + \dfrac{1}{2}\right)\left(1 + \dfrac{1}{3}\right)\left(1 + \dfrac{1}{4}\right)\left(1 + \dfrac{1}{5}\right)$

The dots in the following problems represent factors not written that are in the same pattern as the surrounding factors. Simplify.

49. $\left(1 - \dfrac{1}{2}\right)\left(1 - \dfrac{1}{3}\right)\left(1 - \dfrac{1}{4}\right) \cdots \left(1 - \dfrac{1}{99}\right)\left(1 - \dfrac{1}{100}\right)$

50. $\left(1 - \dfrac{1}{3}\right)\left(1 - \dfrac{1}{4}\right)\left(1 - \dfrac{1}{5}\right) \cdots \left(1 - \dfrac{1}{98}\right)\left(1 - \dfrac{1}{99}\right)$

Applying the Concepts

51. Mount Whitney The top of Mount Whitney, the highest point in California, is 14,494 feet above sea level. Give this height in miles to the nearest tenth of a mile.

52. Motor Displacement The relationship between liters and cubic inches, both of which are measures of volume, is 0.0164 liters = 1 cubic inch. If a Ford Mustang has a motor with a displacement of 4.9 liters, what is the displacement in cubic inches? Round your answer to the nearest cubic inch.

53. Speed of Sound The speed of sound is 1,088 feet per second. Convert the speed of sound to miles per hour. Round your answer to the nearest whole number.

54. Average Speed A car travels 122 miles in 3 hours. Find the average speed of the car in feet per second. Round to the nearest whole number.

55. Ferris Wheel As we mentioned in Problem Set 6.1, the first Ferris wheel was built in 1893. It was a large wheel with a circumference of 785 feet. If one trip around the circumference of the wheel took 20 minutes, find the average speed of a rider in miles per hour. Round to the nearest hundredth.

56. Unit Analysis The photograph shows the Cone Nebula as seen by the Hubble telescope in April 2002. The distance across the photograph is about 2.5 light-years. If we assume light travels 186,000 miles in 1 second, we can find the number of miles in 1 light-year by converting 186,000 miles/second to miles/year. Find the number of miles in 1 light-year. Write your answer in expanded form and in scientific notation.

NASA

57. Ferris Wheel A Ferris wheel called *Colossus* has a circumference of 518 feet. If a trip around the circumference of *Colossus* takes 40 seconds, find the average speed of a rider in miles per hour. Round to the nearest tenth.

58. Average Speed Tina is training for a biathlon. As part of her training, she runs an 8-mile course, 2 miles of which is on level ground and 6 miles of which is downhill. It takes her 20 minutes to run the level part of the course and 40 minutes to run the downhill part of the course. Find her average speed in miles per hour for each part of the course.

59. Running Speed Have you ever wondered about the miles per hour speed of a fast runner? This problem will give you some insight into this question.

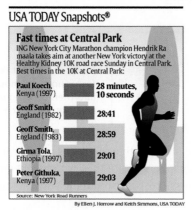

USA TODAY Snapshots®

Fast times at Central Park
ING New York City Marathon champion Hendrik Ra maala takes aim at another New York victory at the Healthy Kidney 10K road race Sunday in Central Park. Best times in the 10K at Central Park:

Paul Koech, Kenya (1997)	**28 minutes, 10 seconds**
Geoff Smith, England (1982)	**28:41**
Geoff Smith, England (1983)	**28:59**
Girma Tola, Ethiopia (1997)	**29:01**
Peter Githuka, Kenya (1997)	**29:03**

Source: New York Road Runners

By Ellen J. Horrow and Keith Simmons, USA TODAY

From *USA Today.* Copyright 2004. Reprinted with permission.

a. Calculate the average speed in kilometers per hour for Paul Koech.

b. If a 10K race is approximately 6.2 miles, convert the answer from part a to miles per hour.

60. Improving Your Quantitative Literacy The guidelines for fitness now indicate that a person who walks 10,000 steps daily is physically fit.

According to *The Walking Site* on the Internet, "The average person's stride length is approximately 2.5 feet long. That means it

Jim Cummings/Getty Images

takes just over 2,000 steps to walk one mile, and 10,000 steps is close to 5 miles." Use your knowledge of unit analysis to determine if these facts are correct.

Maintaining Your Skills

Add the following fractions.

61. $\dfrac{1}{2} + \dfrac{5}{2}$ **62.** $\dfrac{2}{3} + \dfrac{8}{3}$

63. $2 + \dfrac{3}{4}$ **64.** $1 + \dfrac{4}{7}$

Simplify each term, then add.

65. $\dfrac{10x^4}{2x^2} + \dfrac{12x^6}{3x^4}$ **66.** $\dfrac{32x^8}{8x^3} + \dfrac{27x^7}{3x^2}$

67. $\dfrac{12a^2b^5}{3ab^3} + \dfrac{14a^4b^7}{7a^3b^5}$ **68.** $\dfrac{16a^3b^2}{4ab} + \dfrac{25a^6b^5}{5a^4b^4}$

Getting Ready for the Next Section

Perform the indicated operation.

69. $\dfrac{1}{5} + \dfrac{3}{5}$ **70.** $\dfrac{1}{7} + \dfrac{5}{7}$

71. $\dfrac{1}{10} + \dfrac{3}{14}$ **72.** $\dfrac{1}{21} + \dfrac{4}{15}$

73. $\dfrac{1}{10} - \dfrac{3}{14}$ **74.** $\dfrac{1}{21} - \dfrac{4}{15}$

Multiply.

75. $2(x - 3)$ **76.** $x(x + 2)$

77. $(x + 4)(x - 5)$ **78.** $(x + 3)(x - 4)$

Reduce to lowest terms.

79. $\dfrac{x + 3}{x^2 - 9}$ **80.** $\dfrac{x + 7}{x^2 - 49}$

81. $\dfrac{x^2 - x - 30}{2(x + 5)(x - 5)}$

82. $\dfrac{x^2 - x - 20}{2(x + 4)(x - 4)}$

Simplify.

83. $(x + 4)(x - 5) - 10$

84. $(x + 3)(x - 4) - 8$

6.3 Addition and Subtraction of Rational Expressions

OBJECTIVES

A Add and subtract rational expressions that have the same denominators.

B Add and subtract rational expressions that have different denominators.

In Chapter 1 we combined fractions having the same denominator by combining their numerators and putting the result over the common denominator. We use the same process to add two rational expressions with the same denominator.

EXAMPLES

1. Add $\dfrac{5}{x} + \dfrac{3}{x}$.

SOLUTION Adding numerators, we have:

$$\frac{5}{x} + \frac{3}{x} = \frac{8}{x}$$

2. Add $\dfrac{x}{x^2 - 9} + \dfrac{3}{x^2 - 9}$.

SOLUTION Since both expressions have the same denominator, we add numerators and reduce to lowest terms:

$$\frac{x}{x^2 - 9} + \frac{3}{x^2 - 9} = \frac{x + 3}{x^2 - 9}$$

$$= \frac{\cancel{x+3}}{\cancel{(x+3)}(x - 3)} \left.\begin{array}{l} \end{array}\right\} \begin{array}{l}\textbf{Reduce to lowest terms by}\\ \textbf{factoring the denominator}\\ \textbf{and then dividing out the}\\ \textbf{common factor } x + 3\end{array}$$

$$= \frac{1}{x - 3}$$

Remember, it is the distributive property that allows us to add rational expressions by simply adding numerators. Because of this, we must begin all addition problems involving rational expressions by first making sure all the expressions have the same denominator.

> **DEFINITION** The **least common denominator** (LCD) for a set of denominators is the simplest quantity that is exactly divisible by all the denominators.

EXAMPLE 3   Add $\dfrac{1}{10} + \dfrac{3}{14}$.

SOLUTION

Note

If you have had difficulty in the past with addition and subtraction of fractions with different denominators, this is the time to get it straightened out. Go over Example 3 as many times as is necessary for you to understand the process.

Step 1: Find the LCD for 10 and 14. To do so, we factor each denominator and build the LCD from the factors:

$$\left.\begin{array}{l}10 = 2 \cdot 5\\ 14 = 2 \cdot 7\end{array}\right\} \quad \textbf{LCD} = 2 \cdot 5 \cdot 7 = 70$$

We know the LCD is divisible by 10 because it contains the factors 2 and 5. It is also divisible by 14 because it contains the factors 2 and 7.

Step 2: Change to equivalent fractions that each have denominator 70. To accomplish this task, we multiply the numerator and denominator of

43. Work each problem according to the instructions given.

a. Multiply: $\dfrac{4}{9} \cdot \dfrac{1}{6}$

b. Divide: $\dfrac{4}{9} \div \dfrac{1}{6}$

c. Add: $\dfrac{4}{9} + \dfrac{1}{6}$

d. Multiply: $\dfrac{x+2}{x-2} \cdot \dfrac{3x+10}{x^2-4}$

e. Divide: $\dfrac{x+2}{x-2} \div \dfrac{3x+10}{x^2-4}$

f. Subtract: $\dfrac{x+2}{x-2} - \dfrac{3x+10}{x^2-4}$

44. Work each problem according to the instructions given.

a. Multiply: $\dfrac{9}{25} \cdot \dfrac{1}{15}$

b. Divide: $\dfrac{9}{25} \div \dfrac{1}{15}$

c. Subtract: $\dfrac{9}{25} - \dfrac{1}{15}$

d. Multiply: $\dfrac{3x-2}{3x+2} \cdot \dfrac{15x+6}{9x^2-4}$

e. Divide: $\dfrac{3x-2}{3x+2} \div \dfrac{15x+6}{9x^2-4}$

f. Subtract: $\dfrac{3x+2}{3x-2} - \dfrac{15x+6}{9x^2-4}$

Complete the following tables.

45.

Number x	Reciprocal $\dfrac{1}{x}$	Sum $1+\dfrac{1}{x}$	Sum $\dfrac{x+1}{x}$
1			
2			
3			
4			

46.

Number x	Reciprocal $\dfrac{1}{x}$	Difference $1-\dfrac{1}{x}$	Difference $\dfrac{x-1}{x}$
1			
2			
3			
4			

47.

x	$x+\dfrac{4}{x}$	$\dfrac{x^2+4}{x}$	$x+4$
1			
2			
3			
4			

48.

x	$2x+\dfrac{6}{x}$	$\dfrac{2x^2+6}{x}$	$2x+6$
1			
2			
3			
4			

Add or subtract as indicated.

49. $1 + \dfrac{1}{x+2}$

50. $1 - \dfrac{1}{x+2}$

51. $1 - \dfrac{1}{x+3}$

52. $1 + \dfrac{1}{x+3}$

Maintaining Your Skills

Solve each equation.

53. $2x + 3(x-3) = 6$

54. $4x - 2(x-5) = 6$

55. $x - 3(x+3) = x - 3$

56. $x - 4(x+4) = x - 4$

57. $7 - 2(3x+1) = 4x + 3$

58. $8 - 5(2x-1) = 2x + 4$

Solve each quadratic equation.

59. $x^2 + 5x + 6 = 0$

60. $x^2 - 5x + 6 = 0$

61. $x^2 - x = 6$

62. $x^2 + x = 6$

63. $x^2 - 5x = 0$

64. $x^2 - 6x = 0$

69. $\dfrac{5}{0}$

70. $\dfrac{2}{0}$

71. $1 - \dfrac{5}{2}$

72. $1 - \dfrac{5}{3}$

Getting Ready for the Next Section

Simplify.

65. $6\left(\dfrac{1}{2}\right)$

66. $10\left(\dfrac{1}{5}\right)$

67. $\dfrac{0}{5}$

68. $\dfrac{0}{2}$

Use the distributive property to simplify.

73. $6\left(\dfrac{x}{3} + \dfrac{5}{2}\right)$

74. $10\left(\dfrac{x}{2} + \dfrac{3}{5}\right)$

75. $x^2\left(1 - \dfrac{5}{x}\right)$

76. $x^2\left(1 - \dfrac{3}{x}\right)$

Solve.

77. $2x + 15 = 3$

78. $15 = 3x - 3$

79. $-2x - 9 = x - 3$

80. $a^2 - a - 20 = -2a$

6.4 Equations Involving Rational Expressions

OBJECTIVES

A Solve equations that contain rational expressions.

The first step in solving an equation that contains one or more rational expressions is to find the LCD for all denominators in the equation. Once the LCD has been found, we multiply both sides of the equation by it. The resulting equation should be equivalent to the original one (unless we inadvertently multiplied by zero) and free from any denominators except the number 1.

EXAMPLE 1 Solve $\dfrac{x}{3} + \dfrac{5}{2} = \dfrac{1}{2}$ for x.

SOLUTION The LCD for 3 and 2 is 6. If we multiply both sides by 6, we have:

$$6\left(\frac{x}{3} + \frac{5}{2}\right) = 6\left(\frac{1}{2}\right) \qquad \textbf{Multiply both sides by 6}$$

$$6\left(\frac{x}{3}\right) + 6\left(\frac{5}{2}\right) = 6\left(\frac{1}{2}\right) \qquad \textbf{Distributive property}$$

$$2x + 15 = 3$$

$$2x = -12$$

$$x = -6$$

We can check our solution by replacing x with -6 in the original equation:

$$-\frac{6}{3} + \frac{5}{2} \overset{?}{=} \frac{1}{2}$$

$$\frac{1}{2} = \frac{1}{2}$$

Multiplying both sides of an equation containing fractions by the LCD clears the equation of all denominators, because the LCD has the property that all denominators will divide it evenly.

EXAMPLE 2

Solve for x: $\dfrac{3}{x-1} = \dfrac{3}{5}$.

SOLUTION   The LCD for $(x-1)$ and 5 is $5(x-1)$. Multiplying both sides by $5(x-1)$, we have:

$$5(x-1) \cdot \frac{3}{x-1} = 5(x-1) \cdot \frac{3}{5}$$

$$5 \cdot 3 = (x-1) \cdot 3$$

$$15 = 3x - 3$$

$$18 = 3x$$

$$6 = x$$

If we substitute $x = 6$ into the original equation, we have:

$$\frac{3}{6-1} \overset{?}{=} \frac{3}{5}$$

$$\frac{3}{5} = \frac{3}{5}$$

The solution set is {6}.

EXAMPLE 3

Solve $1 - \dfrac{5}{x} = \dfrac{-6}{x^2}$.

SOLUTION   The LCD is x^2. Multiplying both sides by x^2, we have

$$x^2\left(1 - \frac{5}{x}\right) = x^2\left(\frac{-6}{x^2}\right) \qquad \textbf{Multiply both sides by } x^2$$

$$x^2(1) - x^2\left(\frac{5}{x}\right) = x^2\left(\frac{-6}{x^2}\right) \qquad \begin{array}{l}\textbf{Apply distributive property}\\ \textbf{to the left side}\end{array}$$

$$x^2 - 5x = -6 \qquad \textbf{Simplify each side}$$

We have a quadratic equation, which we write in standard form, factor, and solve as we did in Chapter 6.

$$x^2 - 5x + 6 = 0 \qquad \textbf{Standard form}$$

$$(x-2)(x-3) = 0 \qquad \textbf{Factor}$$

$$x - 2 = 0 \quad \text{or} \quad x - 3 = 0 \qquad \textbf{Set factors equal to 0}$$

$$x = 2 \quad \text{or} \quad x = 3$$

The two possible solutions are 2 and 3. Checking each in the original equation, we find they both give true statements. They are both solutions to the original equation:

Check $x = 2$ Check $x = 3$

$$1 - \frac{5}{2} \overset{?}{=} \frac{-6}{4} \qquad\qquad 1 - \frac{5}{3} \overset{?}{=} \frac{-6}{9}$$

$$\frac{2}{2} - \frac{5}{2} = -\frac{3}{2} \qquad\qquad \frac{3}{3} - \frac{5}{3} = -\frac{2}{3}$$

$$-\frac{3}{2} = -\frac{3}{2} \qquad\qquad -\frac{2}{3} = -\frac{2}{3}$$

EXAMPLE 4 Solve $\dfrac{x}{x^2 - 9} - \dfrac{3}{x - 3} = \dfrac{1}{x + 3}$.

SOLUTION The factors of $x^2 - 9$ are $(x + 3)(x - 3)$. The LCD, then, is $(x + 3)(x - 3)$:

$$(x+3)(x-3) \cdot \dfrac{x}{(x+3)(x-3)} + (x+3)(x-3) \cdot \dfrac{-3}{x-3}$$

$$= (x+3)(x - 3) \cdot \dfrac{1}{x+3}$$

$$x + (x + 3)(-3) = (x - 3)1$$

$$x + (-3x) + (-9) = x - 3$$

$$-2x - 9 = x - 3$$

$$-3x = 6$$

$$x = -2$$

The solution is $x = -2$. It checks when substituted for x in the original equation.

 EXAMPLE 5 Solve $\dfrac{x}{x - 3} + \dfrac{3}{2} = \dfrac{3}{x - 3}$.

SOLUTION We begin by multiplying each term on both sides of the equation by $2(x - 3)$:

$$2(x-3) \cdot \dfrac{x}{x-3} + 2(x - 3) \cdot \dfrac{3}{2} = 2(x-3) \cdot \dfrac{3}{x-3}$$

$$2x + (x - 3) \cdot 3 = 2 \cdot 3$$

$$2x + 3x - 9 = 6$$

$$5x - 9 = 6$$

$$5x = 15$$

$$x = 3$$

Our only possible solution is $x = 3$. If we substitute $x = 3$ into our original equation, we get:

$$\dfrac{3}{3 - 3} + \dfrac{3}{2} \overset{?}{=} \dfrac{3}{3 - 3}$$

$$\dfrac{3}{0} + \dfrac{3}{2} = \dfrac{3}{0}$$

Two of the terms are undefined, so the equation is meaningless. What has happened is that we have multiplied both sides of the original equation by zero. The equation produced by doing this is not equivalent to our original equation. We always must check our solution when we multiply both sides of an equation by an expression containing the variable to make sure we have not multiplied both sides by zero.

Our original equation has no solution; that is, there is no real number x such that:

$$\dfrac{x}{x - 3} + \dfrac{3}{2} = \dfrac{3}{x - 3}$$

The solution set is ∅.

Note

There are two things to note about this problem. The first is that to solve the equation $d = r \cdot t$ for t, we divide each side by r, like this:

$$\frac{d}{r} = \frac{r \cdot t}{r}$$

$$\frac{d}{r} = t$$

The second thing is this: The speed of the boat in still water is the rate at which it would be traveling if there were no current; that is, it is the speed of the boat through the water. Since the water itself is moving at 5 miles per hour, the boat is going 5 miles per hour slower when it travels against the current and 5 miles per hour faster when it travels with the current.

The last positions in the table are filled in by using the equation $t = \dfrac{d}{r}$.

	d	r	t
Upstream	30	$x - 5$	$\dfrac{30}{x - 5}$
Downstream	50	$x + 5$	$\dfrac{50}{x + 5}$

Reading the problem again, we find that the time for the trip upstream is equal to the time for the trip downstream. Setting these two quantities equal to each other, we have our equation:

Time (downstream) = time (upstream)

$$\frac{50}{x + 5} = \frac{30}{x - 5}$$

The LCD is $(x + 5)(x - 5)$. We multiply both sides of the equation by the LCD to clear it of all denominators. Here is the solution:

$$(x + 5)(x - 5) \cdot \frac{50}{x + 5} = (x + 5)(x - 5) \cdot \frac{30}{x - 5}$$

$$50x - 250 = 30x + 150$$

$$20x = 400$$

$$x = 20$$

The speed of the boat in still water is 20 miles per hour.

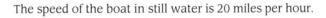

EXAMPLE 3

Tina is training for a biathlon. To train for the bicycle portion, she rides her bike 15 miles up a hill and then 15 miles back down the same hill. The complete trip takes her 2 hours. If her downhill speed is 20 miles per hour faster than her uphill speed, how fast does she ride uphill?

SOLUTION Again, we make a table. As in the previous example, we label the top row with distance, rate, and time. We label the left column with the two trips, uphill and downhill.

	d	r	t
Uphill			
Downhill			

Up and back
total time = 2 hours

Next, we fill in the table with as much information as we can from the problem. We know the distance traveled is 15 miles uphill and 15 miles downhill, which allows us to fill in the distance column. To fill in the rate column, we first note that she rides 20 miles per hour faster downhill than uphill. Therefore, if we let x equal her rate uphill, then her rate downhill is $x + 20$. Filling in the table with this information gives us

	d	r	t
Uphill	15	x	
Downhill	15	x + 20	

Since time is distance divided by rate, $t = d/r$, we can fill in the last column in the table.

	d	r	t
Uphill	15	x	$\dfrac{15}{x}$
Downhill	15	x + 20	$\dfrac{15}{x + 20}$

Rereading the problem, we find that the total time (the time riding uphill plus the time riding downhill) is two hours. We write our equation as follows:

$$\text{Time (uphill)} + \text{time (downhill)} = 2$$

$$\frac{15}{x} + \frac{15}{x + 20} = 2$$

We solve this equation for x by first finding the LCD and then multiplying each term in the equation by it to clear the equation of all denominators. Our LCD is $x(x + 20)$. Here is our solution:

$$x(x + 20)\frac{15}{x} + x(x + 20)\frac{15}{x + 20} = 2 \cdot [x(x + 20)]$$

$$15(x + 20) + 15x = 2x(x + 20)$$

$$15x + 300 + 15x = 2x^2 + 40x$$

$$0 = 2x^2 + 10x - 300$$

$$0 = x^2 + 5x - 150 \quad \textbf{Divide both sides by 2}$$

$$0 = (x + 15)(x - 10)$$

$$x = -15 \quad \text{or} \quad x = 10$$

Since we cannot have a negative speed, our only solution is $x = 10$. Tina rides her bike at a rate of 10 miles per hour when going uphill. (Her downhill speed is $x + 20 = 30$ miles per hour.)

EXAMPLE 4 An inlet pipe can fill a water tank in 10 hours, while an outlet pipe can empty the same tank in 15 hours. By mistake, both pipes are left open. How long will it take to fill the water tank with both pipes open?

SOLUTION Let $x = $ amount of time to fill the tank with both pipes open.

One method of solving this type of problem is to think in terms of how much of the job is done by a pipe in 1 hour.

1. If the inlet pipe fills the tank in 10 hours, then in 1 hour the inlet pipe fills $\frac{1}{10}$ of the tank.

Inlet pipe
10 hours
to fill

Outlet pipe
15 hours
to empty

2. If the outlet pipe empties the tank in 15 hours, then in 1 hour the outlet pipe empties $\frac{1}{15}$ of the tank.

3. If it takes x hours to fill the tank with both pipes open, then in 1 hour the tank is $\frac{1}{x}$ full.

Note

In solving a problem of this type, we have to assume that the thing doing the work (whether it is a pipe, a person, or a machine) is working at a constant rate; that is, as much work gets done in the first hour as is done in the last hour and any other hour in between.

Here is how we set up the equation. *In 1 hour,*

$$\underset{\substack{\text{Amount of water let} \\ \text{in by inlet pipe}}}{\frac{1}{10}} \quad - \quad \underset{\substack{\text{Amount of water let} \\ \text{out by outlet pipe}}}{\frac{1}{15}} \quad = \quad \underset{\substack{\text{Total amount of} \\ \text{water into tank}}}{\frac{1}{x}}$$

The LCD for our equation is $30x$. We multiply both sides by the LCD and solve:

$$30x\left(\frac{1}{10}\right) - 30x\left(\frac{1}{15}\right) = 30x\left(\frac{1}{x}\right)$$

$$3x - 2x = 30$$

$$x = 30$$

It takes 30 hours with both pipes open to fill the tank.

EXAMPLE 5   Graph the equation $y = \frac{1}{x}$.

SOLUTION   Since this is the first time we have graphed an equation of this form, we will make a table of values for x and y that satisfy the equation. Before we do, let's make some generalizations about the graph (Figure 1).

First, notice that since y is equal to 1 divided by x, y will be positive when x is positive. (The quotient of two positive numbers is a positive number.) Likewise, when x is negative, y will be negative. In other words, x and y always will have the same sign. Thus, our graph will appear in quadrants I and III only because in those quadrants x and y have the same sign.

Next, notice that the expression $\frac{1}{x}$ will be undefined when x is 0, meaning that there is no value of y corresponding to $x = 0$. Because of this, the graph will not cross the y-axis. Further, the graph will not cross the x-axis either. If we try to find the x-intercept by letting $y = 0$, we have

$$0 = \frac{1}{x}$$

x	y
-3	$-\frac{1}{3}$
-2	$-\frac{1}{2}$
-1	-1
$-\frac{1}{2}$	-2
$-\frac{1}{3}$	-3
0	Undefined
$\frac{1}{3}$	3
$\frac{1}{2}$	2
1	1
2	$\frac{1}{2}$
3	$\frac{1}{3}$

But there is no value of x to divide into 1 to obtain 0. Therefore, since there is no solution to this equation, our graph will not cross the x-axis.

To summarize, we can expect to find the graph in quadrants I and III only, and the graph will cross neither axis.

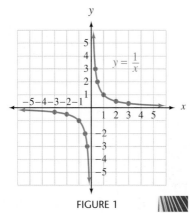

FIGURE 1

EXAMPLE 6 Graph the equation $y = \dfrac{-6}{x}$.

SOLUTION Since y is -6 divided by x, when x is positive, y will be negative (a negative divided by a positive is negative), and when x is negative, y will be positive (a negative divided by a negative). Thus, the graph (Figure 2) will appear in quadrants II and IV only. As was the case in Example 5, the graph will not cross either axis.

x	y
-6	1
-3	2
-2	3
-1	6
0	Undefined
1	-6
2	-3
3	-2
6	-1

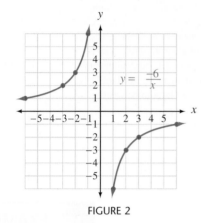

FIGURE 2

LINKING OBJECTIVES AND EXAMPLES

Next to each **objective** we have listed the examples that are best described by that objective.

A 1–4

B 5, 6

GETTING READY FOR CLASS

After reading through the preceding section, respond in your own words and in complete sentences.

1. Write an application problem for which the solution depends on solving the equation $\dfrac{1}{2} + \dfrac{1}{3} = \dfrac{1}{x}$.

2. How does the current of a river affect the speed of a motor boat traveling against the current?

3. How does the current of a river affect the speed of a motor boat traveling in the same direction as the current?

4. What is the relationship between the total number of minutes it takes for a drain to empty a sink and the amount of water that drains out of the sink in 1 minute?

Chapter 6 SUMMARY

Rational Numbers [6.1]

EXAMPLES

1. We can reduce $\frac{6}{8}$ to lowest terms by dividing the numerator and denominator by their greatest common factor 2:

$$\frac{6}{8} = \frac{2 \cdot 3}{2 \cdot 4} = \frac{3}{4}$$

Any number that can be put in the form $\frac{a}{b}$, where a and b are integers ($b \neq 0$), is called a rational number.

Multiplying or dividing the numerator and denominator of a rational number by the same nonzero number never changes the value of the rational number.

Rational Expressions [6.1]

2. We reduce rational expressions to lowest terms by factoring the numerator and denominator and then dividing out any factors they have in common:

$$\frac{x-3}{x^2-9} = \frac{x-3}{(x-3)(x+3)} = \frac{1}{x+3}$$

Any expression of the form $\frac{P}{Q}$, where P and Q are polynomials ($Q \neq 0$), is a rational expression.

Multiplying or dividing the numerator and denominator of a rational expression by the same nonzero quantity always produces a rational expression equivalent to the original one.

Multiplication [6.2]

3. $\dfrac{x-1}{x^2+2x-3} \cdot \dfrac{x^2-9}{x-2}$

$= \dfrac{x-1}{(x+3)(x-1)} \cdot \dfrac{(x-3)(x+3)}{x-2}$

$= \dfrac{x-3}{x-2}$

To multiply two rational numbers or two rational expressions, multiply numerators, multiply denominators, and divide out any factors common to the numerator and denominator:

For rational numbers $\dfrac{a}{b}$ and $\dfrac{c}{d}$, $\dfrac{a}{b} \cdot \dfrac{c}{d} = \dfrac{ac}{bd}$

For rational expressions $\dfrac{P}{Q}$ and $\dfrac{R}{S}$, $\dfrac{P}{Q} \cdot \dfrac{R}{S} = \dfrac{PR}{QS}$

Division [6.2]

4. $\dfrac{2x}{x^2-25} \div \dfrac{4}{x-5}$

$= \dfrac{2x}{(x-5)(x+5)} \cdot \dfrac{(x-5)}{4}$

$= \dfrac{x}{2(x+5)}$

To divide by a rational number or rational expression, simply multiply by its reciprocal:

For rational numbers $\dfrac{a}{b}$ and $\dfrac{c}{d}$, $\dfrac{a}{b} \div \dfrac{c}{d} = \dfrac{a}{b} \cdot \dfrac{d}{c}$

For rational expressions $\dfrac{P}{Q}$ and $\dfrac{R}{S}$, $\dfrac{P}{Q} \div \dfrac{R}{S} = \dfrac{P}{Q} \cdot \dfrac{S}{R}$

Addition [6.3]

5. $\dfrac{3}{x-1} + \dfrac{x}{2}$

$= \dfrac{3}{x-1} \cdot \dfrac{2}{2} + \dfrac{x}{2} \cdot \dfrac{x-1}{x-1}$

$= \dfrac{6}{2(x-1)} + \dfrac{x^2 - x}{2(x-1)}$

$= \dfrac{x^2 - x + 6}{2(x-1)}$

To add two rational numbers or rational expressions, find a common denominator, change each expression to an equivalent expression having the common denominator, then add numerators and reduce if possible:

For rational numbers $\dfrac{a}{c}$ and $\dfrac{b}{c}$,　$\dfrac{a}{c} + \dfrac{b}{c} = \dfrac{a+b}{c}$

For rational expressions $\dfrac{P}{S}$ and $\dfrac{Q}{S}$,　$\dfrac{P}{S} + \dfrac{Q}{S} = \dfrac{P+Q}{S}$

Subtraction [6.3]

6. $\dfrac{x}{x^2 - 4} - \dfrac{2}{x^2 - 4}$

$= \dfrac{x-2}{x^2 - 4}$

$= \dfrac{x-2}{(x-2)(x+2)}$

$= \dfrac{1}{x+2}$

To subtract a rational number or rational expression, simply add its opposite:

For rational numbers $\dfrac{a}{c}$ and $\dfrac{b}{c}$,　$\dfrac{a}{c} - \dfrac{b}{c} = \dfrac{a}{c} + \left(\dfrac{-b}{c}\right)$

For rational expressions $\dfrac{P}{S}$ and $\dfrac{Q}{S}$,　$\dfrac{P}{S} - \dfrac{Q}{S} = \dfrac{P}{S} + \left(\dfrac{-Q}{S}\right)$

Equations [6.4]

7. Solve $\dfrac{1}{2} + \dfrac{3}{x} = 5.$

$2x\left(\dfrac{1}{2}\right) + 2x\left(\dfrac{3}{x}\right) = 2x(5)$

$x + 6 = 10x$

$6 = 9x$

$x = \dfrac{2}{3}$

To solve equations involving rational expressions, first find the least common denominator (LCD) for all denominators. Then multiply both sides by the LCD and solve as usual. Be sure to check all solutions in the original equation to be sure there are no undefined terms.

Complex Fractions [6.6]

8. $\dfrac{1 - \dfrac{4}{x}}{x - \dfrac{16}{x}} = \dfrac{x\left(1 - \dfrac{4}{x}\right)}{x\left(x - \dfrac{16}{x}\right)}$

$= \dfrac{x - 4}{x^2 - 16}$

$= \dfrac{x - 4}{(x - 4)(x + 4)}$

$= \dfrac{1}{x + 4}$

A rational expression that contains a fraction in its numerator or denominator is called a complex fraction. The most common method of simplifying a complex fraction is to multiply the top and bottom by the LCD for all denominators.

Ratio and Proportion [6.1, 6.7]

9. Solve for x: $\dfrac{3}{x} = \dfrac{5}{20}.$

$3 \cdot 20 = 5 \cdot x$

$60 = 5x$

$x = 12$

The ratio of a to b is:

$$\dfrac{a}{b}$$

Two equal ratios form a proportion. In the proportion

$$\dfrac{a}{b} = \dfrac{c}{d}$$

a and d are the *extremes,* and b and c are the *means.* In any proportion the product of the extremes is equal to the product of the means.

The problems below form a comprehensive review of the material in this chapter. They can be used to study for exams. If you would like to take a practice test on this chapter, you can use the odd-numbered problems. Give yourself an hour and work as many of the odd-numbered problems as possible. When you are finished, or when an hour has passed, check your answers with the answers in the back of the book. You can use the even-numbered problems for a second practice test.

The numbers in brackets refer to the sections of the text in which similar problems can be found.

Reduce to lowest terms. Also specify any restriction on the variable. [6.1]

1. $\dfrac{7}{14x - 28}$

2. $\dfrac{a + 6}{a^2 - 36}$

3. $\dfrac{8x - 4}{4x + 12}$

4. $\dfrac{x + 4}{x^2 + 8x + 16}$

5. $\dfrac{3x^3 + 16x^2 - 12x}{2x^3 + 9x^2 - 18x}$

6. $\dfrac{x + 2}{x^4 - 16}$

7. $\dfrac{x^2 + 5x - 14}{x + 7}$

8. $\dfrac{a^2 + 16a + 64}{a + 8}$

9. $\dfrac{xy + bx + ay + ab}{xy + 5x + ay + 5a}$

Multiply or divide as indicated. [6.2]

10. $\dfrac{3x + 9}{x^2} \cdot \dfrac{x^3}{6x + 18}$

11. $\dfrac{x^2 + 8x + 16}{x^2 + x - 12} \div \dfrac{x^2 - 16}{x^2 - x - 6}$

12. $(a^2 - 4a - 12)\left(\dfrac{a - 6}{a + 2}\right)$

13. $\dfrac{3x^2 - 2x - 1}{x^2 + 6x + 8} \div \dfrac{3x^2 + 13x + 4}{x^2 + 8x + 16}$

Find the following sums and differences. [6.3]

14. $\dfrac{2x}{2x + 3} + \dfrac{3}{2x + 3}$

15. $\dfrac{x^2}{x - 9} - \dfrac{18x - 81}{x - 9}$

16. $\dfrac{a + 4}{a + 8} - \dfrac{a - 9}{a + 8}$

17. $\dfrac{x}{x + 9} + \dfrac{5}{x}$

18. $\dfrac{5}{4x + 20} + \dfrac{x}{x + 5}$

19. $\dfrac{3}{x^2 - 36} - \dfrac{2}{x^2 - 4x - 12}$

20. $\dfrac{3a}{a^2 + 8a + 15} - \dfrac{2}{a + 5}$

Solve each equation. [6.4]

21. $\dfrac{3}{x} + \dfrac{1}{2} = \dfrac{5}{x}$

22. $\dfrac{a}{a - 3} = \dfrac{3}{2}$

23. $1 - \dfrac{7}{x} = \dfrac{-6}{x^2}$

24. $\dfrac{3}{x + 6} - \dfrac{1}{x - 2} = \dfrac{-8}{x^2 + 4x - 12}$

25. $\dfrac{2}{y^2 - 16} = \dfrac{10}{y^2 + 4y}$

26. Number Problem The sum of a number and 7 times its reciprocal is $\frac{16}{3}$. Find the number. [6.5]

27. Distance, Rate, and Time A boat travels 48 miles up a river in the same amount of time it takes to travel 72 miles down the same river. If the current is 3 miles per hour, what is the speed of the boat in still water? [6.5]

28. Filling a Pool An inlet pipe can fill a pool in 21 hours, whereas an outlet pipe can empty it in 28 hours. If both pipes are left open, how long will it take to fill the pool? [6.5]

Simplify each complex fraction. [6.6]

29. $\dfrac{\dfrac{x + 4}{x^2 - 16}}{\dfrac{2}{x - 4}}$

30. $\dfrac{1 - \dfrac{9}{y^2}}{1 + \dfrac{4}{y} - \dfrac{21}{y^2}}$

31. $\dfrac{\dfrac{1}{a - 2} + 4}{\dfrac{1}{a - 2} + 1}$

32. Write the ratio of 40 to 100 as a fraction in lowest terms. [6.7]

33. If there are 60 seconds in 1 minute, what is the ratio of 40 seconds to 3 minutes? [6.7]

Solve each proportion. [6.7]

34. $\dfrac{x}{9} = \dfrac{4}{3}$

35. $\dfrac{a}{3} = \dfrac{12}{a}$

36. $\dfrac{8}{x - 2} = \dfrac{x}{6}$

GROUP PROJECT   Rational Expressions

Number of People 3

Time Needed 10–15 minutes

Equipment Pencil and paper

Procedure The four problems shown here all involve the same rational expressions. Often, students who have worked problems successfully on their homework have trouble when they take a test on rational expressions because the problems are mixed up and do not have similar instructions. Noticing similarities and differences between the types of problems involving rational expressions can help with this situation.

1. Which problems here do not require the use of a least common denominator?

2. Which two problems involve multiplying by the least common denominator?

3. Which of the problems will have an answer that is one or two numbers but no variables?

4. Work each of the four problems.

Problem 1: Add: $\dfrac{-2}{x^2 - 2x - 3} + \dfrac{3}{x^2 - 9}$

Problem 2: Divide: $\dfrac{-2}{x^2 - 2x - 3} \div \dfrac{3}{x^2 - 9}$

Problem 3: Solve: $\dfrac{-2}{x^2 - 2x - 3} + \dfrac{3}{x^2 - 9} = -1$

Problem 4: Simplify: $\dfrac{\dfrac{-2}{x^2 - 2x - 3}}{\dfrac{3}{x^2 - 9}}$

Note
Because subtraction is defined in terms of addition and division is defined in terms of multiplication, we do not need to introduce separate properties for subtraction and division. The solution set for an equation will never be changed by subtracting the same amount from both sides or by dividing both sides by the same nonzero quantity.

Our second new property is called the *multiplication property of equality* and is stated as follows.

> **Multiplication Property of Equality** For any three algebraic expressions A, B, and C, where $C \neq 0$,
>
> $$\text{if} \qquad A = B$$
> $$\text{then} \qquad AC = BC$$
>
> *In words:* Multiplying both sides of an equation by the same nonzero quantity will not change the solution set.

EXAMPLE 2   Find the solution set for $3a - 5 = -6a + 1$.

SOLUTION   To solve for a we must isolate it on one side of the equation. Let's decide to isolate a on the left side by adding $6a$ to both sides of the equation.

$$3a - 5 = -6a + 1$$

$$3a + \mathbf{6a} - 5 = -6a + \mathbf{6a} + 1 \qquad \textbf{Add } 6a \textbf{ to both sides}$$

$$9a - 5 = 1$$

$$9a - 5 + \mathbf{5} = 1 + \mathbf{5} \qquad \textbf{Add 5 to both sides}$$

$$9a = 6$$

$$\frac{\mathbf{1}}{\mathbf{9}}(9a) = \frac{\mathbf{1}}{\mathbf{9}}(6) \qquad \textbf{Multiply both sides by } \tfrac{1}{9}$$

$$a = \frac{2}{3} \qquad\qquad \tfrac{1}{9}(6) = \tfrac{6}{9} = \tfrac{2}{3}$$

Note
We know that multiplication by a number and division by its reciprocal always produce the same result. Because of this fact, instead of multiplying each side of our equation by $\frac{1}{9}$, we could just as easily divide each side by 9. If we did so, the last two lines in our solution would look like this:

$$\frac{9a}{9} = \frac{6}{9}$$
$$a = \frac{2}{3}$$

The solution set is $\left\{\frac{2}{3}\right\}$.

We can check our solution in Example 2 by replacing a in the original equation with $\frac{2}{3}$.

When $$a = \frac{2}{3}$$

the equation $$3a - 5 = -6a + 1$$

becomes $$3\left(\frac{2}{3}\right) - 5 = -6\left(\frac{2}{3}\right) + 1$$

$$2 - 5 = -4 + 1$$

$$-3 = -3 \qquad \textbf{A true statement}$$

There will be times when we solve equations and end up with a negative sign in front of the variable. The next example shows how to handle this situation.

EXAMPLE 3   Solve each equation.

    a. $-x = 4$ **b.** $-y = -8$

SOLUTION   Neither equation can be considered solved because of the negative sign in front of the variable. To eliminate the negative signs we simply multiply both sides of each equation by -1.

a.	$-x = 4$	**b.** $-y = -8$
	$-\mathbf{1}(-x) = -\mathbf{1}(4)$	$-\mathbf{1}(-y) = -\mathbf{1}(-8)$ **Multiply each side by −1**
	$x = -4$	$y = 8$

EXAMPLE 4   Solve $\frac{2}{3}x + \frac{1}{2} = -\frac{3}{8}$.

SOLUTION   We can solve this equation by applying our properties and working with fractions, or we can begin by eliminating the fractions. Let's use both methods.

Method 1 Working with the fractions.

$$\frac{2}{3}x + \frac{1}{2} + \left(-\frac{\mathbf{1}}{\mathbf{2}}\right) = -\frac{3}{8} + \left(-\frac{\mathbf{1}}{\mathbf{2}}\right) \quad \text{Add } -\tfrac{1}{2} \text{ to each side}$$

$$\frac{2}{3}x = -\frac{7}{8} \qquad\qquad -\tfrac{3}{8} + \left(-\tfrac{1}{2}\right) = -\tfrac{3}{8} + \left(-\tfrac{4}{8}\right)$$

$$\frac{\mathbf{3}}{\mathbf{2}}\left(\frac{2}{3}x\right) = \frac{\mathbf{3}}{\mathbf{2}}\left(-\frac{7}{8}\right) \qquad \text{Multiply each side by } \tfrac{3}{2}$$

$$x = -\frac{21}{16}$$

Method 2 Eliminating the fractions in the beginning.

Our original equation has denominators of 3, 2, and 8. The least common denominator, abbreviated LCD, for these three denominators is 24, and it has the property that all three denominators will divide it evenly. If we multiply both sides of our equation by 24, each denominator will divide into 24, and we will be left with an equation that does not contain any denominators other than 1.

$$\mathbf{24}\left(\frac{2}{3}x + \frac{1}{2}\right) = \mathbf{24}\left(-\frac{3}{8}\right) \qquad \text{Multiply each side by the LCD 24}$$

$$24\left(\frac{2}{3}x\right) + 24\left(\frac{1}{2}\right) = 24\left(-\frac{3}{8}\right) \qquad \text{Distributive property on the left side}$$

$$16x + 12 = -9 \qquad\qquad \text{Multiply}$$

$$16x = -21 \qquad\qquad \text{Add } -12 \text{ to each side}$$

$$x = -\frac{21}{16} \qquad\qquad \text{Multiply each side by } \tfrac{1}{16}$$

Check To check our solution, we substitute $x = -\frac{21}{16}$ back into our original equation to obtain

$$\frac{2}{3}\left(-\frac{21}{16}\right) + \frac{1}{2} \overset{?}{=} -\frac{3}{8}$$

$$-\frac{7}{8} + \frac{1}{2} \overset{?}{=} -\frac{3}{8}$$

$$-\frac{7}{8} + \frac{4}{8} \overset{?}{=} -\frac{3}{8}$$

$$-\frac{3}{8} = -\frac{3}{8} \qquad \textbf{A true statement}$$

 EXAMPLE 5 Solve the equation $0.06x + 0.05(10{,}000 - x) = 560$.

SOLUTION We can solve the equation in its original form by working with the decimals, or we can eliminate the decimals first by using the multiplication property of equality and solve the resulting equation. Here are both methods.

Method 1 Working with the decimals.

$0.06x + 0.05(10{,}000 - x) = 560$	**Original equation**
$0.06x + 0.05(10{,}000) - 0.05x = 560$	**Distributive property**
$0.01x + 500 = 560$	**Simplify the left side**
$0.01x + 500 + (\mathbf{-500}) = 560 + (\mathbf{-500})$	**Add −500 to each side**
$0.01x = 60$	
$\dfrac{0.01x}{\mathbf{0.01}} = \dfrac{60}{\mathbf{0.01}}$	**Divide each side by 0.01**
$x = 6{,}000$	

Method 2 Eliminating the decimals in the beginning: To move the decimal point two places to the right in $0.06x$ and 0.05, we multiply each side of the equation by 100.

$0.06x + 0.05(10{,}000 - x) = 560$	**Original equation**
$0.06x + 500 - 0.05x = 560$	**Distributive property**
$\mathbf{100}(0.06x) + \mathbf{100}(500) - \mathbf{100}(0.05x) = \mathbf{100}(560)$	**Multiply each side by 100**
$6x + 50{,}000 - 5x = 56{,}000$	
$x + 50{,}000 = 56{,}000$	**Simplify the left side**
$x = 6{,}000$	**Add −50,000 to each side**

Using either method, the solution to our equation is 6,000.

Check We check our work (to be sure we have not made a mistake in applying the properties or in arithmetic) by substituting 6,000 into our original equation and simplifying each side of the result separately, as the following shows.

$$0.06(\mathbf{6{,}000}) + 0.05(10{,}000 - \mathbf{6{,}000}) \overset{?}{=} 560$$

$$0.06(6{,}000) + 0.05(4{,}000) \overset{?}{=} 560$$

$$360 + 200 \overset{?}{=} 560$$

$$560 = 560 \qquad \textbf{A true statement}$$

Here is a list of steps to use as a guideline for solving linear equations in one variable.

> **Strategy for Solving Linear Equations in One Variable**
>
> **Step 1:** **a.** Use the distributive property to separate terms, if necessary.
>
> **b.** If fractions are present, consider multiplying both sides by the LCD to eliminate the fractions. If decimals are present, consider multiplying both sides by a power of 10 to clear the equation of decimals.
>
> **c.** Combine similar terms on each side of the equation.
>
> **Step 2:** Use the addition property of equality to get all variable terms on one side of the equation and all constant terms on the other side. A **variable term** is a term that contains the variable. A **constant term** is a term that does not contain the variable (the number 3, for example).
>
> **Step 3:** Use the multiplication property of equality to get the variable by itself on one side of the equation.
>
> **Step 4:** Check your solution in the original equation to be sure that you have not made a mistake in the solution process.

As you will see as you work through the problems in the problem set, it is not always necessary to use all four steps when solving equations. The number of steps used depends on the equation. In Example 6 there are no fractions or decimals in the original equation, so Step 1b will not be used.

EXAMPLE 6
Solve the equation $8 - 3(4x - 2) + 5x = 35$.

SOLUTION We must begin by distributing the -3 across the quantity $4x - 2$.

Note
It would be a mistake to subtract 3 from 8 first because the rule for order of operations indicates we are to do multiplication before subtraction.

Step 1:	**a.** $8 - 3(4x - 2) + 5x = 35$	**Original equation**
	$8 - 12x + 6 + 5x = 35$	**Distributive property**
	c. $-7x + 14 = 35$	**Simplify**
Step 2:	$-7x = 21$	**Add -14 to each side**
Step 3:	$x = -3$	**Multiply by $-\frac{1}{7}$**

Step 4: When x is replaced by -3 in the original equation, a true statement results. Therefore, -3 is the solution to our equation.

Identities and Equations with No Solution

Two special cases are associated with solving linear equations in one variable, each of which is illustrated in the following examples.

EXAMPLE 7
Solve for x: $2(3x - 4) = 3 + 6x$.

SOLUTION Applying the distributive property to the left side gives us

$6x - 8 = 3 + 6x$ **Distributive property**

Now, if we add $-6x$ to each side, we are left with the following

$$-8 = 3$$

which is a false statement. This means that there is no solution to our equation. Any number we substitute for x in the original equation will lead to a similar false statement.

EXAMPLE 8   Solve for x: $-15 + 3x = 3(x - 5)$.

SOLUTION   We start by applying the distributive property to the right side.

$$-15 + 3x = 3x - 15 \quad \textbf{Distributive property}$$

If we add $-3x$ to each side, we are left with the true statement

$$-15 = -15$$

In this case, our result tells us that any number we use in place of x in the original equation will lead to a true statement. Therefore, all real numbers are solutions to our equation. We say the original equation is an *identity* because the left side is always identically equal to the right side.

Solving Equations by Factoring

Next we will use our knowledge of factoring to solve equations. Most of the equations we will see are *quadratic equations.* Here is the definition of a quadratic equation.

DEFINITION Any equation that can be written in the form

$$ax^2 + bx + c = 0$$

where a, b, and c are constants and a is not 0 ($a \neq 0$) is called a **quadratic equation.** The form $ax^2 + bx + c = 0$ is called **standard form** for quadratic equations.

Each of the following is a quadratic equation:

$$2x^2 = 5x + 3 \qquad 5x^2 = 75 \qquad 4x^2 - 3x + 2 = 0$$

Notation For a quadratic equation written in standard form, the first term ax^2 is called the *quadratic term;* the second term bx is the *linear term;* and the last term c is called the *constant term.*

In the past we have noticed that the number 0 is a special number. There is another property of 0 that is the key to solving quadratic equations. It is called the *zero-factor property.*

Zero-Factor Property For all real numbers r and s,

$$r \cdot s = 0 \qquad \text{if and only if} \qquad r = 0 \qquad \text{or} \qquad s = 0 \quad \text{(or both)}$$

Note

The third equation is clearly a quadratic equation since it is in standard form. (Notice that a is 4, b is -3, and c is 2.) The first two equations are also quadratic because they could be put in the form $ax^2 + bx + c = 0$ by using the addition property of equality.

Note

What the zero-factor property says in words is that we can't multiply and get 0 without multiplying by 0; that is, if we multiply two numbers and get 0, then one or both of the original two numbers we multiplied must have been 0.

 EXAMPLE 9   Solve $x^2 - 2x - 24 = 0$.

SOLUTION   We begin by factoring the left side as $(x - 6)(x + 4)$ and get

$$(x - 6)(x + 4) = 0$$

Now both $(x - 6)$ and $(x + 4)$ represent real numbers. We notice that their product is 0. By the zero-factor property, one or both of them must be 0:

$$x - 6 = 0 \quad \text{or} \quad x + 4 = 0$$

We have used factoring and the zero-factor property to rewrite our original second-degree equation as two first-degree equations connected by the word *or*. Completing the solution, we solve the two first-degree equations:

$$x - 6 = 0 \quad \text{or} \quad x + 4 = 0$$
$$x = 6 \quad \text{or} \quad x = -4$$

We check our solutions in the original equation as follows:

$$\text{Check } x = 6 \qquad\qquad \text{Check } x = -4$$
$$6^2 - 2(6) - 24 \stackrel{?}{=} 0 \quad (-4)^2 - 2(-4) - 24 \stackrel{?}{=} 0$$
$$36 - 12 - 24 \stackrel{?}{=} 0 \qquad\qquad 16 + 8 - 24 \stackrel{?}{=} 0$$
$$0 = 0 \qquad\qquad\qquad\qquad 0 = 0$$

Note

We are placing a question mark over the equal sign because we don't know yet if the expression on the left will be equal to the expression on the right.

In both cases the result is a true statement, which means that both 6 and -4 are solutions to the original equation.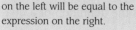

Although the next equation is not quadratic, the method we use is similar.

 EXAMPLE 10   Solve $\frac{1}{3}x^3 = \frac{5}{6}x^2 + \frac{1}{2}x$.

SOLUTION   We can simplify our work if we clear the equation of fractions. Multiplying both sides by the LCD, 6, we have

$$\mathbf{6} \cdot \tfrac{1}{3}x^3 = \mathbf{6} \cdot \tfrac{5}{6}x^2 + \mathbf{6} \cdot \tfrac{1}{2}x$$

$$2x^3 = 5x^2 + 3x$$

Next we add $-5x^2$ and $-3x$ to each side so that the right side will become 0.

$$2x^3 - 5x^2 - 3x = 0 \qquad \textbf{Standard form}$$

We factor the left side and then use the zero-factor property to set each factor to 0.

$$x(2x^2 - 5x - 3) = 0 \qquad\qquad \textbf{Factor out the greatest common factor}$$

$$x(2x + 1)(x - 3) = 0 \qquad\qquad \textbf{Continue factoring}$$

$$x = 0 \quad \text{or} \quad 2x + 1 = 0 \quad \text{or} \quad x - 3 = 0 \qquad \textbf{Zero-factor property}$$

Solving each of the resulting equations, we have

$$x = 0 \quad \text{or} \quad x = -\tfrac{1}{2} \quad \text{or} \quad x = 3$$

To generalize the preceding example, here are the steps used in solving a quadratic equation by factoring.

> **To Solve an Equation by Factoring**
>
> **Step 1:** Write the equation in standard form.
>
> **Step 2:** Factor the left side.
>
> **Step 3:** Use the zero-factor property to set each factor equal to 0.
>
> **Step 4:** Solve the resulting linear equations.

 EXAMPLE 11   Solve $100x^2 = 300x$.

SOLUTION   We begin by writing the equation in standard form and factoring:

$$100x^2 = 300x$$
$$100x^2 - 300x = 0 \quad \textbf{Standard form}$$
$$100x(x - 3) = 0 \quad \textbf{Factor}$$

Using the zero-factor property to set each factor to 0, we have

$$100x = 0 \quad \text{or} \quad x - 3 = 0$$
$$x = 0 \quad \text{or} \quad x = 3$$

The two solutions are 0 and 3.

 EXAMPLE 12   Solve $(x - 2)(x + 1) = 4$.

SOLUTION   We begin by multiplying the two factors on the left side. (Notice that it would be incorrect to set each of the factors on the left side equal to 4. The fact that the product is 4 does not imply that either of the factors must be 4.)

$$(x - 2)(x + 1) = 4$$
$$x^2 - x - 2 = 4 \quad \textbf{Multiply the left side}$$
$$x^2 - x - 6 = 0 \quad \textbf{Standard form}$$
$$(x - 3)(x + 2) = 0 \quad \textbf{Factor}$$
$$x - 3 = 0 \quad \text{or} \quad x + 2 = 0 \quad \textbf{Zero-factor property}$$
$$x = 3 \quad \text{or} \quad x = -2$$

 EXAMPLE 13   Solve for x: $x^3 + 2x^2 - 9x - 18 = 0$.

SOLUTION   We start with factoring by grouping.

$$x^3 + 2x^2 - 9x - 18 = 0$$
$$x^2(x + 2) - 9(x + 2) = 0$$
$$(x + 2)(x^2 - 9) = 0$$
$$(x + 2)(x - 3)(x + 3) = 0 \quad \textbf{The difference of two squares}$$
$$x + 2 = 0 \quad \text{or} \quad x - 3 = 0 \quad \text{or} \quad x + 3 = 0 \quad \textbf{Set factors to 0}$$
$$x = -2 \quad \text{or} \quad x = 3 \quad \text{or} \quad x = -3$$

We have three solutions: -2, 3, and -3.

Application

Recall that the method, or strategy, that we use to solve application problems is called the *Blueprint for Problem Solving*.

 BLUEPRINT FOR PROBLEM SOLVING

> **STEP 1:** **Read** the problem, and then mentally **list** the items that are known and the items that are unknown.
>
> **STEP 2:** **Assign a variable** to one of the unknown items. (In most cases this will amount to letting $x =$ the item that is asked for in the problem.) Then **translate** the other **information** in the problem to expressions involving the variable.
>
> **STEP 3:** **Reread** the problem, and then **write an equation,** using the items and variable listed in steps 1 and 2, that describes the situation.
>
> **STEP 4:** **Solve the equation** found in step 3.
>
> **STEP 5:** **Write your answer** using a complete sentence.
>
> **STEP 6:** **Reread** the problem, and **check** your solution with the original words in the problem.

EXAMPLE 14   The sum of the squares of two consecutive integers is 25. Find the two integers.

SOLUTION   We apply the Blueprint for Problem Solving to solve this application problem. Remember, step 1 in the blueprint is done mentally.

Step 1: **Read and list.**
Known items: Two consecutive integers. If we add their squares, the result is 25.
Unknown items: The two integers

Step 2: **Assign a variable and translate information.**
Let $x =$ the first integer; then $x + 1 =$ the next consecutive integer.

Step 3: **Reread and write an equation.**
Since the sum of the squares of the two integers is 25, the equation that describes the situation is

$$x^2 + (x + 1)^2 = 25$$

Step 4: **Solve the equation.**

$$x^2 + (x + 1)^2 = 25$$
$$x^2 + (x^2 + 2x + 1) = 25$$
$$2x^2 + 2x - 24 = 0$$
$$x^2 + x - 12 = 0 \quad \textbf{Divide each side by 2}$$
$$(x + 4)(x - 3) = 0$$
$$x = -4 \quad \text{or} \quad x = 3$$

Step 5: **Write the answer.**

If $x = -4$, then $x + 1 = -3$. If $x = 3$, then $x + 1 = 4$. The two integers are -4 and -3, or the two integers are 3 and 4.

Step 6: **Reread and check.**

The two integers in each pair are consecutive integers, and the sum of the squares of either pair is 25.

LINKING OBJECTIVES AND EXAMPLES

Next to each **objective** we have listed the examples that are best described by that objective.

A	2–8
B	9–13
C	14

GETTING READY FOR CLASS

After reading through the preceding section, respond in your own words and in complete sentences.

1. What is a solution to an equation?

2. What are equivalent equations?

3. Describe how to eliminate fractions in an equation.

4. What is the zero-factor property?

Problem Set 7.1

Online support materials can be found at academic.cengage.com/login

Solve each of the following equations.

1. $2x - 4 = 6$

2. $3x - 5 = 4$

3. $-300y + 100 = 500$

4. $-20y + 80 = 30$

5. $-x = 2$

6. $-x = \dfrac{1}{2}$

7. $-a = -\dfrac{3}{4}$

8. $-a = -5$

9. $-\dfrac{3}{5}a + 2 = 8$

10. $-\dfrac{5}{3}a + 3 = 23$

11. $2x - 5 = 3x + 2$

12. $5x - 1 = 4x + 3$

13. $5 - 2x = 3x + 1$

14. $7 - 3x = 8x - 4$

15. $5(y + 2) - 4(y + 1) = 3$

16. $6(y - 3) - 5(y + 2) = 8$

17. $6 - 7(m - 3) = -1$

▶ **18.** $3 - 5(2m - 5) = -2$

19. $7 + 3(x + 2) = 4(x - 1)$

20. $5 + 2(4x - 4) = 3(2x - 1)$

▶ **21.** $\dfrac{1}{2}x + \dfrac{1}{4} = \dfrac{1}{3}x + \dfrac{5}{4}$

22. $\dfrac{2}{3}x - \dfrac{3}{4} = \dfrac{1}{6}x + \dfrac{21}{4}$

23. $\dfrac{1}{2}x + \dfrac{1}{3}x + \dfrac{1}{4}x = 13$

24. $\dfrac{1}{3}x + \dfrac{1}{4}x + \dfrac{1}{5}x = 47$

25. $0.08x + 0.09(9,000 - x) = 750$

26. $0.08x + 0.09(9,000 - x) = 500$

▶ **27.** $0.35x - 0.2 = 0.15x + 0.1$

28. $0.25x - 0.05 = 0.2x + 0.15$

29. $x^2 - 5x - 6 = 0$

30. $x^2 + 5x - 6 = 0$

▶ **31.** $x^3 - 5x^2 + 6x = 0$

32. $x^3 + 5x^2 + 6x = 0$

33. $3y^2 + 11y - 4 = 0$

34. $3y^2 - y - 4 = 0$

▶ **35.** $\dfrac{1}{10}t^2 - \dfrac{5}{2} = 0$

36. $\dfrac{2}{7}t^2 - \dfrac{7}{2} = 0$

37. $\dfrac{1}{5}y^2 - 2 = -\dfrac{3}{10}y$

▨ = Videos available by instructor request

▶ = Online student support materials available at academic.cengage.com/login

38. $\frac{1}{2}y^2 + \frac{5}{3} = \frac{17}{6}y$

39. $9x^2 - 12x = 0$

40. $4x^2 + 4x = 0$

41. $0.02r + 0.01 = 0.15r^2$

42. $0.02r - 0.01 = -0.08r^2$

▶ **43.** $9a^3 = 16a$

44. $16a^3 = 25a$

45. $-100x = 10x^2$

46. $800x = 100x^2$

▶ **47.** $(x + 6)(x - 2) = -7$

48. $(x - 7)(x + 5) = -20$

49. $(x + 1)^2 = 3x + 7$

50. $(x + 2)^2 = 9x$

51. $x^3 + 3x^2 - 4x - 12 = 0$

52. $x^3 + 5x^2 - 4x - 20 = 0$

53. $2x^3 + 3x^2 - 8x - 12 = 0$

54. $3x^3 + 2x^2 - 27x - 18 = 0$

55. $3x - 6 = 3(x + 4)$

56. $7x - 14 = 7(x - 2)$

57. $4y + 2 - 3y + 5 = 3 + y + 4$

58. $7y + 5 - 2y - 3 = 6 + 5y - 4$

59. $2(4t - 1) + 3 = 5t + 4 + 3t$

60. $5(2t - 1) + 1 = 2t - 4 + 8t$

Now that you have practiced solving a variety of equations, we can turn our attention to the type of equation you will see as you progress through the book. Each equation appears later in the book exactly as you see it below.

Solve each equation.

61. $0 = 6400a + 70$

62. $0 = 6400a + 60$

63. $x + 2 = 2x$

64. $x + 2 = 7x$

65. $.07x = 1.4$

66. $.02x = 0.3$

67. $5(2x + 1) = 12$

68. $4(3x - 2) = 21$

69. $50 = \frac{K}{48}$

70. $50 = \frac{K}{24}$

71. $100P = 2,400$

72. $3.5d = 16(3.5)^2$

73. $x + (3x + 2) = 26$

74. $2(1) + y = 4$

75. $2x - 3(3x - 5) = -6$

76. $2(2y + 6) + 3y = 5$

77. $5\left(-\frac{19}{15}\right) + 5y = 9$

78. $5\left(-\frac{19}{15}\right) - 2y = 4$

79. $2\left(-\frac{29}{22}\right) - 3y = 4$

80. $2x - 3\left(\frac{5}{11}\right) = 4$

81. $3x^2 + x = 10$

82. $y^2 + y - 20 = 2y$

83. $12(x + 3) + 12(x - 3) = 3(x^2 - 9)$

84. $8(x + 2) + 8(x - 2) = 3(x^2 - 4)$

Applying the Concepts

85. Rectangle A rectangle is twice as long as it is wide. The perimeter is 60 feet. Find the dimensions.

86. Rectangle The length of a rectangle is 5 times the width. The perimeter is 48 inches. Find the dimensions.

87. Livestock Pen A livestock pen is built in the shape of a rectangle that is twice as long as it is wide. The perimeter is 48 feet. If the material used to build the pen is $1.75 per foot for the longer sides and $2.25 per foot for the shorter sides (the shorter sides have gates, which increase the cost per foot), find the cost to build the pen.

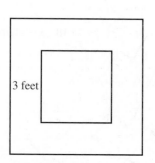

$p = 48$ ft.

88. Garden A garden is in the shape of a square with a perimeter of 42 feet. The garden is surrounded by two fences. One fence is around the perimeter of the garden, whereas the second fence is 3 feet from the first fence, as Figure 2 indicates. If the material used to build the two fences is $1.28 per foot, what was the total cost of the fences?

3 feet

89. Sales Tax A woman owns a small, cash-only business in a state that requires her to charge 6% sales tax on each item she sells. At the beginning of the day, she has $250 in the cash register. At the end of the day, she has $1,204 in the register. How much money should she send to the state government for the sales tax she collected?

90. Sales Tax A store is located in a state that requires 6% tax on all items sold. If the store brings in a total of $3,392 in one day, how much of that total was sales tax?

Maximum Heart Rate In exercise physiology, a person's maximum heart rate, in beats per minute, is found by subtracting his age in years from 220. So, if A represents your age in years, then your maximum heart rate is

$$M = 220 - A$$

Use this formula to complete the following tables.

91.

Age (years)	Maximum Heart Rate (beats per minute)
18	
19	
20	
21	
22	
23	

92.

Age (years)	Maximum Heart Rate (beats per minute)
15	
20	
25	
30	
35	
40	

Training Heart Rate A person's training heart rate, in beats per minute, is the person's resting heart rate plus 60% of the difference between maximum heart rate and their resting heart rate. If resting heart rate is R and maximum heart rate is M, then the formula that gives training heart rate is

$$T = R + 0.6(M - R)$$

Use this formula along with the results of Problems 91 and 92 to fill in the following two tables.

93. For a 20-year-old person

Resting Heart Rate (beats per minute)	Training Heart Rate (beats per minute)
60	
62	
64	
68	
70	
72	

94. For a 40-year-old person

Resting Heart Rate (beats per minute)	Training Heart Rate (beats per minute)
60	
62	
64	
68	
70	
72	

95. Consecutive Integers The sum of the squares of two consecutive odd integers is 34. Find the two integers.

96. Consecutive Integers The sum of the squares of two consecutive even integers is 100. Find the two integers.

Maintaining Your Skills

Simplify.

97. $|-3|$

98. $|3|$

99. $-|-3|$

100. $-(-3)$

101. Give a definition for the absolute value of x that involves the number line. (This is the geometric definition.)

102. Give a definition of the absolute value of x that does not involve the number line. (This is the algebraic definition.)

Getting Ready for the Next Section

Solve each equation.

103. $2a - 1 = -7$

104. $3x - 6 = 9$

105. $\dfrac{2}{3}x - 3 = 7$

106. $\dfrac{2}{3}x - 3 = -7$

107. $x - 5 = x - 7$

108. $x + 3 = x + 8$

109. $x - 5 = -x - 7$

110. $x + 3 = -x + 8$

7.2 Equations with Absolute Value

OBJECTIVES

A Solve equations with absolute value symbols.

Previously we defined the absolute value of x, $|x|$, to be the distance between x and 0 on the number line. The absolute value of a number measures its distance from 0.

EXAMPLE 1 Solve for x: $|x| = 5$.

SOLUTION Using the definition of absolute value, we can read the equation as, "The distance between x and 0 on the number line is 5." If x is 5 units from 0, then x can be 5 or -5:

If $|x| = 5$ then $x = 5$ or $x = -5$

In general, then, we can see that any equation of the form $|a| = b$ is equivalent to the equations $a = b$ or $a = -b$, as long as $b > 0$.

EXAMPLE 2 Solve $|2a - 1| = 7$.

SOLUTION We can read this question as "$2a - 1$ is 7 units from 0 on the number line." The quantity $2a - 1$ must be equal to 7 or -7:

$$|2a - 1| = 7$$
$$2a - 1 = 7 \quad \text{or} \quad 2a - 1 = -7$$

We have transformed our absolute value equation into two equations that do not involve absolute value. We can solve each equation separately.

$2a - 1 = 7$	or	$2a - 1 = -7$
$2a = 8$	or	$2a = -6$	**Add 1 to both sides**
$a = 4$	or	$a = -3$	**Multiply by $\frac{1}{2}$**

Our solution set is $\{4, -3\}$.

To check our solutions, we put them into the original absolute value equation:

When	$a = 4$	When	$a = -3$
the equation	$\lvert 2a - 1 \rvert = 7$	the equation	$\lvert 2a - 1 \rvert = 7$
becomes	$\lvert 2(4) - 1 \rvert = 7$	becomes	$\lvert 2(-3) - 1 \rvert = 7$
	$\lvert 7 \rvert = 7$		$\lvert -7 \rvert = 7$
	$7 = 7$		$7 = 7$

 EXAMPLE 3   Solve $\left\lvert \frac{2}{3}x - 3 \right\rvert + 5 = 12$.

SOLUTION   To use the definition of absolute value to solve this equation, we must isolate the absolute value on the left side of the equal sign. To do so, we add -5 to both sides of the equation to obtain

$$\left\lvert \frac{2}{3}x - 3 \right\rvert = 7$$

Now that the equation is in the correct form, we can write

$$\frac{2}{3}x - 3 = 7 \quad \text{or} \quad \frac{2}{3}x - 3 = -7$$
$$\frac{2}{3}x = 10 \quad \text{or} \quad \frac{2}{3}x = -4 \qquad \textbf{Add 3 to both sides}$$
$$x = 15 \quad \text{or} \quad x = -6 \qquad \textbf{Multiply by } \frac{3}{2}$$

The solution set is $\{15, -6\}$.

 EXAMPLE 4   Solve $\lvert 3a - 6 \rvert = -4$.

SOLUTION   The solution set is $\varnothing$ because the left side cannot be negative and the right side is negative. No matter what we try to substitute for the variable a, the quantity $\lvert 3a - 6 \rvert$ will always be positive or zero. It can never be -4.

Consider the statement $\lvert a \rvert = \lvert b \rvert$. What can we say about a and b? We know they are equal in absolute value. By the definition of absolute value, they are the same distance from 0 on the number line. They must be equal to each other or opposites of each other. In symbols, we write

$$\lvert a \rvert = \lvert b \rvert \quad \Leftrightarrow \quad a = b \quad \text{or} \quad a = -b$$

↑	↑		↑
Equal in absolute value	Equals	or	Opposites

EXAMPLE 5   Solve $\lvert x - 5 \rvert = \lvert x - 7 \rvert$.

SOLUTION   The quantities $x - 5$ and $x - 7$ must be equal or they must be opposites because their absolute values are equal:

Equals		*Opposites*
$x - 5 = x - 7$	or	$x - 5 = -(x - 7)$
$-5 = -7$		$x - 5 = -x + 7$
No solution here		$2x - 5 = 7$
		$2x = 12$
		$x = 6$

Because the first equation leads to a false statement, it will not give us a solution. (If either of the two equations were to reduce to a true statement, it would mean all real numbers would satisfy the original equation.) In this case, our only solution is $x = 6$.

GETTING READY FOR CLASS

After reading through the preceding section, respond in your own words and in complete sentences.

1. Why do some of the equations in this section have two solutions instead of one?
2. Translate $|x| = 6$ into words using the definition of absolute value.
3. Explain in words what the equation $|x - 3| = 4$ means with respect to distance on the number line.
4. When is the statement $|x| = x$ true?

LINKING OBJECTIVES AND EXAMPLES

Next to each **objective** we have listed the examples that are best described by that objective.

A   1–5

Problem Set 7.2

Online support materials can be found at academic.cengage.com/login

Use the definition of absolute value to solve each of the following equations.

1. $|x| = 4$

2. $|x| = 7$

3. $2 = |a|$

4. $5 = |a|$

5. $|x| = -3$

6. $|x| = -4$

7. $|a| + 2 = 3$

8. $|a| - 5 = 2$

9. $|y| + 4 = 3$

10. $|y| + 3 = 1$

11. $4 = |x| - 2$

12. $3 = |x| - 5$

13. $|x - 2| = 5$

14. $|x + 1| = 2$

15. $|a - 4| = \dfrac{5}{3}$

16. $|a + 2| = \dfrac{7}{5}$

17. $1 = |3 - x|$

18. $2 = |4 - x|$

19. $\left|\dfrac{3}{5}a + \dfrac{1}{2}\right| = 1$

20. $\left|\dfrac{2}{7}a + \dfrac{3}{4}\right| = 1$

21. $60 = |20x - 40|$

22. $800 = |400x - 200|$

23. $|2x + 1| = -3$

24. $|2x - 5| = -7$

25. $\left|\dfrac{3}{4}x - 6\right| = 9$

26. $\left|\dfrac{4}{5}x - 5\right| = 15$

27. $\left|1 - \dfrac{1}{2}a\right| = 3$

28. $\left|2 - \dfrac{1}{3}a\right| = 10$

Solve each equation.

29. $|3x + 4| + 1 = 7$

30. $|5x - 3| - 4 = 3$

31. $|3 - 2y| + 4 = 3$

32. $|8 - 7y| + 9 = 1$

33. $3 + |4t - 1| = 8$

34. $2 + |2t - 6| = 10$

35. $\left|9 - \dfrac{3}{5}x\right| + 6 = 12$

36. $\left|4 - \dfrac{2}{7}x\right| + 2 = 14$

37. $5 = \left|\dfrac{2x}{7} + \dfrac{4}{7}\right| - 3$

38. $7 = \left|\dfrac{3x}{5} + \dfrac{1}{5}\right| + 2$

39. $2 = -8 + \left|4 - \dfrac{1}{2}y\right|$

40. $1 = -3 + \left|2 - \dfrac{1}{4}y\right|$

41. $|3a + 1| = |2a - 4|$

42. $|5a + 2| = |4a + 7|$

43. $\left|x - \dfrac{1}{3}\right| = \left|\dfrac{1}{2}x + \dfrac{1}{6}\right|$

44. $\left|\dfrac{1}{10}x - \dfrac{1}{2}\right| = \left|\dfrac{1}{5}x + \dfrac{1}{10}\right|$

▪ = Videos available by instructor request

▶ = Online student support materials available at academic.cengage.com/login

45. $|y - 2| = |y + 3|$ **46.** $|y - 5| = |y - 4|$

47. $|3x - 1| = |3x + 1|$ **48.** $|5x - 8| = |5x + 8|$

49. $|3 - m| = |m + 4|$ **50.** $|5 - m| = |m + 8|$

51. $|0.03 - 0.01x| = |0.04 + 0.05x|$

52. $|0.07 - 0.01x| = |0.08 - 0.02x|$

53. $|x - 2| = |2 - x|$

54. $|x - 4| = |4 - x|$

55. $\left|\dfrac{x}{5} - 1\right| = \left|1 - \dfrac{x}{5}\right|$

56. $\left|\dfrac{x}{3} - 1\right| = \left|1 - \dfrac{x}{3}\right|$

57. Work each problem according to the instructions given.
   a. Solve: $4x - 5 = 0$
   b. Solve: $|4x - 5| = 0$
   c. Solve: $4x - 5 = 3$
   d. Solve: $|4x - 5| = 3$
   e. Solve: $|4x - 5| = |2x + 3|$

58. Work each problem according to the instructions given.
   a. Solve: $3x + 6 = 0$
   b. Solve: $|3x + 6| = 0$
   c. Solve: $3x + 6 = 4$
   d. Solve: $|3x + 6| = 4$
   e. Solve: $|3x + 6| = |7x + 4|$

Applying the Concepts

59. Amtrak Amtrak's annual passenger revenue for the years 1985–1995 is modeled approximately by the formula

$$R = -60|x - 11| + 962$$

where R is the annual revenue in millions of dollars and x is the number of years since January 1, 1980 (Association of American Railroads, Washington, DC, *Railroad Facts, Statistics of Railroads of Class 1,* annual). In what years was the passenger revenue $722 million?

60. Corporate Profits The corporate profits for various U.S. industries vary from year to year. An approximate model for profits of U.S. "communications

companies" during a given year between 1990 and 1997 is given by

$$P = -3,400|x - 5.5| + 36,000$$

where P is the annual profits (in millions of dollars) and x is the number of years since January 1, 1990 (U.S. Bureau of Economic Analysis, Income and Product Accounts of the U.S. (1929–1994), *Survey of Current Business,* September 1998). Use the model to determine the years in which profits of "communication companies" were $31.5 billion ($31,500 million).

Maintaining Your Skills

Simplify each expression.

61. $\dfrac{38}{30}$ **62.** $\dfrac{10}{25}$

63. $\dfrac{240}{6}$ **64.** $\dfrac{39}{13}$

65. $\dfrac{0 + 6}{0 - 3}$ **66.** $\dfrac{6 + 6}{6 - 3}$

67. $\dfrac{4 - 4}{4 - 2}$ **68.** $\dfrac{3 + 6}{3 - 3}$

Getting Ready for the Next Section

Solve each equation.

69. $-2x - 3 = 7$ **70.** $3x + 3 = 2x - 1$

71. $3(2x - 4) - 7x = -3x$ **72.** $3(2x + 5) = -3x$

Extending the Concepts

Solve each formula for x. (Assume a, b, and c are positive.)

73. $|x - a| = b$

74. $|x + a| - b = 0$

75. $|ax + b| = c$

76. $|ax - b| - c = 0$

77. $\left|\dfrac{x}{a} + \dfrac{y}{b}\right| = 1$

78. $\left|\dfrac{x}{a} + \dfrac{y}{b}\right| = c$

OBJECTIVES

A Solve a linear inequality in one variable and graph the solution set.

B Solve a compound inequality and graph the solution set.

C Write solutions to inequalities using interval notation.

D Solve application problems using inequalities

The instrument panel on most cars includes a temperature gauge. The one shown below indicates that the normal operating temperature for the engine is from 50°F to 270°F.

We can represent the same situation with an inequality by writing $50 \leq F \leq 270$, where F is the temperature in degrees Fahrenheit. This inequality is a compound inequality. In this section we present the notation and definitions associated with compound inequalities.

Let's begin with an example reviewing how we solved inequalities in Chapter 2.

 EXAMPLE 1   Solve $3x + 3 < 2x - 1$, and graph the solution.

SOLUTION   We use the addition property for inequalities to write all the variable terms on one side and all constant terms on the other side:

$$3x + 3 < 2x - 1$$
$$3x + (-2x) + 3 < 2x + (-2x) - 1 \qquad \textbf{Add } -2x \textbf{ to each side}$$
$$x + 3 < -1$$
$$x + 3 + (-3) < -1 + (-3) \qquad \textbf{Add } -3 \textbf{ to each side}$$
$$x < -4$$

The solution set is all real numbers that are less than −4. To show this we can use set notation and write

$$\{x \mid x < -4\}$$

We can graph the solution set on the number line using an open circle at −4 to show that −4 is not part of the solution set. This is the format you used when graphing inequalities in Chapter 2.

Here is an equivalent graph that uses a parenthesis opening left, instead of an open circle, to represent the end point of the graph.

Substituting $1{,}300 - 100p$ for x gives us an inequality in the variable p.

$$1{,}300 - 100p \geq 300$$

$$-100p \geq -1{,}000 \qquad \textbf{Add } -\textbf{1,300 to each side}$$

$$p \leq 10 \qquad \textbf{Divide each side by } -\textbf{100, and reverse}$$
$$\textbf{the direction of the inequality symbol}$$

To sell at least 300 cartridges each week, the price per cartridge should be no more than \$10; that is, selling the cartridges for \$10 or less will produce weekly sales of 300 or more cartridges.

EXAMPLE 8   The formula $F = \frac{9}{5}C + 32$ gives the relationship between the Celsius and Fahrenheit temperature scales. If the temperature range on a certain day is 86° to 104° Fahrenheit, what is the temperature range in degrees Celsius?

SOLUTION   From the given information we can write $86 \leq F \leq 104$. But, because F is equal to $\frac{9}{5}C + 32$, we can also write

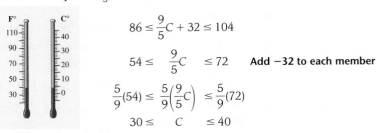

$$86 \leq \frac{9}{5}C + 32 \leq 104$$

$$54 \leq \frac{9}{5}C \leq 72 \qquad \textbf{Add } -\textbf{32 to each member}$$

$$\frac{5}{9}(54) \leq \frac{5}{9}\left(\frac{9}{5}C\right) \leq \frac{5}{9}(72)$$

$$30 \leq C \leq 40$$

A temperature range of 86° to 104° Fahrenheit corresponds to a temperature range of 30° to 40° Celsius.

LINKING OBJECTIVES AND EXAMPLES

Next to each **objective** we have listed the examples that are best described by that objective.

A	1, 2
B	5, 6
C	1–6
D	7, 8

GETTING READY FOR CLASS

After reading through the preceding section, respond in your own words and in complete sentences.

1. What is the addition property for inequalities?
2. When we use interval notation to denote a section of the real number line, when do we use parentheses ( ) and when do we use brackets [ ]?
3. Explain the difference between the multiplication property of equality and the multiplication property for inequalities.
4. When solving an inequality, when do we change the direction of the inequality symbol?

Solve each of the following inequalities and graph each solution.

1. $2x \le 3$

2. $5x \ge -115$

3. $\frac{1}{2}x > 2$

4. $\frac{1}{3}x > 4$

5. $-5x \le 25$

6. $-7x \ge 35$

7. $-\frac{3}{2}x > -6$

8. $-\frac{2}{3}x < -8$

9. $-12 \le 2x$

10. $-20 \ge 4x$

11. $-1 \ge -\frac{1}{4}x$

12. $-1 \le -\frac{1}{5}x$

13. $-3x + 1 > 10$

14. $-2x - 5 \le 15$

15. $\frac{1}{2} - \frac{m}{12} \le \frac{7}{12}$

16. $\frac{1}{2} - \frac{m}{10} > -\frac{1}{5}$

▶ 17. $\frac{1}{2} \ge -\frac{1}{6} - \frac{2}{9}x$

18. $\frac{9}{5} > -\frac{1}{5} - \frac{1}{2}x$

19. $-40 \le 30 - 20y$

20. $-20 > 50 - 30y$

21. $\frac{2}{3}x - 3 < 1$

22. $\frac{3}{4}x - 2 > 7$

23. $10 - \frac{1}{2}y \le 36$

24. $8 - \frac{1}{3}y \ge 20$

Simplify each side first, then solve the following inequalities. Write your answers with interval notation.

25. $2(3y + 1) \le -10$

26. $3(2y - 4) > 0$

27. $-(a + 1) - 4a \le 2a - 8$

28. $-(a - 2) - 5a \le 3a + 7$

29. $\frac{1}{3}t - \frac{1}{2}(5 - t) < 0$

30. $\frac{1}{4}t - \frac{1}{3}(2t - 5) < 0$

31. $-2 \le 5 - 7(2a + 3)$

32. $1 < 3 - 4(3a - 1)$

33. $-\frac{1}{3}(x + 5) \le -\frac{2}{9}(x - 1)$

34. $-\frac{1}{2}(2x + 1) \le -\frac{3}{8}(x + 2)$

Solve each inequality. Write your answer using inequality notation.

35. $20x + 9,300 > 18,000$

36. $20x + 4,800 > 18,000$

Solve the following continued inequalities. Use both a line graph and interval notation to write each solution set.

37. $-2 \le m - 5 \le 2$

38. $-3 \le m + 1 \le 3$

39. $-60 < 20a + 20 < 60$

40. $-60 < 50a - 40 < 60$

41. $0.5 \le 0.3a - 0.7 \le 1.1$

42. $0.1 \le 0.4a + 0.1 \le 0.3$

43. $3 < \frac{1}{2}x + 5 < 6$

44. $5 < \frac{1}{4}x + 1 < 9$

45. $4 < 6 + \frac{2}{3}x < 8$

46. $3 < 7 + \frac{4}{5}x < 15$

Graph the solution sets for the following compound inequalities. Then write each solution set using interval notation.

47. $x + 5 \le -2$ or $x + 5 \ge 2$

48. $3x + 2 < -3$ or $3x + 2 > 3$

49. $5y + 1 \le -4$ or $5y + 1 \ge 4$

50. $7y - 5 \le -2$ or $7y - 5 \ge 2$

51. $2x + 5 < 3x - 1$ or $x - 4 > 2x + 6$

52. $3x - 1 > 2x + 4$ or $5x - 2 < 3x + 4$

Translate each of the following phrases into an equivalent inequality statement.

▶ 53. x is greater than -2 and at most 4.

▶ 54. x is less than 9 and at least -3.

▶ 55. x is less than -4 or at least 1.

▶ 56. x is at most 1 or more than 6.

57. Write each statement using inequality notation.
   a. x is always positive.
   b. x is never negative.
   c. x is greater than or equal to 0.

58. Match each expression on the left with a phrase on the right.
   a. $x^2 \ge 0$ **e.** never true
   b. $x^2 < 0$ **f.** sometimes true
   c. $x^2 \le 0$ **g.** always true

= Videos available by instructor request

▶ = Online student support materials available at academic.cengage.com/login

Solve each inequality by inspection, without showing any work.

59. $x^2 < 0$ $\qquad$ **60.** $x^2 \leq 0$

61. $x^2 \geq 0$

62. $\dfrac{1}{x^2} \geq 0$

63. $\dfrac{1}{x^2} < 0$ $\qquad$ **64.** $\dfrac{1}{x^2} = 0$

The next two problems are intended to give you practice reading, and paying attention to, the instructions that accompany the problems you are working.

65. Work each problem according to the instructions given.
   a. Evaluate when $x = 0$: $-\dfrac{1}{2}x + 1$

   b. Solve: $-\dfrac{1}{2}x + 1 = -7$

   c. Is 0 a solution to $-\dfrac{1}{2}x + 1 < -7$?

   d. Solve: $-\dfrac{1}{2}x + 1 < -7$

66. Work each problem according to the instructions given.
   a. Evaluate when $x = 0$: $-\dfrac{2}{3}x - 5$

   b. Solve: $-\dfrac{2}{3}x - 5 = 1$

   c. Is 0 a solution to $-\dfrac{2}{3}x - 5 > 1$?

   d. Solve: $-\dfrac{2}{3}x - 5 > 1$

Applying the Concepts

A store selling art supplies finds that they can sell x sketch pads each week at a price of p dollars each, according to the formula $x = 900 - 300p$. What price should they charge if they want to sell

67. at least 300 pads each week?

68. more than 600 pads each week?

69. less than 525 pads each week?

70. at most 375 pads each week?

71. **Amtrak** The average number of passengers carried by Amtrak declined each year for the years 1990–1996 (American Association of Railroads, Washington, DC, *Railroad Facts, Statistics of Railroads of Class 1.*) The linear model for the number of passengers carried each year by Amtrak is given by $P = 22,419 - 399x$, where P is the number of passengers, in millions, and x is the number of years after January 1, 1990. In what years did Amtrak have more than 20,500 million passengers?

Courtesy of Amtrak

72. **Student Loan** When considering how much debt to incur in student loans, you learn that it is wise to keep your student loan payment to 8% or less of your starting monthly income. Suppose you anticipate a starting annual salary of $24,000. Set up and solve an inequality that represents the amount of monthly debt for student loans that would be considered manageable.

73. Here is what the U.S. Geological Survey has to say about the survival rates of the Apapane, one of the endemic birds of Hawaii.

> *Annual survival rates based on 1,584 recaptures of 429 banded individuals 0.72 ± 0.11 for adults and 0.13 ± 0.07 for juveniles.*

Write the survival rates using inequalities. Then give the survival rates in terms of percent.

Jack Jeffrey Photography

74. Survival Rates for Sea Gulls Here is part of a report concerning the survival rates of Western Gulls that appeared on the website of Cornell University.

Survival of eggs to hatching is 70%–80%; of hatched chicks to fledglings 50%–70%; of fledglings to age of first breeding <50%.

Write the survival rates using inequalities without percent.

75. Temperature Each of the following temperature ranges is in degrees Fahrenheit. Use the formula $F = \frac{9}{5}C + 32$ to find the corresponding temperature range in degrees Celsius.

a. 95° to 113°
b. 68° to 86°
c. −13° to 14°
d. −4° to 23°

76. Improving Your Quantitative Literacy As you can see, the percents shown in the graph are from a survey of 770 families.

USA TODAY Snapshots®

Luxury isn't just about expensive things

What is luxury?

Happiness and satisfaction from being with family or friends and enjoying good times — **44%**

Having enough time to do whatever you want and being able to afford it — **37%**

19%

Finer things in life that provide supreme comfort, beauty and quality

Source: American Express Platinum
Survey of 770 respondents with household incomes of $125,000 and over.
Margin of error ±4 percentage points

By Darryl Haralson and Dave Merrill, USA TODAY

Data from *USA Today*. Copyright 2005. Reprinted with permission.

a. How many families responded that luxury is happiness and satisfaction from being with friends and family and enjoying good times?
b. If x represents the annual income of any of the families surveyed, write an inequality using x that indicates their level of income.
c. The margin of error for the survey is ±4%. What does this mean in relation to the 19% category?

Maintaining Your Skills

Reduce each rational expression to lowest terms.

77. $\dfrac{x^2 - 16}{x + 4}$

78. $\dfrac{x - 5}{x^2 - 25}$

79. $\dfrac{10a + 20}{5a^2 - 20}$

80. $\dfrac{8a - 16}{4a^2 - 16}$

81. $\dfrac{2x^2 - 5x - 3}{x^2 - 3x}$

82. $\dfrac{x^2 - 5x}{3x^2 - 13x - 10}$

83. $\dfrac{xy + 3x + 2y + 6}{xy + 3x + ay + 3a}$

84. $\dfrac{xy + 5x + 4y + 20}{x^2 + bx + 4x + 4b}$

Getting Ready for the Next Section

To understand all of the explanations and examples in the next section, you must be able to work the problems below.

Solve each inequality. Do not graph the solution set.

85. $2x - 5 < 3$

86. $-3 < 2x - 5$

87. $-4 \le 3a + 7$

88. $3a + 7 \le 4$

89. $4t - 3 \le -9$

90. $4t - 3 \ge 9$

Checking Solutions We can check our solution by substituting it into both of our equations.

Substituting $x = 1$ and $y = 2$ into $4x + 3y = 10$, we have

$$4(1) + 3(2) \stackrel{?}{=} 10$$

$$4 + 6 \stackrel{?}{=} 10$$

$$10 = 10 \qquad \textbf{A true statement}$$

Substituting $x = 1$ and $y = 2$ into $2x + y = 4$, we have

$$2(1) + 2 \stackrel{?}{=} 4$$

$$2 + 2 \stackrel{?}{=} 4$$

$$4 = 4 \qquad \textbf{A true statement}$$

Our solution satisfies both equations; therefore, it is a solution to our system of equations.

 EXAMPLE 2   Solve the system.

$$3x - 5y = -2$$

$$2x - 3y = 1$$

SOLUTION We can eliminate either variable. Let's decide to eliminate the variable x. We can do so by multiplying the top equation by 2 and the bottom equation by -3, and then adding the left and right sides of the resulting equations:

$$3x - 5y = -2 \xrightarrow{\text{Multiply by 2}} 6x - 10y = -4$$

$$2x - 3y = 1 \xrightarrow[\text{Multiply by } -3]{} \quad \underline{-6x + 9y = -3}$$

$$-y = -7$$

$$y = 7$$

The y-coordinate of the solution to the system is 7. Substituting this value of y into any of the equations with both x- and y-variables gives $x = 11$. The solution to the system is $(11, 7)$. It is the only ordered pair that satisfies both equations.

Checking Solutions Checking $(11, 7)$ in each equation looks like this

Substituting $x = 11$ and $y = 7$ into $3x - 5y = -2$, we have

$$3(11) - 5(7) \stackrel{?}{=} -2$$

$$33 - 35 \stackrel{?}{=} -2$$

$$-2 = -2 \qquad \textbf{A true statement}$$

Substituting $x = 11$ and $y = 7$ into $2x - 3y = 1$, we have

$$2(11) - 3(7) \stackrel{?}{=} 1$$

$$22 - 21 \stackrel{?}{=} 1$$

$$1 = 1 \qquad \textbf{A true statement}$$

Our solution satisfies both equations; therefore, $(11, 7)$ is a solution to our system.

EXAMPLE 3   Solve the system.

$$2x - 3y = 4$$

$$4x + 5y = 3$$

SOLUTION We can eliminate x by multiplying the top equation by -2 and adding it to the bottom equation:

$$2x - 3y = 4 \xrightarrow{\text{Multiply by } -2} -4x + 6y = -8$$

$$4x + 5y = 3 \xrightarrow[\text{No change}]{} \quad \underline{4x + 5y = 3}$$

$$11y = -5$$

$$y = -\frac{5}{11}$$

The y-coordinate of our solution is $-\frac{5}{11}$. If we were to substitute this value of y back into either of our original equations, we would find the arithmetic necessary to solve for x cumbersome. For this reason, it is probably best to go back to the original system and solve it a second time—for x instead of y. Here is how we do that:

$$2x - 3y = 4 \xrightarrow{\text{Multiply by 5}} 10x - 15y = 20$$

$$4x + 5y = 3 \xrightarrow[\text{Multiply by 3}]{} \quad \underline{12x + 15y = 9}$$

$$22x = 29$$

$$x = \frac{29}{22}$$

The solution to our system is $\left(\frac{29}{22}, -\frac{5}{11}\right)$.

The main idea in solving a system of linear equations by the elimination method is to use the multiplication property of equality on one or both of the original equations, if necessary, to make the coefficients of either variable opposites. The following box shows some steps to follow when solving a system of linear equations by the elimination method.

Strategy for Solving a System of Linear Equations by the Elimination Method

Step 1: Decide which variable to eliminate. (In some cases, one variable will be easier to eliminate than the other. With some practice, you will notice which one it is.)

Step 2: Use the multiplication property of equality on each equation separately to make the coefficients of the variable that is to be eliminated opposites.

Step 3: Add the respective left and right sides of the system together.

Step 4: Solve for the remaining variable.

Step 5: Substitute the value of the variable from step 4 into an equation containing both variables and solve for the other variable. (Or repeat steps 2–4 to eliminate the other variable.)

Step 6: Check your solution in both equations, if necessary.

EXAMPLE 4 Solve the system.

$$5x - 2y = 5$$

$$-10x + 4y = 15$$

SOLUTION We can eliminate y by multiplying the first equation by 2 and adding the result to the second equation:

$$5x - 2y = 5 \xrightarrow{\text{Multiply by 2}} 10x - 4y = 10$$

$$-10x + 4y = 15 \xrightarrow[\text{No change}]{} \underline{-10x + 4y = 15}$$
$$0 = 25$$

The result is the false statement $0 = 25$, which indicates there is no solution to the system. If we were to graph the two lines, we would find that they are parallel. In a case like this, we say the system is *inconsistent*. Whenever both variables have been eliminated and the resulting statement is false, the solution set for the system will be the empty set, $\varnothing$.

EXAMPLE 5   Solve the system.

$$4x + 3y = 2$$
$$8x + 6y = 4$$

SOLUTION Multiplying the top equation by -2 and adding, we can eliminate the variable x:

$$4x + 3y = 2 \xrightarrow{\text{Multiply by } -2} -8x - 6y = -4$$

$$8x + 6y = 4 \xrightarrow[\text{No change}]{} \underline{8x + 6y = 4}$$
$$0 = 0$$

Both variables have been eliminated and the resulting statement $0 = 0$ is true. In this case, the lines coincide and the equations are said to be *dependent*. The solution set consists of all ordered pairs that satisfy either equation. We can write the solution set as $\{(x, y) \mid 4x + 3y = 2\}$ or $\{(x, y) \mid 8x + 6y = 4\}$.

Special Cases

The previous two examples illustrate the two special cases in which the graphs of the equations in the system either coincide or are parallel. In both cases the left-hand sides of the equations were multiples of each other. In the case of the dependent equations the right-hand sides were also multiples. We can generalize these observations for the system

$$a_1 x + b_1 y = c_1$$
$$a_2 x + b_2 y = c_2$$

Inconsistent System

What Happens	*Geometric Interpretation*	*Algebraic Interpretation*
Both variables are eliminated, and the resulting statement is false.	The lines are parallel, and there is no solution to the system.	$\dfrac{a_1}{a_2} = \dfrac{b_1}{b_2} \neq \dfrac{c_1}{c_2}$

Dependent Equations

What Happens	*Geometric Interpretation*	*Algebraic Interpretation*
Both variables are eliminated, and the resulting statement is true.	The lines coincide, and there are an infinite number of solutions to the system.	$\dfrac{a_1}{a_2} = \dfrac{b_1}{b_2} = \dfrac{c_1}{c_2}$

EXAMPLE 6   Solve the system.

$$\frac{1}{2}x - \frac{1}{3}y = 2$$

$$\frac{1}{4}x + \frac{2}{3}y = 6$$

SOLUTION   Although we could solve this system without clearing the equations of fractions, there is probably less chance for error if we have only integer coefficients to work with. So let's begin by multiplying both sides of the top equation by 6 and both sides of the bottom equation by 12 to clear each equation of fractions:

$$\frac{1}{2}x - \frac{1}{3}y = 2 \xrightarrow{\text{Multiply by 6}} 3x - 2y = 12$$

$$\frac{1}{4}x + \frac{2}{3}y = 6 \xrightarrow[\text{Multiply by 12}]{} 3x + 8y = 72$$

Now we can eliminate x by multiplying the top equation by -1 and leaving the bottom equation unchanged:

$$
\begin{array}{lll}
3x - 2y = 12 & \xrightarrow{\text{Multiply by } -1} & -3x + 2y = -12 \\
3x + 8y = 72 & \xrightarrow[\text{No change}]{} & \underline{3x + 8y = 72} \\
& & 10y = 60 \\
& & y = 6
\end{array}
$$

We can substitute $y = 6$ into any equation that contains both x and y. Let's use $3x - 2y = 12$.

$$3x - 2(6) = 12$$

$$3x - 12 = 12$$

$$3x = 24$$

$$x = 8$$

The solution to the system is $(8, 6)$.

The Substitution Method

EXAMPLE 7   Solve the system.

$$2x - 3y = -6$$

$$y = 3x - 5$$

SOLUTION The second equation tells us y is $3x - 5$. Substituting the expression $3x - 5$ for y in the first equation, we have

$$2x - 3(3x - 5) = -6$$

The result of the substitution is the elimination of the variable y. Solving the resulting linear equation in x as usual, we have

$$2x - 9x + 15 = -6$$
$$-7x + 15 = -6$$
$$-7x = -21$$
$$x = 3$$

Putting $x = 3$ into the second equation in the original system, we have

$$y = 3(3) - 5$$
$$= 9 - 5$$
$$= 4$$

The solution to the system is $(3, 4)$.

Checking Solutions Checking $(3, 4)$ in each equation looks like this

Substituting $x = 3$ and $y = 4$ into
$2x - 3y = -6$, we have

$2(3) - 3(4) \overset{?}{=} -6$

$6 - 12 \overset{?}{=} -6$

$-6 = -6$ **A true statement**

Substituting $x = 3$ and $y = 4$ into
$y = 3x - 5$, we have

$4 \overset{?}{=} 3(3) - 5$

$4 \overset{?}{=} 9 - 5$

$4 = 4$ **A true statement**

Our solution satisfies both equations; therefore, $(3, 4)$ is a solution to our system.

> **Strategy for Solving a System of Equations by the Substitution Method**
>
> **Step 1:** Solve either one of the equations for x or y. (This step is not necessary if one of the equations is already in the correct form, as in Example 7.)
>
> **Step 2:** Substitute the expression for the variable obtained in step 1 into the other equation and solve it.
>
> **Step 3:** Substitute the solution for step 2 into any equation in the system that contains both variables and solve it.
>
> **Step 4:** Check your results, if necessary.

 EXAMPLE 8   Solve by substitution.

$$2x + 3y = 5$$
$$x - 2y = 6$$

SOLUTION To use the substitution method, we must solve one of the two equations for x or y. We can solve for x in the second equation by adding $2y$ to both sides:

$$x - 2y = 6$$
$$x = 2y + 6 \qquad \text{**Add 2y to both sides.**}$$

Note

Both the substitution method and the elimination method can be used to solve any system of linear equations in two variables. Systems like the one in Example 7, however, are easier to solve using the substitution method because one of the variables is already written in terms of the other. A system like the one in Example 2 is easier to solve using the elimination method because solving for one of the variables would lead to an expression involving fractions. The system in Example 8 could be solved easily by either method because solving the second equation for x is a one-step process.

Substituting the expression $2y + 6$ for x in the first equation of our system, we have

$$2(2y + 6) + 3y = 5$$
$$4y + 12 + 3y = 5$$
$$7y + 12 = 5$$
$$7y = -7$$
$$y = -1$$

Using $y = -1$ in either equation in the original system, we get $x = 4$. The solution is $(4, -1)$.

Application

To review, here is our Blueprint for Problem Solving, modified to fit the application problems that depend on a system of equations.

> **BLUEPRINT FOR PROBLEM SOLVING USING A SYSTEM OF EQUATIONS**
>
> **STEP 1: Read** the problem, and then mentally **list** the items that are known and the items that are unknown.
>
> **STEP 2: Assign variables** to each of the unknown items; that is, let $x =$ one of the unknown items and $y =$ the other unknown item (and $z =$ the third unknown item, if there is a third one). Then **translate** the other **information** in the problem to expressions involving the two (or three) variables.
>
> **STEP 3: Reread** the problem, and then **write a system of equations,** using the items and variables listed in steps 1 and 2, that describes the situation.
>
> **STEP 4: Solve the system** found in step 3.
>
> **STEP 5: Write your answers** using complete sentences.
>
> **STEP 6: Reread** the problem, and **check** your solution with the original words in the problem.

EXAMPLE 9   It takes 2 hours for a boat to travel 28 miles downstream (with the current). The same boat can travel 18 miles upstream (against the current) in 3 hours. What is the speed of the boat in still water, and what is the speed of the current of the river?

SOLUTION

> **Step 1: Read and list.**
> A boat travels 18 miles upstream and 28 miles downstream. The trip upstream takes 3 hours. The trip downstream takes 2 hours. We don't know the speed of the boat or the speed of the current.
>
> **Step 2: Assign variables and translate information.**
> Let $x =$ the speed of the boat in still water and let $y =$ the speed of the current. The average speed (rate) of the boat upstream is $x - y$ because it is traveling against the current. The rate of the boat downstream is $x + y$ because the boat is traveling with the current.

Step 3: Write a system of equations.

Putting the information into a table, we have

	d (Distance, miles)	r (Rate, mph)	t (Time, h)
Upstream	18	$x - y$	3
Downstream	28	$x + y$	2

The formula for the relationship between distance d, rate r, and time t is $d = rt$ (the rate equation). Because $d = r \cdot t$, the system we need to solve the problem is

$$18 = (x - y) \cdot 3$$
$$28 = (x + y) \cdot 2$$

which is equivalent to

$$6 = x - y$$
$$14 = x + y$$

Step 4: Solve the system.

Adding the two equations, we have

$$20 = 2x$$
$$x = 10$$

Substituting $x = 10$ into $14 = x + y$, we see that

$$y = 4$$

Step 5: Write answers.

The speed of the boat in still water is 10 miles per hour; the speed of the current is 4 miles per hour.

Step 6: Reread and check.

The boat travels at $10 + 4 = 14$ miles per hour downstream, so in 2 hours it will travel $14 \cdot 2 = 28$ miles. The boat travels at $10 - 4 = 6$ miles per hour upstream, so in 3 hours it will travel $6 \cdot 3 = 18$ miles.

LINKING OBJECTIVES AND EXAMPLES

Next to each **objective** we have listed the examples that are best described by that objective.

A	None
B	1–6
C	7, 8

GETTING READY FOR CLASS

After reading through the preceding section, respond in your own words and in complete sentences.

1. Two linear equations, each with the same two variables, form a system of equations. How do we define a solution to this system? That is, what form will a solution have, and what properties does a solution possess?

2. When would substitution be more efficient than the elimination method in solving two linear equations?

3. Explain what an inconsistent system of linear equations looks like graphically and what would result algebraically when attempting to solve the system.

4. When might the graphing method of solving a system of equations be more desirable than the other techniques, and when might it be less desirable?

Solve each system by graphing both equations on the same set of axes and then reading the solution from the graph.

1. $3x - 2y = 6$
$\quad x - y = 1$

2. $5x - 2y = 10$
$\quad x - y = -1$

3. $\quad y = \dfrac{3}{5}x - 3$
$\quad 2x - y = -4$

4. $\quad y = \dfrac{1}{2}x - 2$
$\quad 2x - y = -1$

5. $y = \dfrac{1}{2}x$
$\quad y = -\dfrac{3}{4}x + 5$

6. $y = \dfrac{2}{3}x$
$\quad y = -\dfrac{1}{3}x + 6$

7. $3x + 3y = -2$
$\quad y = -x + 4$

8. $2x - 2y = 6$
$\quad y = x - 3$

9. $2x - y = 5$
$\quad y = 2x - 5$

10. $x + 2y = 5$
$\quad y = -\dfrac{1}{2}x + 3$

Solve each of the following systems by the elimination method.

11. $\quad x + y = 5$
$\quad 3x - y = 3$

12. $\quad x - y = 4$
$\quad -x + 2y = -3$

13. $3x + y = 4$
$\quad 4x + y = 5$

14. $6x - 2y = -10$
$\quad 6x + 3y = -15$

15. $3x - 2y = 6$
$\quad 6x - 4y = 12$

16. $\quad 4x + 5y = -3$
$\quad -8x - 10y = 3$

17. $\quad x + 2y = 0$
$\quad 2x - 6y = 5$

18. $\quad x + 3y = 3$
$\quad 2x - 9y = 1$

19. $2x - 5y = 16$
$\quad 4x - 3y = 11$

20. $5x - 3y = -11$
$\quad 7x + 6y = -12$

21. $6x + 3y = -1$
$\quad 9x + 5y = 1$

22. $5x + 4y = -1$
$\quad 7x + 6y = -2$

23. $4x + 3y = 14$
$\quad 9x - 2y = 14$

24. $7x - 6y = 13$
$\quad 6x - 5y = 11$

25. $\quad 2x - 5y = 3$
$\quad -4x + 10y = 3$

26. $\quad 3x - 2y = 1$
$\quad -6x + 4y = -2$

27. $\quad \dfrac{1}{4}x - \dfrac{1}{6}y = -2$
$\quad -\dfrac{1}{6}x + \dfrac{1}{5}y = 4$

28. $\quad -\dfrac{1}{3}x + \dfrac{1}{4}y = 0$
$\quad \dfrac{1}{5}x - \dfrac{1}{10}y = 1$

29. $\dfrac{1}{2}x + \dfrac{1}{3}y = 13$
$\quad \dfrac{2}{5}x + \dfrac{1}{4}y = 10$

30. $\dfrac{1}{2}x + \dfrac{1}{3}y = \dfrac{2}{3}$
$\quad \dfrac{2}{3}x + \dfrac{2}{5}y = \dfrac{14}{15}$

Solve each of the following systems by the substitution method.

31. $7x - y = 24$
$\quad x = 2y + 9$

32. $3x - y = -8$
$\quad y = 6x + 3$

33. $6x - y = 10$
$\quad y = -\dfrac{3}{4}x - 1$

34. $2x - y = 6$
$\quad y = -\dfrac{4}{3}x + 1$

35. $3y + 4z = 23$
$\quad 6y + z = 32$

36. $\quad 2x - y = 650$
$\quad 3.5x - y = 1,400$

37. $y = 3x - 2$
$\quad y = 4x - 4$

38. $y = 5x - 2$
$\quad y = -2x + 5$

39. $\quad 2x - y = 5$
$\quad 4x - 2y = 10$

40. $-10x + 8y = -6$
$\quad y = \dfrac{5}{4}x$

41. $\dfrac{1}{3}x - \dfrac{1}{2}y = 0$
$\quad x = \dfrac{3}{2}y$

42. $\dfrac{2}{5}x - \dfrac{2}{3}y = 0$
$\quad y = \dfrac{3}{5}x$

You may want to read Example 3 again before solving the systems that follow.

43. $4x - 7y = 3$
$\quad 5x + 2y = -3$

44. $3x - 4y = 7$
$\quad 6x - 3y = 5$

45. $9x - 8y = 4$
$\quad 2x + 3y = 6$

46. $\quad 4x - 7y = 10$
$\quad -3x + 2y = -9$

47. $3x - 5y = 2$
$\quad 7x + 2y = 1$

48. $4x - 3y = -1$
$\quad 5x + 8y = 2$

Solve each of the following systems by using either the elimination or substitution method. Choose the method that is most appropriate for the problem.

49. $x - 3y = 7$
$\quad 2x + y = -6$

50. $2x - y = 9$
$\quad x + 2y = -11$

51. $y = \dfrac{1}{2}x + \dfrac{1}{3}$
$\quad y = -\dfrac{1}{3}x + 2$

52. $y = \dfrac{3}{4}x - \dfrac{4}{5}$
$\quad y = \dfrac{1}{2}x - \dfrac{1}{2}$

53. $3x - 4y = 12$
$\quad x = \dfrac{2}{3}y - 4$

54. $-5x + 3y = -15$
$\quad x = \dfrac{4}{5}y - 2$

55. $\quad 4x - 3y = -7$
$\quad -8x + 6y = -11$

56. $3x - 4y = 8$
$\quad y = \dfrac{3}{4}x - 2$

= Videos available by instructor request

▶ = Online student support materials available at academic.cengage.com/login

▶ **57.** $3y + z = 17$
$5y + 20z = 65$

▶ **58.** $x + y = 850$
$1.5x + y = 1{,}100$

▶ **59.** $\dfrac{3}{4}x - \dfrac{1}{3}y = 1$
$y = \dfrac{1}{4}x$

▶ **60.** $-\dfrac{2}{3}x + \dfrac{1}{2}y = -1$
$y = -\dfrac{1}{3}x$

▶ **61.** $\dfrac{1}{4}x - \dfrac{1}{2}y = \dfrac{1}{3}$
$\dfrac{1}{3}x - \dfrac{1}{4}y = -\dfrac{2}{3}$

▶ **62.** $\dfrac{1}{5}x - \dfrac{1}{10}y = -\dfrac{1}{5}$
$\dfrac{2}{3}x - \dfrac{1}{2}y = -\dfrac{1}{6}$

The next two problems are intended to give you practice reading, and paying attention to, the instructions that accompany the problems you are working.

63. Work each problem according to the instructions given.
  a. Simplify: $(3x - 4y) - 3(x - y)$
  b. Find y when x is 0 in $3x - 4y = 8$.
  c. Find the y-intercept: $3x - 4y = 8$
  d. Graph: $3x - 4y = 8$
  e. Find the point where the graphs of $3x - 4y = 8$ and $x - y = 2$ cross.

64. Work each problem according to the instructions given.
  a. Solve: $4x - 5 = 20$
  b. Solve for y: $4x - 5y = 20$
  c. Solve for x: $x - y = 5$
  d. Solve the system: $4x - 5y = 20$
                   $x - y = 5$

Applying the Concepts

65. Number Problem One number is 3 more than twice another. The sum of the numbers is 18. Find the two numbers.

66. Number Problem The sum of two numbers is 32. One of the numbers is 4 less than 5 times the other. Find the two numbers.

67. Ticket Problem A total of 925 tickets were sold for a game for a total of $1,150. If adult tickets sold for $2.00 and children's tickets sold for $1.00, how many of each kind of ticket were sold?

68. Internet Problem If tickets for a show cost $2.00 for adults and $1.50 for children, how many of each kind of ticket were sold if a total of 300 tickets were sold for $525?

69. Mixture Problem A mixture of 16% disinfectant solution is to be made from 20% and 14% disinfectant solutions. How much of each solution should be used if 15 gallons of the 16% solution are needed?

70. Mixture Problem Paul mixes nuts worth $1.55 per pound with oats worth $1.35 per pound to get 25 pounds of trail mix worth $1.45 per pound. How many pounds of nuts and how many pounds of oats did he use?

71. Rate Problem It takes about 2 hours to travel 24 miles downstream and 3 hours to travel 18 miles upstream. What is the speed of the boat in still water? What is the speed of the current of the river?

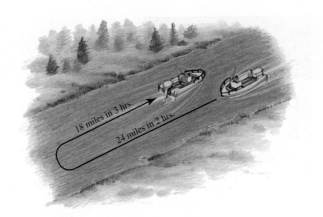

18 miles in 3 hrs.
24 miles in 2 hrs.

72. Rate Problem A boat on a river travels 20 miles downstream in only 2 hours. It takes the same boat 6 hours to travel 12 miles upstream. What are the speed of the boat and the speed of the current?

73. Rate Problem An airplane flying with the wind can cover a certain distance in 2 hours. The return trip against the wind takes $2\frac{1}{2}$ hours. How fast is the plane and what is the speed of the air, if the distance is 600 miles?

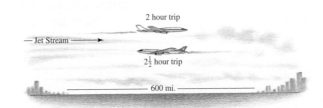

2 hour trip

Jet Stream

$2\frac{1}{2}$ hour trip

600 mi.

74. Rate Problem An airplane covers a distance of 1,500 miles in 3 hours when it flies with the wind and $3\frac{1}{3}$ hours when it flies against the wind. What is the speed of the plane in still air?

Maintaining Your Skills

Add or subtract as indicated.

75. $\dfrac{x^2}{x+5} + \dfrac{10x+25}{x+5}$ **76.** $\dfrac{x^2}{x-3} - \dfrac{9}{x-3}$

77. $\dfrac{a}{3} + \dfrac{2}{5}$ **78.** $\dfrac{4}{a} + \dfrac{2}{3}$

79. $\dfrac{6}{a^2-9} - \dfrac{5}{a^2-a-6}$

80. $\dfrac{4a}{a^2+6a+5} - \dfrac{3a}{a^2+5a+4}$

Getting Ready for the Next Section

Simplify.

81. $(x+3y) - 1(x-2z)$

82. $(x+y+z) + (2x-y+z)$

Solve.

83. $-9y = -9$ **84.** $30x = 38$

85. $3(1) + 2z = 9$ **86.** $4\left(\dfrac{19}{15}\right) - 2y = 4$

Apply the distributive property, then simplify if possible.

87. $2(5x - z)$ **88.** $-1(x - 2z)$

89. $3(3x + y - 2z)$

90. $2(2x - y + z)$

7.6   Systems of Linear Equations in Three Variables

OBJECTIVES

A   Solve systems of linear equations in three variables.

A solution to an equation in three variables such as

$$2x + y - 3z = 6$$

Is an ordered triple of numbers (x, y, z). For example, the ordered triples $(0, 0, -2)$, $(2, 2, 0)$, and $(0, 9, 1)$ are solutions to the equation $2x + y - 3z = 6$ since they produce a true statement when their coordinates are substituted for $x, y,$ and z in the equation.

> **DEFINITION** The **solution set** for a system of three linear equations in three variables is the set of ordered triples that satisfy all three equations.

 EXAMPLE 1   Solve the system.

$$\begin{array}{rcl} x + y + z &=& 6 \quad \textbf{(1)} \\ 2x - y + z &=& 3 \quad \textbf{(2)} \\ x + 2y - 3z &=& -4 \quad \textbf{(3)} \end{array}$$

SOLUTION   We want to find the ordered triple (x, y, z) that satisfies all three equations. We have numbered the equations so it will be easier to keep track of where they are and what we are doing.

    There are many ways to proceed. The main idea is to take two different pairs of equations and eliminate the same variable from each pair. We begin by adding equations (1) and (2) to eliminate the y-variable. The resulting equation is numbered (4):

$$
\begin{array}{ll}
x + y + z = 6 & \textbf{(1)} \\
\underline{2x - y + z = 3} & \textbf{(2)} \\
3x \quad\; + 2z = 9 & \textbf{(4)}
\end{array}
$$

Adding twice equation (2) to equation (3) will also eliminate the variable y. The resulting equation is numbered (5):

$$
\begin{array}{ll}
4x - 2y + 2z = \;\; 6 & \textbf{Twice (2)} \\
\underline{\;\;x + 2y - 3z = -4} & \textbf{(3)} \\
5x \qquad\; - z = \;\; 2 & \textbf{(5)}
\end{array}
$$

Equations (4) and (5) form a linear system in two variables. By multiplying equation (5) by 2 and adding the result to equation (4), we succeed in eliminating the variable z from the new pair of equations:

$$
\begin{array}{ll}
3x + 2z = \;\; 9 & \textbf{(4)} \\
\underline{10x - 2z = \;\; 4} & \textbf{Twice (5)} \\
13x \qquad = 13 & \\
\quad\;\; x = \;\; 1 &
\end{array}
$$

Substituting $x = 1$ into equation (4), we have

$$
3(1) + 2z = 9
$$
$$
2z = 6
$$
$$
z = 3
$$

Using $x = 1$ and $z = 3$ in equation (1) gives us

$$
1 + y + 3 = 6
$$
$$
y + 4 = 6
$$
$$
y = 2
$$

The solution is the ordered triple $(1, 2, 3)$.

EXAMPLE 2   Solve the system.

$$
\begin{array}{ll}
2x + \;\; y - z = 3 & \textbf{(1)} \\
3x + 4y + z = 6 & \textbf{(2)} \\
2x - 3y + z = 1 & \textbf{(3)}
\end{array}
$$

SOLUTION   It is easiest to eliminate z from the equations. The equation produced by adding (1) and (2) is

$$
5x + 5y = 9 \qquad \textbf{(4)}
$$

The equation that results from adding (1) and (3) is

$$
4x - 2y = 4 \qquad \textbf{(5)}
$$

Equations (4) and (5) form a linear system in two variables. We can eliminate the variable y from this system as follows:

$$5x + 5y = 9 \xrightarrow{\text{Multiply by 2}} 10x + 10y = 18$$

$$4x - 2y = 4 \xrightarrow[\text{Multiply by 5}]{} \underline{20x - 10y = 20}$$

$$30x \qquad = 38$$

$$x = \frac{38}{30}$$

$$= \frac{19}{15}$$

Substituting $x = \frac{19}{15}$ into equation (5) or equation (4) and solving for y gives

$$y = \frac{8}{15}$$

Using $x = \frac{19}{15}$ and $y = \frac{8}{15}$ in equation (1), (2), or (3) and solving for z results in

$$z = \frac{1}{15}$$

The ordered triple that satisfies all three equations is $(\frac{19}{15}, \frac{8}{15}, \frac{1}{15})$.

 EXAMPLE 3   Solve the system.

$$2x + 3y - z = 5 \quad \textbf{(1)}$$
$$4x + 6y - 2z = 10 \quad \textbf{(2)}$$
$$x - 4y + 3z = 5 \quad \textbf{(3)}$$

SOLUTION   Multiplying equation (1) by -2 and adding the result to equation (2) looks like this:

$$-4x - 6y + 2z = -10 \qquad \textbf{-2 times (1)}$$
$$\underline{4x + 6y - 2z = 10} \qquad \textbf{(2)}$$
$$0 = 0$$

Note
On the last page of this section, you will find a discussion of the geometric interpretations associated with systems of equations in three variables.

All three variables have been eliminated, and we are left with a true statement. As was the case in the previous section, this implies that the two equations are dependent. With a system of three equations in three variables, however, a system such as this one can have no solution or an infinite number of solutions. In either case, we have no unique solution, meaning there is no single ordered triple that is the only solution to the system.

 EXAMPLE 4   Solve the system.

$$x - 5y + 4z = 8 \quad \textbf{(1)}$$
$$3x + y - 2z = 7 \quad \textbf{(2)}$$
$$-9x - 3y + 6z = 5 \quad \textbf{(3)}$$

Method 3: Using Lists

On the TI-83/84 you can set Y_1 as follows

$$Y_1 = X + \{-3, -2, -1, 0, 1, 2, 3\}$$

When you press $\boxed{\text{GRAPH}}$, the calculator will graph each line from $y = x + (-3)$ to $y = x + 3$.

Each of the three methods will produce graphs similar to those in Figure 10.

LINKING OBJECTIVES AND EXAMPLES

Next to each **objective** we have listed the examples that are best described by that objective.

A	1
B	2, 6
C	3–5

GETTING READY FOR CLASS

After reading through the preceding section, respond in your own words and in complete sentences.

1. If you were looking at a graph that described the performance of a stock you had purchased, why would it be better if the slope of the line were positive, rather than negative?
2. Describe the behavior of a line with a negative slope.
3. Would you rather climb a hill with a slope of $\frac{1}{2}$ or a slope of 3? Explain why.
4. Describe how to obtain the slope of a line if you know the coordinates of two points on the line.

Problem Set 8.1

Online support materials can be found at academic.cengage.com/login

Find the slope of each of the following lines from the given graph.

1.

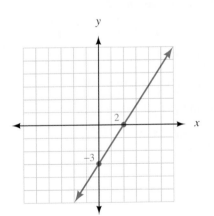

2.

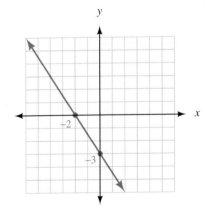

= Videos available by instructor request

▶ = Online student support materials available at academic.cengage.com/login

3.

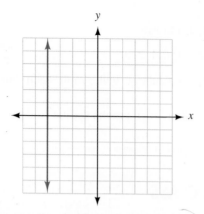

6.

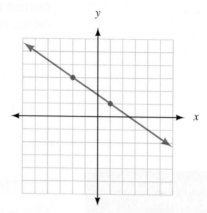

4.

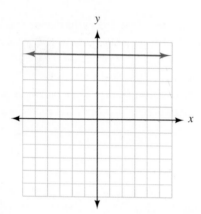

5.

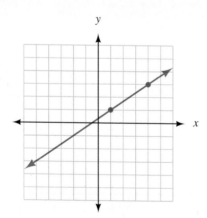

Find the slope of the line through each of the following pairs of points. Then, plot each pair of points, draw a line through them, and indicate the rise and run in the graph in the manner shown in Example 3.

7. $(2, 1)$, $(4, 4)$ ▶ **8.** $(3, 1)$, $(5, 4)$

9. $(1, 4)$, $(5, 2)$ **10.** $(1, 3)$, $(5, 2)$

11. $(1, -3)$, $(4, 2)$ **12.** $(2, -3)$, $(5, 2)$

13. $(-3, -2)$, $(1, 3)$ **14.** $(-3, -1)$, $(1, 4)$

▶ **15.** $(-3, 2)$, $(3, -2)$ **16.** $(-3, 3)$, $(3, -1)$

17. $(2, -5)$, $(3, -2)$ **18.** $(2, -4)$, $(3, -1)$

Solve for the indicated variable if the line through the two given points has the given slope.

▶ **19.** $(5, 2)$ and $(x, -2)$, $m = 2$.

▶ **20.** $(-1, -4)$ and $(x, 5)$, $m = 3$.

▶ **21.** $(a, 3)$ and $(2, 6)$, $m = -1$.

▶ **22.** $(a, -2)$ and $(4, -6)$, $m = -3$.

▶ **23.** $(2, b)$ and $(-1, 4b)$, $m = -2$.

▶ **24.** $(-4, y)$ and $(-1, 6y)$, $m = 2$.

For each equation below, complete the table, and then use the results to find the slope of the graph of the equation.

25. $2x + 3y = 6$

x	y
0	
	0

26. $3x - 2y = 6$

x	y
0	
	0

27 $y = \frac{2}{3}x - 5$

x	y
0	
3	

28. $y = -\frac{3}{4}x + 2$

x	y
0	
4	

▶ **29. Finding Slope from Intercepts** Graph the line with x-intercept 4 and y-intercept 2. What is the slope of this line?

30. Finding Slope from Intercepts Graph the line with x-intercept −4 and y-intercept −2. What is the slope of this line?

31. Parallel Lines Find the slope of any line parallel to the line through (2, 3) and (−8, 1).

32. Parallel Lines Find the slope of any line parallel to the line through (2, 5) and (5, −3).

33. Perpendicular Lines Line l contains the points (5, −6) and (5, 2). Give the slope of any line perpendicular to l.

34. Perpendicular Lines Line l contains the points (3, 4) and (−3, 1). Give the slope of any line perpendicular to l.

▶ **35. Parallel Lines** Line l contains the points (−2, 1) and (4, −5). Find the slope of any line parallel to l.

▶ **36. Parallel Lines** Line l contains the points (3, −4) and (−2, −6). Find the slope of any line parallel to l.

▶ **37. Perpendicular Lines** Line l contains the points (−2, −5) and (1, −3). Find the slope of any line perpendicular to l.

▶ **38. Perpendicular Lines** Line l contains the points (6, −3) and (−2, 7). Find the slope of any line perpendicular to l.

39. Determine if each of the following tables could represent ordered pairs from an equation of a line.

a.

x	y
0	5
1	7
2	9
3	11

b.

x	y
−2	−5
0	−2
2	0
4	1

40. The following lines have slope 2, $\frac{1}{2}$, 0, and −1. Match each line to its slope value.

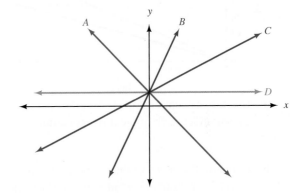

Applying the Concepts

41. An object is traveling at a constant speed. The distance and time data are shown on the given graph. Use the graph to find the speed of the object

a.

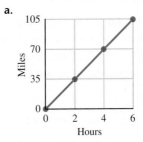

b.

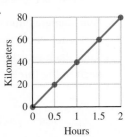

c.

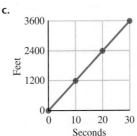

d.

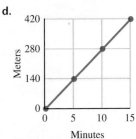

42. Improving Your Quantitative Literacy A cyclist is traveling at a constant speed. The graph shows the distance the cyclist travels over time when there is no wind present, when she travels against the wind, and when she travels with the wind.

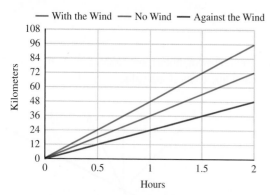

a.   Use the concept of slope to find the speed of the cyclist under each of the three conditions.

b.   Compare the speed of the cyclist when she is traveling without any wind to when she is riding against the wind. How do the two speeds differ?

c.   Compare the speed of the cyclist when she is traveling without any wind to when she is riding with the wind. How do the two speeds differ?

d.   What is the speed of the wind?

▶ **43. Non–Camera Phone Sales** The table and line graph here each show the projected non–camera phone sales each year through 2010. Find the slope of each of the three line segments, *A*, *B*, and *C*.

Year	2006	2007	2008	2009	2010
Sales (in millions)	300	250	175	150	125

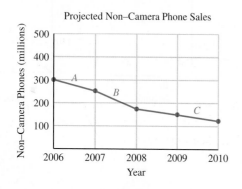

▶ **44. Camera Phone Sales** The table and line graph here each show the projected camera phone sales each year through 2010. Find the slope of each of the three line segments, *A*, *B*, and *C*.

Year	2006	2007	2008	2009	2010
Sales (in millions)	500	650	750	875	900

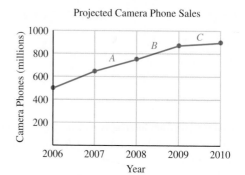

45. Slope of a Highway A sign at the top of the Cuesta Grade, outside of San Luis Obispo, reads "7% downgrade next 3 miles." The diagram shown here is a model of the Cuesta Grade that takes into account the information on that sign.

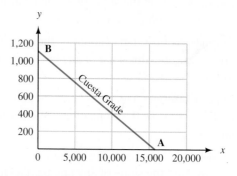

a.   At point B, the graph crosses the *y*-axis at 1,106 feet. How far is it from the origin to point A?

b.   What is the slope of the Cuesta Grade?

46. Heating a Block of Ice A block of ice with an initial temperature of −20°C is heated at a steady rate. The graph shows how the temperature changes as the ice melts to become water and the water boils to become steam and water.

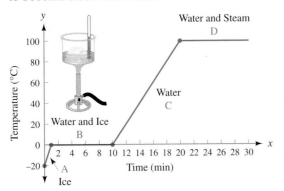

a. How long does it take all the ice to melt?

b. From the time the heat is applied to the block of ice, how long is it before the water boils?

c. Find the slope of the line segment labeled A. What units would you attach to this number?

d. Find the slope of the line segment labeled C. Be sure to attach units to your answer.

e. Is the temperature changing faster during the 1st minute or the 16th minute?

47. Slope and Rate of Change Find the slope of the line connecting the first and last points on the chart. Explain in words what the slope represents.

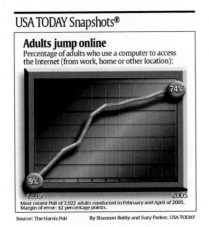

USA TODAY Snapshots®

Adults jump online
Percentage of adults who use a computer to access the Internet (from work, home or other location):

74%

9%

1995   2005

Most recent Poll of 2,022 adults conducted in February and April of 2005. Margin of error: ±2 percentage points.

Source: The Harris Poll   By Shannon Reilly and Suzy Parker, USA TODAY

From *USA Today.* Copyright 2005. Reprinted with permission.

48. Bridal Books Find the slope of the line that connects the points (1999, 118) and (2002, 75). Round to the nearest tenth. Explain in words what the slope represents.

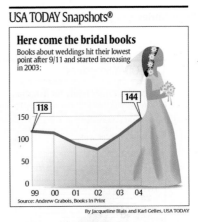

USA TODAY Snapshots®

Here come the bridal books
Books about weddings hit their lowest point after 9/11 and started increasing in 2003:

118   144

150
100
50
0
99 00 01 02 03 04

Source: Andrew Grabois, Books In Print

By Jacqueline Blais and Karl Gelles, USA TODAY

From *USA Today.* Copyright 2005. Reprinted with permission.

49. Health Care Use the chart to work the following problems involving slope.

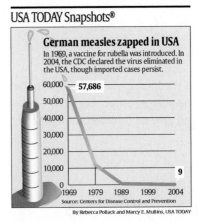

USA TODAY Snapshots®

German measles zapped in USA
In 1969, a vaccine for rubella was introduced. In 2004, the CDC declared the virus eliminated in the USA, though imported cases persist.

60,000 — 57,686
50,000
40,000
30,000
20,000
10,000
0
1969 1979 1989 1999 2004

9

Source: Centers for Disease Control and Prevention

By Rebecca Pollack and Marcy E. Mullins, USA TODAY

From *USA Today.* Copyright 2005. Reprinted with permission.

a. Find the slope of the line from 1969 to 1979, and then give a written explanation of the significance of that number.

b. Find the slope of the line from 1979 to 1989, and then give a written explanation of the significance of that number.

c. What is the slope of the line connecting (1989, 9) and (2004, 9)? What does this tell us about German measles?

50. Triathlon Entries Use the chart to work the following problems.

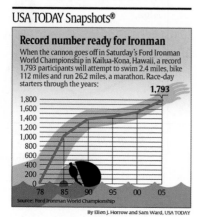

USA TODAY Snapshots®

Record number ready for Ironman
When the cannon goes off in Saturday's Ford Ironman World Championship in Kailua-Kona, Hawaii, a record 1,793 participants will attempt to swim 2.4 miles, bike 112 miles and run 26.2 miles, a marathon. Race-day starters through the years:

Source: Ford Ironman World Championship

By Ellen J. Horrow and Sam Ward, USA TODAY

From *USA Today.* Copyright 2005. Reprinted with permission.

a. Find the slope of the line that connects the points (1978, 0) and (2005, 1,793). Round your answer to the nearest tenth. Explain in words what the slope represents.

b. Of the line segments shown in the chart, one has a larger slope than the rest. Find that slope and give an interpretation of its meaning.

Maintaining Your Skills

The problems that follow review some of the more important skills you have learned in previous sections and chapters.

51. If $3x + 2y = 12$, find y when x is 4.

52. If $y = 3x - 1$, find x when y is 0.

53. Solve the formula $3x + 2y = 12$ for y.

54. Solve the formula $y = 3x - 1$ for x.

55. Solve the formula $A = P + Prt$ for t.

56. Solve the formula $S = \pi r^2 + 2\pi rh$ for h.

Getting Ready for the Next Section

Simplify.

57. $2\left(-\dfrac{1}{2}\right)$

58. $\dfrac{3 - (-1)}{-3 - 3}$

59. $\dfrac{5 - (-3)}{2 - 6}$

60. $3\left(-\dfrac{2}{3}x + 1\right)$

Solve for y.

61. $\dfrac{y - b}{x - 0} = m$

62. $2x + 3y = 6$

63. $y - 3 = -2(x + 4)$

64. $y + 1 = -\dfrac{2}{3}(x - 3)$

65. If $y = -\dfrac{4}{3}x + 5$, find y when x is 0.

66. If $y = -\dfrac{4}{3}x + 5$, find y when x is 3.

Extending the Concepts

67. Find a point P on the line $y = 3x - 2$ such that the slope of the line through (1, 2) and P is equal to -1.

68. Find a point P on the line $y = -2x + 3$ such that the slope of the line through (2, 3) and P is equal to 2.

69. Use your Y-variables list or write a program to graph the family of curves $Y = -2X + B$ for $B = -3, -2, -1, 0, 1, 2,$ and 3.

70. Use your Y-variables list or write a program to graph the family of curves $Y = -2X - B$ for $B = -3, -2, -1, 0, 1, 2,$ and 3.

71. Use your Y-variables list or write a program to graph the family of curves $Y = AX$ for $A = -3, -2, -1, 0, 1, 2,$ and 3.

72. Use your Y-variables list or write a program to graph the family of curves $Y = AX$ for $A = \dfrac{1}{4}, \dfrac{1}{3}, \dfrac{1}{2}, 1, 2,$ and 3.

73. Use your Y-variables list or write a program to graph the family of curves $Y = AX + 2$ for $A = -3, -2, -1, 0, 1, 2,$ and 3.

74. Use your Y-variables list or write a program to graph the family of curves $Y = AX - 2$ for $A = \dfrac{1}{4}, \dfrac{1}{3}, \dfrac{1}{2}, 1, 2,$ and 3.

OBJECTIVES

A Find the equation of a line given its slope and y-intercept.

B Find the slope and y-intercept from the equation of a line.

C Find the equation of a line given the slope and a point on the line.

D Find the equation of a line given two points on the line.

The table and illustrations show some corresponding temperatures on the Fahrenheit and Celsius temperature scales. For example, water freezes at 32°F and 0°C, and boils at 212°F and 100°C.

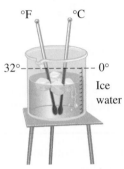

Degrees Celsius	Degrees Fahrenheit
0	32
25	77
50	122
75	167
100	212

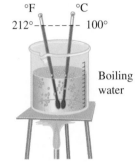

If we plot all the points in the table using the x-axis for temperatures on the Celsius scale and the y-axis for temperatures on the Fahrenheit scale, we see that they line up in a straight line (Figure 1). This means that a linear equation in two variables will give a perfect description of the relationship between the two scales. That equation is

$$F = \frac{9}{5}C + 32$$

The techniques we use to find the equation of a line from a set of points is what this section is all about.

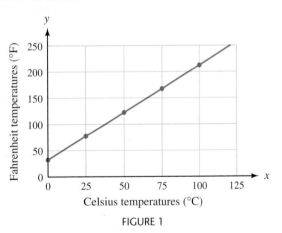

FIGURE 1

Suppose line *l* has slope *m* and y-intercept *b*. What is the equation of *l*? Because the y-intercept is *b*, we know the point (0, *b*) is on the line. If (*x*, *y*) is any other point on *l*, then using the definition for slope, we have

$$\frac{y - b}{x - 0} = m \qquad \textbf{Definition of slope}$$

$$y - b = mx \qquad \textbf{Multiply both sides by } x$$

$$y = mx + b \qquad \textbf{Add } b \textbf{ to both sides}$$

This last equation is known as the *slope-intercept form* of the equation of a straight line.

> **Slope-Intercept Form of the Equation of a Line** The equation of any line with slope *m* and *y*-intercept *b* is given by
>
> $$y = mx + b$$
>
> Slope   *y*-intercept

When the equation is in this form, the *slope* of the line is always the *coefficent of x*, and the *y-intercept* is always the *constant term*.

 EXAMPLE 1   Find the equation of the line with slope $-\frac{4}{3}$ and *y*-intercept 5. Then graph the line.

SOLUTION   Substituting $m = -\frac{4}{3}$ and $b = 5$ into the equation $y = mx + b$, we have

$$y = -\frac{4}{3}x + 5$$

Finding the equation from the slope and *y*-intercept is just that easy. If the slope is *m* and the *y*-intercept is *b*, then the equation is always $y = mx + b$. Now, let's graph the line.

Because the *y*-intercept is 5, the graph goes through the point (0, 5). To find a second point on the graph, we start at (0, 5) and move 4 units down (that's a rise of −4) and 3 units to the right (a run of 3). The point we end up at is (3, 1). Drawing a line that passes through (0, 5) and (3, 1), we have the graph of our equation. (Note that we could also let the rise = 4 and the run = −3 and obtain the same graph.) The graph is shown in Figure 2.

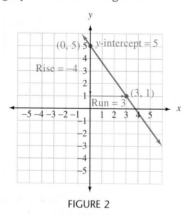

FIGURE 2

 EXAMPLE 2   Give the slope and *y*-intercept for the line $2x - 3y = 5$.

SOLUTION   To use the slope-intercept form, we must solve the equation for *y* in terms of *x*.

$$2x - 3y = 5$$

$$-3y = -2x + 5 \qquad \textbf{Add } -2x \textbf{ to both sides}$$

$$y = \frac{2}{3}x - \frac{5}{3} \qquad \textbf{Divide by } -3$$

The form $y = mx + b$. The slope must be $m = \frac{2}{3}$, and the y-intercept is $b = -\frac{5}{3}$.

EXAMPLE 3   Graph the equation $2x + 3y = 6$ using the slope and y-intercept.

SOLUTION   Although we could graph this equation by finding ordered pairs that are solutions to the equation and drawing a line through their graphs, it is sometimes easier to graph a line using the slope-intercept form of the equation.
Solving the equation for y, we have,

$$2x + 3y = 6$$

$$3y = -2x + 6 \qquad \textbf{Add } -2x \textbf{ to both sides}$$

$$y = -\frac{2}{3}x + 2 \qquad \textbf{Divide by 3}$$

The slope is $m = -\frac{2}{3}$ and the y-intercept is $b = 2$. Therefore, the point $(0, 2)$ is on the graph and the ratio of rise to run going from $(0, 2)$ to any other point on the line is $-\frac{2}{3}$. If we start at $(0, 2)$ and move 2 units up (that's a rise of 2) and 3 units to the left (a run of -3), we will be at another point on the graph. (We could also go down 2 units and right 3 units and still be assured of ending up at another point on the line because $\frac{2}{-3}$ is the same as $\frac{-2}{3}$.)

Note

As we mentioned earlier, the rectangular coordinate system is the tool we use to connect algebra and geometry. Example 3 illustrates this connection, as do the many other examples in this chapter. In Example 3, Descartes's rectangular coordinate system allows us to associate the equation $2x + 3y = 6$ (an algebraic concept) with the straight line (a geometric concept) shown in Figure 3.

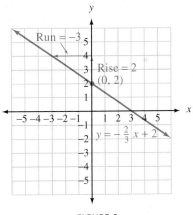

FIGURE 3

A second useful form of the equation of a straight line is the point-slope form.
Let line l contain the point (x_1, y_1) and have slope m. If (x, y) is any other point on l, then by the definition of slope we have

$$\frac{y - y_1}{x - x_1} = m$$

Multiplying both sides by $(x - x_1)$ gives us

$$(\boldsymbol{x - x_1}) \cdot \frac{y - y_1}{x - x_1} = \boldsymbol{m(x - x_1)}$$

$$y - y_1 = m(x - x_1)$$

This last equation is known as the *point-slope form* of the equation of a straight line.

> **Point-Slope Form of the Equation of a Line** The equation of the line through (x_1, y_1) with slope m is given by
> $$y - y_1 = m(x - x_1)$$

This form of the equation of a straight line is used to find the equation of a line, either given one point on the line and the slope, or given two points on the line.

EXAMPLE 4   Find the equation of the line with slope -2 that contains the point $(-4, 3)$. Write the answer in slope-intercept form.

SOLUTION

Using $(x_1, y_1) = (-4, 3)$ and $m = -2$
in $y - y_1 = m(x - x_1)$ **Point-slope form**
gives us $y - 3 = -2(x + 4)$ **Note: $x - (-4) = x + 4$**
                     $y - 3 = -2x - 8$ **Multiply out right side**
                          $y = -2x - 5$ **Add 3 to each side**

Figure 4 is the graph of the line that contains $(-4, 3)$ and has a slope of -2. Notice that the y-intercept on the graph matches that of the equation we found.

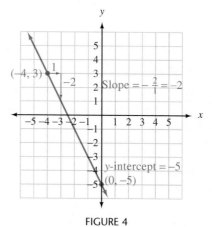

FIGURE 4

EXAMPLE 5   Find the equation of the line that passes through the points $(-3, 3)$ and $(3, -1)$.

SOLUTION   We begin by finding the slope of the line:
$$m = \frac{3 - (-1)}{-3 - 3} = \frac{4}{-6} = -\frac{2}{3}$$

Using $(x_1, y_1) = (3, -1)$ and $m = -\frac{2}{3}$ in $y - y_1 = m(x - x_1)$ yields

$$y + 1 = -\frac{2}{3}(x - 3)$$

$$y + 1 = -\frac{2}{3}x + 2 \qquad \textbf{Multiply out right side}$$

$$y = -\frac{2}{3}x + 1 \qquad \textbf{Add } -1 \textbf{ to each side}$$

Note

In Example 5 we could have used the point $(-3, 3)$ instead of $(3, -1)$ and obtained the same equation; that is, using $(x_1, y_1) = (-3, 3)$ and $m = -\frac{2}{3}$ in $y - y_1 = m(x - x_1)$ gives us

$$y - 3 = -\frac{2}{3}(x + 3)$$

$$y - 3 = -\frac{2}{3}x - 2$$

$$y = -\frac{2}{3}x + 1$$

which is the same result we obtained using $(3, -1)$.

Figure 5 shows the graph of the line that passes through the points $(-3, 3)$ and $(3, -1)$. As you can see, the slope and y-intercept are $-\frac{2}{3}$ and 1, respectively.

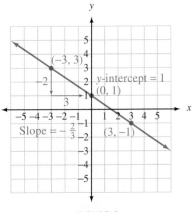

FIGURE 5

The last form of the equation of a line that we will consider in this section is called the standard form. It is used mainly to write equations in a form that is free of fractions and is easy to compare with other equations.

> **Standard Form for the Equation of a Line** If a, b, and c are integers, then the equation of a line is in standard form when it has the form
>
> $$ax + by = c$$

If we were to write the equation

$$y = -\frac{2}{3}x + 1$$

in standard form, we would first multiply both sides by 3 to obtain

$$3y = -2x + 3$$

Then we would add $2x$ to each side, yielding

$$2x + 3y = 3$$

which is a linear equation in standard form.

 EXAMPLE 6   Give the equation of the line through $(-1, 4)$ whose graph is perpendicular to the graph of $2x - y = -3$. Write the answer in standard form.

SOLUTION   To find the slope of $2x - y = -3$, we solve for y:

$$2x - y = -3$$

$$y = 2x + 3$$

The slope of this line is 2. The line we are interested in is perpendicular to the line with slope 2 and must, therefore, have a slope of $-\frac{1}{2}$.

Using $(x_1, y_1) = (-1, 4)$ and $m = -\frac{1}{2}$, we have

$$y - y_1 = m(x - x_1)$$

$$y - 4 = -\frac{1}{2}(x + 1)$$

Because we want our answer in standard form, we multiply each side by 2.

$$2y - 8 = -1(x + 1)$$

$$2y - 8 = -x - 1$$

$$x + 2y - 8 = -1$$

$$x + 2y = 7$$

The last equation is in standard form.

As a final note, the summary reminds us that all horizontal lines have equations of the form $y = b$ and slopes of 0. Because they cross the y-axis at b, the y-intercept is b; there is no x-intercept. Vertical lines have undefined slope and equations of the form $x = a$. Each will have an x-intercept at a and no y-intercept. Finally, equations of the form $y = mx$ have graphs that pass through the origin. The slope is always m and both the x-intercept and the y-intercept are 0.

FACTS FROM GEOMETRY

Special Equations and Their Graphs, Slopes, and Intercepts

For the equations below, m, a, and b are real numbers.

Through the Origin	*Vertical Line*	*Horizontal Line*
Equation: $y = mx$	Equation: $x = a$	Equation: $y = b$
Slope $= m$	Slope is undefined	Slope $= 0$
x-intercept $= 0$	x-intercept $= a$	No x-intercept
y-intercept $= 0$	No y-intercept	y-intercept $= b$

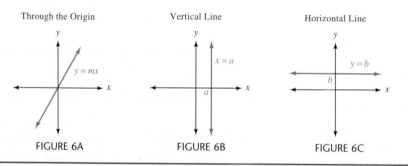

FIGURE 6A   FIGURE 6B   FIGURE 6C

USING TECHNOLOGY

Graphing Calculators

One advantage of using a graphing calculator to graph lines is that a calculator does not care whether the equation has been simplified or not. To illustrate, in Example 5 we found that the equation of the line with slope $-\frac{2}{3}$ that passes through the point $(3, -1)$ is

$$y + 1 = -\frac{2}{3}(x - 3)$$

Normally, to graph this equation we would simplify it first. With a graphing calculator, we add -1 to each side and enter the equation this way:

$$Y_1 = -(2/3)(X - 3) - 1$$

LINKING OBJECTIVES AND EXAMPLES

Next to each objective we have listed the examples that are best described by that objective.

A	1
B	2, 3
C	4, 6
D	5

GETTING READY FOR CLASS

After reading through the preceding section, respond in your own words and in complete sentences.

1. How would you graph the line $y = \frac{1}{2}x + 3$?
2. What is the slope-intercept form of the equation of a line?
3. Describe how you would find the equation of a line if you knew the slope and the y-intercept of the line.
4. If you had the graph of a line, how would you use it to find the equation of the line?

Problem Set 8.2

Online support materials can be found at academic.cengage.com/login

Write the equation of the line with the given slope and y-intercept.

1. $m = 2, b = 3$

2. $m = -4, b = 2$

3. $m = 1, b = -5$

4. $m = -5, b = -3$

5. $m = \frac{1}{2}, b = \frac{3}{2}$ **6.** $m = \frac{2}{3}, b = \frac{5}{6}$

7. $m = 0, b = 4$ **8.** $m = 0, b = -2$

Find the slope of a line (a) parallel and (b) perpendicular to the given line.

▶ **9.** $y = 3x - 4$ ▶ **10.** $y = -4x + 1$

▶ **11.** $3x + y = -2$ ▶ **12.** $2x - y = -4$

▶ **13.** $2x + 5y = -11$ ▶ **14.** $3x - 5y = -4$

Give the slope and y-intercept for each of the following equations. Sketch the graph using the slope and y-intercept. Give the slope of any line perpendicular to the given line.

15. $y = 3x - 2$

16. $y = 2x + 3$

17. $2x - 3y = 12$

18. $3x - 2y = 12$

▶ **19.** $4x + 5y = 20$

20. $5x - 4y = 20$

For each of the following lines, name the slope and y-intercept. Then write the equation of the line in slope-intercept form.

21.
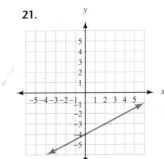

22.
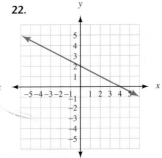

   ▮ = Videos available by instructor request
   ▶ = Online student support materials available at academic.cengage.com/login

23.

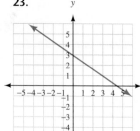

24.

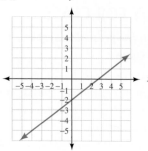

For each of the following problems, the slope and one point on a line are given. In each case, find the equation of that line. (Write the equation for each line in slope-intercept form.)

▶ **25.** $(-2, -5)$; $m = 2$

26. $(-1, -5)$; $m = 2$

27. $(-4, 1)$; $m = -\frac{1}{2}$

28. $(-2, 1)$; $m = -\frac{1}{2}$

29. $\left(-\frac{1}{3}, 2\right)$; $m = -3$

30. $\left(-\frac{2}{3}, 5\right)$; $m = -3$

▶ **31.** $(-4, -2)$; $m = \frac{2}{3}$

▶ **32.** $(3, 4)$; $m = -\frac{1}{3}$

▶ **33.** $(-5, 2)$; $m = -\frac{1}{4}$

▶ **34.** $(-4, 3)$; $m = -\frac{1}{6}$

Find the equation of the line that passes through each pair of points. Write your answers in standard form.

35. $(-2, -4)$, $(1, -1)$

36. $(2, 4)$, $(-3, -1)$

37. $(-1, -5)$, $(2, 1)$

38. $(-1, 6)$, $(1, 2)$

39. $\left(\frac{1}{3}, -\frac{1}{5}\right)$, $\left(-\frac{1}{3}, -1\right)$

40. $\left(-\frac{1}{2}, -\frac{1}{2}\right)$, $\left(\frac{1}{2}, \frac{1}{10}\right)$

41. The equation $3x - 2y = 10$ is a linear equation in standard form. From this equation, answer the following:

  a. Find the x- and y-intercepts.

  b. Find a solution to this equation other than the intercepts in part a.

  c. Write this equation in slope-intercept form.

  d. Is the point $(2, 2)$ a solution to the equation?

42. The equation $4x + 3y = 8$ is a linear equation in standard form. From this equation, answer the following:

  a. Find the x- and y-intercepts.

  b. Find a solution to this equation other than the intercepts in part a.

  c. Write this equation in slope-intercept form.

  d. Is the point $(-3, 2)$ a solution to the equation?

43. The equation $\frac{3x}{4} - \frac{y}{2} = 1$ is a linear equation. From this equation, answer the following:

  a. Find the x- and y-intercepts.

  b. Find a solution to this equation other than the intercepts in part a.

  c. Write this equation in slope-intercept form.

  d. Is the point $(1, 2)$ a solution to the equation?

44. The equation $\frac{3x}{5} + \frac{2y}{3} = 1$ is a linear equation. From this equation, answer the following:

  a. Find the x- and y-intercepts.

  b. Find a solution to this equation other than the intercepts in part a.

  c. Write this equation in slope-intercept form.

  d. Is the point $(-5, 3)$ a solution to the equation?

The next two problems are intended to give you practice reading, and paying attention to, the instructions that accompany the problems you are working. Working these problems is an excellent way to get ready for a test or a quiz.

45. Work each problem according to the instructions given.

  a. Solve: $-2x + 1 = -3$

  b. Write in slope-intercept form:
    $-2x + y = -3$

  c. Find the y-intercept: $-2x + y = -3$

  d. Find the slope: $-2x + y = -3$

  e. Graph: $-2x + y = -3$

46. Work each problem according to the instructions given.

  a. Solve: $\frac{x}{3} + \frac{1}{4} = 1$

  b. Write in slope-intercept form: $\frac{x}{3} + \frac{y}{4} = 1$

c. Rosemary works 30 hours to earn $360.

d. Marcus works 40 hours to earn $320.

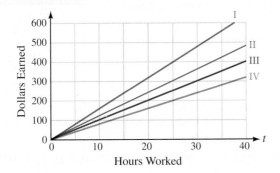

Hours Worked

46. Find an equation for each of the functions shown in Problem 45. Show dollars earned, E, as a function of hours worked, t. Then, indicate the domain and range of each function.

  a. Graph I: $E =$

        Domain = $\{t \mid \qquad \}$

        Range = $\{E \mid \qquad \}$

  b. Graph II: $E =$

        Domain = $\{t \mid \qquad \}$

        Range = $\{E \mid \qquad \}$

  c. Graph III: $E =$

        Domain = $\{t \mid \qquad \}$

        Range = $\{E \mid \qquad \}$

  d. Graph IV: $E =$

        Domain = $\{t \mid \qquad \}$

        Range = $\{E \mid \qquad \}$

Maintaining Your Skills

The problems that follow review some of the more important skills you have learned in previous sections and chapters.

For the equation $y = 3x - 2$:

47. Find y if x is 4.

48. Find y if x is 0.

49. Find y if x is -4.

50. Find y if x is -2.

For the equation $y = x^2 - 3$:

51. Find y if x is 2.

52. Find y if x is -2.

53. Find y if x is 0.

54. Find y if x is -4.

55. If $x - 2y = 4$ and $x = \dfrac{8}{5}$, find y.

56. If $5x - 10y = 15$, find y when x is 3.

57. Let $x = 0$ and $y = 0$ in $y = a(x - 80)^2 + 70$ and solve for a.

58. Find R if $p = 2.5$ and $R = (900 - 300p)p$.

Getting Ready for the Next Section

Simplify. Round to the nearest whole number if necessary.

59. $7.5(20)$

60. $60 \div 7.5$

61. $4(3.14)(9)$

62. $\dfrac{4}{3}(3.14) \cdot 3^3$

63. $4(-2) - 1$

64. $3(3)^2 + 2(3) - 1$

65. If $s = \dfrac{60}{t}$, find s when

  a. $t = 10$ **b.** $t = 8$

66. If $y = 3x^2 + 2x - 1$, find y when

  a. $x = 0$ **b.** $x = -2$

67. Find the value of $x^2 + 2$ for

  a. $x = 5$ **b.** $x = -2$

68. Find the value of $125 \cdot 2^t$ for

  a. $t = 0$ **b.** $t = 1$

Extending the Concepts

Graph each of the following relations. In each case, use the graph to find the domain and range, and indicate whether the graph is the graph of a function.

69. $y = 5 - |x|$

70. $y = |x| - 3$

71. $x = |y| + 3$

72. $x = 2 - |y|$

73. $|x| + |y| = 4$

74. $2|x| + |y| = 6$

8.4 Function Notation

Let's return to the discussion that introduced us to functions. If a job pays $7.50 per hour for working from 0 to 40 hours a week, then the amount of money y earned in 1 week is a function of the number of hours worked, x. The exact relationship between x and y is written

$$y = 7.5x \quad \text{for} \quad 0 \le x \le 40$$

Because the amount of money earned y depends on the number of hours worked x, we call y the *dependent variable* and x the *independent variable*. Furthermore, if we let f represent all the ordered pairs produced by the equation, then we can write

$$f = \{(x, y) \mid y = 7.5x \text{ and } 0 \le x \le 40\}$$

> **312 Help Wanted**
>
> **YARD PERSON**
> Full-time 40 hrs. with weekend work required. Cleaning & loading trucks. $7.50/hr. Valid CDL with clean record & drug screen required. Submit current MVR to KCI, 225 Suburban Rd., SLO. 805-555-3304.

Once we have named a function with a letter, we can use an alternative notation to represent the dependent variable y. The alternative notation for y is $f(x)$. It is read "f of x" and can be used instead of the variable y when working with functions. The notation y and the notation $f(x)$ are equivalent—that is,

$$y = 7.5x \Leftrightarrow f(x) = 7.5x$$

When we use the notation $f(x)$ we are using *function notation.* The benefit of using function notation is that we can write more information with fewer symbols than we can by using just the variable y. For example, asking how much money a person will make for working 20 hours is simply a matter of asking for $f(20)$. Without function notation, we would have to say, "Find the value of y that corresponds to a value of $x = 20$." To illustrate further, using the variable y, we can say, "y is 150 when x is 20." Using the notation $f(x)$, we simply say, "$f(20) = 150$." Each expression indicates that you will earn $150 for working 20 hours.

 EXAMPLE 1 If $f(x) = 7.5x$, find $f(0), f(10),$ and $f(20)$.

SOLUTION To find $f(0)$ we substitute 0 for x in the expression $7.5x$ and simplify. We find $f(10)$ and $f(20)$ in a similar manner—by substitution.

$$\text{If} \qquad f(x) = 7.5x$$

$$\text{then} \qquad f(\mathbf{0}) = 7.5(\mathbf{0}) = 0$$

$$f(\mathbf{10}) = 7.5(\mathbf{10}) = 75$$

$$f(\mathbf{20}) = 7.5(\mathbf{20}) = 150 \qquad \text{}$$

If we changed the example in the discussion that opened this section so that the hourly wage was $6.50 per hour, we would have a new equation to work with:

$$y = 6.5x \quad \text{for} \quad 0 \le x \le 40$$

Suppose we name this new function with the letter g. Then

$$g = \{(x, y) \mid y = 6.5x \text{ and } 0 \le x \le 40\}$$

and

$$g(x) = 6.5x$$

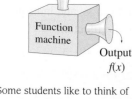

Input x

Function machine

Output $f(x)$

Some students like to think of functions as machines. Values of x are put into the machine, which transforms them into values of $f(x)$, which then are output by the machine.

EXAMPLE 7   If the function f is given by

$$f = \{(-2, 0), (3, -1), (2, 4), (7, 5)\}$$

then $f(-2) = 0, f(3) = -1, f(2) = 4$, and $f(7) = 5$.

 EXAMPLE 8   If $f(x) = 2x^2$ and $g(x) = 3x - 1$, find
a. $f[g(2)]$ **b.** $g[f(2)]$

SOLUTION   The expression $f[g(2)]$ is read "f of g of 2."
a. Since $g(2) = 3(2) - 1 = 5$,

$$f[g(2)] = f(5) = 2(5)^2 = 50$$

b. Since $f(2) = 2(2)^2 = 8$,

$$g[f(2)] = g(8) = 3(8) - 1 = 23$$

LINKING OBJECTIVES AND EXAMPLES

Next to each **objective** we have listed the examples that are best described by that objective.

A   1–8

GETTING READY FOR CLASS

After reading through the preceding section, respond in your own words and in complete sentences.

1. Explain what you are calculating when you find f(2) for a given function f.

2. If $s(t) = \dfrac{60}{t}$, how do you find s(10)?

3. If f(2) = 3 for a function f, what is the relationship between the numbers 2 and 3 and the graph of f?

4. If f(6) = 0 for a particular function f, then you can immediately graph one of the intercepts. Explain.

Problem Set 8.4

 Online support materials can be found at academic.cengage.com/login

Let $f(x) = 2x - 5$ and $g(x) = x^2 + 3x + 4$. Evaluate the following.

1. $f(2)$

2. $f(3)$

3. $f(-3)$

4. $g(-2)$

5. $g(-1)$

6. $f(-4)$

7. $g(-3)$

8. $g(2)$

9. $g(4) + f(4)$

10. $f(2) - g(3)$

11. $f(3) - g(2)$

12. $g(-1) + f(-1)$

Let $f(x) = 3x^2 - 4x + 1$ and $g(x) = 2x - 1$. Evaluate the following.

13. $f(0)$

14. $g(0)$

15. $g(-4)$

16. $f(1)$

17. $f(-1)$

18. $g(-1)$

19. $g(10)$

20. $f(10)$

21. $f(3)$

22. $g(3)$

23. $g\left(\frac{1}{2}\right)$

24. $g\left(\frac{1}{4}\right)$

25. $f(a)$

26. $g(b)$

If $f = \{(1, 4), (-2, 0), (3, \frac{1}{2}), (\pi, 0)\}$ and $g = \{(1, 1), (-2, 2), (\frac{1}{2}, 0)\}$, find each of the following values of f and g.

27. $f(1)$ **28.** $g(1)$

29. $g(\frac{1}{2})$ **30.** $f(3)$

31. $g(-2)$ **32.** $f(\pi)$

Let $f(x) = 2x^2 - 8$ and $g(x) = \frac{1}{2}x + 1$. Evaluate each of the following.

▶ **33.** $f(0)$ **34.** $g(0)$

35. $g(-4)$ **36.** $f(1)$

▶ **37.** $f(a)$ **38.** $g(z)$

39. $f(b)$ **40.** $g(t)$

41. $f[g(2)]$ **42.** $g[f(2)]$

43. $g[f(-1)]$ **44.** $f[g(-2)]$

45. $g[f(0)]$ **46.** $f[g(0)]$

47. Graph the function $f(x) = \frac{1}{2}x + 2$. Then draw and label the line segments that represent $x = 4$ and $f(4)$.

48. Graph the function $f(x) = -\frac{1}{2}x + 6$. Then draw and label the line segments that represent $x = 4$ and $f(4)$.

49. For the function $f(x) = \frac{1}{2}x + 2$, find the value of x for which $f(x) = x$.

50. For the function $f(x) = -\frac{1}{2}x + 6$, find the value of x for which $f(x) = x$.

▶ **51.** Graph the function $f(x) = x^2$. Then draw and label the line segments that represent $x = 1$ and $f(1)$, $x = 2$ and $f(2)$, and, finally, $x = 3$ and $f(3)$.

52. Graph the function $f(x) = x^2 - 2$. Then draw and label the line segments that represent $x = 2$ and $f(2)$, and the line segments corresponding to $x = 3$ and $f(3)$.

Applying the Concepts

53. Investing in Art A painting is purchased as an investment for $150. If its value increases continuously so that it doubles every 3 years, then its value is given by the function

$$V(t) = 150 \cdot 2^{t/3} \quad \text{for} \quad t \geq 0$$

where t is the number of years since the painting was purchased, and $V(t)$ is its value (in dollars) at time t. Find $V(3)$ and $V(6)$, and then explain what they mean.

54. Average Speed If it takes Minke t minutes to run a mile, then her average speed $s(t)$, in miles per hour, is given by the formula

$$s(t) = \frac{60}{t} \quad \text{for} \quad t > 0$$

Find $s(4)$ and $s(5)$, and then explain what they mean.

55. Antidepressant Sales Suppose x represents one of the years in the chart. Suppose further that we have three functions f, g, and h that do the following:

    f pairs each year with the total sales of Zoloft in billions of dollars for that year.

    g pairs each year with the total sales of Effexor in billions of dollars for that year.

    h pairs each year with the total sales of Wellbutrin in billions of dollars for that year.

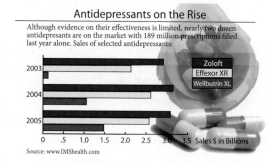

Antidepressants on the Rise

Although evidence on their effectiveness is limited, nearly two dozen antidepressants are on the market with 189 million prescriptions filled last year alone. Sales of selected antidepressants:

Source: www.IMShealth.com

For each statement below, indicate whether the statement is true or false.

a. The domain of g is {2003, 2004, 2005}.

b. $f(2003) < f(2004)$

c. $f(2004) > g(2004)$

d. $h(2005) > 1.5$

e. $h(2005) > h(2004) > h(2003)$

LINKING OBJECTIVES AND EXAMPLES

Next to each **objective** we have listed the examples that are best described by that objective.

A	1–3
B	4

GETTING READY FOR CLASS

After reading through the preceding section, respond in your own words and in complete sentences.

1. How are profit, revenue, and cost related?
2. How do you find the maximum heart rate?
3. For functions f and g, how do you find the composition of f with g?
4. For functions f and g, how do you find the composition of g with f?

Problem Set 8.5

Online support materials can be found at academic.cengage.com/login

Let $f(x) = 4x - 3$ and $g(x) = 2x + 5$. Write a formula for each of the following functions.

1. $f + g$

2. $f - g$

3. $g - f$

4. $g + f$

If the functions f, g, and h are defined by $f(x) = 3x - 5$, $g(x) = x - 2$, and $h(x) = 3x^2$, write a formula for each of the following functions.

▶ **5.** $g + f$

▶ **6.** $f + h$

▶ **7.** $g + h$

▶ **8.** $f - g$

▶ **9.** $g - f$

▶ **10.** $h - g$

▶ **11.** fh

▶ **12.** gh

▶ **13.** h/f

▶ **14.** h/g

▶ **15.** f/h

▶ **16.** g/h

▶ **17.** $f + g + h$

▶ **18.** $h - g + f$

Let $f(x) = 2x + 1$, $g(x) = 4x + 2$, and $h(x) = 4x^2 + 4x + 1$, and find the following.

19. $(f + g)(2)$

20. $(f - g)(-1)$

21. $(fg)(3)$

22. $(f/g)(-3)$

23. $(h/g)(1)$

24. $(hg)(1)$

25. $(fh)(0)$

26. $(h - g)(-4)$

27. $(f + g + h)(2)$

28. $(h - f + g)(0)$

29. $(h + fg)(3)$

30. $(h - fg)(5)$

31. Let $f(x) = x^2$ and $g(x) = x + 4$, and find
   a. $(f \circ g)(5)$
   b. $(g \circ f)(5)$
   c. $(f \circ g)(x)$
   d. $(g \circ f)(x)$

32. Let $f(x) = 3 - x$ and $g(x) = x^3 - 1$, and find
   a. $(f \circ g)(0)$
   b. $(g \circ f)(0)$
   c. $(f \circ g)(x)$

33. Let $f(x) = x^2 + 3x$ and $g(x) = 4x - 1$, and find
   a. $(f \circ g)(0)$
   b. $(g \circ f)(0)$
   c. $(g \circ f)(x)$

34. Let $f(x) = (x - 2)^2$ and $g(x) = x + 1$, and find the following
   a. $(f \circ g)(-1)$
   b. $(g \circ f)(-1)$

For each of the following pairs of functions f and g, show that $(f \circ g)(x) = (g \circ f)(x) = x$.

35. $f(x) = 5x - 4$ and $g(x) = \dfrac{x + 4}{5}$

36. $f(x) = \dfrac{x}{6} - 2$ and $g(x) = 6x + 12$

	= Videos available by instructor request
▶	= Online student support materials available at academic.cengage.com/login

Use the graph to answer problems 37–44.

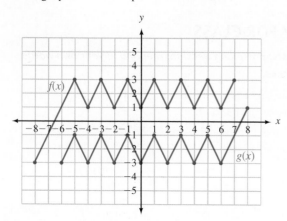

Evaluate.

37. $f(2) + 5$

38. $g(-2) - 5$

39. $f(-3) + g(-3)$

40. $f(5) - g(5)$

41. $(f \circ g)(0)$

42. $(g \circ f)(0)$

43. Find x if $f(x) = -3$.

44. Find x if $g(x) = 1$.

Use the graph to answer problems 45–52.

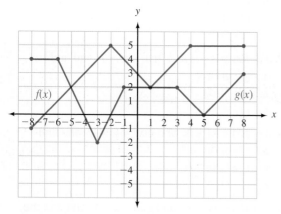

Evaluate.

45. $f(-3) + 2$

46. $g(3) - 3$

47. $f(2) + g(2)$

48. $f(-5) - g(-5)$

49. $(f \circ g)(0)$

50. $(g \circ f)(0)$

51. Find x if $f(x) = 1$.

52. Find x if $g(x) = -2$.

Applying the Concepts

53. Profit, Revenue, and Cost A company manufactures and sells prerecorded videotapes. Here are the equations they use in connection with their business.

Number of tapes sold each day: $n(x) = x$

Selling price for each tape: $p(x) = 11.5 - 0.05x$

Daily fixed costs: $f(x) = 200$

Daily variable costs: $v(x) = 2x$

Find the following functions.

a. Revenue = $R(x)$ = the product of the number of tapes sold each day and the selling price of each tape.

b. Cost = $C(x)$ = the sum of the fixed costs and the variable costs.

c. Profit = $P(x)$ = the difference between revenue and cost.

d. Average cost = $\overline{C}(x)$ = the quotient of cost and the number of tapes sold each day.

54. Profit, Revenue, and Cost A company manufactures and sells diskettes for home computers. Here are the equations they use in connection with their business.

Number of diskettes sold each day: $n(x) = x$

Selling price for each diskette: $p(x) = 3 - \dfrac{1}{300}x$

Daily fixed costs: $f(x) = 200$

Daily variable costs: $v(x) = 2x$

Find the following functions.

a. Revenue = $R(x)$ = the product of the number of diskettes sold each day and the selling price of each diskette.

b. Cost = $C(x)$ = the sum of the fixed costs and the variable costs.

c. Profit = $P(x)$ = the difference between revenue and cost.

d. Average cost = $\overline{C}(x)$ = the quotient of cost and the number of diskettes sold each day.

55. Training Heart Rate Find the training heart rate function, $T(M)$, for a person with a resting heart rate of 62 beats per minute, then find the following.

a. Find the maximum heart rate function, $M(x)$, for a person x years of age.

b. What is the maximum heart rate for a 24-year-old person?

c. What is the training heart rate for a 24-year-old person with a resting heart rate of 62 beats per minute?

d. What is the training heart rate for a 36-year-old person with a resting heart rate of 62 beats per minute?

e. What is the training heart rate for a 48-year-old person with a resting heart rate of 62 beats per minute?

56. **Training Heart Rate** Find the training heart rate function, $T(M)$, for a person with a resting heart rate of 72 beats per minute, then finding the following to the nearest whole number.

   a. Find the maximum heart rate function, $M(x)$, for a person x years of age.

   b. What is the maximum heart rate for a 20-year-old person?

   c. What is the training heart rate for a 20-year-old person with a resting heart rate of 72 beats per minute?

   d. What is the training heart rate for a 30-year-old person with a resting heart rate of 72 beats per minute?

   e. What is the training heart rate for a 40-year-old person with a resting heart rate of 72 beats per minute?

Maintaining Your Skills

Solve each inequality.

57. $|x - 3| < 1$

58. $|x - 3| > 1$

59. $|6 - x| > 2$

60. $\left| 1 - \frac{1}{2}x \right| > 2$

61. $|7x - 1| \le 6$

62. $|7x - 1| \ge 6$

Getting Ready for the Next Section

Simplify.

63. $16(3.5)^2$

64. $\dfrac{2{,}400}{100}$

65. $\dfrac{180}{45}$

66. $4(2)(4)^2$

67. $\dfrac{0.0005(200)}{(0.25)^2}$

68. $\dfrac{0.2(0.5)^2}{100}$

69. If $y = Kx$, find K if $x = 5$ and $y = 15$.

70. If $d = Kt^2$, find K if $t = 2$ and $d = 64$.

71. If $V = \dfrac{K}{P}$, find K if $P = 48$ and $V = 50$.

72. If $y = Kxz^2$, find K if $x = 5$, $z = 3$, and $y = 180$.

8.6 Variation

OBJECTIVES

A Set up and solve problems with direct, inverse, or joint variation.

If you are a runner and you average t minutes for every mile you run during one of your workouts, then your speed s in miles per hour is given by the equation and graph shown here. Figure 1 is shown in the first quadrant only because both t and s are positive.

$$s = \frac{60}{t}$$

Input t	Output s
4	15.0
6	10.0
8	7.5
10	6.0
12	5.0
14	4.3

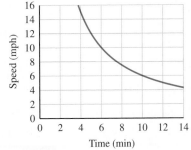

FIGURE 1

You know intuitively that as your average time per mile (t) increases, your speed (s) decreases. Likewise, lowering your time per mile will increase your speed. The equation and Figure 1 also show this to be true: increasing t decreases s, and decreasing t increases s. Quantities that are connected in this way are said to *vary inversely* with each other. Inverse variation is one of the topics we will study in this section.

There are two main types of variation: *direct variation* and *inverse variation*. Variation problems are most common in the sciences, particularly in chemistry and physics.

Direct Variation

When we say the variable *y varies directly* with the variable *x*, we mean that the relationship can be written in symbols as $y = Kx$, where K is a nonzero constant called the *constant of variation* (or *proportionality constant*). Another way of saying *y* varies directly with *x* is to say *y* is *directly proportional* to *x*.

Study the following list. It gives the mathematical equivalent of some direct variation statements.

English Phrase	Algebraic Equation
y varies directly with *x*	$y = Kx$
s varies directly with the square of *t*	$s = Kt^2$
y is directly proportional to the cube of *z*	$y = Kz^3$
u is directly proportional to the square root of *v*	$u = K\sqrt{v}$

EXAMPLE 1 *y* varies directly with *x*. If *y* is 15 when *x* is 5, find *y* when *x* is 7.

SOLUTION The first sentence gives us the general relationship between *x* and *y*. The equation equivalent to the statement "*y* varies directly with *x*" is

$$y = Kx$$

The first part of the second sentence in our example gives us the information necessary to evaluate the constant K:

When	$y = 15$	
and	$x = 5$	
the equation	$y = Kx$	
becomes	$15 = K \cdot 5$	
or	$K = 3$	

The equation now can be written specifically as

$$y = 3x$$

Letting $x = 7$, we have

$$y = 3 \cdot 7$$
$$y = 21$$

EXAMPLE 2 A skydiver jumps from a plane. As with any object that falls toward Earth, the distance the skydiver falls is directly proportional to the square of the time he has been falling until he reaches his terminal velocity. If the skydiver falls 64 feet in the first 2 seconds of the jump, then

    a. How far will he have fallen after 3.5 seconds?

    b. Graph the relationship between distance and time.

    c. How long will it take him to fall 256 feet?

SOLUTION We let t represent the time the skydiver has been falling. Then we can let $d(t)$ represent the distance he has fallen.

  a. Because $d(t)$ is directly proportional to the square of t, we have the general function that describes this situation:

  $$d(t) = Kt^2$$

  Next, we use the fact that $d(2) = 64$ to find K.

  $$64 = K(2^2)$$

  $$K = 16$$

  The specific equation that describes this situation is

  $$d(t) = 16t^2$$

  To find how far a skydiver will have fallen after 3.5 seconds, we find $d(3.5)$:

  $$d(3.5) = 16(3.5^2)$$

  $$d(3.5) = 196$$

  A skydiver will have fallen 196 feet after 3.5 seconds.

  b. To graph this equation, we use a table:

Input t	Output $d(t)$
0	0
1	16
2	64
3	144
4	256
5	400

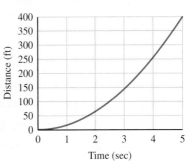

FIGURE 2

  c. From the table or the graph (Figure 2), we see that it will take 4 seconds for the skydiver to fall 256 feet.

Inverse Variation

Running From the introduction to this section, we know that the relationship between the number of minutes (t) it takes a person to run a mile and his or her average speed in miles per hour (s) can be described with the following equation, table, and Figure 3.

$$s = \frac{60}{t}$$

Input t	Output s
4	15.0
6	10.0
8	7.5
10	6.0
12	5.0
14	4.3

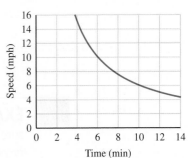

FIGURE 3

If t decreases, then s will increase, and if t increases, then s will decrease. The variable s is *inversely proportional* to the variable t. In this case, the *constant of proportionality* is 60.

Photography If you are familiar with the terminology and mechanics associated with photography, you know that the f-stop for a particular lens will increase as the aperture (the maximum diameter of the opening of the lens) decreases. In mathematics we say that f-stop and aperture vary inversely with each other. The diagram illustrates this relationship.

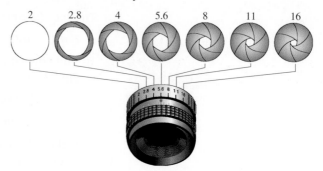

If f is the f-stop and d is the aperture, then their relationship can be written

$$f = \frac{K}{d}$$

In this case, K is the constant of proportionality. (Those of you familiar with photography know that K is also the focal length of the camera lens.)

In General We generalize this discussion of inverse variation as follows: If y varies inversely with x, then

$$y = K\frac{1}{x} \quad \text{or} \quad y = \frac{K}{x}$$

We can also say y is inversely proportional to x. The constant K is again called the constant of variation or proportionality constant.

English Phrase	Algebraic Equation
y is inversely proportional to x	$y = \dfrac{K}{x}$
s varies inversely with the square of t	$s = \dfrac{K}{t^2}$
y is inversely proportional to x^4	$y = \dfrac{K}{x^4}$
z varies inversely with the cube root of t	$z = \dfrac{K}{\sqrt[3]{t}}$

EXAMPLE 3   The volume of a gas is inversely proportional to the pressure of the gas on its container. If a pressure of 48 pounds per square inch corresponds to a volume of 50 cubic feet, what pressure is needed to produce a volume of 100 cubic feet?

SOLUTION   We can represent volume with V and pressure with P:

$$V = \frac{K}{P}$$

Using $P = 48$ and $V = 50$, we have

$$50 = \frac{K}{48}$$

$$K = 50(48)$$

$$K = 2{,}400$$

The equation that describes the relationship between P and V is

$$V = \frac{2{,}400}{P}$$

Here is a graph of this relationship.

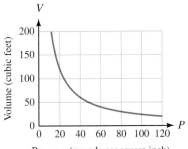

Substituting $V = 100$ into our last equation, we get

$$100 = \frac{2{,}400}{P}$$

$$100P = 2{,}400$$

$$P = \frac{2{,}400}{100}$$

$$P = 24$$

A volume of 100 cubic feet is produced by a pressure of 24 pounds per square inch.

Joint Variation and Other Variation Combinations

Many times relationships among different quantities are described in terms of more than two variables. If the variable y varies directly with *two* other variables, say x and z, then we say y varies *jointly* with x and z. In addition to joint variation, there are many other combinations of direct and inverse variation involving more than two variables. The following table is a list of some variation statements and their equivalent mathematical forms:

English Phrase	Algebraic Equation
y varies jointly with x and z	$y = Kxz$
z varies jointly with r and the square of s	$z = Krs^2$
V is directly proportional to T and inversely proportional to P	$V = \dfrac{KT}{P}$
F varies jointly with m_1 and m_2 and inversely with the square of r	$F = \dfrac{Km_1 m_2}{r^2}$

EXAMPLE 4 y varies jointly with x and the square of z. When x is 5 and z is 3, y is 180. Find y when x is 2 and z is 4.

SOLUTION The general equation is given by

$$y = Kxz^2$$

Substituting $x = 5$, $z = 3$, and $y = 180$, we have

$$180 = K(5)(3)^2$$

$$180 = 45K$$

$$K = 4$$

The specific equation is

$$y = 4xz^2$$

When $x = 2$ and $z = 4$, the last equation becomes

$$y = 4(2)(4)^2$$

$$y = 128$$

 EXAMPLE 5 In electricity, the resistance of a cable is directly proportional to its length and inversely proportional to the square of the diameter. If a 100-foot cable 0.5 inch in diameter has a resistance of 0.2 ohm, what will be the resistance of a cable made from the same material if it is 200 feet long with a diameter of 0.25 inch?

SOLUTION Let R = resistance, l = length, and d = diameter. The equation is

$$R = \frac{Kl}{d^2}$$

When $R = 0.2$, $l = 100$, and $d = 0.5$, the equation becomes

$$0.2 = \frac{K(100)}{(0.5)^2}$$

or

$$K = 0.0005$$

Using this value of K in our original equation, the result is

$$R = \frac{0.0005l}{d^2}$$

When $l = 200$ and $d = 0.25$, the equation becomes

$$R = \frac{0.0005(200)}{(0.25)^2}$$

$$R = 1.6 \text{ ohms}$$

LINKING OBJECTIVES AND EXAMPLES

Next to each **objective** we have listed the examples that are best described by that objective.

A   1–5

GETTING READY FOR CLASS

After reading through the preceding section, respond in your own words and in complete sentences.

1. Give an example of a direct variation statement, and then translate it into symbols.
2. Translate the equation $y = \frac{K}{x}$ into words.
3. For the inverse variation equation $y = \frac{3}{x}$, what happens to the values of y as x gets larger?
4. How are direct variation statements and linear equations in two variables related?

Problem Set 8.6

Online support materials can be found at academic.cengage.com/login

Describe each relationship below as direct or inverse.

▶ **1.** The number of homework problems assigned each night and the time it takes to complete the assignment

▶ **2.** The number of years of school and starting salary

▶ **3.** The length of the line for an amusement park ride and the number of people in line

▶ **4.** The age of a home computer and its relative speed

▶ **5.** The age of a bottle of wine and its price

▶ **6.** The speed of an Internet connection and the time to download a song

▶ **7.** The length of a song and the time it takes to download

8. The speed of a rollercoaster and your willingness to ride it

9. The diameter of a Ferris wheel and circumference

10. The measure angle α to the measure of angle β if $\alpha + \beta = 90°$ ($\alpha = 0$)

11. The altitude above sea level and the air temperature

12. The uncommonness of a search term and the number of search results

For the following problems, y varies directly with x.

▶ **13.** If y is 10 when x is 2, find y when x is 6.

14. If y is 20 when x is 5, find y when x is 3.

15. If y is −32 when x is 4, find x when y is −40.

16. If y is −50 when x is 5, find x when y is −70.

For the following problems, r is inversely proportional to s.

17. If r is −3 when s is 4, find r when s is 2.

18. If r is −10 when s is 6, find r when s is −5.

19. If r is 8 when s is 3, find s when r is 48.

20. If r is 12 when s is 5, find s when r is 30.

For the following problems, d varies directly with the square of r.

21. If $d = 10$ when $r = 5$, find d when $r = 10$.

22. If $d = 12$ when $r = 6$, find d when $r = 9$.

23. If $d = 100$ when $r = 2$, find d when $r = 3$.

24. If $d = 50$ when $r = 5$, find d when $r = 7$.

For the following problems, y varies inversely with the absolute value of x.

▶ **25.** If $y = 6$ when $x = 3$, find y when $x = 9$.

▶ **26.** If $y = 6$ when $x = −3$, find y when $x = −9$.

▶ **27.** If $y = 20$ when $x = −5$, find y when $x = 10$.

= Videos available by instructor request

▶ = Online student support materials available at academic.cengage.com/login

▶ **28.** If $y = 20$ when $x = 5$, find y when $x = 10$.

For the following problems, y varies inversely with the square of x.

29. If $y = 45$ when $x = 3$, find y when x is 5.

30. If $y = 12$ when $x = 2$, find y when x is 6.

31. If $y = 18$ when $x = 3$, find y when x is 2.

32. If $y = 45$ when $x = 4$, find y when x is 5.

For the following problems, z varies jointly with x and the square of y.

33. If z is 54 when x and y are 3, find z when $x = 2$ and $y = 4$.

34. If z is 80 when x is 5 and y is 2, find z when $x = 2$ and $y = 5$.

35. If z is 64 when $x = 1$ and $y = 4$, find x when $z = 32$ and $y = 1$.

36. If z is 27 when $x = 6$ and $y = 3$, find x when $z = 50$ and $y = 4$.

Applying the Concepts

37. Length of a Spring The length a spring stretches is directly proportional to the force applied. If a force of 5 pounds stretches a spring 3 inches, how much force is necessary to stretch the same spring 10 inches?

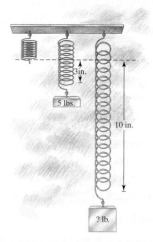

38. Weight and Surface Area The weight of a certain material varies directly with the surface area of that material. If 8 square feet weighs half a pound, how much will 10 square feet weigh?

39. Pressure and Temperature The temperature of a gas varies directly with its pressure. A temperature of 200 K produces a pressure of 50 pounds per square inch.
  a. Find the equation that relates pressure and temperature.
  b. Graph the equation from part (a) in the first quadrant only.
  c. What pressure will the gas have at 280° K?

40. Circumference and Diameter The circumference of a wheel is directly proportional to its diameter. A wheel has a circumference of 8.5 feet and a diameter of 2.7 feet.
  a. Find the equation that relates circumference and diameter.
  b. Graph the equation from part (a) in the first quadrant only.
  c. What is the circumference of a wheel that has a diameter of 11.3 feet?

41. Volume and Pressure The volume of a gas is inversely proportional to the pressure. If a pressure of 36 pounds per square inch corresponds to a volume of 25 cubic feet, what pressure is needed to produce a volume of 75 cubic feet?

42. Wave Frequency The frequency of an electromagnetic wave varies inversely with the wavelength. If a wavelength of 200 meters has a frequency of 800 kilocycles per second, what frequency will be associated with a wavelength of 500 meters?

43. f-Stop and Aperture Diameter The relative aperture or f-stop for a camera lens is inversely proportional to the diameter of the aperture. An f-stop of 2 corresponds to an aperture diameter of 40 millimeters for the lens on an automatic camera.
  a. Find the equation that relates f-stop and diameter.
  b. Graph the equation from part (a) in the first quadrant only.
  c. What is the f-stop of this camera when the aperture diameter is 10 millimeters?

44. f-Stop and Aperture Diameter The relative aperture or f-stop for a camera lens is inversely proportional to the diameter of the aperture. An f-stop of 2.8

corresponds to an aperture diameter of 75 millimeters for a certain telephoto lens.

a. Find the equation that relates f-stop and diameter.

b. Graph the equation from part (a) in the first quadrant only.

c. What aperture diameter corresponds to an f-stop of 5.6?

45. **Surface Area of a Cylinder** The surface area of a hollow cylinder varies jointly with the height and radius of the cylinder. If a cylinder with radius 3 inches and height 5 inches has a surface area of 94 square inches, what is the surface area of a cylinder with radius 2 inches and height 8 inches?

46. **Capacity of a Cylinder** The capacity of a cylinder varies jointly with its height and the square of its radius. If a cylinder with a radius of 3 centimeters and a height of 6 centimeters has a capacity of 169.56 cubic centimeters, what will be the capacity of a cylinder with radius 4 centimeters and height 9 centimeters?

47. **Electrical Resistance** The resistance of a wire varies directly with its length and inversely with the square of its diameter. If 100 feet of wire with diameter 0.01 inch has a resistance of 10 ohms, what is the resistance of 60 feet of the same type of wire if its diameter is 0.02 inch?

48. **Volume and Temperature** The volume of a gas varies directly with its temperature and inversely with the pressure. If the volume of a certain gas is 30 cubic feet at a temperature of 300 K and a pressure of 20 pounds per square inch, what is the volume of the same gas at 340 K when the pressure is 30 pounds per square inch?

49. **Music** A musical tone's pitch varies inversely with its wavelength. If one tone has a pitch of 420 vibrations each second and a wavelength of 2.2 meters, find the wavelength of a tone that has a pitch of 720 vibrations each second.

50. **Hooke's Law** Hooke's law states that the stress (force per unit area) placed on a solid object varies directly with the strain (deformation) produced.

a. Using the variables S_1 for stress and S_2 for strain, state this law in algebraic form.

b. Find the constant, K, if for one type of material $S_1 = 24$ and $S_2 = 72$.

51. **Gravity** In Book Three of his *Principia,* Isaac Newton (depicted on the postage stamp) states that there is a single force in the universe that holds everything together, called the force of universal gravity. Newton stated that the force of universal gravity, F, is directly proportional with the product of two masses, m_1 and m_2, and inversely proportional with the square of the distance d between them. Write the equation for Newton's force of universal gravity, using the symbol G as the constant of proportionality.

52. **Boyle's Law and Charles's Law** Boyle's law states that for low pressures, the pressure of an ideal gas kept at a constant temperature varies inversely with the volume of the gas. Charles's law states that for low pressures, the density of an ideal gas kept at a constant pressure varies inversely with the absolute temperature of the gas.

a. State Boyle's law as an equation using the symbols P, K, and V.

b. State Charles's law as an equation using the symbols D, K, and T.

Maintaining Your Skills

Solve the following equations.

53. $x - 5 = 7$

54. $3y = -4$

55. $5 - \frac{4}{7}a = -11$

56. $\frac{1}{5}x - \frac{1}{2} - \frac{1}{10}x + \frac{2}{5} = \frac{3}{10}x + \frac{1}{2}$

57. $5(x - 1) - 2(2x + 3) = 5x - 4$

58. $0.07 - 0.02(3x + 1) = -0.04x + 0.01$

Solve for the indicated variable.

59. $P = 2l + 2w$ for w

60. $A = \frac{1}{2}h(b + B)$ for B

Solve the following inequalities. Write the solution set using interval notation, then graph the solution set.

61. $-5t \leq 30$

62. $5 - \frac{3}{2}x > -1$

63. $1.6x - 2 < 0.8x + 2.8$

64. $3(2y + 4) \geq 5(y - 8)$

Solve the following equations.

65. $\left|\frac{1}{4}x - 1\right| = \frac{1}{2}$ **66.** $\left|\frac{2}{3}a + 4\right| = 6$

67. $\left|3 - 2x\right| + 5 = 2$ **68.** $5 = \left|3y + 6\right| - 4$

Solve each inequality and graph the solution set.

69. $\left|\frac{x}{5} + 1\right| \geq \frac{4}{5}$

70. $\left|2 - 6t\right| < -5$

71. $\left|3 - 4t\right| > -5$

72. $\left|6y - 1\right| - 4 \leq 2$

Extending the Concepts

73. Human Cannonball A circus company is deciding where to position the net for the human cannonball so that he will land safely during the act. They do this by firing a 100-pound sack of potatoes out of the cannon at different speeds and then measuring how far from the cannon the sack lands. The results are shown in the table.

Speed in Miles/Hour	Distance in Feet
40	108
50	169
60	243
70	331

a. Does distance vary directly with the speed, or directly with the square of the speed?

b. Write the equation that describes the relationship between speed and distance.

c. If the cannon will fire a human safely at 55 miles/hour, where should they position the net so the cannonball has a safe landing?

d. How much farther will he land if his speed out of the cannon is 56 miles/hour?

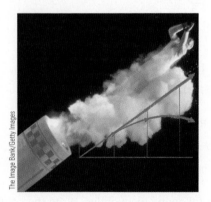

The Image Bank/Getty Images

Linear Equations in Two Variables [8.1, 8.3]

EXAMPLES

1. The equation $3x + 2y = 6$ is an example of a linear equation in two variables.

A *linear equation in two variables* is any equation that can be put in *standard form* $ax + by = c$. The graph of every linear equation is a straight line.

Intercepts [8.1]

2. To find the *x*-intercept for $3x + 2y = 6$, we let $y = 0$ and get

$$3x = 6$$
$$x = 2$$

In this case the *x*-intercept is 2, and the graph crosses the *x*-axis at (2, 0).

The *x-intercept* of an equation is the *x-coordinate* of the point where the graph crosses the *x*-axis. The *y-intercept* is the *y*-coordinate of the point where the graph crosses the *y*-axis. We find the *y*-intercept by substituting $x = 0$ into the equation and solving for *y*. The *x*-intercept is found by letting $y = 0$ and solving for *x*.

The Slope of a Line [8.1]

3. The slope of the line through (6, 9) and (1, −1) is

$$m = \frac{9 - (-1)}{6 - 1} = \frac{10}{5} = 2$$

The *slope* of the line containing points (x_1, y_1) and (x_2, y_2) is given by

$$\text{Slope} = m = \frac{\text{Rise}}{\text{Run}} = \frac{y_2 - y_1}{x_2 - x_1}$$

Horizontal lines have 0 slope, and vertical lines have no slope.
Parallel lines have equal slopes, and perpendicular lines have slopes that are negative reciprocals.

The Slope Intercept Form of a Line [8.2]

4. The equation of the line with slope 5 and *y*-intercept 3 is

$$y = 5x + 3$$

The equation of a line with slope *m* and *y*-intercept *b* is given by

$$y = mx + b$$

The Point Slope Form of a Line [8.2]

5. The equation of the line through (3, 2) with slope −4 is

$$y - 2 = -4(x - 3)$$

which can be simplified to

$$y = -4x + 14$$

The equation of the line through (x_1, y_1) that has slope *m* can be written as

$$y - y_1 = m(x - x_1)$$

Relations and Functions [8.3]

6. The relation

$$\{(8, 1), (6, 1), (-3, 0)\}$$

is also a function because no ordered pairs have the same first coordinates. The domain is $\{8, 6, -3\}$ and the range is $\{1, 0\}$.

A *function* is a rule that pairs each element in one set, called the *domain*, with exactly one element from a second set, called the *range*.

A *relation* is any set of ordered pairs. The set of all first coordinates is called the *domain* of the relation, and the set of all second coordinates is the *range* of the relation. A function is a relation in which no two different ordered pairs have the same first coordinates.

Vertical Line Test [8.4]

7. The graph of $x = y^2$ shown in Figure 5 in Section 3.5 fails the vertical line test. It is not the graph of a function.

If a vertical line crosses the graph of a relation in more than one place, the relation cannot be a function. If no vertical line can be found that crosses the graph in more than one place, the relation must be a function.

Function Notation [8.4]

8. If $f(x) = 5x - 3$ then
$$f(0) = 5(0) - 3$$
$$= -3$$
$$f(1) = 5(1) - 3$$
$$= 2$$
$$f(-2) = 5(-2) - 3$$
$$= -13$$
$$f(a) = 5a - 3$$

The alternative notation for y is $f(x)$. It is read "f of x" and can be used instead of the variable y when working with functions. The notation y and the notation $f(x)$ are equivalent; that is, $y = f(x)$.

Algebra with Functions [8.5]

9. If $f(x) = 4x$ and
$g(x) = x^2 - 3$, then
$(f + g)(x) = x^2 + 4x - 3$
$(f - g)(x) = -x^2 + 4x + 3$

$(fg)(x) = 4x^3 - 12x$
$$\dfrac{f}{g}(x) = \dfrac{4x}{x^2 - 3}$$

If f and g are any two functions with a common domain, then:

$(f + g)(x) = f(x) + g(x)$ — The function $f + g$ is the sum of the functions f and g.

$(f - g)(x) = f(x) - g(x)$ — The function $f - g$ is the difference of the functions f and g.

$(fg)(x) = f(x)g(x)$ — The function fg is the product of the functions f and g.

$\dfrac{f}{g}(x) = \dfrac{f(x)}{g(x)}$ — The function $\dfrac{f}{g}$ is the quotient of the functions f and g,

where $g(x) \neq 0$

Composition of Functions [8.5]

10. If $f(x) = 4x$ and $g(x) = x^2 - 3$, then

$$(f \circ g)(x) = f[g(x)] = f(x^2 - 3)$$
$$= 4x^2 - 12$$
$$(g \circ f)(x) = g[f(x)] = g(4x)$$
$$= 16x^2 - 3$$

If f and g are two functions for which the range of each has numbers in common with the domain of the other, then we have the following definitions:

The composition of f with g: $(f \circ g)(x) = f[g(x)]$

The composition of g with f: $(g \circ f)(x) = g[f(x)]$

Variation [8.6]

11. If y varies directly with x, then

$$y = Kx$$

Then if y is 18 when x is 6,

$$18 = K \cdot 6$$

or

$$K = 3$$

So the equation can be written more specifically as

$$y = 3x$$

If we want to know what y is when x is 4, we simply substitute:

$$y = 3 \cdot 4$$
$$y = 12$$

If y varies *directly* with x (y is directly proportional to x), then

$$y = Kx$$

If y varies *inversely* with x (y is inversely proportional to x), then

$$y = \frac{K}{x}$$

If z varies *jointly* with x and y (z is directly proportional to both x and y), then
$$z = Kxy$$

In each case, K is called the *constant of variation*.

! COMMON MISTAKES

1. When graphing ordered pairs, the most common mistake is to associate the first coordinate with the y-axis and the second with the x-axis. If you make this mistake you would graph (3, 1) by going up 3 and to the right 1, which is just the reverse of what you should do. Remember, the first coordinate is always associated with the horizontal axis, and the second coordinate is always associated with the vertical axis.

2. The two most common mistakes students make when first working with the formula for the slope of a line are the following:
   a. Putting the difference of the x-coordinates over the difference of the y-coordinates.
   b. Subtracting in one order in the numerator and then subtracting in the opposite order in the denominator.

The problems below form a comprehensive review of the material in this chapter. They can be used to study for exams. If you would like to take a practice test on this chapter, you can use the odd-numbered problems. Give yourself an hour and work as many of the odd-numbered problems as possible. When you are finished, or when an hour has passed, check your answers with the answers in the back of the book. You can use the even-numbered problems for a second practice test.

Graph each line. [8.1]

1. $3x + 2y = 6$

2. $y = -\frac{3}{2}x + 1$

3. $x = 3$

Find the slope of the line through the following pairs of points. [8.1]

4. $(5, 2), (3, 6)$

5. $(-4, 2), (3, 2)$

Find x if the line through the two given points has the given slope. [8.1]

6. $(4, x), (1, -3); m = 2$

7. $(-4, 7), (2, x); m = -\frac{1}{3}$

8. Find the slope of any line parallel to the line through $(3, 8)$ and $(5, -2)$. [8.1]

9. The line through $(5, 3y)$ and $(2, y)$ is parallel to a line with slope 4. What is the value of y? [8.1]

Give the equation of the line with the following slope and y-intercept. [8.2]

10. $m = 3, b = 5$

11. $m = -2, b = 0$

Give the slope and y-intercept of each equation. [8.2]

12. $3x - y = 6$

13. $2x - 3y = 9$

Find the equation of the line that contains the given point and has the given slope. [8.2]

14. $(2, 4), m = 2$

15. $(-3, 1), m = -\frac{1}{3}$

Find the equation of the line that contains the given pair of points. [8.2]

16. $(2, 5), (-3, -5)$

17. $(-3, 7), (4, 7)$

18. $(-5, -1), (-3, -4)$

19. Find the equation of the line that is parallel to $2x - y = 4$ and contains the point $(2, -3)$. [8.2]

20. Find the equation of the line perpendicular to $y = -3x + 1$ that has an x-intercept of 2. [8.2]

State the domain and range of each relation, and then indicate which relations are also functions. [8.3]

21. $\{(2, 4), (3, 3), (4, 2)\}$

22. $\{(6, 3), (-4, 3), (-2, 0)\}$

If $f = \{(2, -1), (-3, 0), (4, \frac{1}{2}), (\pi, 2)\}$ and $g = \{(2, 2), (-1, 4), (0, 0)\}$, find the following. [8.4]

23. $f(\pi)$

24. $f(-3) + g(-1)$

25. $f(-3)$

26. $f(2) + g(2)$

Let $f(x) = 2x^2 - 4x + 1$ and $g(x) = 3x + 2$, and evaluate each of the following. [8.4]

27. $f(0)$

28. $g(a)$

29. $f[g(0)]$

30. $f[g(1)]$

For the following problems, y varies directly with x. [8.6]

31. If y is 6 when x is 2, find y when x is 8.

32. If y is -3 when x is 5, find y when x is -10.

For the following problems, y varies inversely with the square of x. [8.6]

33. If y is 9 when x is 2, find y when x is 3.

34. If y is 4 when x is 5, find y when x is 2.

Solve each application problem. [8.6]

35. Tension in a Spring The tension t in a spring varies directly with the distance d the spring is stretched. If the tension is 42 pounds when the spring is stretched 2 inches, find the tension when the spring is stretched twice as far.

36. Light Intensity The intensity of a light source varies inversely with the square of the distance from the source. Four feet from the source the intensity is 9 foot-candles. What is the intensity 3 feet from the source?

GROUP PROJECT   Light Intensity

Number of People 2–3

Time Needed 15 minutes

Equipment Paper and pencil

Background I found the following diagram while shopping for some track lighting for my home. I was impressed by the diagram because it displays a lot of useful information in a very efficient manner. As the diagram indicates, the amount of light that falls on a surface depends on how far above the surface the light is placed and how much the light spreads out on the surface. Assume that this light illuminates a circle on a flat surface, and work the following problems.

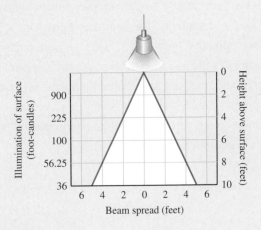

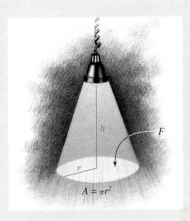

Procedure

a. Fill in each table.

Height Above Surface (ft)	Illumination (foot-candles)
2	
4	
6	
8	
10	

Distance Above Surface (ft)	Area of Illuminated Region (ft²)
2	
4	
6	
8	
10	

b. Construct line graphs from the data in the tables.

c. Which of the relationships is direct variation, and which is inverse variation?

d. Let F represent the number of foot-candles that fall on the surface, h the distance the light source is above the surface, and A the area of the illuminated region. Write an equation that shows the relationship between A and h, then write another equation that gives the relationship between F and h.

Descartes and Pascal

René Descartes, 1596–1650

Blaise Pascal, 1623–1662

In this chapter, we mentioned that René Descartes, the inventor of the rectangular coordinate system, is the person who made the statement, "I think, therefore, I am." Blaise Pascal, another French philosopher, is responsible for the statement, "The heart has its reasons which reason does not know." Although Pascal and Descartes were contemporaries, the philosophies of the two men differed greatly. Research the philosophy of both Descartes and Pascal, and then write an essay that gives the main points of each man's philosophy. In the essay, show how the quotations given here fit in with the philosophy of the man responsible for the quotation.

Rational Exponents and Roots

David Woodfall/Getty Images

Ecology and conservation are topics that interest most college students. If our rivers and oceans are to be preserved for future generations, we need to work to eliminate pollution from our waters. If a river is flowing at 1 meter per second and a pollutant is entering the river at a constant rate, the shape of the pollution plume can often be modeled by the simple equation

$$y = \sqrt{x}$$

The following table and graph were produced from the equation.

Width of a Pollutant Plume

Distance from Source (meters) x	Width of Plume (meters) y
0	0
1	1
4	2
9	3
16	4

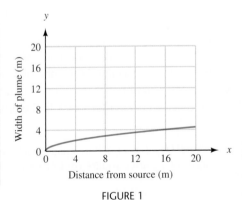

FIGURE 1

To visualize how Figure 1 models the pollutant plume, imagine that the river is flowing from left to right, parallel to the x-axis, with the x-axis as one of its banks. The pollutant is entering the river from the bank at (0, 0).

By modeling pollution with mathematics, we can use our knowledge of mathematics to help control and eliminate pollution.

▶ Improve your grade and save time!
Go online to **academic.cengage.com/login** where you can
- Watch videos of instructors working through the in-text examples
- Follow step-by-step online tutorials of in-text examples and review questions
- Work practice problems
- Check your readiness for an exam by taking a pre-test and exploring the modules recommended in your Personalized Study plan
- Receive help from a live tutor online through vMentor™
Try it out! Log in with an access code or purchase access at **www.ichapters.com**.

Rational Exponents

A Simplify radical expressions using the definition for roots.

B Simplify expressions with rational exponents.

Figure 1 shows a square in which each of the four sides is 1 inch long. To find the square of the length of the diagonal c, we apply the Pythagorean theorem:

$$c^2 = 1^2 + 1^2$$
$$c^2 = 2$$

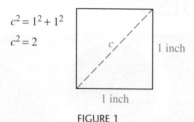

FIGURE 1

Because we know that c is positive and that its square is 2, we call c the *positive square root* of 2, and we write $c = \sqrt{2}$. Associating numbers, such as $\sqrt{2}$, with the diagonal of a square or rectangle allows us to analyze some interesting items from geometry. One particularly interesting geometric object that we will study in this section is shown in Figure 2. It is constructed from a right triangle, and the length of the diagonal is found from the Pythagorean theorem. We will come back to this figure at the end of this section.

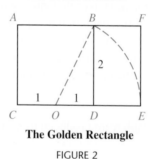

The Golden Rectangle

FIGURE 2

Previously, we developed notation (exponents) to give us the square, cube, or any other power of a number. For instance, if we wanted the square of 3, we wrote $3^2 = 9$. If we wanted the cube of 3, we wrote $3^3 = 27$. In this section, we will develop notation that will take us in the reverse direction—that is, from the square of a number, say 25, back to the original number, 5.

Note

It is a common mistake to assume that an expression like $\sqrt{25}$ indicates both square roots, 5 and −5. The expression $\sqrt{25}$ indicates only the positive square root of 25, which is 5. If we want the negative square root, we must use a negative sign: $-\sqrt{25} = -5$.

DEFINITION If x is a nonnegative real number, then the expression $\sqrt{x}$ is called the **positive square root** of x and is the nonnegative number such that

$$(\sqrt{x})^2 = x$$

In words: $\sqrt{x}$ is the nonnegative number we square to get x.

The negative square root of x, $-\sqrt{x}$, is defined in a similar manner.

EXAMPLE 1 The positive square root of 64 is 8 because 8 is the positive number with the property $8^2 = 64$. The negative square root of 64 is −8 since −8 is the negative number whose square is 64. We can summarize both of these facts by saying

$$\sqrt{64} = 8 \qquad \text{and} \qquad -\sqrt{64} = -8$$

The higher roots, cube roots, fourth roots, and so on are defined by definitions similar to that of square roots.

> **DEFINITION** If x is a real number and n is a positive integer, then
>
> Positive square root of x, $\sqrt{x}$, is such that $(\sqrt{x})^2 = x$ $x \geq 0$
>
> Cube root of x, $\sqrt[3]{x}$, is such that $(\sqrt[3]{x})^3 = x$
>
> Positive fourth root of x, $\sqrt[4]{x}$, is such that $(\sqrt[4]{x})^4 = x$ $x \geq 0$
>
> Fifth root of x, $\sqrt[5]{x}$, is such that $(\sqrt[5]{x})^5 = x$
>
> $\vdots$
>
> The **nth root of x**, $\sqrt[n]{x}$, is such that $(\sqrt[n]{x})^n = x$ $x \geq 0$ if n is even

The following is a table of the most common roots used in this book. Any of the roots that are unfamiliar should be memorized.

Square Roots		Cube Roots	Fourth Roots
$\sqrt{0} = 0$	$\sqrt{49} = 7$	$\sqrt[3]{0} = 0$	$\sqrt[4]{0} = 0$
$\sqrt{1} = 1$	$\sqrt{64} = 8$	$\sqrt[3]{1} = 1$	$\sqrt[4]{1} = 1$
$\sqrt{4} = 2$	$\sqrt{81} = 9$	$\sqrt[3]{8} = 2$	$\sqrt[4]{16} = 2$
$\sqrt{9} = 3$	$\sqrt{100} = 10$	$\sqrt[3]{27} = 3$	$\sqrt[4]{81} = 3$
$\sqrt{16} = 4$	$\sqrt{121} = 11$	$\sqrt[3]{64} = 4$	
$\sqrt{25} = 5$	$\sqrt{144} = 12$	$\sqrt[3]{125} = 5$	
$\sqrt{36} = 6$	$\sqrt{169} = 13$		

Notation An expression like $\sqrt[3]{8}$ that involves a root is called a *radical expression*. In the expression $\sqrt[3]{8}$, the 3 is called the *index*, the $\sqrt{}$ is the *radical sign*, and 8 is called the *radicand*. The index of a radical must be a positive integer greater than 1. If no index is written, it is assumed to be 2.

Roots and Negative Numbers

When dealing with negative numbers and radicals, the only restriction concerns negative numbers under even roots. We can have negative signs in front of radicals and negative numbers under odd roots and still obtain real numbers. Here are some examples to help clarify this. In the last section of this chapter we will see how to deal with even roots of negative numbers.

EXAMPLES   Simplify each expression, if possible.

2. $\sqrt[3]{-8} = -2$ because $(-2)^3 = -8$.

3. $\sqrt{-4}$ is not a real number since there is no real number whose square is -4.

4. $-\sqrt{25} = -5$ is the negative square root of 25.

5. $\sqrt[5]{-32} = -2$ because $(-2)^5 = -32$.

6. $\sqrt[4]{-81}$ is not a real number since there is no real number we can raise to the fourth power and obtain -81.

Variables Under a Radical

From the preceding examples it is clear that we must be careful that we do not try to take an even root of a negative number. For this reason, we will assume that all variables appearing under a radical sign represent nonnegative numbers.

EXAMPLES Assume all variables represent nonnegative numbers and simplify each expression as much as possible.

7. $\sqrt{25a^4b^6} = 5a^2b^3$ because $(5a^2b^3)^2 = 25a^4b^6$.

8. $\sqrt[3]{x^6y^{12}} = x^2y^4$ because $(x^2y^4)^3 = x^6y^{12}$.

9. $\sqrt[4]{81r^8s^{20}} = 3r^2s^5$ because $(3r^2s^5)^4 = 81r^8s^{20}$.

Rational Numbers as Exponents

Next we develop a second kind of notation involving exponents that will allow us to designate square roots, cube roots, and so on in another way.

Consider the equation $x = 8^{1/3}$. Although we have not encountered fractional exponents before, let's assume that all the properties of exponents hold in this case. Cubing both sides of the equation, we have

$$x^3 = (8^{1/3})^3$$

$$x^3 = 8^{(1/3)(3)}$$

$$x^3 = 8^1$$

$$x^3 = 8$$

The last line tells us that x is the number whose cube is 8. It must be true, then, that x is the cube root of 8, $x = \sqrt[3]{8}$. Since we started with $x = 8^{1/3}$, it follows that

$$8^{1/3} = \sqrt[3]{8}$$

It seems reasonable, then, to define fractional exponents as indicating roots. Here is the formal definition.

> **DEFINITION** If x is a real number and n is a positive integer greater than 1, then
>
> $$x^{1/n} = \sqrt[n]{x} \qquad (x \geq 0 \text{ when } n \text{ is even})$$
>
> *In words:* The quantity $x^{1/n}$ is the nth root of x.

With this definition we have a way of representing roots with exponents. Here are some examples.

EXAMPLES Write each expression as a root and then simplify, if possible.

10. $8^{1/3} = \sqrt[3]{8} = 2$

11. $36^{1/2} = \sqrt{36} = 6$

12. $-25^{1/2} = -\sqrt{25} = -5$

13. $(-25)^{1/2} = \sqrt{-25}$, which is not a real number

14. $\left(\dfrac{4}{9}\right)^{1/2} = \sqrt{\dfrac{4}{9}} = \dfrac{2}{3}$

The properties of exponents developed in a previous chapter were applied to integer exponents only. We will now extend these properties to include rational exponents also. We do so without proof.

Properties of Exponents If a and b are real numbers and r and s are rational numbers, and a and b are nonnegative whenever r and s indicate even roots, then

1. $a^r \cdot a^s = a^{r+s}$ **4.** $a^{-r} = \dfrac{1}{a^r}$ $(a \neq 0)$

2. $(a^r)^s = a^{rs}$ **5.** $\left(\dfrac{a}{b}\right)^r = \dfrac{a^r}{b^r}$ $(b \neq 0)$

3. $(ab)^r = a^r b^r$ **6.** $\dfrac{a^r}{a^s} = a^{r-s}$ $(a \neq 0)$

There are times when rational exponents can simplify our work with radicals. Here are Examples 8 and 9 again, but this time we will work them using rational exponents.

EXAMPLES   Write each radical with a rational exponent and then simplify.

15. $\sqrt[3]{x^6 y^{12}} = (x^6 y^{12})^{1/3}$

$\qquad\qquad = (x^6)^{1/3}(y^{12})^{1/3}$

$\qquad\qquad = x^2 y^4$

16. $\sqrt[4]{81 r^8 s^{20}} = (81 r^8 s^{20})^{1/4}$

$\qquad\qquad\quad = 81^{1/4}(r^8)^{1/4}(s^{20})^{1/4}$

$\qquad\qquad\quad = 3 r^2 s^5$

So far, the numerators of all the rational exponents we have encountered have been 1. The next theorem extends the work we can do with rational exponents to rational exponents with numerators other than 1.

We can extend our properties of exponents with the following theorem.

The Rational Exponent Theorem If a is a nonnegative real number, m is an integer, and n is a positive integer, then

$$a^{m/n} = (a^{1/n})^m = (a^m)^{1/n}$$

Proof We can prove this theorem using the properties of exponents. Since $\dfrac{m}{n} = m\left(\dfrac{1}{n}\right)$ we have

$$a^{m/n} = a^{m(1/n)} \qquad\qquad a^{m/n} = a^{(1/n)(m)}$$

$$= (a^m)^{1/n} \qquad\qquad\quad = (a^{1/n})^m$$

Here are some examples that illustrate how we use this theorem.

Note

On a scientific calculator, Example 17 would look like this:

$8 \boxed{y^x} \boxed{(} \boxed{2} \boxed{\div} \boxed{3} \boxed{)} \boxed{=}$

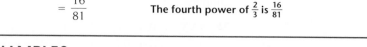
EXAMPLES   Simplify as much as possible.

17. $8^{2/3} = (8^{1/3})^2$ **Rational exponent theorem**

     $= 2^2$ **Definition of fractional exponents**

     $= 4$ **The square of 2 is 4**

18. $25^{3/2} = (25^{1/2})^3$ **Rational exponent theorem**

     $= 5^3$ **Definition of fractional exponents**

     $= 125$ **The cube of 5 is 125**

19. $9^{-3/2} = (9^{1/2})^{-3}$ **Rational exponent theorem**

     $= 3^{-3}$ **Definition of fractional exponents**

     $= \dfrac{1}{3^3}$ **Property 4 for exponents**

     $= \dfrac{1}{27}$ **The cube of 3 is 27**

20. $\left(\dfrac{27}{8}\right)^{-4/3} = \left[\left(\dfrac{27}{8}\right)^{1/3}\right]^{-4}$ **Rational exponent theorem**

     $= \left(\dfrac{3}{2}\right)^{-4}$ **Definition of fractional exponents**

     $= \left(\dfrac{2}{3}\right)^{4}$ **Property 4 for exponents**

     $= \dfrac{16}{81}$ **The fourth power of $\frac{2}{3}$ is $\frac{16}{81}$**

EXAMPLES   Assume all variables represent positive quantities and simplify as much as possible.

21. $x^{1/3} \cdot x^{5/6} = x^{1/3+5/6}$ **Property 1**

     $= x^{2/6+5/6}$ **LCD is 6**

     $= x^{7/6}$ **Add fractions**

22. $(y^{2/3})^{3/4} = y^{(2/3)(3/4)}$ **Property 2**

     $= y^{1/2}$ **Multiply fractions: $\frac{2}{3} \cdot \frac{3}{4} = \frac{6}{12} = \frac{1}{2}$**

23. $\dfrac{z^{1/3}}{z^{1/4}} = z^{1/3-1/4}$ **Property 6**

     $= z^{4/12-3/12}$ **LCD is 12**

     $= z^{1/12}$ **Subtract fractions**

24. $\dfrac{(x^{-3}y^{1/2})^4}{x^{10}y^{3/2}} = \dfrac{(x^{-3})^4(y^{1/2})^4}{x^{10}y^{3/2}}$ **Property 3**

     $= \dfrac{x^{-12}y^2}{x^{10}y^{3/2}}$ **Property 2**

     $= x^{-22}y^{1/2}$ **Property 6**

     $= \dfrac{y^{1/2}}{x^{22}}$ **Property 4**

⚡ FACTS FROM GEOMETRY

The Pythagorean Theorem (Again) and the Golden Rectangle

Now that we have had some experience working with square roots, we can rewrite the Pythagorean theorem using a square root. If triangle ABC is a right triangle with $C = 90°$, then the length of the longest side is the *positive square root* of the sum of the squares of the other two sides (see Figure 3).

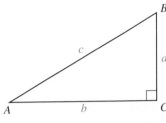

$$c = \sqrt{a^2 + b^2}$$

FIGURE 3

Constructing a Golden Rectangle from a Square of Side 2

Step 1: Draw a square with a side of length two. Connect the midpoint of side CD to corner B. (Note that we have labeled the midpoint of segment CD with the letter O.)

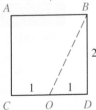

Step 2: Drop the diagonal from step 1 down so it aligns with side CD.

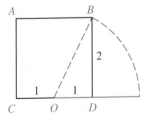

Step 3: Form rectangle $ACEF$. This is a golden rectangle.

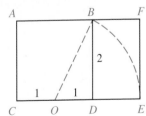

> **Note**
>
> In the introduction to this section we mentioned the golden rectangle. Its origins can be traced back more than 2,000 years to the Greek civilization that produced Pythagoras, Socrates, Plato, Aristotle, and Euclid. The most important mathematical work to come from that Greek civilization was Euclid's *Elements,* an elegantly written summary of all that was known about geometry at that time in history. Euclid's *Elements,* according to Howard Eves, an authority on the history of mathematics, exercised a greater influence on scientific thinking than any other work. Here is how we construct a golden rectangle from a square of side 2, using the same method that Euclid used in his *Elements.*

All golden rectangles are constructed from squares. Every golden rectangle, no matter how large or small it is, will have the same shape. To associate a number with the shape of the golden rectangle, we use the ratio of its length to its width. This ratio is called the *golden ratio.* To calculate the golden ratio, we must first find the length of the diagonal we used to construct the golden rectangle. Figure 4 shows the golden rectangle that we constructed from a square of side 2. The length of the diagonal OB is found by applying the Pythagorean theorem to triangle OBD.

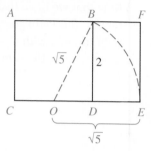

FIGURE 4

The length of segment OE is equal to the length of diagonal OB; both are $\sqrt{5}$. Since the distance from C to O is 1, the length CE of the golden rectangle is $1 + \sqrt{5}$. Now we can find the golden ratio:

$$\text{Golden ratio} = \frac{\text{length}}{\text{width}} = \frac{CE}{EF} = \frac{1 + \sqrt{5}}{2}$$

USING TECHNOLOGY

Graphing Calculators—A Word of Caution

Some graphing calculators give surprising results when evaluating expressions such as $(-8)^{2/3}$. As you know from reading this section, the expression $(-8)^{2/3}$ simplifies to 4, either by taking the cube root first and then squaring the result, or by squaring the base first and then taking the cube root of the result. Here are three different ways to evaluate this expression on your calculator:

1. $(-8)\wedge(2/3)$ **To evaluate $(-8)^{2/3}$**
2. $((-8)\wedge2)\wedge(1/3)$ **To evaluate $((-8)^2)^{1/3}$**
3. $((-8)\wedge(1/3))\wedge2$ **To evaluate $((-8)^{1/3})^2$**

Note any difference in the results.

Next, graph each of the following functions, one at a time.

1. $Y_1 = X^{2/3}$ **2.** $Y_2 = (X^2)^{1/3}$ **3.** $Y_3 = (X^{1/3})^2$

The correct graph is shown in Figure 5. Note which of your graphs match the correct graph.

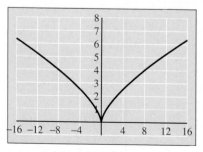

FIGURE 5

Different calculators evaluate exponential expressions in different ways. You should use the method (or methods) that gave you the correct graph.

GETTING READY FOR CLASS

After reading through the preceding section, respond in your own words and in complete sentences.

1. Every real number has two square roots. Explain the notation we use to tell them apart. Use the square roots of 3 for examples.
2. Explain why a square root of -4 is not a real number.
3. We use the notation $\sqrt{2}$ to represent the positive square root of 2. Explain why there isn't a simpler way to express the positive square root of 2.
4. For the expression $a^{m/n}$, explain the significance of the numerator m and the significance of the denominator n in the exponent.

Problem Set 9.1

Online support materials can be found at academic.cengage.com/login

Find each of the following roots, if possible.

1. $\sqrt{144}$

2. $-\sqrt{144}$

3. $\sqrt{-144}$

4. $\sqrt{-49}$

5. $-\sqrt{49}$

6. $\sqrt{49}$

7. $\sqrt[3]{-27}$

8. $-\sqrt[3]{27}$

9. $\sqrt[4]{16}$

10. $-\sqrt[4]{16}$

11. $\sqrt[4]{-16}$

12. $-\sqrt[4]{-16}$

13. $\sqrt{0.04}$

14. $\sqrt{0.81}$

15. $\sqrt[3]{0.008}$

16. $\sqrt[3]{0.125}$

17. $\sqrt[3]{125}$

18. $\sqrt[3]{-125}$

▶ 19. $-\sqrt[3]{216}$

▶ 20. $-\sqrt[3]{-216}$

▶ 21. $\sqrt{\dfrac{1}{36}}$

▶ 22. $\sqrt{\dfrac{9}{25}}$

▶ 23. $\sqrt[3]{\dfrac{8}{125}}$

▶ 24. $\sqrt[3]{-\dfrac{27}{216}}$

Simplify each expression. Assume all variables represent nonnegative numbers.

25. $\sqrt{36a^8}$

26. $\sqrt{49a^{10}}$

▶ 27. $\sqrt[3]{27a^{12}}$

28. $\sqrt[3]{8a^{15}}$

29. $\sqrt[5]{32x^{10}y^5}$

30. $\sqrt[5]{32x^5y^{10}}$

▶ 31. $\sqrt[4]{16a^{12}b^{20}}$

32. $\sqrt[4]{81a^{24}b^8}$

Use the definition of rational exponents to write each of the following with the appropriate root. Then simplify.

33. $36^{1/2}$

34. $49^{1/2}$

35. $-9^{1/2}$

36. $-16^{1/2}$

37. $8^{1/3}$

38. $-8^{1/3}$

▶ 39. $(-8)^{1/3}$

40. $-27^{1/3}$

41. $32^{1/5}$

42. $81^{1/4}$

43. $\left(\dfrac{81}{25}\right)^{1/2}$

44. $\left(\dfrac{64}{125}\right)^{1/3}$

Use the rational exponent theorem to simplify each of the following as much as possible.

45. $27^{2/3}$

46. $8^{4/3}$

▶ 47. $25^{3/2}$

48. $81^{3/4}$

Simplify each expression. Remember, negative exponents give reciprocals.

49. $27^{-1/3}$

50. $9^{-1/2}$

▶ 51. $81^{-3/4}$

52. $4^{-3/2}$

53. $\left(\dfrac{25}{36}\right)^{-1/2}$

54. $\left(\dfrac{16}{49}\right)^{-1/2}$

55. $\left(\dfrac{81}{16}\right)^{-3/4}$

56. $\left(\dfrac{27}{8}\right)^{-2/3}$

57. $16^{1/2} + 27^{1/3}$

58. $25^{1/2} + 100^{1/2}$

59. $8^{-2/3} + 4^{-1/2}$

60. $49^{-1/2} + 25^{-1/2}$

Use the properties of exponents to simplify each of the following as much as possible. Assume all bases are positive.

61. $x^{3/5} \cdot x^{1/5}$

62. $x^{3/4} \cdot x^{5/4}$

63. $(a^{3/4})^{4/3}$

64. $(a^{2/3})^{3/4}$

65. $\dfrac{x^{1/5}}{x^{3/5}}$

66. $\dfrac{x^{2/7}}{x^{5/7}}$

67. $\dfrac{x^{5/6}}{x^{2/3}}$

68. $\dfrac{x^{7/8}}{x^{8/7}}$

69. $(x^{3/5}y^{5/6}z^{1/3})^{3/5}$

70. $(x^{3/4}y^{1/8}z^{5/6})^{4/5}$

71. $\dfrac{a^{3/4}b^2}{a^{7/8}b^{1/4}}$

72. $\dfrac{a^{1/3}b^4}{a^{3/5}b^{1/3}}$

73. $\dfrac{(y^{2/3})^{3/4}}{(y^{1/3})^{3/5}}$

74. $\dfrac{(y^{5/4})^{2/5}}{(y^{1/4})^{4/3}}$

75. $\left(\dfrac{a^{-1/4}}{b^{1/2}}\right)^8$

76. $\left(\dfrac{a^{-1/5}}{b^{1/3}}\right)^{15}$

77. $\dfrac{(r^{-2}s^{1/3})^6}{r^8s^{3/2}}$

78. $\dfrac{(r^{-5}s^{1/2})^4}{r^{12}s^{5/2}}$

79. $\dfrac{(25a^6b^4)^{1/2}}{(8a^{-9}b^3)^{-1/3}}$

80. $\dfrac{(27a^3b^6)^{1/3}}{(81a^8b^{-4})^{1/4}}$

Applying the Concepts

81. Maximum Speed The maximum speed (v) that an automobile can travel around a curve of radius r without skidding is given by the equation

$$v = \left(\dfrac{5r}{2}\right)^{1/2}$$

where v is in miles per hour and r is measured in feet. What is the maximum speed a car can travel around a curve with a radius of 250 feet without skidding?

82. Golden Ratio The golden ratio is the ratio of the length to the width in any golden rectangle. The exact value of this number is $\dfrac{1 + \sqrt{5}}{2}$. Use a calculator to find a decimal approximation to this number and round it to the nearest thousandth.

83. Chemistry Figure 6 shows part of a model of a magnesium oxide (MgO) crystal. Each corner of the square is at the center of one oxygen ion (O^{2-}), and the center of the middle ion is at the center of the square. The radius for each oxygen ion is 126 picometers (pm), and the radius for each magnesium ion (Mg^{2+}) is 86 picometers.

a. Find the length of the side of the square. Write your answer in picometers.

b. Find the length of the diagonal of the square to the nearest picometer.

c. If 1 meter is 10^{12} picometers, give the length of the diagonal of the square in meters.

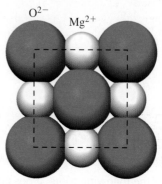
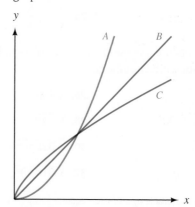

FIGURE 6   (Susan M. Young)

84. Geometry The length of each side of the cube shown in Figure 7 is 1 inch.

a. Find the length of the diagonal CH.

b. Find the length of the diagonal CF.

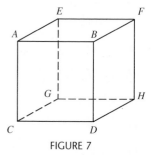

FIGURE 7

85. Comparing Graphs Identify the graph with the correct equation.

a. $y = x$

b. $y = x^2$

c. $y = x^{2/3}$

d. What are the two points of intersection of all three graphs?

86. Falling Objects The time t in seconds it takes an object to fall d feet is given by the equation

$$t = \frac{1}{4}\sqrt{d}$$

a. The Sears Tower in Chicago is 1,450 feet tall. How long would it take a penny to fall to the ground from the top of the Sears Tower? Round to the nearest hundredth.

b. An object took 30 seconds to fall to the ground. From what distance must it have been dropped?

Maintaining Your Skills

Multiply.

87. $x^2(x^4 - x)$

88. $5x^2(2x^3 - x)$

89. $(x - 3)(x + 5)$

90. $(x - 2)(x + 2)$

91. $(x^2 - 5)^2$

92. $(x^2 + 5)^2$

93. $(x - 3)(x^2 + 3x + 9)$

94. $(x + 3)(x^2 - 3x + 9)$

Getting Ready for the Next Section

Simplify.

95. $x^2(x^4 - x^3)$

96. $(x^2 - 3)(x^2 + 5)$

97. $(3a - 2b)(4a - b)$

98. $(x^2 + 3)^2$

99. $(x^3 - 2)(x^3 + 2)$

100. $(a - b)(a^2 + ab + b^2)$

101. $\dfrac{15x^2y - 20x^4y^2}{5xy}$

102. $\dfrac{12a^3b^2 - 24a^2b^4}{3ab}$

Factor.

103. $x^2 - 3x - 10$

104. $x^2 + x - 12$

105. $6x^2 + 11x - 10$

106. $10x^2 - x - 3$

Use rules of exponents to simplify.

107. $x^{2/3} \cdot x^{4/3}$

108. $x^{1/4} \cdot x^{3/4}$

109. $(t^{1/2})^2$

110. $(x^{3/2})^2$

111. $\dfrac{x^{2/3}}{x^{1/3}}$

112. $\dfrac{x^{1/2}}{x^{1/2}}$

Extending the Concepts

113. Show that the expression $(a^{1/2} + b^{1/2})^2$ is not equal to $a + b$ by replacing a with 9 and b with 4 in both expressions and then simplifying each.

114. Show that the statement $(a^2 + b^2)^{1/2} = a + b$ is not, in general, true by replacing a with 3 and b with 4 and then simplifying both sides.

115. You may have noticed, if you have been using a calculator to find roots, that you can find the fourth root of a number by pressing the square root button twice. Written in symbols, this fact looks like this:

$$\sqrt{\sqrt{a}} = \sqrt[4]{a} \qquad (a \geq 0)$$

Show that this statement is true by rewriting each side with exponents instead of radical notation and then simplifying the left side.

116. Show that the following statement is true by rewriting each side with exponents instead of radical notation and then simplifying the left side.

$$\sqrt[3]{\sqrt{a}} = \sqrt[6]{a} \qquad (a \geq 0)$$

9.2 More Expressions Involving Rational Exponents

OBJECTIVES

A Multiply expressions with rational exponents.

B Divide expressions with rational exponents.

C Factor expressions with rational exponents.

D Add and subtract expressions with rational exponents.

Suppose you purchased 10 silver proof coin sets in 1997 for $21 each, for a total investment of $210. Three years later, in 2000, you find that each set is worth $30, which means that your 10 sets have a total value of $300.

United States Mint Proof Set

You can calculate the annual rate of return on this investment using a formula that involves rational exponents. The annual rate of return will tell you at what interest rate you would have to invest your original $210 for it to be worth $300 three years later. As you will see at the end of this section, the annual rate of return on this investment is 12.6%, which is a good return on your money.

In this section we will look at multiplication, division, factoring, and simplification of some expressions that resemble polynomials but contain rational exponents. The problems in this section will be of particular interest to you if you are planning to take either an engineering calculus class or a business calculus class. As was the case in the previous section, we will assume all variables represent nonnegative real numbers. That way, we will not have to worry about the possibility of introducing undefined terms—even roots of negative numbers—into any of our examples. Let's begin this section with a look at multiplication of expressions containing rational exponents.

EXAMPLE 1   Multiply $x^{2/3}(x^{4/3} - x^{1/3})$.

SOLUTION   Applying the distributive property and then simplifying the resulting terms, we have:

$$x^{2/3}(x^{4/3} - x^{1/3}) = x^{2/3}x^{4/3} - x^{2/3}x^{1/3} \qquad \text{Distributive property}$$

$$= x^{6/3} - x^{3/3} \qquad \text{Add exponents}$$

$$= x^2 - x \qquad \text{Simplify}$$

EXAMPLE 2   Multiply $(x^{2/3} - 3)(x^{2/3} + 5)$.

SOLUTION   Applying the FOIL method, we multiply as if we were multiplying two binomials:

$$(x^{2/3} - 3)(x^{2/3} + 5) = x^{2/3}x^{2/3} + 5x^{2/3} - 3x^{2/3} - 15$$

$$= x^{4/3} + 2x^{2/3} - 15$$

EXAMPLE 3 Multiply $(3a^{1/3} - 2b^{1/3})(4a^{1/3} - b^{1/3})$.

SOLUTION Again, we use the FOIL method to multiply:

$$(3a^{1/3} - 2b^{1/3})(4a^{1/3} - b^{1/3})$$

$$= 3a^{1/3}4a^{1/3} - 3a^{1/3}b^{1/3} - 2b^{1/3}4a^{1/3} + 2b^{1/3}b^{1/3}$$

$$= 12a^{2/3} - 11a^{1/3}b^{1/3} + 2b^{2/3}$$

 EXAMPLE 4 Expand $(t^{1/2} - 5)^2$.

SOLUTION We can use the definition of exponents and the FOIL method:

$$(t^{1/2} - 5)^2 = (t^{1/2} - 5)(t^{1/2} - 5)$$

$$= t^{1/2}t^{1/2} - 5t^{1/2} - 5t^{1/2} + 25$$

$$= t - 10t^{1/2} + 25$$

We can obtain the same result by using the formula for the square of a binomial, $(a - b)^2 = a^2 - 2ab + b^2$.

$$(t^{1/2} - 5)^2 = (t^{1/2})^2 - 2t^{1/2} \cdot 5 + 5^2$$

$$= t - 10t^{1/2} + 25$$

 EXAMPLE 5 Multiply $(x^{3/2} - 2^{3/2})(x^{3/2} + 2^{3/2})$.

SOLUTION This product has the form $(a - b)(a + b)$, which will result in the difference of two squares, $a^2 - b^2$:

$$(x^{3/2} - 2^{3/2})(x^{3/2} + 2^{3/2}) = (x^{3/2})^2 - (2^{3/2})^2$$

$$= x^3 - 2^3$$

$$= x^3 - 8$$

 EXAMPLE 6 Multiply $(a^{1/3} - b^{1/3})(a^{2/3} + a^{1/3}b^{1/3} + b^{2/3})$.

SOLUTION We can find this product by multiplying in columns:

$$
\begin{array}{r}
a^{2/3} + a^{1/3}b^{1/3} + b^{2/3} \\
a^{1/3} - b^{1/3} \\
\hline
a + a^{2/3}b^{1/3} + a^{1/3}b^{2/3} \\
- a^{2/3}b^{1/3} - a^{1/3}b^{2/3} - b \\
\hline
a \qquad\qquad\qquad - b
\end{array}
$$

The product is $a - b$.

Our next example involves division with expressions that contain rational exponents. As you will see, this kind of division is very similar to division of a polynomial by a monomial.

EXAMPLE 7 Divide $\dfrac{15x^{2/3}y^{1/3} - 20x^{4/3}y^{2/3}}{5x^{1/3}y^{1/3}}$.

Note
For Example 7, assume x and y are positive numbers.

SOLUTION We can approach this problem in the same way we approached division by a monomial. We simply divide each term in the numerator by the term in the denominator:

$$\frac{15x^{2/3}y^{1/3} - 20x^{4/3}y^{2/3}}{5x^{1/3}y^{1/3}} = \frac{15x^{2/3}y^{1/3}}{5x^{1/3}y^{1/3}} - \frac{20x^{4/3}y^{2/3}}{5x^{1/3}y^{1/3}}$$

$$= 3x^{1/3} - 4xy^{1/3}$$

The next three examples involve factoring. In the first example, we are told what to factor from each term of an expression.

EXAMPLE 8 Factor $3(x - 2)^{1/3}$ from $12(x - 2)^{4/3} - 9(x - 2)^{1/3}$, and then simplify, if possible.

SOLUTION This solution is similar to factoring out the greatest common factor:

$$12(x - 2)^{4/3} - 9(x - 2)^{1/3} = 3(x - 2)^{1/3}[4(x - 2) - 3]$$

$$= 3(x - 2)^{1/3}(4x - 11)$$

Although an expression containing rational exponents is not a polynomial—remember, a polynomial must have exponents that are whole numbers—we are going to treat the expressions that follow as if they were polynomials.

EXAMPLE 9 Factor $x^{2/3} - 3x^{1/3} - 10$ as if it were a trinomial.

SOLUTION We can think of $x^{2/3} - 3x^{1/3} - 10$ as if it is a trinomial in which the variable is $x^{1/3}$. To see this, replace $x^{1/3}$ with y to get

$$y^2 - 3y - 10$$

Since this trinomial in y factors as $(y - 5)(y + 2)$, we can factor our original expression similarly:

$$x^{2/3} - 3x^{1/3} - 10 = (x^{1/3} - 5)(x^{1/3} + 2)$$

Remember, with factoring, we can always multiply our factors to check that we have factored correctly.

EXAMPLE 10 Factor $6x^{2/5} + 11x^{1/5} - 10$ as if it were a trinomial.

SOLUTION We can think of the expression in question as a trinomial in $x^{1/5}$.

$$6x^{2/5} + 11x^{1/5} - 10 = (3x^{1/5} - 2)(2x^{1/5} + 5)$$

In our next example, we combine two expressions by applying the methods we used to add and subtract fractions or rational expressions.

EXAMPLE 11 Subtract $(x^2 + 4)^{1/2} - \dfrac{x^2}{(x^2 + 4)^{1/2}}$.

SOLUTION To combine these two expressions, we need to find a least common denominator, change to equivalent fractions, and subtract numerators. The least common denominator is $(x^2 + 4)^{1/2}$.

$$(x^2 + 4)^{1/2} - \frac{x^2}{(x^2 + 4)^{1/2}} = \frac{(x^2 + 4)^{1/2}}{1} \cdot \frac{(x^2 + 4)^{1/2}}{(x^2 + 4)^{1/2}} - \frac{x^2}{(x^2 + 4)^{1/2}}$$

$$= \frac{x^2 + 4 - x^2}{(x^2 + 4)^{1/2}}$$

$$= \frac{4}{(x^2 + 4)^{1/2}}$$

EXAMPLE 12 If you purchase an investment for P dollars and t years later it is worth A dollars, then the annual rate of return r on that investment is given by the formula

$$r = \left(\frac{A}{P}\right)^{1/t} - 1$$

Find the annual rate of return on a coin collection that was purchased for $210 and sold 3 years later for $300.

SOLUTION Using $A = 300$, $P = 210$, and $t = 3$ in the formula, we have

$$r = \left(\frac{300}{210}\right)^{1/3} - 1$$

The easiest way to simplify this expression is with a calculator.

$$\boxed{(}\ 300\ \boxed{\div}\ 210\ \boxed{)}\ \boxed{\wedge}\ \boxed{(}\ 1\ \boxed{\div}\ 3\ \boxed{)}\ \boxed{-}\ 1\ \boxed{=}$$

Allowing three decimal places, the result is 0.126. The annual return on the coin collection is approximately 12.6%. To do as well with a savings account, we would have to invest the original $210 in an account that paid 12.6%, compounded annually.

LINKING OBJECTIVES AND EXAMPLES

Next to each objective we have listed the examples that are best described by that objective.

A	1–6
B	7
C	8–10
D	11

GETTING READY FOR CLASS

After reading through the preceding section, respond in your own words and in complete sentences.

1. When multiplying expressions with fractional exponents, when do we add the fractional exponents?

2. Is it possible to multiply two expressions with fractional exponents and end up with an expression containing only integer exponents? Support your answer with examples.

3. Write an application modeled by the equation $r = \left(\dfrac{1,000}{600}\right)^{1/8} - 1$.

4. When can you use the FOIL method with expressions that contain rational exponents?

Multiply. (Assume all variables in this problem set represent nonnegative real numbers.)

1. $x^{2/3}(x^{1/3} + x^{4/3})$

2. $x^{2/5}(x^{3/5} - x^{8/5})$

▶ **3.** $a^{1/2}(a^{3/2} - a^{1/2})$

4. $a^{1/4}(a^{3/4} + a^{7/4})$

5. $2x^{1/3}(3x^{8/3} - 4x^{5/3} + 5x^{2/3})$

6. $5x^{1/2}(4x^{5/2} + 3x^{3/2} + 2x^{1/2})$

7. $4x^{1/2}y^{3/5}(3x^{3/2}y^{-3/5} - 9x^{-1/2}y^{7/5})$

8. $3x^{4/5}y^{1/3}(4x^{6/5}y^{-1/3} - 12x^{-4/5}y^{5/3})$

9. $(x^{2/3} - 4)(x^{2/3} + 2)$

10. $(x^{2/3} - 5)(x^{2/3} + 2)$

11. $(a^{1/2} - 3)(a^{1/2} - 7)$

12. $(a^{1/2} - 6)(a^{1/2} - 2)$

13. $(4y^{1/3} - 3)(5y^{1/3} + 2)$

14. $(5y^{1/3} - 2)(4y^{1/3} + 3)$

15. $(5x^{2/3} + 3y^{1/2})(2x^{2/3} + 3y^{1/2})$

16. $(4x^{2/3} - 2y^{1/2})(5x^{2/3} - 3y^{1/2})$

17. $(t^{1/2} + 5)^2$

18. $(t^{1/2} - 3)^2$

▶ **19.** $(x^{3/2} + 4)^2$

20. $(x^{3/2} - 6)^2$

21. $(a^{1/2} - b^{1/2})^2$

22. $(a^{1/2} + b^{1/2})^2$

23. $(2x^{1/2} - 3y^{1/2})^2$

24. $(5x^{1/2} + 4y^{1/2})^2$

25. $(a^{1/2} - 3^{1/2})(a^{1/2} + 3^{1/2})$

26. $(a^{1/2} - 5^{1/2})(a^{1/2} + 5^{1/2})$

27. $(x^{3/2} + y^{3/2})(x^{3/2} - y^{3/2})$

28. $(x^{5/2} + y^{5/2})(x^{5/2} - y^{5/2})$

29. $(t^{1/2} - 2^{3/2})(t^{1/2} + 2^{3/2})$

30. $(t^{1/2} - 5^{3/2})(t^{1/2} + 5^{3/2})$

31. $(2x^{3/2} + 3^{1/2})(2x^{3/2} - 3^{1/2})$

32. $(3x^{1/2} + 2^{3/2})(3x^{1/2} - 2^{3/2})$

33. $(x^{1/3} + y^{1/3})(x^{2/3} - x^{1/3}y^{1/3} + y^{2/3})$

34. $(x^{1/3} - y^{1/3})(x^{2/3} + x^{1/3}y^{1/3} + y^{2/3})$

35. $(a^{1/3} - 2)(a^{2/3} + 2a^{1/3} + 4)$

36. $(a^{1/3} + 3)(a^{2/3} - 3a^{1/3} + 9)$

37. $(2x^{1/3} + 1)(4x^{2/3} - 2x^{1/3} + 1)$

38. $(3x^{1/3} - 1)(9x^{2/3} + 3x^{1/3} + 1)$

39. $(t^{1/4} - 1)(t^{1/4} + 1)(t^{1/2} + 1)$

40. $(t^{1/4} - 2)(t^{1/4} + 2)(t^{1/2} + 4)$

Divide. (Assume all variables represent positive numbers.)

▶ **41.** $\dfrac{18x^{3/4} + 27x^{1/4}}{9x^{1/4}}$

42. $\dfrac{25x^{1/4} + 30x^{3/4}}{5x^{1/4}}$

43. $\dfrac{12x^{2/3}y^{1/3} - 16x^{1/3}y^{2/3}}{4x^{1/3}y^{1/3}}$

44. $\dfrac{12x^{4/3}y^{1/3} - 18x^{1/3}y^{4/3}}{6x^{1/3}y^{1/3}}$

45. $\dfrac{21a^{7/5}b^{3/5} - 14a^{2/5}b^{8/5}}{7a^{2/5}b^{3/5}}$

46. $\dfrac{24a^{9/5}b^{3/5} - 16a^{4/5}b^{8/5}}{8a^{4/5}b^{3/5}}$

47. Factor $3(x - 2)^{1/2}$ from $12(x - 2)^{3/2} - 9(x - 2)^{1/2}$.

48. Factor $4(x + 1)^{1/3}$ from $4(x + 1)^{4/3} + 8(x + 1)^{1/3}$.

49. Factor $5(x - 3)^{7/5}$ from $5(x - 3)^{12/5} - 15(x - 3)^{7/5}$.

50. Factor $6(x + 3)^{8/7}$ from $6(x + 3)^{15/7} - 12(x + 3)^{8/7}$.

51. Factor $3(x + 1)^{1/2}$ from $9x(x + 1)^{3/2} + 6(x + 1)^{1/2}$.

52. Factor $4x(x + 1)^{1/2}$ from $4x^2(x + 1)^{1/2} + 8x(x + 1)^{3/2}$.

Factor each of the following as if it were a trinomial.

53. $x^{2/3} - 5x^{1/3} + 6$

54. $x^{2/3} - x^{1/3} - 6$

55. $a^{2/5} - 2a^{1/5} - 8$

56. $a^{2/5} + 2a^{1/5} - 8$

57. $2y^{2/3} - 5y^{1/3} - 3$

58. $3y^{2/3} + 5y^{1/3} - 2$

59. $9t^{2/5} - 25$

= Videos available by instructor request

▶ = Online student support materials available at academic.cengage.com/login

60. $16t^{2/5} - 49$

61. $4x^{2/7} + 20x^{1/7} + 25$

62. $25x^{2/7} - 20x^{1/7} + 4$

Evaluate the following functions for the given value.

▶ **63.** If $f(x) = x - 2\sqrt{x} - 8$, find $f(4)$.

▶ **64.** If $g(x) = x - 2\sqrt{x} - 3$, find $g(9)$.

▶ **65.** If $f(x) = 2x + 9\sqrt{x} - 5$, find $f(25)$.

▶ **66.** If $f(t) = t - 2\sqrt{t} - 15$, find $f(9)$.

▶ **67.** If $g(x) = 2x - \sqrt{x} - 6$, find $g\left(\dfrac{9}{4}\right)$.

▶ **68.** If $f(x) = 2x + \sqrt{x} - 15$, find $f(9)$.

▶ **69.** If $f(x) = x^{2/3} - 2x^{1/3} - 8$, find $f(-8)$.

▶ **70.** If $g(x) = x^{2/3} + 4x^{1/3} - 12$, find $g(8)$.

Simplify each of the following to a single fraction. (Assume all variables represent positive numbers.)

71. $\dfrac{3}{x^{1/2}} + x^{1/2}$

72. $\dfrac{2}{x^{1/2}} - x^{1/2}$

73. $x^{2/3} + \dfrac{5}{x^{1/3}}$

74. $x^{3/4} - \dfrac{7}{x^{1/4}}$

75. $\dfrac{3x^2}{(x^3 + 1)^{1/2}} + (x^3 + 1)^{1/2}$

76. $\dfrac{x^3}{(x^2 - 1)^{1/2}} + 2x(x^2 - 1)^{1/2}$

77. $\dfrac{x^2}{(x^2 + 4)^{1/2}} - (x^2 + 4)^{1/2}$

78. $\dfrac{x^5}{(x^2 - 2)^{1/2}} + 4x^3(x^2 - 2)^{1/2}$

Applying the Concepts

79. Investing A coin collection is purchased as an investment for $500 and sold 4 years later for $900. Find the annual rate of return on the investment.

80. Investing An investor buys stock in a company for $800. Five years later, the same stock is worth $1,600. Find the annual rate of return on the stocks.

Maintaining Your Skills

81. Find the slope of the line that contains $(-4, -1)$ and $(-2, 5)$.

82. A line has a slope of $\dfrac{2}{3}$. Find the slope of any line
  a. Parallel to it.
  b. Perpendicular to it.

83. Give the slope and y-intercept of the line $2x - 3y = 6$.

84. Give the equation of the line with slope -3 and y-intercept 5.

85. Find the equation of the line with slope $\dfrac{2}{3}$ that contains the point $(-6, 2)$.

86. Find the equation of the line through $(1, 3)$ and $(-1, -5)$.

87. Find the equation of the line with x-intercept 3 and y-intercept -2.

88. Find the equation of the line through $(-1, 4)$ whose graph is perpendicular to the graph of $y = 2x + 3$.

Getting Ready for the Next Section

Simplify. Assume all variable are positive real numbers.

89. $\sqrt{25}$

90. $\sqrt{4}$

91. $\sqrt{6^2}$

92. $\sqrt{3^2}$

93. $\sqrt{16x^4y^2}$

94. $\sqrt{4x^6y^8}$

95. $\sqrt{(5y)^2}$

96. $\sqrt{(8x^3)^2}$

97. $\sqrt[3]{27}$

98. $\sqrt[3]{-8}$

99. $\sqrt[3]{2^3}$

100. $\sqrt[3]{(-5)^3}$

101. $\sqrt[3]{8a^3b^3}$

102. $\sqrt[3]{64a^6b^3}$

Fill in the blank.

103. $50 = \underline{\hspace{2cm}} \cdot 2$

104. $12 = \underline{\hspace{2cm}} \cdot 3$

105. $48x^4y^3 = \underline{\hspace{2cm}} \cdot y$

106. $40a^5b^4 = \underline{\hspace{2cm}} \cdot 5a^2b$

107. $12x^7y^6 = \underline{\hspace{2cm}} \cdot 3x$

108. $54a^6b^2c^4 = \underline{\hspace{2cm}} \cdot 2b^2c$

9.3 Simplified Form for Radicals

OBJECTIVES

A Write radical expressions in simplified form.

B Rationalize a denominator that contains only one term.

Earlier in this chapter, we showed how the Pythagorean theorem can be used to construct a golden rectangle. In a similar manner, the Pythagorean theorem can be used to construct the attractive spiral shown here.

The Spiral of Roots

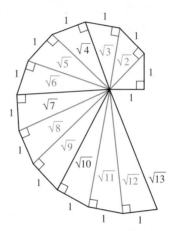

This spiral is called the Spiral of Roots because each of the diagonals is the positive square root of one of the positive integers. At the end of this section, we will use the Pythagorean theorem and some of the material in this section to construct this spiral.

In this section we will use radical notation instead of rational exponents. We will begin by stating two properties of radicals. Following this, we will give a definition for simplified form for radical expressions. The examples in this section show how we use the properties of radicals to write radical expressions in simplified form.

There are two properties of radicals. For these two properties, we will assume a and b are nonnegative real numbers whenever n is an even number.

Property 1 for Radicals
$$\sqrt[n]{ab} = \sqrt[n]{a}\ \sqrt[n]{b}$$

In words: The nth root of a product is the product of the nth roots.

Proof of Property 1

$\sqrt[n]{ab} = (ab)^{1/n}$ **Definition of fractional exponents**

$\quad\quad = a^{1/n}b^{1/n}$ **Exponents distribute over products**

$\quad\quad = \sqrt[n]{a}\ \sqrt[n]{b}$ **Definition of fractional exponents**

Property 2 for Radicals
$$\sqrt[n]{\frac{a}{b}} = \frac{\sqrt[n]{a}}{\sqrt[n]{b}} \quad (b \neq 0)$$

In words: The nth root of a quotient is the quotient of the nth roots.

Note

There is no property for radicals that says the nth root of a sum is the sum of the nth roots; that is, in general

$$\sqrt[n]{a + b} \neq \sqrt[n]{a} + \sqrt[n]{b}$$

The proof of property 2 is similar to the proof of property 1.

The two properties of radicals allow us to change the form of and simplify radical expressions without changing their value.

> **Simplified Form for Radical Expressions** A radical expression is in *simplified form* if
>
> **1.** None of the factors of the radicand (the quantity under the radical sign) can be written as powers greater than or equal to the index—that is, no perfect squares can be factors of the quantity under a square root sign, no perfect cubes can be factors of what is under a cube root sign, and so forth;
>
> **2.** There are no fractions under the radical sign; and
>
> **3.** There are no radicals in the denominator.

Satisfying the first condition for simplified form actually amounts to taking as much out from under the radical sign as possible. The following examples illustrate the first condition for simplified form.

EXAMPLE 1   Write $\sqrt{50}$ in simplified form.

SOLUTION   The largest perfect square that divides 50 is 25. We write 50 as $25 \cdot 2$ and apply property 1 for radicals:

$$\sqrt{50} = \sqrt{25 \cdot 2} \qquad \textbf{50 = 25} \cdot \textbf{2}$$
$$= \sqrt{25}\,\sqrt{2} \qquad \textbf{Property 1}$$
$$= 5\sqrt{2} \qquad \sqrt{\textbf{25}} = \textbf{5}$$

We have taken as much as possible out from under the radical sign—in this case, factoring 25 from 50 and then writing $\sqrt{25}$ as 5.

EXAMPLE 2   Write in simplified form: $\sqrt{48x^4y^3}$, where $x, y \geq 0$.

SOLUTION   The largest perfect square that is a factor of the radicand is $16x^4y^2$. Applying property 1 again, we have

$$\sqrt{48x^4y^3} = \sqrt{16x^4y^2 \cdot 3y}$$
$$= \sqrt{16x^4y^2}\,\sqrt{3y}$$
$$= 4x^2y\sqrt{3y}$$

EXAMPLE 3   Write $\sqrt[3]{40a^5b^4}$ in simplified form.

SOLUTION   We now want to factor the largest perfect cube from the radicand. We write $40a^5b^4$ as $8a^3b^3 \cdot 5a^2b$ and proceed as we did in Examples 1 and 2.

$$\sqrt[3]{40a^5b^4} = \sqrt[3]{8a^3b^3 \cdot 5a^2b}$$
$$= \sqrt[3]{8a^3b^3}\,\sqrt[3]{5a^2b}$$
$$= 2ab\sqrt[3]{5a^2b}$$

Here are some further examples concerning the first condition for simplified form.

EXAMPLES   Write each expression in simplified form.

4. $\sqrt{12x^7y^6} = \sqrt{4x^6y^6 \cdot 3x}$

    $= \sqrt{4x^6y^6} \sqrt{3x}$

    $= 2x^3y^3\sqrt{3x}$

5. $\sqrt[3]{54a^6b^2c^4} = \sqrt[3]{27a^6c^3 \cdot 2b^2c}$

    $= \sqrt[3]{27a^6c^3} \sqrt[3]{2b^2c}$

    $= 3a^2c\sqrt[3]{2b^2c}$

The second property of radicals is used to simplify a radical that contains a fraction.

EXAMPLE 6   Simplify $\sqrt{\dfrac{3}{4}}$.

SOLUTION   Applying property 2 for radicals, we have

$$\sqrt{\frac{3}{4}} = \frac{\sqrt{3}}{\sqrt{4}}   \quad\textbf{Property 2}$$

$$= \frac{\sqrt{3}}{2}   \quad\sqrt{4} = 2$$

The last expression is in simplified form because it satisfies all three conditions for simplified form.

EXAMPLE 7   Write $\sqrt{\dfrac{5}{6}}$ in simplified form.

SOLUTION   Proceeding as in Example 6, we have

$$\sqrt{\frac{5}{6}} = \frac{\sqrt{5}}{\sqrt{6}}$$

> **Note**
> The idea behind rationalizing the denominator is to produce a perfect square under the square root sign in the denominator. This is accomplished by multiplying both the numerator and denominator by the appropriate radical.

The resulting expression satisfies the second condition for simplified form since neither radical contains a fraction. It does, however, violate condition 3 since it has a radical in the denominator. Getting rid of the radical in the denominator is called *rationalizing the denominator* and is accomplished, in this case, by multiplying the numerator and denominator by $\sqrt{6}$:

$$\frac{\sqrt{5}}{\sqrt{6}} = \frac{\sqrt{5}}{\sqrt{6}} \cdot \frac{\sqrt{6}}{\sqrt{6}}$$

$$= \frac{\sqrt{30}}{\sqrt{6^2}}$$

$$= \frac{\sqrt{30}}{6}$$

EXAMPLES   Rationalize the denominator.

8. $\dfrac{4}{\sqrt{3}} = \dfrac{4}{\sqrt{3}} \cdot \dfrac{\sqrt{3}}{\sqrt{3}}$

$\qquad = \dfrac{4\sqrt{3}}{\sqrt{3^2}}$

$\qquad = \dfrac{4\sqrt{3}}{3}$

9. $\dfrac{2\sqrt{3x}}{\sqrt{5y}} = \dfrac{2\sqrt{3x}}{\sqrt{5y}} \cdot \dfrac{\sqrt{5y}}{\sqrt{5y}}$

$\qquad = \dfrac{2\sqrt{15xy}}{\sqrt{(5y)^2}}$

$\qquad = \dfrac{2\sqrt{15xy}}{5y}$

When the denominator involves a cube root, we must multiply by a radical that will produce a perfect cube under the cube root sign in the denominator, as our next example illustrates.

 EXAMPLE 10   Rationalize the denominator in $\dfrac{7}{\sqrt[3]{4}}$.

SOLUTION   Since $4 = 2^2$, we can multiply both numerator and denominator by $\sqrt[3]{2}$ and obtain $\sqrt[3]{2^3}$ in the denominator.

$$\dfrac{7}{\sqrt[3]{4}} = \dfrac{7}{\sqrt[3]{2^2}}$$

$$= \dfrac{7}{\sqrt[3]{2^2}} \cdot \dfrac{\sqrt[3]{2}}{\sqrt[3]{2}}$$

$$= \dfrac{7\sqrt[3]{2}}{\sqrt[3]{2^3}}$$

$$= \dfrac{7\sqrt[3]{2}}{2}$$

 EXAMPLE 11   Simplify $\sqrt{\dfrac{12x^5y^3}{5z}}$.

SOLUTION   We use property 2 to write the numerator and denominator as two separate radicals:

$$\sqrt{\dfrac{12x^5y^3}{5z}} = \dfrac{\sqrt{12x^5y^3}}{\sqrt{5z}}$$

Simplifying the numerator, we have

$$\dfrac{\sqrt{12x^5y^3}}{\sqrt{5z}} = \dfrac{\sqrt{4x^4y^2}\,\sqrt{3xy}}{\sqrt{5z}}$$

$$= \dfrac{2x^2y\sqrt{3xy}}{\sqrt{5z}}$$

To rationalize the denominator, we multiply the numerator and denominator by $\sqrt{5z}$:

$$\frac{2x^2y\sqrt{3xy}}{\sqrt{5z}} \cdot \frac{\sqrt{5z}}{\sqrt{5z}} = \frac{2x^2y\sqrt{15xyz}}{\sqrt{(5z)^2}}$$

$$= \frac{2x^2y\sqrt{15xyz}}{5z}$$

The Square Root of a Perfect Square

So far in this chapter we have assumed that all our variables are nonnegative when they appear under a square root symbol. There are times, however, when this is not the case.

Consider the following two statements:

$$\sqrt{3^2} = \sqrt{9} = 3 \quad \text{and} \quad \sqrt{(-3)^2} = \sqrt{9} = 3$$

Whether we operate on 3 or -3, the result is the same: Both expressions simplify to 3. The other operation we have worked with in the past that produces the same result is absolute value; that is,

$$|3| = 3 \quad \text{and} \quad |-3| = 3$$

This leads us to the next property of radicals.

> **Property 3 for Radicals** If a is a real number, then $\sqrt{a^2} = |a|$.

The result of this discussion and property 3 is simply this:

If we know a is positive, then $\sqrt{a^2} = a$.

If we know a is negative, then $\sqrt{a^2} = |a|$.

If we don't know if a is positive or negative, then $\sqrt{a^2} = |a|$.

 EXAMPLES   Simplify each expression. Do *not* assume the variables represent positive numbers.

12. $\sqrt{9x^2} = 3|x|$

13. $\sqrt{x^3} = |x|\sqrt{x}$

14. $\sqrt{x^2 - 6x + 9} = \sqrt{(x-3)^2} = |x-3|$

15. $\sqrt{x^3 - 5x^2} = \sqrt{x^2(x-5)} = |x|\sqrt{x-5}$

As you can see, we must use absolute value symbols when we take a square root of a perfect square, unless we know the base of the perfect square is a positive number. The same idea holds for higher even roots, but not for odd roots. With odd roots, no absolute value symbols are necessary.

 EXAMPLES   Simplify each expression.

16. $\sqrt[3]{(-2)^3} = \sqrt[3]{-8} = -2$

17. $\sqrt[3]{(-5)^3} = \sqrt[3]{-125} = -5$

We can extend this discussion to all roots as follows:

> **Extending Property 3 for Radicals** If a is a real number, then
>
> $$\sqrt[n]{a^n} = |a| \qquad \text{if} \qquad n \text{ is even}$$
>
> $$\sqrt[n]{a^n} = a \qquad \text{if} \qquad n \text{ is odd}$$

LINKING OBJECTIVES AND EXAMPLES

Next to each **objective** we have listed the examples that are best described by that objective.

A 1–6, 12–17

B 7–11

GETTING READY FOR CLASS

After reading through the preceding section, respond in your own words and in complete sentences.

1. Explain why this statement is false: "The square root of a sum is the sum of the square roots."
2. What is simplified form for an expression that contains a square root?
3. Why is it not necessarily true that $\sqrt{a^2} = a$?
4. What does it mean to rationalize the denominator in an expression?

Problem Set 9.3

Online support materials can be found at academic.cengage.com/login

Use property 1 for radicals to write each of the following expressions in simplified form. (Assume all variables are nonnegative through Problem 84.)

1. $\sqrt{8}$
2. $\sqrt{32}$
3. $\sqrt{98}$
4. $\sqrt{75}$
5. $\sqrt{288}$
6. $\sqrt{128}$
7. $\sqrt{80}$
8. $\sqrt{200}$
9. $\sqrt{48}$
10. $\sqrt{27}$
11. $\sqrt{675}$
12. $\sqrt{972}$
13. $\sqrt[3]{54}$
14. $\sqrt[3]{24}$
15. $\sqrt[3]{128}$
16. $\sqrt[3]{162}$
17. $\sqrt[3]{432}$
18. $\sqrt[3]{1,536}$
19. $\sqrt[5]{64}$
20. $\sqrt[4]{48}$
21. $\sqrt{18x^3}$
22. $\sqrt{27x^5}$
23. $\sqrt[4]{32y^7}$
24. $\sqrt[5]{32y^7}$
25. $\sqrt[3]{40x^4y^7}$
26. $\sqrt[3]{128x^6y^2}$
27. $\sqrt{48a^2b^3c^4}$
28. $\sqrt{72a^4b^3c^2}$
29. $\sqrt[3]{48a^2b^3c^4}$
30. $\sqrt[3]{72a^4b^3c^2}$

31. $\sqrt[5]{64x^8y^{12}}$
32. $\sqrt[4]{32x^9y^{10}}$
33. $\sqrt[5]{243x^7y^{10}z^5}$
34. $\sqrt[5]{64x^8y^4z^{11}}$

Substitute the given number into the expression $\sqrt{b^2 - 4ac}$, and then simplify.

35. $a = 2, b = -6, c = 3$
36. $a = 6, b = 7, c = -5$
37. $a = 1, b = 2, c = 6$
38. $a = 2, b = 5, c = 3$
39. $a = \dfrac{1}{2}, b = -\dfrac{1}{2}, c = -\dfrac{5}{4}$
40. $a = \dfrac{7}{4}, b = -\dfrac{3}{4}, c = -2$

Rationalize the denominator in each of the following expressions.

41. $\dfrac{2}{\sqrt{3}}$
42. $\dfrac{3}{\sqrt{2}}$
43. $\dfrac{5}{\sqrt{6}}$
44. $\dfrac{7}{\sqrt{5}}$

= Videos available by instructor request

▶ = Online student support materials available at academic.cengage.com/login

45. $\sqrt{\dfrac{1}{2}}$

46. $\sqrt{\dfrac{1}{3}}$

47. $\sqrt{\dfrac{1}{5}}$

48. $\sqrt{\dfrac{1}{6}}$

49. $\dfrac{4}{\sqrt[3]{2}}$

50. $\dfrac{5}{\sqrt[3]{3}}$

51. $\dfrac{2}{\sqrt[3]{9}}$

52. $\dfrac{3}{\sqrt[3]{4}}$

53. $\sqrt[4]{\dfrac{3}{2x^2}}$

54. $\sqrt[4]{\dfrac{5}{3x^2}}$

55. $\sqrt[4]{\dfrac{8}{y}}$

56. $\sqrt[4]{\dfrac{27}{y}}$

57. $\sqrt[3]{\dfrac{4x}{3y}}$

58. $\sqrt[3]{\dfrac{7x}{6y}}$

59. $\sqrt[3]{\dfrac{2x}{9y}}$

60. $\sqrt[3]{\dfrac{5x}{4y}}$

61. $\sqrt[4]{\dfrac{1}{8x^3}}$

62. $\sqrt[4]{\dfrac{8}{9x^3}}$

Write each of the following in simplified form.

63. $\sqrt{\dfrac{27x^3}{5y}}$

64. $\sqrt{\dfrac{12x^5}{7y}}$

65. $\sqrt{\dfrac{75x^3y^2}{2z}}$

66. $\sqrt{\dfrac{50x^2y^3}{3z}}$

67. $\sqrt[3]{\dfrac{16a^4b^3}{9c}}$

68. $\sqrt[3]{\dfrac{54a^5b^4}{25c^2}}$

69. $\sqrt[3]{\dfrac{8x^3y^6}{9z}}$

70. $\sqrt[3]{\dfrac{27x^6y^3}{2z^2}}$

71. $\sqrt{\sqrt{x^2}}$

72. $\sqrt{\sqrt{2x^3}}$

73. $\sqrt[3]{\sqrt{xy}}$

74. $\sqrt{\sqrt{4x}}$

75. $\sqrt[3]{\sqrt[4]{a}}$

76. $\sqrt[6]{\sqrt[4]{x}}$

77. $\sqrt[3]{\sqrt[3]{6x^{10}}}$

78. $\sqrt[5]{\sqrt{x^{14}y^{11}z}}$

79. $\sqrt[4]{\sqrt[3]{a^{12}b^{24}c^{14}}}$

80. $\sqrt{\sqrt[3]{4a^{17}}}$

81. $\sqrt[3]{\sqrt[5]{3a^{17}b^{16}c^{30}}}$

82. $\left(\sqrt{\sqrt[4]{x^4y^8z^9}}\right)^2$

83. $\left(\sqrt{\sqrt[3]{8ab^6}}\right)^2$

84. $\left(\sqrt[4]{\sqrt[3]{16x^8y^{12}z^3}}\right)^3$

Simplify each expression. Do *not* assume the variables represent positive numbers.

85. $\sqrt{25x^2}$

86. $\sqrt{49x^2}$

87. $\sqrt{27x^3y^2}$

88. $\sqrt{40x^3y^2}$

89. $\sqrt{x^2 - 10x + 25}$

90. $\sqrt{x^2 - 16x + 64}$

91. $\sqrt{4x^2 + 12x + 9}$

92. $\sqrt{16x^2 + 40x + 25}$

93. $\sqrt{4a^4 + 16a^3 + 16a^2}$

94. $\sqrt{9a^4 + 18a^3 + 9a^2}$

95. $\sqrt{4x^3 - 8x^2}$

96. $\sqrt{18x^3 - 9x^2}$

97. Show that the statement $\sqrt{a + b} = \sqrt{a} + \sqrt{b}$ is not true by replacing a with 9 and b with 16 and simplifying both sides.

98. Find a pair of values for a and b that will make the statement $\sqrt{a + b} = \sqrt{a} + \sqrt{b}$ true.

Applying the Concepts

99. **Diagonal Distance** The distance d between opposite corners of a rectangular room with length l and width w is given by

$$d = \sqrt{l^2 + w^2}$$

How far is it between opposite corners of a living room that measures 10 by 15 feet?

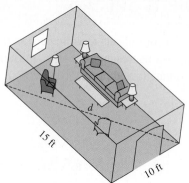

100. **Radius of a Sphere** The radius r of a sphere with volume V can be found by using the formula

$$r = \sqrt[3]{\dfrac{3V}{4\pi}}$$

Find the radius of a sphere with volume 9 cubic feet. Write your answer in simplified form. (Use $\dfrac{22}{7}$ for π.)

Volume

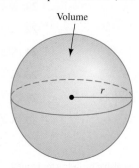

101. Diagonal of a Box The length of the diagonal of a rectangular box with length l, width w, and height h is given by $d = \sqrt{l^2 + w^2 + h^2}$.

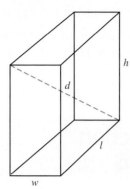

a. Find the length of the diagonal of a rectangular box that is 3 feet wide, 4 feet long, and 12 feet high.

b. Find the length of the diagonal of a rectangular box that is 2 feet wide, 4 feet high, and 6 feet long.

102. Distance to the Horizon If you are at a point k miles above the surface of the Earth, the distance you can see, in miles, is approximated by the equation $d = \sqrt{8000k + k^2}$.

a. How far can you see from a point that is 1 mile above the surface of the Earth?

b. How far can you see from a point that is 2 miles above the surface of the Earth?

c. How far can you see from a point that is 3 miles above the surface of the Earth?

103. Spiral of Roots Construct your own spiral of roots by using a ruler. Draw the first triangle by using two 1-inch lines. The first diagonal will have a length of $\sqrt{2}$ inches. Each new triangle will be formed by drawing a 1-inch line segment at the end of the previous diagonal so that the angle formed is 90°.

104. Spiral of Roots Construct a spiral of roots by using line segments of length 2 inches. The length of the first diagonal will be $2\sqrt{2}$ inches. The length of the second diagonal will be $2\sqrt{3}$ inches.

Maintaining Your Skills

Let $f(x) = \frac{1}{2}x + 3$ and $g(x) = x^2 - 4$, and find

105. $f(0)$ **106.** $g(0)$

107. $g(2)$ **108.** $f(2)$

109. $f(-4)$ **110.** $g(-6)$

111. $f[g(2)]$ **112.** $g[f(2)]$

Getting Ready for the Next Section

Simplify the following.

113. $35xy^2 - 8xy^2$ **114.** $20a^2b + 33a^2b$

115. $\frac{1}{2}x + \frac{1}{3}x$ **116.** $\frac{2}{3}x + \frac{5}{8}x$

Write in simplified form for radicals.

117. $\sqrt{18}$ **118.** $\sqrt{8}$

119. $\sqrt{75xy^3}$ **120.** $\sqrt{12xy}$

121. $\sqrt[3]{8a^4b^2}$ **122.** $\sqrt[3]{27ab^2}$

Extending the Concepts

Factor each radicand into the product of prime factors. Then simplify each radical.

123. $\sqrt[3]{8,640}$ **124.** $\sqrt{8,640}$

125. $\sqrt[3]{10,584}$ **126.** $\sqrt{10,584}$

Assume a is a positive number, and rationalize each denominator.

127. $\dfrac{1}{\sqrt[10]{a^3}}$ **128.** $\dfrac{1}{\sqrt[12]{a^7}}$

129. $\dfrac{1}{\sqrt[20]{a^{11}}}$ **130.** $\dfrac{1}{\sqrt[15]{a^{13}}}$

131. Show that the two expressions $\sqrt{x^2 + 1}$ and $x + 1$ are not, in general, equal to each other by graphing $y = \sqrt{x^2 + 1}$ and $y = x + 1$ in the same viewing window.

132. Show that the two expressions $\sqrt{x^2 + 9}$ and $x + 3$ are not, in general, equal to each other by graphing $y = \sqrt{x^2 + 9}$ and $y = x + 3$ in the same viewing window.

133. Approximately how far apart are the graphs in Problem 131 when $x = 2$?

134. Approximately how far apart are the graphs in Problem 132 when $x = 2$?

135. For what value of x are the expressions $\sqrt{x^2 + 1}$ and $x + 1$ in Problem 131 equal?

136. For what value of x are the expressions $\sqrt{x^2 + 9}$ and $x + 3$ in Problem 132 equal?

9.4 Addition and Subtraction of Radical Expressions

OBJECTIVES

A Add and subtract radicals.

B Construct golden rectangles from squares.

We have been able to add and subtract polynomials by combining similar terms. The same idea applies to addition and subtraction of radical expressions.

> **DEFINITION** Two radicals are said to be **similar radicals** if they have the same index and the same radicand.

The expressions $5\sqrt[3]{7}$ and $-8\sqrt[3]{7}$ are similar since the index is 3 in both cases and the radicands are 7. The expressions $3\sqrt[4]{5}$ and $7\sqrt[3]{5}$ are not similar since they have different indices, and the expressions $2\sqrt[5]{8}$ and $3\sqrt[5]{9}$ are not similar because the radicands are not the same.

 EXAMPLE 1   Combine $5\sqrt{3} - 4\sqrt{3} + 6\sqrt{3}$.

SOLUTION   All three radicals are similar. We apply the distributive property to get

$$5\sqrt{3} - 4\sqrt{3} + 6\sqrt{3} = (5 - 4 + 6)\sqrt{3}$$
$$= 7\sqrt{3}$$

 EXAMPLE 2   Combine $3\sqrt{8} + 5\sqrt{18}$.

SOLUTION   The two radicals do not seem to be similar. We must write each in simplified form before applying the distributive property.

$$3\sqrt{8} + 5\sqrt{18} = 3\sqrt{4 \cdot 2} + 5\sqrt{9 \cdot 2}$$
$$= 3\sqrt{4}\,\sqrt{2} + 5\sqrt{9}\,\sqrt{2}$$
$$= 3 \cdot 2\sqrt{2} + 5 \cdot 3\sqrt{2}$$
$$= 6\sqrt{2} + 15\sqrt{2}$$
$$= (6 + 15)\sqrt{2}$$
$$= 21\sqrt{2}$$

The result of Example 2 can be generalized to the following rule for sums and differences of radical expressions.

> **Rule** To add or subtract radical expressions, put each in simplified form and apply the distributive property if possible. We can add only similar radicals. We must write each expression in simplified form for radicals before we can tell if the radicals are similar.

 EXAMPLE 3   Combine $7\sqrt{75xy^3} - 4y\sqrt{12xy}$, where $x, y \geq 0$.

SOLUTION   We write each expression in simplified form and combine similar radicals:

$$7\sqrt{75xy^3} - 4y\sqrt{12xy} = 7\sqrt{25y^2}\,\sqrt{3xy} - 4y\sqrt{4}\,\sqrt{3xy}$$
$$= 35y\sqrt{3xy} - 8y\sqrt{3xy}$$
$$= (35y - 8y)\sqrt{3xy}$$
$$= 27y\sqrt{3xy}$$

 EXAMPLE 4   Combine $10\sqrt[3]{8a^4b^2} + 11a\sqrt[3]{27ab^2}$.

SOLUTION   Writing each radical in simplified form and combining similar terms, we have

$$10\sqrt[3]{8a^4b^2} + 11a\sqrt[3]{27ab^2} = 10\sqrt[3]{8a^3}\,\sqrt[3]{ab^2} + 11a\sqrt[3]{27}\,\sqrt[3]{ab^2}$$
$$= 20a\sqrt[3]{ab^2} + 33a\sqrt[3]{ab^2}$$
$$= 53a\sqrt[3]{ab^2}$$

 EXAMPLE 5   Combine $\dfrac{\sqrt{3}}{2} + \dfrac{1}{\sqrt{3}}$.

SOLUTION   We begin by writing the second term in simplified form.

$$\frac{\sqrt{3}}{2} + \frac{1}{\sqrt{3}} = \frac{\sqrt{3}}{2} + \frac{1}{\sqrt{3}} \cdot \frac{\sqrt{3}}{\sqrt{3}}$$
$$= \frac{\sqrt{3}}{2} + \frac{\sqrt{3}}{3}$$
$$= \frac{1}{2}\sqrt{3} + \frac{1}{3}\sqrt{3}$$
$$= \left(\frac{1}{2} + \frac{1}{3}\right)\sqrt{3}$$
$$= \frac{5}{6}\sqrt{3} = \frac{5\sqrt{3}}{6}$$

FIGURE 1

EXAMPLE 6   Construct a golden rectangle from a square of side 4. Then show that the ratio of the length to the width is the golden ratio $\dfrac{1 + \sqrt{5}}{2}$.

SOLUTION   Figure 1 shows the golden rectangle constructed from a square of side 4.

The length of the diagonal OB is found from the Pythagorean theorem.

$$OB = \sqrt{2^2 + 4^2} = \sqrt{4 + 16} = \sqrt{20} = 2\sqrt{5}$$

The ratio of the length to the width for the rectangle is the golden ratio.

$$\text{Golden ratio} = \frac{CE}{EF} = \frac{2 + 2\sqrt{5}}{4} = \frac{2(1 + \sqrt{5})}{2 \cdot 2} = \frac{1 + \sqrt{5}}{2}$$

LINKING OBJECTIVES AND EXAMPLES

Next to each objective we have listed the examples that are best described by that objective.

A   1–5

B   6

GETTING READY FOR CLASS

After reading through the preceding section, respond in your own words and in complete sentences.

1. What are similar radicals?
2. When can we add two radical expressions?
3. What is the first step when adding or subtracting expressions containing radicals?
4. What is the golden ratio, and where does it come from?

Problem Set 9.4

Online support materials can be found at academic.cengage.com/login

Combine the following expressions. (Assume any variables under an even root are nonnegative.)

1. $3\sqrt{5} + 4\sqrt{5}$ **2.** $6\sqrt{3} - 5\sqrt{3}$

3. $3x\sqrt{7} - 4x\sqrt{7}$ **4.** $6y\sqrt{a} + 7y\sqrt{a}$

5. $5\sqrt[3]{10} - 4\sqrt[3]{10}$ **6.** $6\sqrt[4]{2} + 9\sqrt[4]{2}$

7. $8\sqrt[5]{6} - 2\sqrt[5]{6} + 3\sqrt[5]{6}$

8. $7\sqrt[6]{7} - \sqrt[6]{7} + 4\sqrt[6]{7}$

9. $3x\sqrt{2} - 4x\sqrt{2} + x\sqrt{2}$

10. $5x\sqrt{6} - 3x\sqrt{6} - 2x\sqrt{6}$

11. $\sqrt{20} - \sqrt{80} + \sqrt{45}$

12. $\sqrt{8} - \sqrt{32} - \sqrt{18}$

13. $4\sqrt{8} - 2\sqrt{50} - 5\sqrt{72}$

14. $\sqrt{48} - 3\sqrt{27} + 2\sqrt{75}$

15. $5x\sqrt{8} + 3\sqrt{32x^2} - 5\sqrt{50x^2}$

16. $2\sqrt{50x^2} - 8x\sqrt{18} - 3\sqrt{72x^2}$

17. $5\sqrt[3]{16} - 4\sqrt[3]{54}$

18. $\sqrt[3]{81} + 3\sqrt[3]{24}$

19. $\sqrt[3]{x^4y^2} + 7x\sqrt[3]{xy^2}$

20. $2\sqrt[3]{x^8y^6} - 3y^2\sqrt[3]{8x^8}$

21. $5a^2\sqrt{27ab^3} - 6b\sqrt{12a^5b}$

22. $9a\sqrt{20a^3b^2} + 7b\sqrt{45a^5}$

23. $b\sqrt[3]{24a^5b} + 3a\sqrt[3]{81a^2b^4}$

24. $7\sqrt[3]{a^4b^3c^2} - 6ab\sqrt[3]{ac^2}$

25. $5x\sqrt[4]{3y^5} + y\sqrt[4]{243x^4y} + \sqrt[4]{48x^4y^5}$

26. $x\sqrt[4]{5xy^8} + y\sqrt[4]{405x^5y^4} + y^2\sqrt[4]{80x^5}$

27. $\dfrac{\sqrt{2}}{2} + \dfrac{1}{\sqrt{2}}$ **28.** $\dfrac{\sqrt{3}}{3} + \dfrac{1}{\sqrt{3}}$

29. $\dfrac{\sqrt{5}}{3} + \dfrac{1}{\sqrt{5}}$ **30.** $\dfrac{\sqrt{6}}{2} + \dfrac{1}{\sqrt{6}}$

31. $\sqrt{x} - \dfrac{1}{\sqrt{x}}$ **32.** $\sqrt{x} + \dfrac{1}{\sqrt{x}}$

33. $\dfrac{\sqrt{18}}{6} + \sqrt{\dfrac{1}{2}} + \dfrac{\sqrt{2}}{2}$

34. $\dfrac{\sqrt{12}}{6} + \sqrt{\dfrac{1}{3}} + \dfrac{\sqrt{3}}{3}$

35. $\sqrt{6} - \sqrt{\dfrac{2}{3}} + \sqrt{\dfrac{1}{6}}$

36. $\sqrt{15} - \sqrt{\dfrac{3}{5}} + \sqrt{\dfrac{5}{3}}$

= Videos available by instructor request

▶ = Online student support materials available at academic.cengage.com/login

37. $\sqrt[3]{25} + \dfrac{3}{\sqrt[3]{5}}$

38. $\sqrt[4]{8} + \dfrac{1}{\sqrt[4]{2}}$

The following problems apply to what you have learned with the algebra of functions.

▶ **39.** Let $f(x) = \sqrt{8x}$ and $g(x) = \sqrt{72x}$, then find
  a. $f(x) + g(x)$
  b. $f(x) - g(x)$

▶ **40.** Let $f(x) = x + \sqrt{3}$ and $g(x) = x + 2\sqrt{3}$, then find
  a. $f(x) + g(x)$
  b. $f(x) - g(x)$

▶ **41.** Let $f(x) = 3\sqrt{2x}$ and $g(x) = \sqrt{2x}$, then find
  a. $f(x) + g(x)$
  b. $f(x) - g(x)$

▶ **42.** Let $f(x) = x\sqrt[3]{64}$ and $g(x) = x\sqrt{81}$, then find
  a. $f(x) + g(x)$
  b. $f(x) - g(x)$

▶ **43.** Let $f(x) = x\sqrt{2}$ and $g(x) = 2x\sqrt{2}$, then find
  a. $f(x) + g(x)$
  b. $f(x) - g(x)$

▶ **44.** Let $f(x) = 5 + 2\sqrt{5x}$ and $g(x) = 3\sqrt{5x}$, then find
  a. $f(x) + g(x)$
  b. $f(x) - g(x)$

▶ **45.** Let $f(x) = \sqrt{2x} - 2$ and $g(x) = 2\sqrt{2x} + 5$, then find
  a. $f(x) + g(x)$
  b. $f(x) - g(x)$

▶ **46.** Let $f(x) = 1 - 2\sqrt[3]{3x}$ and $g(x) = 2 + 3\sqrt[3]{3x}$, then find
  a. $f(x) + g(x)$
  b. $f(x) - g(x)$

47. Use a calculator to find a decimal approximation for $\sqrt{12}$ and for $2\sqrt{3}$.

48. Use a calculator to find decimal approximations for $\sqrt{50}$ and $5\sqrt{2}$.

49. Use a calculator to find a decimal approximation for $\sqrt{8} + \sqrt{18}$. Is it equal to the decimal approximation for $\sqrt{26}$ or $\sqrt{50}$?

50. Use a calculator to find a decimal approximation for $\sqrt{3} + \sqrt{12}$. Is it equal to the decimal approximation for $\sqrt{15}$ or $\sqrt{27}$?

Applying the Concepts

51. Golden Rectangle Construct a golden rectangle from a square of side 8. Then show that the ratio of the length to the width is the golden ratio $\dfrac{1 + \sqrt{5}}{2}$.

52. Golden Rectangle Construct a golden rectangle from a square of side 10. Then show that the ratio of the length to the width is the golden ratio $\dfrac{1 + \sqrt{5}}{2}$.

53. Golden Rectangle To show that all golden rectangles have the same ratio of length to width, construct a golden rectangle from a square of side $2x$. Then show that the ratio of the length to the width is the golden ratio.

54. Golden Rectangle To show that all golden rectangles have the same ratio of length to width, construct a golden rectangle from a square of side x. Then show that the ratio of the length to the width is the golden ratio.

55. Equilateral Triangles A triangle is equilateral if it has three equal sides. The triangle in the figure is equilateral with each side of length $2x$. Find the ratio of the height to a side.

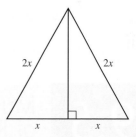

56. Pyramids Refer to the diagram of a square pyramid below. Find the ratio of the height h of the pyramid to the altitude a.

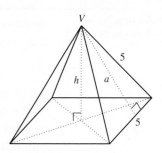

Maintaining Your Skills

Graph each inequality.

57. $2x + 3y < 6$

58. $2x + y < -5$

59. $y \geq -3x - 4$

60. $y \geq 2x - 1$

61. $x \geq 3$

62. $y > -5$

Solve each inequality.

63. $-2 \leq 5x - 1 \leq 2$

64. $2x - 3 < 7$ or $2x + 1 > 15$

Getting Ready for the Next Section

Simplify the following.

65. $3 \cdot 2$

66. $5 \cdot 7$

67. $(x + y)(4x - y)$

68. $(2x + y)(x - y)$

69. $(x + 3)^2$

70. $(3x - 2y)^2$

71. $(x - 2)(x + 2)$

72. $(2x + 5)(2x - 5)$

Simplify the following expressions.

73. $2\sqrt{18}$

74. $5\sqrt{36}$

75. $(\sqrt{6})^2$

76. $(\sqrt{2})^2$

77. $(3\sqrt{x})^2$

78. $(2\sqrt{y})^2$

Rationalize the denominator.

79. $\dfrac{\sqrt{3}}{\sqrt{2}}$

80. $\dfrac{\sqrt{5}}{\sqrt{6}}$

Extending the Concepts

Assume all variables represent positive numbers. Simplify.

81. $\sqrt[5]{32x^5y^5} - y\sqrt[3]{27x^3}$

82. $\sqrt[6]{x^4} + 4\sqrt[3]{8x^2}$

83. $3\sqrt[9]{x^9y^{18}z^{27}} - 4\sqrt[6]{x^6y^{12}z^{18}}$

84. $4a\sqrt{b^4c^6} + 3b\sqrt[3]{a^3b^3c^9}$

85. $3c\sqrt[8]{4a^6b^{18}} + b\sqrt[4]{32a^3b^5c^4}$

86. $4x\sqrt[6]{16y^6z^8} - y\sqrt[3]{32x^3z^4}$

87. $3\sqrt[9]{8a^{12}b^9} + b\sqrt[3]{16a^4} - 8\sqrt[6]{4a^8b^6}$

88. $-ac\sqrt{108bc^3} - 4\sqrt[6]{27a^6b^3c^{15}} + 3\sqrt[4]{9a^4b^2c^{10}}$

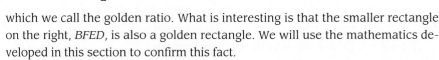

9.5 Multiplication and Division of Radical Expressions

OBJECTIVES

A Multiply expressions containing radicals.

B Rationalize a denominator containing two terms.

We have worked with the golden rectangle more than once in this chapter. The following is one such golden rectangle.

By now you know that in any golden rectangle constructed from a square (of any size) the ratio of the length to the width will be

$$\frac{1 + \sqrt{5}}{2}$$

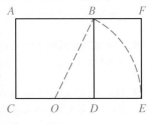

which we call the golden ratio. What is interesting is that the smaller rectangle on the right, *BFED*, is also a golden rectangle. We will use the mathematics developed in this section to confirm this fact.

In this section we will look at multiplication and division of expressions that contain radicals. As you will see, multiplication of expressions that contain radicals is very similar to multiplication of polynomials. The division problems in this section are just an extension of the work we did previously when we rationalized denominators.

 EXAMPLE 1   Multiply $(3\sqrt{5})(2\sqrt{7})$.

SOLUTION   We can rearrange the order and grouping of the numbers in this product by applying the commutative and associative properties. Following this, we apply property 1 for radicals and multiply:

$$(3\sqrt{5})(2\sqrt{7}) = (3 \cdot 2)(\sqrt{5}\,\sqrt{7}) \quad \textbf{Commutative and associative properties}$$

$$= (3 \cdot 2)(\sqrt{5 \cdot 7}) \quad \textbf{Property 1 for radicals}$$

$$= 6\sqrt{35} \quad \textbf{Multiplication}$$

In practice, it is not necessary to show the first two steps.

 EXAMPLE 2   Multiply $\sqrt{3}(2\sqrt{6} - 5\sqrt{12})$.

SOLUTION   Applying the distributive property, we have

$$\sqrt{3}(2\sqrt{6} - 5\sqrt{12}) = \sqrt{3} \cdot 2\sqrt{6} - \sqrt{3} \cdot 5\sqrt{12}$$

$$= 2\sqrt{18} - 5\sqrt{36}$$

Writing each radical in simplified form gives

$$2\sqrt{18} - 5\sqrt{36} = 2\sqrt{9}\,\sqrt{2} - 5\sqrt{36}$$

$$= 6\sqrt{2} - 30$$

 EXAMPLE 3   Multiply $(\sqrt{3} + \sqrt{5})(4\sqrt{3} - \sqrt{5})$.

SOLUTION   The same principle that applies when multiplying two binomials applies to this product. We must multiply each term in the first expression by each term in the second one. Any convenient method can be used. Let's use the FOIL method.

$$(\sqrt{3} + \sqrt{5})(4\sqrt{3} - \sqrt{5}) = \overset{F}{\sqrt{3} \cdot 4\sqrt{3}} - \overset{O}{\sqrt{3}\sqrt{5}} + \overset{I}{\sqrt{5} \cdot 4\sqrt{3}} - \overset{L}{\sqrt{5}\sqrt{5}}$$

$$= 4 \cdot 3 - \sqrt{15} + 4\sqrt{15} - 5$$

$$= 12 + 3\sqrt{15} - 5$$

$$= 7 + 3\sqrt{15}$$

 EXAMPLE 4   Expand and simplify $(\sqrt{x} + 3)^2$.

SOLUTION 1   We can write this problem as a multiplication problem and proceed as we did in Example 3:

$$(\sqrt{x} + 3)^2 = (\sqrt{x} + 3)(\sqrt{x} + 3)$$

$$= \overset{F}{\sqrt{x} \cdot \sqrt{x}} + \overset{O}{3\sqrt{x}} + \overset{I}{3\sqrt{x}} + \overset{L}{3 \cdot 3}$$

$$= x + 3\sqrt{x} + 3\sqrt{x} + 9$$

$$= x + 6\sqrt{x} + 9$$

SOLUTION 2 We can obtain the same result by applying the formula for the square of a sum: $(a + b)^2 = a^2 + 2ab + b^2$.

$$(\sqrt{x} + 3)^2 = (\sqrt{x})^2 + 2(\sqrt{x})(3) + 3^2$$
$$= x + 6\sqrt{x} + 9$$

 EXAMPLE 5 Expand $(3\sqrt{x} - 2\sqrt{y})^2$ and simplify the result.

SOLUTION Let's apply the formula for the square of a difference, $(a - b)^2 = a^2 - 2ab + b^2$.

$$(3\sqrt{x} - 2\sqrt{y})^2 = (3\sqrt{x})^2 - 2(3\sqrt{x})(2\sqrt{y}) + (2\sqrt{y})^2$$
$$= 9x - 12\sqrt{xy} + 4y$$

 EXAMPLE 6 Expand and simplify $(\sqrt{x + 2} - 1)^2$.

SOLUTION Applying the formula $(a - b)^2 = a^2 - 2ab + b^2$, we have

$$(\sqrt{x + 2} - 1)^2 = (\sqrt{x + 2})^2 - 2\sqrt{x + 2}(1) + 1^2$$
$$= x + 2 - 2\sqrt{x + 2} + 1$$
$$= x + 3 - 2\sqrt{x + 2}$$

 EXAMPLE 7 Multiply $(\sqrt{6} + \sqrt{2})(\sqrt{6} - \sqrt{2})$.

SOLUTION We notice the product is of the form $(a + b)(a - b)$, which always gives the difference of two squares, $a^2 - b^2$:

$$(\sqrt{6} + \sqrt{2})(\sqrt{6} - \sqrt{2}) = (\sqrt{6})^2 - (\sqrt{2})^2$$
$$= 6 - 2$$
$$= 4$$

Note

We can prove that conjugates always multiply to yield a rational number as follows: If a and b are positive integers, then
$(\sqrt{a} + \sqrt{b})(\sqrt{a} - \sqrt{b})$
$= \sqrt{a}\sqrt{a} - \sqrt{a}\sqrt{b} + \sqrt{a}\sqrt{b} - \sqrt{b}\sqrt{b}$
$= a - \sqrt{ab} + \sqrt{ab} - b$
$= a - b$
which is rational if a and b are rational.

The two expressions $(\sqrt{6} + \sqrt{2})$ and $(\sqrt{6} - \sqrt{2})$ are called *conjugates*. In general, the conjugate of $\sqrt{a} + \sqrt{b}$ is $\sqrt{a} - \sqrt{b}$. If a and b are integers, multiplying njugates of this form always produces a rational number.

Division with radical expressions is the same as rationalizing the denomina-r. In a previous section we were able to divide $\sqrt{3}$ by $\sqrt{2}$ by rationalizing the nominator:

$$\frac{\sqrt{3}}{\sqrt{2}} = \frac{\sqrt{3}}{\sqrt{2}} \cdot \frac{\sqrt{2}}{\sqrt{2}} = \frac{\sqrt{6}}{2}$$

We can accomplish the same result with expressions such as

$$\frac{6}{\sqrt{5} - \sqrt{3}}$$

by multiplying the numerator and denominator by the conjugate of the denominator.

EXAMPLE 8 Divide $\dfrac{6}{\sqrt{5} - \sqrt{3}}$. (Rationalize the denominator.)

SOLUTION Since the product of two conjugates is a rational number, we multiply the numerator and denominator by the conjugate of the denominator.

$$\frac{6}{\sqrt{5} - \sqrt{3}} = \frac{6}{\sqrt{5} - \sqrt{3}} \cdot \frac{(\sqrt{5} + \sqrt{3})}{(\sqrt{5} + \sqrt{3})}$$

$$= \frac{6\sqrt{5} + 6\sqrt{3}}{(\sqrt{5})^2 - (\sqrt{3})^2}$$

$$= \frac{6\sqrt{5} + 6\sqrt{3}}{5 - 3}$$

$$= \frac{6\sqrt{5} + 6\sqrt{3}}{2}$$

The numerator and denominator of this last expression have a factor of 2 in common. We can reduce to lowest terms by factoring 2 from the numerator and then dividing both the numerator and denominator by 2:

$$= \frac{\cancel{2}(3\sqrt{5} + 3\sqrt{3})}{\cancel{2}}$$

$$= 3\sqrt{5} + 3\sqrt{3}$$

EXAMPLE 9 Rationalize the denominator $\dfrac{\sqrt{5} - 2}{\sqrt{5} + 2}$.

SOLUTION To rationalize the denominator, we multiply the numerator and denominator by the conjugate of the denominator:

$$\frac{\sqrt{5} - 2}{\sqrt{5} + 2} = \frac{\sqrt{5} - 2}{\sqrt{5} + 2} \cdot \frac{(\sqrt{5} - 2)}{(\sqrt{5} - 2)}$$

$$= \frac{5 - 2\sqrt{5} - 2\sqrt{5} + 4}{(\sqrt{5})^2 - 2^2}$$

$$= \frac{9 - 4\sqrt{5}}{5 - 4}$$

$$= \frac{9 - 4\sqrt{5}}{1}$$

$$= 9 - 4\sqrt{5}$$

EXAMPLE 10 A golden rectangle constructed from a square of side 2 is shown in Figure 1. Show that the smaller rectangle *BDEF* is also a golden rectangle by finding the ratio of its length to its width.

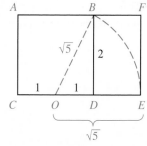

FIGURE 1

SOLUTION First we find expressions for the length and width of the smaller rectangle.

$$\text{Length} = EF = 2$$

$$\text{Width} = DE = \sqrt{5} - 1$$

Next, we find the ratio of length to width.

$$\text{Ratio of length to width} = \frac{EF}{DE} = \frac{2}{\sqrt{5} - 1}$$

To show that the small rectangle is a golden rectangle, we must show that the ratio of length to width is the golden ratio. We do so by rationalizing the denominator.

$$\frac{2}{\sqrt{5} - 1} = \frac{2}{\sqrt{5} - 1} \cdot \frac{\sqrt{5} + 1}{\sqrt{5} + 1}$$

$$= \frac{2(\sqrt{5} + 1)}{5 - 1}$$

$$= \frac{2(\sqrt{5} + 1)}{4}$$

$$= \frac{\sqrt{5} + 1}{2} \qquad \textbf{Divide out common factor 2}$$

Since addition is commutative, this last expression is the golden ratio. Therefore, the small rectangle in Figure 1 is a golden rectangle.

LINKING OBJECTIVES AND EXAMPLES

Next to each objective we have listed the examples that are best described by that objective.

A	1–7
B	8–10

GETTING READY FOR CLASS

After reading through the preceding section, respond in your own words and in complete sentences.

1. Explain why $(\sqrt{5} + \sqrt{2})^2 \neq 5 + 2$.
2. Explain in words how you would rationalize the denominator in the expression $\dfrac{\sqrt{3}}{\sqrt{5} - \sqrt{2}}$
3. What are conjugates?
4. What result is guaranteed when multiplying radical expressions that are conjugates?

EXAMPLE 3 Solve $\sqrt{5x - 1} + 3 = 7$.

SOLUTION We must isolate the radical on the left side of the equation. If we attempt to square both sides without doing so, the resulting equation will also contain a radical. Adding -3 to both sides, we have

$$\sqrt{5x - 1} + 3 = 7$$

$$\sqrt{5x - 1} = 4$$

We can now square both sides and proceed as usual:

$$(\sqrt{5x - 1})^2 = 4^2$$

$$5x - 1 = 16$$

$$5x = 17$$

$$x = \frac{17}{5}$$

Checking $x = \frac{17}{5}$, we have

$$\sqrt{5\left(\frac{17}{5}\right) - 1} + 3 \stackrel{?}{=} 7$$

$$\sqrt{17 - 1} + 3 \stackrel{?}{=} 7$$

$$\sqrt{16} + 3 \stackrel{?}{=} 7$$

$$4 + 3 \stackrel{?}{=} 7$$

$$7 = 7$$

 EXAMPLE 4 Solve $t + 5 = \sqrt{t + 7}$.

SOLUTION This time, squaring both sides of the equation results in a quadratic equation:

$$(t + 5)^2 = (\sqrt{t + 7})^2 \qquad \text{**Square both sides**}$$

$$t^2 + 10t + 25 = t + 7$$

$$t^2 + 9t + 18 = 0 \qquad \text{**Standard form**}$$

$$(t + 3)(t + 6) = 0 \qquad \text{**Factor the left side**}$$

$$t + 3 = 0 \quad \text{or} \quad t + 6 = 0 \qquad \text{**Set factors equal to 0**}$$

$$t = -3 \quad \text{or} \qquad t = -6$$

We must check each solution in the original equation:

Check $t = -3$	Check $t = -6$
$-3 + 5 \stackrel{?}{=} \sqrt{-3 + 7}$	$-6 + 5 \stackrel{?}{=} \sqrt{-6 + 7}$
$2 \stackrel{?}{=} \sqrt{4}$	$-1 \stackrel{?}{=} \sqrt{1}$
$2 = 2$	$-1 = 1$
A true statement	A false statement

Since $t = -6$ does not check, our only solution is $t = -3$.

EXAMPLE 5 Solve $\sqrt{x-3} = \sqrt{x} - 3$.

SOLUTION We begin by squaring both sides. Note what happens when we square the right side of the equation, and compare the square of the right side with the square of the left side. You must convince yourself that these results are correct. (The note in the margin will help if you are having trouble convincing yourself that what is written below is true.)

$$(\sqrt{x-3})^2 = (\sqrt{x}-3)^2$$

$$x - 3 = x - 6\sqrt{x} + 9$$

Now we still have a radical in our equation, so we will have to square both sides again. Before we do, though, let's isolate the remaining radical.

$$x - 3 = x - 6\sqrt{x} + 9$$

$-3 = -6\sqrt{x} + 9$	Add $-x$ to each side
$-12 = -6\sqrt{x}$	Add -9 to each side
$2 = \sqrt{x}$	Divide each side by -6
$4 = x$	Square each side

Our only possible solution is $x = 4$, which we check in our original equation as follows:

$$\sqrt{4-3} \stackrel{?}{=} \sqrt{4} - 3$$

$$\sqrt{1} \stackrel{?}{=} 2 - 3$$

$$1 = -1 \qquad \textbf{A false statement}$$

Substituting 4 for x in the original equation yields a false statement. Since 4 was our only possible solution, there is no solution to our equation.

Here is another example of an equation for which we must apply our squaring property twice before all radicals are eliminated.

 EXAMPLE 6 Solve $\sqrt{x+1} = 1 - \sqrt{2x}$.

SOLUTION This equation has two separate terms involving radical signs. Squaring both sides gives

$$x + 1 = 1 - 2\sqrt{2x} + 2x$$

$-x = -2\sqrt{2x}$	Add $-2x$ and -1 to both sides
$x^2 = 4(2x)$	Square both sides
$x^2 - 8x = 0$	Standard form

Our equation is a quadratic equation in standard form. To solve for x, we factor the left side and set each factor equal to 0.

$x(x - 8) = 0$	Factor left side
$x = 0 \quad \text{or} \quad x - 8 = 0$	Set factors equal to 0
$x = 8$	

Note

It is very important that you realize that the square of $(\sqrt{x} - 3)$ is not $x + 9$. Remember, when we square a difference with two terms, we use the formula

$$(a - b)^2 = a^2 - 2ab + b^2$$

Applying this formula to $(\sqrt{x} - 3)^2$, we have

$$(\sqrt{x} - 3)^2 = (\sqrt{x})^2 - 2(\sqrt{x})(3) + 3^2$$
$$= x - 6\sqrt{x} + 9$$

Since we squared both sides of our equation, we have the possibility that one or both of the solutions are extraneous. We must check each one in the original equation:

Check $x = 8$

$\sqrt{8 + 1} \overset{?}{=} 1 - \sqrt{2 \cdot 8}$

$\sqrt{9} \overset{?}{=} 1 - \sqrt{16}$

$3 \overset{?}{=} 1 - 4$

$3 = -3$

A false statement

Check $x = 0$

$\sqrt{0 + 1} \overset{?}{=} 1 - \sqrt{2 \cdot 0}$

$\sqrt{1} \overset{?}{=} 1 - \sqrt{0}$

$1 \overset{?}{=} 1 - 0$

$1 = 1$

A true statement

Since $x = 8$ does not check, it is an extraneous solution. Our only solution is $x = 0$.

 EXAMPLE 7   Solve $\sqrt{x + 1} = \sqrt{x + 2} - 1$.

SOLUTION   Squaring both sides we have

$$(\sqrt{x + 1})^2 = (\sqrt{x + 2} - 1)^2$$

$$x + 1 = x + 2 - 2\sqrt{x + 2} + 1$$

Once again we are left with a radical in our equation. Before we square each side again, we must isolate the radical on the right side of the equation.

$x + 1 = x + 3 - 2\sqrt{x + 2}$	**Simplify the right side**
$1 = 3 - 2\sqrt{x + 2}$	**Add $-x$ to each side**
$-2 = -2\sqrt{x + 2}$	**Add -3 to each side**
$1 = \sqrt{x + 2}$	**Divide each side by -2**
$1 = x + 2$	**Square both sides**
$-1 = x$	**Add -2 to each side**

Checking our only possible solution, $x = -1$, in our original equation, we have

$$\sqrt{-1 + 1} \overset{?}{=} \sqrt{-1 + 2} - 1$$

$$\sqrt{0} \overset{?}{=} \sqrt{1} - 1$$

$$0 \overset{?}{=} 1 - 1$$

$$0 = 0 \qquad \textbf{A true statement}$$

Our solution checks.

It is also possible to raise both sides of an equation to powers greater than 2. We only need to check for extraneous solutions when we raise both sides of an equation to an even power. Raising both sides of an equation to an odd power will not produce extraneous solutions.

EXAMPLE 8   Solve $\sqrt[3]{4x + 5} = 3$.

SOLUTION   Cubing both sides we have

$$(\sqrt[3]{4x + 5})^3 = 3^3$$

$$4x + 5 = 27$$

$$4x = 22$$

$$x = \frac{22}{4}$$

$$x = \frac{11}{2}$$

We do not need to check $x = \frac{11}{2}$ since we raised both sides to an odd power.

We end this section by looking at graphs of some equations that contain radicals.

 EXAMPLE 9   Graph $y = \sqrt{x}$ and $y = \sqrt[3]{x}$.

SOLUTION   The graphs are shown in Figures 1 and 2. Notice that the graph of $y = \sqrt{x}$ appears in the first quadrant only because in the equation $y = \sqrt{x}$, x and y cannot be negative.

The graph of $y = \sqrt[3]{x}$ appears in quadrants 1 and 3 since the cube root of a positive number is also a positive number and the cube root of a negative number is a negative number; that is, when x is positive, y will be positive and when x is negative, y will be negative.

The graphs of both equations will contain the origin since $y = 0$ when $x = 0$ in both equations.

x	y
-4	undefined
-1	undefined
0	0
1	1
4	2
9	3
16	4

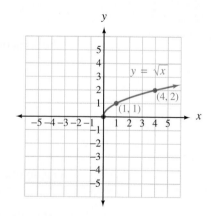

FIGURE 1

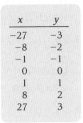

x	y
-27	-3
-8	-2
-1	-1
0	0
1	1
8	2
27	3

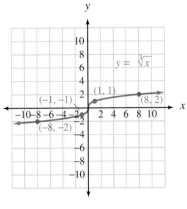

FIGURE 2

LINKING OBJECTIVES AND EXAMPLES

Next to each **objective** we have listed the examples that are best described by that objective.

A 1–8

B 9

GETTING READY FOR CLASS

After reading through the preceding section, respond in your own words and in complete sentences.

1. What is the squaring property of equality?
2. Under what conditions do we obtain extraneous solutions to equations that contain radical expressions?
3. If we have raised both sides of an equation to a power, when is it not necessary to check for extraneous solutions?
4. When will you need to apply the squaring property of equality twice in the process of solving an equation containing radicals?

Problem Set 9.6

Online support materials can be found at academic.cengage.com/login

Solve each of the following equations.

1. $\sqrt{2x + 1} = 3$

2. $\sqrt{3x + 1} = 4$

3. $\sqrt{4x + 1} = -5$

4. $\sqrt{6x + 1} = -5$

5. $\sqrt{2y - 1} = 3$

6. $\sqrt{3y - 1} = 2$

7. $\sqrt{5x - 7} = -1$

8. $\sqrt{8x + 3} = -6$

9. $\sqrt{2x - 3} - 2 = 4$

10. $\sqrt{3x + 1} - 4 = 1$

11. $\sqrt{4x + 1} + 3 = 2$

12. $\sqrt{5a - 3} + 6 = 2$

13. $\sqrt[4]{3x + 1} = 2$

14. $\sqrt[4]{4x + 1} = 3$

15. $\sqrt[3]{2x - 5} = 1$

16. $\sqrt[3]{5x + 7} = 2$

17. $\sqrt[3]{3a + 5} = -3$

18. $\sqrt[3]{2a + 7} = -2$

19. $\sqrt{y - 3} = y - 3$

20. $\sqrt{y + 3} = y - 3$

21. $\sqrt{a + 2} = a + 2$

22. $\sqrt{a + 10} = a - 2$

▶ **23.** $\sqrt{2x + 3} = \dfrac{2x - 7}{3}$

▶ **24.** $\sqrt{3x - 2} = \dfrac{2x - 3}{3}$

▶ **25.** $\sqrt{4x - 3} = \dfrac{x + 3}{2}$

▶ **26.** $\sqrt{4x + 5} = \dfrac{x + 3}{2}$

▶ **27.** $\sqrt{7x + 2} = \dfrac{2x + 2}{3}$

▶ **28.** $\sqrt{7x - 3} = \dfrac{4x - 1}{3}$

29. $\sqrt{2x + 4} = \sqrt{1 - x}$

30. $\sqrt{3x + 4} = -\sqrt{2x + 3}$

░ = Videos available by instructor request

▶ = Online student support materials available at academic.cengage.com/login

31. $\sqrt{4a + 7} = -\sqrt{a + 2}$ **32.** $\sqrt{7a - 1} = \sqrt{2a + 4}$

33. $\sqrt[4]{5x - 8} = \sqrt[4]{4x - 1}$ **34.** $\sqrt[4]{6x + 7} = \sqrt[4]{x + 2}$

35. $x + 1 = \sqrt{5x + 1}$

36. $x - 1 = \sqrt{6x + 1}$

37. $t + 5 = \sqrt{2t + 9}$ **38.** $t + 7 = \sqrt{2t + 13}$

39. $\sqrt{y - 8} = \sqrt{8 - y}$ **40.** $\sqrt{2y + 5} = \sqrt{5y + 2}$

41. $\sqrt[3]{3x + 5} = \sqrt[3]{5 - 2x}$ **42.** $\sqrt[3]{4x + 9} = \sqrt[3]{3 - 2x}$

The following equations will require that you square both sides twice before all the radicals are eliminated. Solve each equation using the methods shown in Examples 5, 6, and 7.

43. $\sqrt{x - 8} = \sqrt{x} - 2$ **44.** $\sqrt{x + 3} = \sqrt{x} - 3$

45. $\sqrt{x + 1} = \sqrt{x} + 1$ **46.** $\sqrt{x - 1} = \sqrt{x} - 1$

47. $\sqrt{x + 8} = \sqrt{x - 4} + 2$ **48.** $\sqrt{x + 5} = \sqrt{x - 3} + 2$

49. $\sqrt{x - 5} - 3 = \sqrt{x - 8}$

50. $\sqrt{x - 3} - 4 = \sqrt{x - 3}$

51. $\sqrt{x + 4} = 2 - \sqrt{2x}$

52. $\sqrt{5x + 1} = 1 + \sqrt{5x}$

53. $\sqrt{2x + 4} = \sqrt{x + 3} + 1$

54. $\sqrt{2x - 1} = \sqrt{x - 4} + 2$

Let $f(x) = \sqrt{2x - 1}$. Find all values for the variable x that produce the following values of $f(x)$.

▶ **55.** $f(x) = 0$ ▶ **56.** $f(x) = -10$

▶ **57.** $f(x) = 2x - 1$ ▶ **58.** $f(x) = \sqrt{5x - 10}$

▶ **59.** $f(x) = \sqrt{x - 4} + 2$ ▶ **60.** $f(x) = 9$

Let $g(x) = \sqrt{2x + 3}$. Find all values for the variable x that produce the following values of $g(x)$.

61. $g(x) = 0$

62. $g(x) = x$

63. $g(x) = -\sqrt{5x}$

64. $g(x) = \sqrt{x^2 - 5}$

Let $f(x) = \sqrt{2x} - 1$. Find all values for the variable x for which $f(x) = g(x)$.

▶ **65.** $g(x) = 0$ ▶ **66.** $g(x) = 5$

67. $g(x) = \sqrt{2x + 5}$

68. $g(x) = x - 1$

Let $h(x) = \sqrt[3]{3x + 5}$. Find all values for the variable x for which $h(x) = f(x)$.

▶ **69.** $f(x) = 2$ ▶ **70.** $f(x) = -1$

Let $h(x) = \sqrt[3]{5 - 2x}$. Find all values for the variable x for which $h(x) = f(x)$.

▶ **71.** $f(x) = 3$ ▶ **72.** $f(x) = -1$

Graph each equation.

73. $y = 2\sqrt{x}$ **74.** $y = -2\sqrt{x}$

75. $y = \sqrt{x} - 2$ **76.** $y = \sqrt{x} + 2$

77. $y = \sqrt{x - 2}$ **78.** $y = \sqrt{x + 2}$

79. $y = 3\sqrt[3]{x}$ **80.** $y = -3\sqrt[3]{x}$

81. $y = \sqrt[3]{x} + 3$ **82.** $y = \sqrt[3]{x} - 3$

83. $y = \sqrt[3]{x + 3}$ **84.** $y = \sqrt[3]{x - 3}$

Applying the Concepts

85. Solving a Formula Solve the following formula for h:
$$t = \frac{\sqrt{100 - h}}{4}$$

86. Solving a Formula Solve the following formula for h:
$$t = \sqrt{\frac{2h - 40t}{g}}$$

87. Pendulum Clock The length of time (T) in seconds it takes the pendulum of a grandfather clock to swing through one complete cycle is given by the formula
$$T = 2\pi\sqrt{\frac{L}{32}}$$
where L is the length, in feet, of the pendulum, and π is approximately $\frac{22}{7}$. How long must the pendulum be if one complete cycle takes 2 seconds?

88. Pendulum Clock Solve the formula in Problem 87 for L.

1 sec

Pollution A long straight river, 100 meters wide, is flowing at 1 meter per second. A pollutant is entering the river at a constant rate from one of its banks. As the pollutant disperses in the water, it forms a plume that is modeled by the equation $y = \sqrt{x}$. Use this information to answer the following questions.

89. How wide is the plume 25 meters down river from the source of the pollution?

90. How wide is the plume 100 meters down river from the source of the pollution?

91. How far down river from the source of the pollution does the plume reach halfway across the river?

92. How far down river from the source of the pollution does the plume reach the other side of the river?

Maintaining Your Skills

Multiply.

93. $\sqrt{2}(\sqrt{3} - \sqrt{2})$

94. $(\sqrt{x} - 4)(\sqrt{x} + 5)$

95. $(\sqrt{x} + 5)^2$

96. $(\sqrt{5} + \sqrt{3})(\sqrt{5} - \sqrt{3})$

Rationalize the denominator.

97. $\dfrac{\sqrt{x}}{\sqrt{x} + 3}$

98. $\dfrac{\sqrt{5} - \sqrt{3}}{\sqrt{5} + \sqrt{3}}$

Getting Ready for the Next Section

Simplify.

99. $\sqrt{25}$

100. $\sqrt{49}$

101. $\sqrt{12}$

102. $\sqrt{50}$

103. $(-1)^{15}$

104. $(-1)^{20}$

105. $(-1)^{50}$

106. $(-1)^5$

Solve.

107. $3x = 12$

108. $4 = 8y$

109. $4x - 3 = 5$

110. $7 = 2y - 1$

Perform the indicated operation.

111. $(3 + 4x) + (7 - 6x)$

112. $(2 - 5x) + (-1 + 7x)$

113. $(7 + 3x) - (5 + 6x)$

114. $(5 - 2x) - (9 - 4x)$

115. $(3 - 4x)(2 + 5x)$

116. $(8 + x)(7 - 3x)$

117. $2x(4 - 6x)$

118. $3x(7 + 2x)$

119. $(2 + 3x)^2$

120. $(3 + 5x)^2$

121. $(2 - 3x)(2 + 3x)$

122. $(4 - 5x)(4 + 5x)$

9.7 Complex Numbers

OBJECTIVES

A Simplify square roots of negative numbers.

B Simplify powers of i.

C Solve for unknown variables by equating real parts and equating imaginary parts of two complex numbers.

D Add and subtract complex numbers.

E Multiply complex numbers.

F Divide complex numbers.

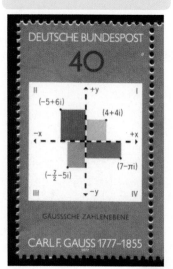

The stamp shown in the margin was issued by Germany in 1977 to commemorate the 200th anniversary of the birth of mathematician Carl Gauss. The number $-5 + 6i$ shown on the stamp is a complex number, as are the other numbers on the stamp. Working with complex numbers gives us a way to solve a wider variety of equations. For example, the equation $x^2 = -9$ has no real number solutions since the square of a real number is always positive. We have been unable to work with square roots of negative numbers like $\sqrt{-25}$ and $\sqrt{-16}$ for the same reason. Complex numbers allow us to expand our work with radicals to include square roots of negative numbers and solve equations like $x^2 = -9$ and $x^2 = -64$. Our work with complex numbers is based on the following definition.

> **DEFINITION** The number i is such that $i = \sqrt{-1}$ (which is the same as saying $i^2 = -1$).

The number i, as we have defined it here, is not a real number. Because of the way we have defined i, we can use it to simplify square roots of negative numbers.

> **Square Roots of Negative Numbers** If a is a positive number, then $\sqrt{-a}$ can always be written as $i\sqrt{a}$; that is,
> $$\sqrt{-a} = i\sqrt{a} \text{ if } a \text{ is a positive number}$$

To justify our rule, we simply square the quantity $i\sqrt{a}$ to obtain $-a$. Here is what it looks like when we do so:

$$(i\sqrt{a})^2 = i^2 \cdot (\sqrt{a})^2$$
$$= -1 \cdot a$$
$$= -a$$

Here are some examples that illustrate the use of our new rule.

EXAMPLES Write each square root in terms of the number i.

1. $\sqrt{-25} = i\sqrt{25} = i \cdot 5 = 5i$

2. $\sqrt{-49} = i\sqrt{49} = i \cdot 7 = 7i$

3. $\sqrt{-12} = i\sqrt{12} = i \cdot 2\sqrt{3} = 2i\sqrt{3}$

4. $\sqrt{-17} = i\sqrt{17}$

Note

In Examples 3 and 4 we wrote i before the radical simply to avoid confusion. If we were to write the answer to 3 as $2\sqrt{3}i$, some people would think the i was under the radical sign and it is not.

If we assume all the properties of exponents hold when the base is i, we can write any power of i as i, -1, $-i$, or 1. Using the fact that $i^2 = -1$, we have

$$i^1 = i$$
$$i^2 = -1$$
$$i^3 = i^2 \cdot i = -1(i) = -i$$
$$i^4 = i^2 \cdot i^2 = -1(-1) = 1$$

Since $i^4 = 1$, i^5 will simplify to i, and we will begin repeating the sequence i, -1, $-i$, 1 as we simplify higher powers of i: Any power of i simplifies to i, -1, $-i$, or 1. The easiest way to simplify higher powers of i is to write them in terms of i^2. For instance, to simplify i^{21}, we would write it as

$$(i^2)^{10} \cdot i \qquad \text{because } 2 \cdot 10 + 1 = 21$$

Then, since $i^2 = -1$, we have

$$(-1)^{10} \cdot i = 1 \cdot i = i$$

EXAMPLES   Simplify as much as possible.

5. $i^{30} = (i^2)^{15} = (-1)^{15} = -1$

6. $i^{11} = (i^2)^5 \cdot i = (-1)^5 \cdot i = (-1)i = -i$

7. $i^{40} = (i^2)^{20} = (-1)^{20} = 1$

> **DEFINITION** A **complex number** is any number that can be put in the form
>
> $$a + bi$$
>
> where a and b are real numbers and $i = \sqrt{-1}$. The form $a + bi$ is called **standard form** for complex numbers. The number a is called the **real part** of the complex number. The number b is called the **imaginary part** of the complex number.

Every real number is a complex number. For example, 8 can be written as $8 + 0i$. Likewise, $-\frac{1}{2}$, π, $\sqrt{3}$, and -9 are complex numbers because they can all be written in the form $a + bi$:

$$-\frac{1}{2} = -\frac{1}{2} + 0i \qquad \pi = \pi + 0i$$

$$\sqrt{3} = \sqrt{3} + 0i \qquad -9 = -9 + 0i$$

The real numbers occur when $b = 0$. When $b \neq 0$, we have complex numbers that contain i, such as $2 + 5i$, $6 - i$, $4i$, and $\frac{1}{2}i$. These numbers are called *imaginary numbers.* The diagram explains this further.

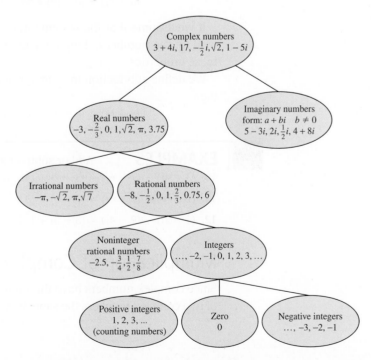

Equality for Complex Numbers

Two complex numbers are equal if and only if their real parts are equal and their imaginary parts are equal; that is, for real numbers a, b, c, and d,

$$a + bi = c + di \quad \text{if and only if} \quad a = c \quad \text{and} \quad b = d$$

 EXAMPLE 8 Find x and y if $3x + 4i = 12 - 8yi$.

SOLUTION Since the two complex numbers are equal, their real parts are equal and their imaginary parts are equal:

$$3x = 12 \quad \text{and} \quad 4 = -8y$$

$$x = 4 \qquad\qquad y = -\frac{1}{2}$$

 EXAMPLE 9 Find x and y if $(4x - 3) + 7i = 5 + (2y - 1)i$.

SOLUTION The real parts are $4x - 3$ and 5. The imaginary parts are 7 and $2y - 1$:

$$4x - 3 = 5 \quad \text{and} \quad 7 = 2y - 1$$

$$4x = 8 \qquad\qquad 8 = 2y$$

$$x = 2 \qquad\qquad y = 4$$

Addition and Subtraction of Complex Numbers

To add two complex numbers, add their real parts and add their imaginary parts; that is, if a, b, c, and d are real numbers, then

$$(a + bi) + (c + di) = (a + c) + (b + d)i$$

If we assume that the commutative, associative, and distributive properties hold for the number i, then the definition of addition is simply an extension of these properties.

We define subtraction in a similar manner. If a, b, c, and d are real numbers, then

$$(a + bi) - (c + di) = (a - c) + (b - d)i$$

 EXAMPLES Add or subtract as indicated.

10. $(3 + 4i) + (7 - 6i) = (3 + 7) + (4 - 6)i = 10 - 2i$

11. $(7 + 3i) - (5 + 6i) = (7 - 5) + (3 - 6)i = 2 - 3i$

12. $(5 - 2i) - (9 - 4i) = (5 - 9) + (-2 + 4)i = -4 + 2i$

Multiplication of Complex Numbers

Since complex numbers have the same form as binomials, we find the product of two complex numbers the same way we find the product of two binomials.

EXAMPLE 13   Multiply $(3 - 4i)(2 + 5i)$.

SOLUTION   Multiplying each term in the second complex number by each term in the first, we have

$$(3 - 4i)(2 + 5i) = 3 \cdot 2 + 3 \cdot 5i - 2 \cdot 4i - 5i(4i)$$
$$= 6 + 15i - 8i - 20i^2$$

Combining similar terms and using the fact that $i^2 = -1$, we can simplify as follows:

$$6 + 15i - 8i - 20i^2 = 6 + 7i - 20(-1)$$
$$= 6 + 7i + 20$$
$$= 26 + 7i$$

The product of the complex numbers $3 - 4i$ and $2 + 5i$ is the complex number $26 + 7i$.

 EXAMPLE 14   Multiply $2i(4 - 6i)$.

SOLUTION   Applying the distributive property gives us

$$2i(4 - 6i) = 2i \cdot 4 - 2i \cdot 6i$$
$$= 8i - 12i^2$$
$$= 12 + 8i$$

 EXAMPLE 15   Expand $(3 + 5i)^2$.

SOLUTION   We treat this like the square of a binomial. Remember, $(a + b)^2 = a^2 + 2ab + b^2$.

$$(3 + 5i)^2 = 3^2 + 2(3)(5i) + (5i)^2$$
$$= 9 + 30i + 25i^2$$
$$= 9 + 30i - 25$$
$$= -16 + 30i$$

> ### Note
> We can obtain the same result by writing $(3 + 5i)^2$ as $(3 + 5i)$ times $(3 + 5i)$ and applying the FOIL method as we did with the problem in Example 13.

EXAMPLE 16   Multiply $(2 - 3i)(2 + 3i)$.

SOLUTION   This product has the form $(a - b)(a + b)$, which we know results in the difference of two squares, $a^2 - b^2$:

$$(2 - 3i)(2 + 3i) = 2^2 - (3i)^2$$
$$= 4 - 9i^2$$
$$= 4 + 9$$
$$= 13$$

The product of the two complex numbers $2 - 3i$ and $2 + 3i$ is the real number 13. The two complex numbers $2 - 3i$ and $2 + 3i$ are called complex conjugates. The fact that their product is a real number is very useful.

> **DEFINITION** The complex numbers $a + bi$ and $a - bi$ are called **complex conjugates.** One important property they have is that their product is the real number $a^2 + b^2$. Here's why:
>
> $$(a + bi)(a - bi) = a^2 - (bi)^2$$
> $$= a^2 - b^2i^2$$
> $$= a^2 - b^2(-1)$$
> $$= a^2 + b^2$$

Division with Complex Numbers

The fact that the product of two complex conjugates is a real number is the key to division with complex numbers.

 EXAMPLE 17   Divide $\dfrac{2 + i}{3 - 2i}$.

SOLUTION   We want a complex number in standard form that is equivalent to the quotient $\dfrac{2 + i}{3 - 2i}$. We need to eliminate i from the denominator. Multiplying the numerator and denominator by $3 + 2i$ will give us what we want:

$$\frac{2 + i}{3 - 2i} = \frac{2 + i}{3 - 2i} \cdot \frac{(3 + 2i)}{(3 + 2i)}$$

$$= \frac{6 + 4i + 3i + 2i^2}{9 - 4i^2}$$

$$= \frac{6 + 7i - 2}{9 + 4}$$

$$= \frac{4 + 7i}{13}$$

$$= \frac{4}{13} + \frac{7}{13}i$$

Dividing the complex number $2 + i$ by $3 - 2i$ gives the complex number $\frac{4}{13} + \frac{7}{13}i$.

 EXAMPLE 18   Divide $\dfrac{7 - 4i}{i}$.

SOLUTION   The conjugate of the denominator is $-i$. Multiplying numerator and denominator by this number, we have

$$\frac{7 - 4i}{i} = \frac{7 - 4i}{i} \cdot \frac{-i}{-i}$$

$$= \frac{-7i + 4i^2}{-i^2}$$

$$= \frac{-7i + 4(-1)}{-(-1)}$$

$$= -4 - 7i$$

GETTING READY FOR CLASS

After reading through the preceding section, respond in your own words and in complete sentences.

1. What is the number *i*?
2. What is a complex number?
3. What kind of number results when we multiply complex conjugates?
4. Explain how to divide complex numbers.

Problem Set 9.7

Online support materials can be found at academic.cengage.com/login

Write the following in terms of *i*, and simplify as much as possible.

1. $\sqrt{-36}$

2. $\sqrt{-49}$

3. $-\sqrt{-25}$

4. $-\sqrt{-81}$

▶ **5.** $\sqrt{-72}$

6. $\sqrt{-48}$

7. $-\sqrt{-12}$

8. $-\sqrt{-75}$

Write each of the following as *i*, −1, −*i*, or 1.

9. i^{28}

10. i^{31}

11. i^{26}

12. i^{37}

13. i^{75}

14. i^{42}

Find *x* and *y* so each of the following equations is true.

15. $2x + 3yi = 6 - 3i$

16. $4x - 2yi = 4 + 8i$

17. $2 - 5i = -x + 10yi$

18. $4 + 7i = 6x - 14yi$

19. $2x + 10i = -16 - 2yi$

20. $4x - 5i = -2 + 3yi$

21. $(2x - 4) - 3i = 10 - 6yi$

22. $(4x - 3) - 2i = 8 + yi$

23. $(7x - 1) + 4i = 2 + (5y + 2)i$

24. $(5x + 2) - 7i = 4 + (2y + 1)i$

Combine the following complex numbers.

25. $(2 + 3i) + (3 + 6i)$

26. $(4 + i) + (3 + 2i)$

27. $(3 - 5i) + (2 + 4i)$

28. $(7 + 2i) + (3 - 4i)$

29. $(5 + 2i) - (3 + 6i)$

▶ **30.** $(6 + 7i) - (4 + i)$

31. $(3 - 5i) - (2 + i)$

32. $(7 - 3i) - (4 + 10i)$

33. $[(3 + 2i) - (6 + i)] + (5 + i)$

34. $[(4 - 5i) - (2 + i)] + (2 + 5i)$

35. $[(7 - i) - (2 + 4i)] - (6 + 2i)$

36. $[(3 - i) - (4 + 7i)] - (3 - 4i)$

= Videos available by instructor request

▶ = Online student support materials available at academic.cengage.com/login

37. $(3 + 2i) - [(3 - 4i) - (6 + 2i)]$

38. $(7 - 4i) - [(-2 + i) - (3 + 7i)]$

39. $(4 - 9i) + [(2 - 7i) - (4 + 8i)]$

40. $(10 - 2i) - [(2 + i) - (3 - i)]$

Find the following products.

41. $3i(4 + 5i)$ **42.** $2i(3 + 4i)$

▶ **43.** $6i(4 - 3i)$ **44.** $11i(2 - i)$

45. $(3 + 2i)(4 + i)$ **46.** $(2 - 4i)(3 + i)$

47. $(4 + 9i)(3 - i)$ **48.** $(5 - 2i)(1 + i)$

49. $(1 + i)^3$ **50.** $(1 - i)^3$

51. $(2 - i)^3$ **52.** $(2 + i)^3$

▶ **53.** $(2 + 5i)^2$ **54.** $(3 + 2i)^2$

55. $(1 - i)^2$ **56.** $(1 + i)^2$

57. $(3 - 4i)^2$ **58.** $(6 - 5i)^2$

59. $(2 + i)(2 - i)$ **60.** $(3 + i)(3 - i)$

61. $(6 - 2i)(6 + 2i)$ **62.** $(5 + 4i)(5 - 4i)$

63. $(2 + 3i)(2 - 3i)$ **64.** $(2 - 7i)(2 + 7i)$

65. $(10 + 8i)(10 - 8i)$ **66.** $(11 - 7i)(11 + 7i)$

Find the following quotients. Write all answers in standard form for complex numbers.

67. $\dfrac{2 - 3i}{i}$ **68.** $\dfrac{3 + 4i}{i}$

69. $\dfrac{5 + 2i}{-i}$ **70.** $\dfrac{4 - 3i}{-i}$

▶ **71.** $\dfrac{4}{2 - 3i}$ **72.** $\dfrac{3}{4 - 5i}$

73. $\dfrac{6}{-3 + 2i}$ **74.** $\dfrac{-1}{-2 - 5i}$

75. $\dfrac{2 + 3i}{2 - 3i}$ **76.** $\dfrac{4 - 7i}{4 + 7i}$

77. $\dfrac{5 + 4i}{3 + 6i}$ **78.** $\dfrac{2 + i}{5 - 6i}$

Applying the Concepts

79. Electric Circuits Complex numbers may be applied to electrical circuits. Electrical engineers use the fact

that resistance R to electrical flow of the electrical current I and the voltage V are related by the formula $V = RI$. (Voltage is measured in volts, resistance in ohms, and current in amperes.) Find the resistance to electrical flow in a circuit that has a voltage $V = (80 + 20i)$ volts and current $I = (-6 + 2i)$ amps.

80. Electric Circuits Refer to the information about electrical circuits in Problem 79, and find the current in a circuit that has a resistance of $(4 + 10i)$ ohms and a voltage of $(5 - 7i)$ volts.

Maintaining Your Skills

For each relation that follows, state the domain and the range, and indicate which are also functions.

81. $\{(1, 2), (3, 4), (4, 2)\}$

82. $\{(0, 0), (1, 1), (0, 1)\}$

83. $\{(3, 1), (2, 3), (1, 2)\}$

84. $\{(-1, 1), (2, -2), (-3, -3)\}$

State whether each of the following graphs is the graph of a function

85.

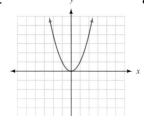

86.

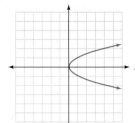

87.

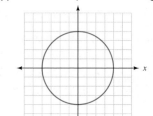

88.

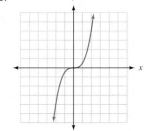

Square Roots [9.1]

1. The number 49 has two square roots, 7 and −7. They are written like this:
$$\sqrt{49} = 7 \qquad -\sqrt{49} = -7$$

Every positive real number x has two square roots. The *positive square root* of x is written $\sqrt{x}$, and the *negative square root* of x is written $-\sqrt{x}$. Both the positive and the negative square roots of x are numbers we square to get x; that is,

$$\left. \begin{array}{c} (\sqrt{x})^2 = x \\ \text{and} \qquad (-\sqrt{x})^2 = x \end{array} \right\} \quad \text{for } x \geq 0$$

Higher Roots [9.1]

2. $\sqrt[3]{8} = 2$
$\sqrt[3]{-27} = -3$

In the expression $\sqrt[n]{a}$, n is the *index*, a is the *radicand*, and $\sqrt{}$ is the *radical sign*. The expression $\sqrt[n]{a}$ is such that

$$(\sqrt[n]{a})^n = a \qquad a \geq 0 \text{ when } n \text{ is even}$$

Rational Exponents [9.1, 9.2]

3. $25^{1/2} = \sqrt{25} = 5$
$8^{2/3} = (\sqrt[3]{8})^2 = 2^2 = 4$
$9^{3/2} = (\sqrt{9})^3 = 3^3 = 27$

Rational exponents are used to indicate roots. The relationship between rational exponents and roots is as follows:

$$a^{1/n} = \sqrt[n]{a} \qquad \text{and} \qquad a^{m/n} = (a^{1/n})^m = (a^m)^{1/n}$$

$$a \geq 0 \text{ when } n \text{ is even}$$

Properties of Radicals [9.3]

4. $\sqrt{4 \cdot 5} = \sqrt{4}\,\sqrt{5} = 2\sqrt{5}$
$\sqrt{\dfrac{7}{9}} = \dfrac{\sqrt{7}}{\sqrt{9}} = \dfrac{\sqrt{7}}{3}$

If a and b are nonnegative real numbers whenever n is even, then

1. $\sqrt[n]{ab} = \sqrt[n]{a}\,\sqrt[n]{b}$

2. $\sqrt[n]{\dfrac{a}{b}} = \dfrac{\sqrt[n]{a}}{\sqrt[n]{b}} \qquad (b \neq 0)$

Simplified Form for Radicals [9.3]

5. $\sqrt{\dfrac{4}{5}} = \dfrac{\sqrt{4}}{\sqrt{5}}$

$\phantom{\sqrt{\dfrac{4}{5}}} = \dfrac{2}{\sqrt{5}} \cdot \dfrac{\sqrt{5}}{\sqrt{5}}$

$\phantom{\sqrt{\dfrac{4}{5}}} = \dfrac{2\sqrt{5}}{5}$

A radical expression is said to be in *simplified form*

1. If there is no factor of the radicand that can be written as a power greater than or equal to the index;

2. If there are no fractions under the radical sign; and

3. If there are no radicals in the denominator.

Addition and Subtraction of Radical Expressions [9.4]

6. $5\sqrt{3} - 7\sqrt{3} = (5 - 7)\sqrt{3}$
$= -2\sqrt{3}$
$\sqrt{20} + \sqrt{45} = 2\sqrt{5} + 3\sqrt{5}$
$= (2 + 3)\sqrt{5}$
$= 5\sqrt{5}$

We add and subtract radical expressions by using the distributive property to combine similar radicals. Similar radicals are radicals with the same index and the same radicand.

Multiplication of Radical Expressions [9.5]

7. $(\sqrt{x} + 2)(\sqrt{x} + 3)$
$= \sqrt{x}\,\sqrt{x} + 3\sqrt{x} + 2\sqrt{x} + 2 \cdot 3$
$= x + 5\sqrt{x} + 6$

We multiply radical expressions in the same way that we multiply polynomials. We can use the distributive property and the FOIL method.

Rationalizing the Denominator [9.3, 9.5]

8. $\dfrac{3}{\sqrt{2}} = \dfrac{3}{\sqrt{2}} \cdot \dfrac{\sqrt{2}}{\sqrt{2}} = \dfrac{3\sqrt{2}}{2}$

$\dfrac{3}{\sqrt{5} - \sqrt{3}} = \dfrac{3}{\sqrt{5} - \sqrt{3}} \cdot \dfrac{\sqrt{5} + \sqrt{3}}{\sqrt{5} + \sqrt{3}}$

$= \dfrac{3\sqrt{5} + 3\sqrt{3}}{5 - 3}$

$= \dfrac{3\sqrt{5} + 3\sqrt{3}}{2}$

When a fraction contains a square root in the denominator, we rationalize the denominator by multiplying numerator and denominator by

1. The square root itself if there is only one term in the denominator, or
2. The conjugate of the denominator if there are two terms in the denominator.

Rationalizing the denominator is also called division of radical expressions.

Squaring Property of Equality [9.6]

9. $\sqrt{2x + 1} = 3$
$(\sqrt{2x + 1})^2 = 3^2$
$2x + 1 = 9$
$x = 4$

We may square both sides of an equation any time it is convenient to do so, as long as we check all resulting solutions in the original equation.

Complex Numbers [9.7]

10. $3 + 4i$ is a complex number.
Addition
$(3 + 4i) + (2 - 5i) = 5 - i$
Multiplication
$(3 + 4i)(2 - 5i)$
$= 6 - 15i + 8i - 20i^2$
$= 6 - 7i + 20$
$= 26 - 7i$

A *complex number* is any number that can be put in the form

$$a + bi$$

where a and b are real numbers and $i = \sqrt{-1}$. The *real part* of the complex number is a, and b is the *imaginary part*.

If $a, b, c,$ and d are real numbers, then we have the following definitions associated with complex numbers:

1. Equality

$$a + bi = c + di \quad \text{if and only if} \quad a = c \text{ and } b = d$$

Division

$$\frac{2}{3 + 4i} = \frac{2}{3 + 4i} \cdot \frac{3 - 4i}{3 - 4i}$$

$$= \frac{6 - 8i}{9 + 16}$$

$$= \frac{6}{25} - \frac{8}{25}i$$

2. Addition and subtraction

$$(a + bi) + (c + di) = (a + c) + (b + d)i$$

$$(a + bi) - (c + di) = (a - c) + (b - d)i$$

3. Multiplication

$$(a + bi)(c + di) = (ac - bd) + (ad + bc)i$$

4. Division is similar to rationalizing the denominator.

Chapter 9 Review Test

The problems below form a comprehensive review of the material in this chapter. They can be used to study for exams. If you would like to take a practice test on this chapter, you can use the odd-numbered problems. Give yourself an hour and work as many of the odd-numbered problems as possible. When you are finished, or when an hour has passed, check your answers with the answers in the back of the book. You can use the even-numbered problems for a second practice test.

Simplify each expression as much as possible. [9.1]

1. $\sqrt{49}$

2. $(-27)^{1/3}$

3. $16^{1/4}$

4. $9^{3/2}$

5. $\sqrt[5]{32x^{15}y^{10}}$

6. $8^{-4/3}$

Use the properties of exponents to simplify each expression. Assume all bases represent positive numbers. [9.1]

7. $x^{2/3} \cdot x^{4/3}$

8. $(a^{2/3}b^{4/3})^3$

9. $\dfrac{a^{3/5}}{a^{1/4}}$

10. $\dfrac{a^{2/3}b^3}{a^{1/4}b^{1/3}}$

Multiply. [9.2]

11. $(3x^{1/2} + 5y^{1/2})(4x^{1/2} - 3y^{1/2})$

12. $(a^{1/3} - 5)^2$

13. Divide: $\dfrac{28x^{5/6} + 14x^{7/6}}{7x^{1/3}}$. (Assume $x > 0$.) [9.2]

14. Factor $2(x-3)^{1/4}$ from $8(x-3)^{5/4} - 2(x-3)^{1/4}$. [9.2]

15. Simplify $x^{3/4} + \dfrac{5}{x^{1/4}}$ into a single fraction.

(Assume $x > 0$.) [9.2]

Write each expression in simplified form for radicals. (Assume all variables represent nonnegative numbers.) [9.3]

16. $\sqrt{12}$

17. $\sqrt{50}$

18. $\sqrt[3]{16}$

19. $\sqrt{18x^2}$

20. $\sqrt{80a^3b^4c^2}$

21. $\sqrt[4]{32a^4b^5c^6}$

Rationalize the denominator in each expression. [9.3]

22. $\dfrac{3}{\sqrt{2}}$

23. $\dfrac{6}{\sqrt[3]{2}}$

Write each expression in simplified form. (Assume all variables represent positive numbers.) [9.3]

24. $\sqrt{\dfrac{48x^3}{7y}}$

25. $\sqrt[3]{\dfrac{40x^2y^3}{3z}}$

Combine the following expressions. (Assume all variables represent positive numbers.) [9.4]

26. $5x\sqrt{6} + 2x\sqrt{6} - 9x\sqrt{6}$

27. $\sqrt{12} + \sqrt{3}$

28. $\dfrac{3}{\sqrt{5}} + \sqrt{5}$

29. $3\sqrt{8} - 4\sqrt{72} + 5\sqrt{50}$

30. $3b\sqrt{27a^5b} + 2a\sqrt{3a^3b^3}$

31. $2x\sqrt[3]{xy^3z^2} - 6y\sqrt[3]{x^4z^2}$

Multiply. [9.5]

32. $\sqrt{2}(\sqrt{3} - 2\sqrt{2})$

33. $(\sqrt{x} - 2)(\sqrt{x} - 3)$

Rationalize the denominator. (Assume $x, y > 0$.) [9.5]

34. $\dfrac{3}{\sqrt{5} - 2}$

35. $\dfrac{\sqrt{7} + \sqrt{5}}{\sqrt{7} - \sqrt{5}}$

36. $\dfrac{3\sqrt{7}}{3\sqrt{7} - 4}$

Solve each equation. [9.6]

37. $\sqrt{4a + 1} = 1$

38. $\sqrt[3]{3x - 8} = 1$

39. $\sqrt{3x + 1} - 3 = 1$

40. $\sqrt{x + 4} = \sqrt{x} - 2$

Graph each equation. [9.6]

41. $y = 3\sqrt{x}$

42. $y = \sqrt[3]{x} + 2$

Write each of the following as i, -1, $-i$, or 1. [9.7]

43. i^{24}

44. i^{27}

Find x and y so that each of the following equations is true. [9.7]

45. $3 - 4i = -2x + 8yi$

46. $(3x + 2) - 8i = -4 + 2yi$

Combine the following complex numbers. [9.7]

47. $(3 + 5i) + (6 - 2i)$

48. $(2 + 5i) - [(3 + 2i) + (6 - i)]$

Multiply. [9.7]

49. $3i(4 + 2i)$

50. $(2 + 3i)(4 + i)$

51. $(4 + 2i)^2$

52. $(4 + 3i)(4 - 3i)$

Divide. Write all answers in standard form for complex numbers. [9.7]

53. $\dfrac{3 + i}{i}$

54. $\dfrac{-3}{2 + i}$

55. Construction The roof of the house shown in Figure 1 is to extend up 13.5 feet above the ceiling, which is 36 feet across. Find the length of one side of the roof.

56. Surveying A surveyor is attempting to find the distance across a pond. From a point on one side of the pond he walks 25 yards to the end of the pond and then makes a 90-degree turn and walks another 60 yards before coming to a point directly across the pond from the point at which he started. What is the distance across the pond? (See Figure 2.)

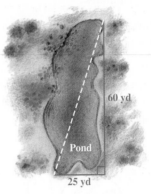

FIGURE 2

FIGURE 1

GROUP PROJECT   Constructing the Spiral of Roots

Number of People 3

Time Needed 20 minutes

Equipment Two sheets of graph paper (4 or 5 squares per inch) and pencils.

Background The spiral of roots gives us a way to visualize the positive square roots of the counting numbers, and in so doing, we see many line segments whose lengths are irrational numbers.

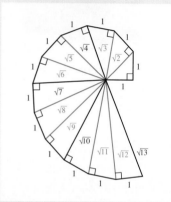

Procedure You are to construct a spiral of roots from a line segment 1 inch long. The graph paper you have contains either 4 or 5 squares per inch, allowing you to accurately draw 1-inch line segments. Because the lines on the graph paper are perpendicular to one another, if you are careful, you can use the graph paper to connect one line segment to another so that they form a right angle.

1. Fold one of the pieces of graph paper so it can be used as a ruler.

2. Use the folded paper to draw a line segment 1-inch long, just to the right of the middle of the unfolded paper. On the end of this segment, attach another segment of 1-inch length at a right angle to the first one. Connect the end points of the segments to form a right triangle. Label each side of this triangle. When you are finished, your work should resemble Figure 1.

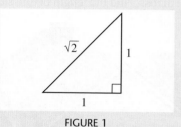

FIGURE 1

3. On the end of the hypotenuse of the triangle, attach a 1-inch line segment so that the two segments form a right angle. (Use the folded paper to do this.) Draw the hypotenuse of this triangle. Label all the sides of this second triangle. Your work should resemble Figure 2.

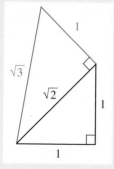

FIGURE 2

4. Continue to draw a new right triangle by attaching 1-inch line segments at right angles to the previous hypotenuse. Label all the sides of each triangle.

5. Stop when you have drawn a hypotenuse $\sqrt{8}$ inches long.

Connections

Although it may not look like it, the three items shown here are related very closely to one another. Your job is to find the connection.

A Continued Fraction

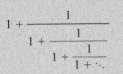

$$1 + \cfrac{1}{1 + \cfrac{1}{1 + \cfrac{1}{1 + \cdots}}}$$

The Fibonacci Sequence

1, 1, 2, 3, 5, . . .

The Golden Rectangle

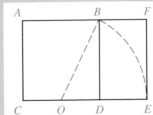

Step 1: The dots in the continued fraction indicate that the pattern shown continues indefinitely. This means that there is no way for us to simplify this expression, as we have simplified the expressions in this chapter. However, we can begin to understand the continued fraction, and what it simplifies to, by working with the following sequence of expressions. Simplify each expression. Write each answer as a fraction, in lowest terms.

$$1 + \cfrac{1}{1+1} \qquad 1 + \cfrac{1}{1 + \cfrac{1}{1+1}} \qquad 1 + \cfrac{1}{1 + \cfrac{1}{1 + \cfrac{1}{1+1}}} \qquad 1 + \cfrac{1}{1 + \cfrac{1}{1 + \cfrac{1}{1 + \cfrac{1}{1+1}}}}$$

Step 2: Compare the fractional answers to step 1 with the numbers in the Fibonacci sequence. Based on your observation, give the answer to the following problem, without actually doing any arithmetic.

$$1 + \cfrac{1}{1 + \cfrac{1}{1 + \cfrac{1}{1 + \cfrac{1}{1 + \cfrac{1}{1+1}}}}}$$

Step 3: Continue the sequence of simplified fractions you have written in steps 1 and 2, until you have nine numbers in the sequence. Convert each of these numbers to a decimal, accurate to four places past the decimal point.

Step 4: Find a decimal approximation to the golden ratio $\dfrac{1 + \sqrt{5}}{2}$, accurate to four places past the decimal point.

Step 5: Compare the results in steps 3 and 4, and then make a conjecture about what number the continued fraction would simplify to, if it was actually possible to simplify it.

Quadratic Functions

10

age fotostock/SuperStock

If you have been to the circus or the county fair, you may have witnessed one of the more spectacular acts, the human cannonball. The human cannonball shown in the photograph will reach a height of 70 feet and travel a distance of 160 feet before landing in a safety net. In this chapter, we use this information to derive the equation

$$f(x) = -\frac{7}{640}(x - 80)^2 + 70 \quad \text{for } 0 \leq x \leq 160$$

which describes the path flown by this particular cannonball. The table and graph below were constructed from this equation.

Path of a Human Cannonball	
x (feet)	$f(x)$ (nearest foot)
0	0
40	53
80	70
120	53
160	0

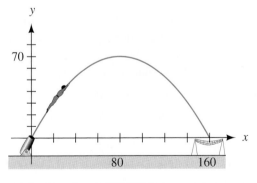

FIGURE 1

▶ Improve your grade and save time!
Go online to **academic.cengage.com/login** where you can
- Watch videos of instructors working through the in-text examples
- Follow step-by-step online tutorials of in-text examples and review questions
- Work practice problems
- Check your readiness for an exam by taking a pre-test and exploring the modules recommended in your Personalized Study plan
- Receive help from a live tutor online through vMentor™
Try it out! Log in with an access code or purchase access at **www.ichapters.com**.

All objects that are projected into the air, whether they are basketballs, bullets, arrows, or coins, follow parabolic paths like the one shown in Figure 1. Studying the material in this chapter will give you a more mathematical hold on the world around you.

OBJECTIVES

A Solve quadratic equations by taking the square root of both sides.

B Solve quadratic equations by completing the square.

C Use quadratic equations to solve for missing parts of right triangles.

Table 1 is taken from the trail map given to skiers at the Northstar at Tahoe Ski Resort in Lake Tahoe, California. The table gives the length of each chair lift at Northstar, along with the change in elevation from the beginning of the lift to the end of the lift.

Right triangles are good mathematical models for chair lifts. In this section, we will use our knowledge of right triangles, along with the new material developed in the section, to solve problems involving chair lifts and a variety of other examples.

TABLE 1
From the Trail Map for the Northstar at Tahoe Ski Resort

Lift Information		
Lift	Vertical Rise (feet)	Length (feet)
Big Springs Gondola	480	4,100
Bear Paw Double	120	790
Echo Triple	710	4,890
Aspen Express Quad	900	5,100
Forest Double	1,170	5,750
Lookout Double	960	4,330
Comstock Express Quad	1,250	5,900
Rendezvous Triple	650	2,900
Schaffer Camp Triple	1,860	6,150
Chipmunk Tow Lift	28	280
Bear Cub Tow Lift	120	750

In this section, we will develop the first of our new methods of solving quadratic equations. The new method is called *completing the square*. Completing the square on a quadratic equation allows us to obtain solutions, regardless of whether the equation can be factored. Before we solve equations by completing the square, we need to learn how to solve equations by taking square roots of both sides.

Consider the equation

$$x^2 = 16$$

We could solve it by writing it in standard form, factoring the left side, and proceeding as we have done previously. We can shorten our work considerably, however, if we simply notice that x must be either the positive square root of 16 or the negative square root of 16; that is,

If $x^2 = 16$

then $x = \sqrt{16}$ or $x = -\sqrt{16}$

$x = 4$ or $x = -4$

We can generalize this result into a theorem as follows.

Theorem 1 If $a^2 = b$ where b is a real number, then $a = \sqrt{b}$ or $a = -\sqrt{b}$.

Notation The expression $a = \sqrt{b}$ or $a = -\sqrt{b}$ can be written in shorthand form as $a = \pm\sqrt{b}$. The symbol $\pm$ is read "plus or minus."

We can apply Theorem 1 to some fairly complicated quadratic equations.

 EXAMPLE 1   Solve $(2x - 3)^2 = 25$.

SOLUTION

$$(2x - 3)^2 = 25$$

$$2x - 3 = \pm\sqrt{25} \qquad \text{Theorem 1}$$

$$2x - 3 = \pm 5 \qquad \sqrt{25} = 5$$

$$2x = 3 \pm 5 \qquad \text{Add 3 to both sides}$$

$$x = \frac{3 \pm 5}{2} \qquad \text{Divide both sides by 2}$$

The last equation can be written as two separate statements:

$$x = \frac{3 + 5}{2} \qquad \text{or} \qquad x = \frac{3 - 5}{2}$$

$$= \frac{8}{2} \qquad\qquad\qquad = \frac{-2}{2}$$

$$= 4 \qquad \text{or} \qquad = -1$$

The solution set is $\{4, -1\}$.

Notice that we could have solved the equation in Example 1 by expanding the left side, writing the resulting equation in standard form, and then factoring. The problem would look like this:

$$(2x - 3)^2 = 25 \qquad \text{Original equation}$$

$$4x^2 - 12x + 9 = 25 \qquad \text{Expand the left side}$$

$$4x^2 - 12x - 16 = 0 \qquad \text{Add } -25 \text{ to each side}$$

$$4(x^2 - 3x - 4) = 0 \qquad \text{Begin factoring}$$

$$4(x - 4)(x + 1) = 0 \qquad \text{Factor completely}$$

$$x - 4 = 0 \quad \text{or} \quad x + 1 = 0 \qquad \text{Set variable factors equal to 0}$$

$$x = 4 \quad \text{or} \qquad x = -1$$

As you can see, solving the equation by factoring leads to the same two solutions.

Note

We cannot solve the equation in Example 2 by factoring. If we expand the left side and write the resulting equation in standard form, we are left with a quadratic equation that does not factor:

$(3x - 1)^2 = -12$

$9x^2 - 6x + 1 = -12$

$9x^2 - 6x + 13 = 0$

EXAMPLE 2   Solve for x: $(3x - 1)^2 = -12$.

SOLUTION

$$(3x - 1)^2 = -12$$

$$3x - 1 = \pm\sqrt{-12} \qquad \text{Theorem 1}$$

$$3x - 1 = \pm 2i\sqrt{3} \qquad \sqrt{-12} = 2i\sqrt{3}$$

$$3x = 1 \pm 2i\sqrt{3} \qquad \text{Add 1 to both sides}$$

$$x = \frac{1 \pm 2i\sqrt{3}}{3} \qquad \text{Divide both sides by 3}$$

The solution set is $\left\{ \dfrac{1 + 2i\sqrt{3}}{3}, \dfrac{1 - 2i\sqrt{3}}{3} \right\}$

Both solutions are complex. Here is a check of the first solution:

$$\text{When} \qquad x = \frac{1 + 2i\sqrt{3}}{3}$$

$$\text{the equation} \qquad (3x - 1)^2 = -12$$

$$\text{becomes} \qquad \left(3 \cdot \frac{1 + 2i\sqrt{3}}{3} - 1\right)^2 \overset{?}{=} -12$$

$$\text{or} \qquad (1 + 2i\sqrt{3} - 1)^2 \overset{?}{=} -12$$

$$(2i\sqrt{3})^2 \overset{?}{=} -12$$

$$4 \cdot i^2 \cdot 3 \overset{?}{=} -12$$

$$12(-1) \overset{?}{=} -12$$

$$-12 = -12 \qquad$$

 EXAMPLE 3   Solve $x^2 + 6x + 9 = 12$.

SOLUTION   We can solve this equation as we have the equations in Examples 1 and 2 if we first write the left side as $(x + 3)^2$.

$x^2 + 6x + 9 = 12$	**Original equation**
$(x + 3)^2 = 12$	**Write $x^2 + 6x + 9$ as $(x + 3)^2$**
$x + 3 = \pm 2\sqrt{3}$	**Theorem 1**
$x = -3 \pm 2\sqrt{3}$	**Add -3 to each side**

We have two irrational solutions: $-3 + 2\sqrt{3}$ and $-3 - 2\sqrt{3}$. What is important about this problem, however, is the fact that the equation was easy to solve because the left side was a perfect square trinomial.

Completing the Square

The method of completing the square is simply a way of transforming any quadratic equation into an equation of the form found in the preceding three examples.

The key to understanding the method of completing the square lies in recognizing the relationship between the last two terms of any perfect square trinomial whose leading coefficient is 1.

Consider the following list of perfect square trinomials and their corresponding binomial squares:

$$x^2 - 6x + 9 = (x - 3)^3$$

$$x^2 + 8x + 16 = (x + 4)^2$$

$$x^2 - 10x + 25 = (x - 5)^2$$

$$x^2 + 12x + 36 = (x + 6)^2$$

In each case the leading coefficient is 1. A more important observation comes from noticing the relationship between the linear and constant terms (middle and last terms) in each trinomial. Observe that the constant term in each case is the square of half the coefficient of x in the middle term. For example, in the last expression, the constant term 36 is the square of half of 12, where 12 is the coefficient of x in the middle term. (Notice also that the second terms in all the binomials on the right side are half the coefficients of the middle terms of the trinomials on the left side.) We can use these observations to build our own perfect square trinomials and, in doing so, solve some quadratic equations.

Consider the following equation:

$$x^2 + 6x = 3$$

We can think of the left side as having the first two terms of a perfect square trinomial. We need only add the correct constant term. If we take half the coefficient of x, we get 3. If we then square this quantity, we have 9. Adding the 9 to both sides, the equation becomes

$$x^2 + 6x + \mathbf{9} = 3 + \mathbf{9}$$

Note

This is the step in which we actually complete the square.

The left side is the perfect square $(x + 3)^2$; the right side is 12:

$$(x + 3)^2 = 12$$

The equation is now in the correct form. We can apply Theorem 1 and finish the solution:

$$(x + 3)^2 = 12$$
$$x + 3 = \pm\sqrt{12} \qquad \textbf{Theorem 1}$$
$$x + 3 = \pm 2\sqrt{3}$$
$$x = -3 \pm 2\sqrt{3}$$

The solution set is $\{-3 + 2\sqrt{3},\ -3 - 2\sqrt{3}\}$. The method just used is called *completing the square* since we complete the square on the left side of the original equation by adding the appropriate constant term.

EXAMPLE 4   Solve by completing the square: $x^2 + 5x - 2 = 0$.

SOLUTION   We must begin by adding 2 to both sides. (The left side of the equation, as it is, is not a perfect square because it does not have the correct constant term. We will simply "move" that term to the other side and use our own constant term.)

$$x^2 + 5x = 2 \qquad \textbf{Add 2 to each side}$$

We complete the square by adding the square of half the coefficient of the linear term to both sides:

$$x^2 + 5x + \frac{\mathbf{25}}{\mathbf{4}} = 2 + \frac{\mathbf{25}}{\mathbf{4}} \qquad \textbf{Half of 5 is } \tfrac{5}{2}\textbf{, the square of which is } \tfrac{25}{4}$$

$$\left(x + \frac{5}{2}\right)^2 = \frac{33}{4} \qquad 2 + \tfrac{25}{4} = \tfrac{8}{4} + \tfrac{25}{4} = \tfrac{33}{4}$$

$$x + \frac{5}{2} = \pm\sqrt{\frac{33}{4}} \qquad \textbf{Theorem 1}$$

Note

We can use a calculator to get decimal approximations to these solutions. If $\sqrt{33} \approx 5.74$, then

$$\frac{-5 + 5.74}{2} = 0.37$$

$$\frac{-5 - 5.74}{2} = -5.37$$

$$x + \frac{5}{2} = \pm\frac{\sqrt{33}}{2} \qquad \textbf{Simplify the radical}$$

$$x = -\frac{5}{2} \pm \frac{\sqrt{33}}{2} \qquad \textbf{Add } -\tfrac{5}{2} \textbf{ to both sides}$$

$$= \frac{-5 \pm \sqrt{33}}{2}$$

The solution set is $\left\{\dfrac{-5 + \sqrt{33}}{2},\ \dfrac{-5 - \sqrt{33}}{2}\right\}$.

EXAMPLE 5   Solve for x: $3x^2 - 8x + 7 = 0$.

SOLUTION

$$3x^2 - 8x + 7 = 0$$

$$3x^2 - 8x = -7 \qquad \textbf{Add } -7 \textbf{ to both sides}$$

We cannot complete the square on the left side because the leading coefficient is not 1. We take an extra step and divide both sides by 3:

$$\frac{3x^2}{3} - \frac{8x}{3} = -\frac{7}{3}$$

$$x^2 - \frac{8}{3}x = -\frac{7}{3}$$

Half of $\frac{8}{3}$ is $\frac{4}{3}$, the square of which is $\frac{16}{9}$.

$$x^2 - \frac{8}{3}x + \mathbf{\frac{16}{9}} = -\frac{7}{3} + \mathbf{\frac{16}{9}} \qquad \textbf{Add } \frac{16}{9} \textbf{ to both sides}$$

$$\left(x - \frac{4}{3}\right)^2 = -\frac{5}{9} \qquad \textbf{Simplify right side}$$

$$x - \frac{4}{3} = \pm\sqrt{-\frac{5}{9}} \qquad \textbf{Theorem 1}$$

$$x - \frac{4}{3} = \pm\frac{i\sqrt{5}}{3} \qquad \sqrt{-\frac{5}{9}} = \frac{\sqrt{-5}}{3} = \frac{i\sqrt{5}}{3}$$

$$x = \frac{4}{3} \pm \frac{i\sqrt{5}}{3} \qquad \textbf{Add } \frac{4}{3} \textbf{ to both sides}$$

$$x = \frac{4 \pm i\sqrt{5}}{3}$$

The solution set is $\left\{ \dfrac{4 + i\sqrt{5}}{3}, \dfrac{4 - i\sqrt{5}}{3} \right\}$.

To Solve a Quadratic Equation by Completing the Square To summarize the method used in the preceding two examples, we list the following steps:

Step 1: Write the equation in the form $ax^2 + bx = c$.

Step 2: If the leading coefficient is not 1, divide both sides by the coefficient so that the resulting equation has a leading coefficient of 1; that is, if $a \neq 1$, then divide both sides by a.

Step 3: Add the square of half the coefficient of the linear term to both sides of the equation.

Step 4: Write the left side of the equation as the square of a binomial, and simplify the right side if possible.

Step 5: Apply Theorem 1, and solve as usual.

FACTS FROM GEOMETRY

More Special Triangles

The triangles shown in Figures 1 and 2 occur frequently in mathematics.

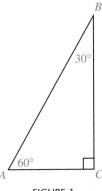

FIGURE 1

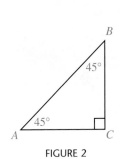

FIGURE 2

Note that both of the triangles are right triangles. We refer to the triangle in Figure 1 as a 30°–60°–90° triangle and the triangle in Figure 2 as a 45°–45°–90° triangle.

EXAMPLE 6   If the shortest side in a 30°–60°–90° triangle is 1 inch, find the lengths of the other two sides.

SOLUTION   In Figure 3 triangle ABC is a 30°–60°–90° triangle in which the shortest side AC is 1 inch long. Triangle DBC is also a 30°–60°–90° triangle in which the shortest side DC is 1 inch long.

FIGURE 3

Notice that the large triangle ABD is an equilateral triangle because each of its interior angles is 60°. Each side of triangle ABD is 2 inches long. Side AB in triangle ABC is therefore 2 inches. To find the length of side BC, we use the Pythagorean theorem.

$$BC^2 + AC^2 = AB^2$$

$$x^2 + 1^2 = 2^2$$

$$x^2 + 1 = 4$$

$$x^2 = 3$$

$$x = \sqrt{3} \text{ inches}$$

Note that we write only the positive square root because x is the length of a side in a triangle and is therefore a positive number.

 EXAMPLE 7   Table 1 in the introduction to this section gives the vertical rise of the Forest Double chair lift as 1,170 feet and the length of the chair lift as 5,750 feet. To the nearest foot, find the horizontal distance covered by a person riding this lift.

SOLUTION   Figure 4 is a model of the Forest Double chair lift. A rider gets on the lift at point A and exits at point B. The length of the lift is AB.

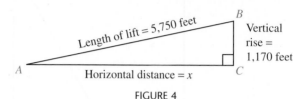

FIGURE 4

To find the horizontal distance covered by a person riding the chair lift, we use the Pythagorean theorem.

$$5,750^2 = x^2 + 1,170^2 \qquad \textbf{Pythagorean theorem}$$

$$33,062,500 = x^2 + 1,368,900 \qquad \textbf{Simplify squares}$$

$$x^2 = 33,062,500 - 1,368,900 \qquad \textbf{Solve for } x^2$$

$$x^2 = 31,693,600 \qquad \textbf{Simplify the right side}$$

$$x = \sqrt{31,693,600} \qquad \textbf{Theorem 1}$$

$$= 5,630 \text{ feet} \ \ (\text{to the nearest foot})$$

A rider getting on the lift at point A and riding to point B will cover a horizontal distance of approximately 5,630 feet.

LINKING OBJECTIVES AND EXAMPLES

Next to each **objective** we have listed the examples that are best described by that objective.

A	1–3
B	4, 5
C	6, 7

GETTING READY FOR CLASS

After reading through the preceding section, respond in your own words and in complete sentences.

1. What kind of equation do we solve using the method of completing the square?

2. Explain in words how you would complete the square on $x^2 - 16x = 4$.

3. What is the relationship between the shortest side and the longest side in a 30°–60°–90° triangle?

4. What two expressions together are equivalent to $x = \pm 4$?

Solve the following equations.

1. $x^2 = 25$ **2.** $x^2 = 16$

3. $y^2 = \dfrac{3}{4}$ **4.** $y^2 = \dfrac{5}{9}$

5. $x^2 + 12 = 0$ **6.** $x^2 + 8 = 0$

7. $4a^2 - 45 = 0$ **8.** $9a^2 - 20 = 0$

9. $(2y - 1)^2 = 25$ **10.** $(3y + 7)^2 = 1$

11. $(2a + 3)^2 = -9$ **12.** $(3a - 5)^2 = -49$

13. $x^2 + 8x + 16 = -27$

14. $x^2 - 12x + 36 = -8$

15. $4a^2 - 12a + 9 = -4$

16. $9a^2 - 12a + 4 = -9$

Copy each of the following, and fill in the blanks so that the left side of each is a perfect square trinomial; that is, complete the square.

17. $x^2 + 12x + \underline{\hspace{1cm}} = (x + \underline{\hspace{1cm}})^2$

18. $x^2 + 6x + \underline{\hspace{1cm}} = (x + \underline{\hspace{1cm}})^2$

19. $x^2 - 4x + \underline{\hspace{1cm}} = (x - \underline{\hspace{1cm}})^2$

20. $x^2 - 2x + \underline{\hspace{1cm}} = (x - \underline{\hspace{1cm}})^2$

21. $a^2 - 10a + \underline{\hspace{1cm}} = (a - \underline{\hspace{1cm}})^2$

22. $a^2 - 8a + \underline{\hspace{1cm}} = (a - \underline{\hspace{1cm}})^2$

23. $x^2 + 5x + \underline{\hspace{1cm}} = (x + \underline{\hspace{1cm}})^2$

24. $x^2 + 3x + \underline{\hspace{1cm}} = (x + \underline{\hspace{1cm}})^2$

25. $y^2 - 7y + \underline{\hspace{1cm}} = (y - \underline{\hspace{1cm}})^2$

26. $y^2 - y + \underline{\hspace{1cm}} = (y - \underline{\hspace{1cm}})^2$

27. $x^2 + \dfrac{1}{2}x + \underline{\hspace{1cm}} = (x + \underline{\hspace{1cm}})^2$

28. $x^2 - \dfrac{3}{4}x + \underline{\hspace{1cm}} = (x - \underline{\hspace{1cm}})^2$

29. $x^2 + \dfrac{2}{3}x + \underline{\hspace{1cm}} = (x + \underline{\hspace{1cm}})^2$

30. $x^2 - \dfrac{4}{5}x + \underline{\hspace{1cm}} = (x - \underline{\hspace{1cm}})^2$

Solve each of the following quadratic equations by completing the square.

31. $x^2 + 12x = -27$ **32.** $x^2 - 6x = 16$

33. $a^2 - 2a + 5 = 0$

34. $a^2 + 10a + 22 = 0$

35. $y^2 - 8y + 1 = 0$

36. $y^2 + 6y - 1 = 0$

37. $x^2 - 5x - 3 = 0$

38. $x^2 - 5x - 2 = 0$

39. $2x^2 - 4x - 8 = 0$

40. $3x^2 - 9x - 12 = 0$

41. $3t^2 - 8t + 1 = 0$

42. $5t^2 + 12t - 1 = 0$

43. $4x^2 - 3x + 5 = 0$

44. $7x^2 - 5x + 2 = 0$

45. For the equation $x^2 = -9$
  a. Can it be solved by factoring?
  b. Solve it.

46. For the equation $x^2 - 10x + 18 = 0$
  a. Can it be solved by factoring?
  b. Solve it.

47. Solve each equation below by the indicated method.
  a. $x^2 - 6x = 0$ Factoring
  b. $x^2 - 6x = 0$ Completing the square

48. Solve each equation below by the indicated method.
  a. $x^2 + ax = 0$ Factoring
  b. $x^2 + ax = 0$ Completing the square

49. Solve the equation $x^2 + 2x = 35$
  a. By factoring.
  b. By completing the square.

50. Solve the equation $8x^2 - 10x - 25 = 0$
  a. By factoring.
  b. By completing the square.

51. Is $x = -3 + \sqrt{2}$ a solution to $x^2 - 6x = 7$?

52. Is $x = 2 - \sqrt{5}$ a solution to $x^2 - 4x = 1$?

53. Solve each equation
  a. $5x - 7 = 0$
  b. $5x - 7 = 8$
  c. $(5x - 7)^2 = 8$
  d. $\sqrt{5x - 7} = 8$
  e. $\dfrac{5}{2} - \dfrac{7}{2x} = \dfrac{4}{x}$

54. Solve each equation

    a. $5x + 11 = 0$

    b. $5x + 11 = 9$

    c. $(5x + 11)^2 = 9$

    d. $\sqrt{5x + 11} = 9$

    e. $\dfrac{5}{3} - \dfrac{11}{3x} = \dfrac{3}{x}$

Simplify the left side of each equation, and then solve for x.

55. $(x + 5)^2 + (x - 5)^2 = 52$

56. $(2x + 1)^2 + (2x - 1)^2 = 10$

57. $(2x + 3)^2 + (2x - 3)^2 = 26$

58. $(3x + 2)^2 + (3x - 2)^2 = 26$

59. $(3x + 4)(3x - 4) - (x + 2)(x - 2) = -4$

60. $(5x + 2)(5x - 2) - (x + 3)(x - 3) = 29$

61. Fill in the table below given the following functions.
$f(x) = (2x - 3)^2, g(x) = 4x^2 - 12x + 9, h(x) = 4x^2 + 9$

x	$f(x)$	$g(x)$	$h(x)$
-2			
-1			
0			
1			
2			

62. Fill in the table below given the following functions. $f(x) = \left(x + \dfrac{1}{2}\right)^2, g(x) = x^2 + x + \dfrac{1}{4}, h(x) = x^2 + \dfrac{1}{4}$

x	$f(x)$	$g(x)$	$h(x)$
-2			
-1			
0			
1			
2			

▶ **63.** If $f(x) = (x - 3)^2$, find x if $f(x) = 0$.

▶ **64.** If $f(x) = (3x - 5)^2$, find x if $f(x) = 0$.

▶ **65.** If $f(x) = x^2 - 5x - 6$ find

    a. The x-intercepts

    b. The value of x for which $f(x) = 0$

    c. $f(0)$

    d. $f(1)$

▶ **66.** If $f(x) = 9x^2 - 12x + 4$ find

    a. The x-intercepts

    b. The value of x for which $f(x) = 0$

    c. $f(0)$

    d. $f\left(\dfrac{2}{3}\right)$

Applying the Concepts

67. Geometry If the shortest side in a 30°–60°–90° triangle is $\frac{1}{2}$ inch long, find the lengths of the other two sides.

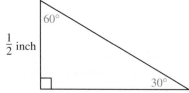

68. Geometry If the length of the longest side of a 30°–60°–90° triangle is x, find the length of the other two sides in terms of x.

69. Geometry If the length of the shorter sides of a 45°–45°–90° triangle is 1 inch, find the length of the hypotenuse.

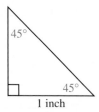

70. Geometry If the length of the shorter sides of a 45°–45°–90° triangle is x, find the length of the hypotenuse, in terms of x.

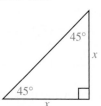

EXAMPLE 3 Solve $\dfrac{1}{x+2} - \dfrac{1}{x} = \dfrac{1}{3}$.

SOLUTION To solve this equation, we must first put it in standard form. To do so, we must clear the equation of fractions by multiplying each side by the LCD for all the denominators, which is $3x(x+2)$. Multiplying both sides by the LCD, we have

$$3x(x+2)\left(\frac{1}{x+2} - \frac{1}{x}\right) = \frac{1}{3} \cdot 3x(x+2) \qquad \text{Multiply each by the LCD}$$

$$3x(x+2) \cdot \frac{1}{x+2} - 3x(x+2) \cdot \frac{1}{x} = \frac{1}{3} \cdot 3x(x+2)$$

$$3x - 3(x+2) = x(x+2)$$

$$3x - 3x - 6 = x^2 + 2x \qquad \text{Multiplication}$$

$$-6 = x^2 + 2x \qquad \text{Simplify left side}$$

$$0 = x^2 + 2x + 6 \qquad \text{Add 6 to each side}$$

Since the right side of our last equation is not factorable, we use the quadratic formula. From our last equation, we have $a = 1$, $b = 2$, and $c = 6$. Using these numbers for a, b, and c in the quadratic formula gives us

$$x = \frac{-2 \pm \sqrt{4 - 4(1)(6)}}{2(1)}$$

$$= \frac{-2 \pm \sqrt{4 - 24}}{2} \qquad \text{Simplify inside the radical}$$

$$= \frac{-2 \pm \sqrt{-20}}{2} \qquad 4 - 24 = -20$$

$$= \frac{-2 \pm 2i\sqrt{5}}{2} \qquad -20 = i\sqrt{20} = i\sqrt{4}\,\sqrt{5} = 2i\sqrt{5}$$

$$= \frac{2(-1 \pm i\sqrt{5})}{2} \qquad \text{Factor 2 from the numerator}$$

$$= -1 \pm i\sqrt{5} \qquad \text{Divide numerator and denominator by 2}$$

Since neither of the two solutions, $-1 + i\sqrt{5}$ nor $-1 - i\sqrt{5}$, will make any of the denominators in our original equation 0, they are both solutions.

Although the equation in our next example is not a quadratic equation, we solve it by using both factoring and the quadratic formula.

EXAMPLE 4 Solve $27t^3 - 8 = 0$.

SOLUTION It would be a mistake to add 8 to each side of this equation and then take the cube root of each side because we would lose two of our solutions. Instead, we factor the left side, and then set the factors equal to 0:

$$27t^3 - 8 = 0 \qquad \text{Equation in standard form}$$

$$(3t - 2)(9t^2 + 6t + 4) = 0 \qquad \text{Factor as the difference of two cubes}$$

$$3t - 2 = 0 \quad \text{or} \quad 9t^2 + 6t + 4 = 0 \qquad \text{Set each factor equal to 0}$$

The first equation leads to a solution of $t = \frac{2}{3}$. The second equation does not factor, so we use the quadratic formula with $a = 9$, $b = 6$, and $c = 4$:

$$t = \frac{-6 \pm \sqrt{36 - 4(9)(4)}}{2(9)}$$

$$= \frac{-6 \pm \sqrt{36 - 144}}{18}$$

$$= \frac{-6 \pm \sqrt{-108}}{18}$$

$$= \frac{-6 \pm 6i\sqrt{3}}{18} \qquad \sqrt{-108} = i\sqrt{36 \cdot 3} = 6i\sqrt{3}$$

$$= \frac{6(-1 \pm i\sqrt{3})}{6 \cdot 3} \qquad \textbf{Factor 6 from the numerator and denominator}$$

$$= \frac{-1 \pm i\sqrt{3}}{3} \qquad \textbf{Divide out common factor 6}$$

The three solutions to our original equation are

$$\frac{2}{3}, \qquad \frac{-1 + i\sqrt{3}}{3}, \qquad \text{and} \qquad \frac{-1 - i\sqrt{3}}{3}$$

EXAMPLE 5   If an object is thrown downward with an initial velocity of 20 feet per second, the distance $s(t)$, in feet, it travels in t seconds is given by the function $s(t) = 20t + 16t^2$. How long does it take the object to fall 40 feet?

SOLUTION   We let $s(t) = 40$, and solve for t:

When $\qquad\qquad s(t) = 40$

the function $\qquad s(t) = 20t + 16t^2$

becomes $\qquad\quad 40 = 20t + 16t^2$

or $\quad 16t^2 + 20t - 40 = 0$

$\qquad\qquad 4t^2 + 5t - 10 = 0 \qquad$ **Divide by 4**

Using the quadratic formula, we have

$$t = \frac{-5 \pm \sqrt{25 - 4(4)(-10)}}{2(4)}$$

$$= \frac{-5 \pm \sqrt{185}}{8}$$

$$= \frac{-5 + \sqrt{185}}{8} \qquad \text{or} \qquad \frac{-5 - \sqrt{185}}{8}$$

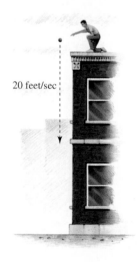

20 feet/sec

The second solution is impossible since it is a negative number and time t must be positive. It takes

$$t = \frac{-5 + \sqrt{185}}{8} \quad \text{or approximately} \quad \frac{-5 + 13.60}{8} \approx 1.08 \text{ seconds}$$

for the object to fall 40 feet.

Recall that the relationship between profit, revenue, and cost is given by the formula

$$P(x) = R(x) - C(x)$$

where $P(x)$ is the profit, $R(x)$ is the total revenue, and $C(x)$ is the total cost of producing and selling x items.

EXAMPLE 6   A company produces and sells copies of an accounting program for home computers. The total weekly cost (in dollars) to produce x copies of the program is $C(x) = 8x + 500$, and the weekly revenue for selling all x copies of the program is $R(x) = 35x - 0.1x^2$. How many programs must be sold each week for the weekly profit to be $1,200?

SOLUTION   Substituting the given expressions for $R(x)$ and $C(x)$ in the equation $P(x) = R(x) - C(x)$, we have a polynomial in x that represents the weekly profit $P(x)$:

$$P(x) = R(x) - C(x)$$
$$= 35x - 0.1x^2 - (8x + 500)$$
$$= 35x - 0.1x^2 - 8x - 500$$
$$= -500 + 27x - 0.1x^2$$

Setting this expression equal to 1,200, we have a quadratic equation to solve that gives us the number of programs x that need to be sold each week to bring in a profit of $1,200:

$$1,200 = -500 + 27x - 0.1x^2$$

We can write this equation in standard form by adding the opposite of each term on the right side of the equation to both sides of the equation. Doing so produces the following equation:

$$0.1x^2 - 27x + 1,700 = 0$$

Applying the quadratic formula to this equation with $a = 0.1$, $b = -27$, and $c = 1,700$, we have

$$x = \frac{27 \pm \sqrt{(-27)^2 - 4(0.1)(1,700)}}{2(0.1)}$$
$$= \frac{27 \pm \sqrt{729 - 680}}{0.2}$$
$$= \frac{27 \pm \sqrt{49}}{0.2}$$
$$= \frac{27 \pm 7}{0.2}$$

Writing this last expression as two separate expressions, we have our two solutions:

$$x = \frac{27 + 7}{0.2} \quad \text{or} \quad x = \frac{27 - 7}{0.2}$$
$$= \frac{34}{0.2} \qquad\qquad = \frac{20}{0.2}$$
$$= 170 \qquad\qquad\quad = 100$$

The weekly profit will be $1,200 if the company produces and sells 100 programs or 170 programs.

What is interesting about the equation we solved in Example 6 is that it has rational solutions, meaning it could have been solved by factoring. But looking back at the equation, factoring does not seem like a reasonable method of solution because the coefficients are either very large or very small. So, there are times when using the quadratic formula is a faster method of solution, even though the equation you are solving is factorable.

USING TECHNOLOGY

Graphing Calculators

More About Example 5

We can solve the problem discussed in Example 5 by graphing the function $Y_1 = 20X + 16X^2$ in a window with X from 0 to 2 (because X is taking the place of t and we know t is a positive quantity) and Y from 0 to 50 (because we are looking for X when Y_1 is 40). Graphing

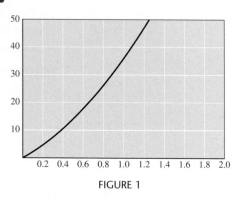

FIGURE 1

Y_1 gives a graph similar to the graph in Figure 1. Using the Zoom and Trace features at $Y_1 = 40$ gives us X = 1.08 to the nearest hundredth, matching the results we obtained by solving the original equation algebraically.

More About Example 6

To visualize the functions in Example 6, we set up our calculator this way:

$$Y_1 = 35X - .1X^2 \qquad \textbf{Revenue function}$$
$$Y_2 = 8X + 500 \qquad \textbf{Cost function}$$
$$Y_3 = Y_1 - Y_2 \qquad \textbf{Profit function}$$

Window: X from 0 to 350, Y from 0 to 3500

Graphing these functions produces graphs similar to the ones shown in Figure 2. The lower graph is the graph of the profit function. Using the Zoom and Trace features on the lower graph at $Y_3 = 1,200$ produces two corresponding values of X, 170 and 100, which match the results in Example 6.

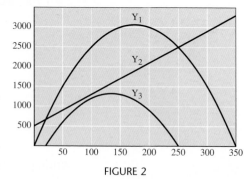

FIGURE 2

We will continue this discussion of the relationship between graphs of functions and solutions to equations in the Using Technology material in the next section.

LINKING OBJECTIVES AND EXAMPLES

Next to each objective we have listed the examples that are best described by that objective.

A 1–4

B 5, 6

GETTING READY FOR CLASS

After reading through the preceding section, respond in your own words and in complete sentences.

1. What is the quadratic formula?

2. Under what circumstances should the quadratic formula be applied?

3. When would the quadratic formula result in complex solutions?

4. When will the quadratic formula result in only one solution?

Problem Set 10.2

Online support materials can be found at academic.cengage.com/login

Solve each equation in each problem using the quadratic formula.

1. a. $3x^2 + 4x - 2 = 0$

b. $3x^2 - 4x - 2 = 0$

c. $3x^2 + 4x + 2 = 0$

d. $2x^2 + 4x - 3 = 0$

e. $2x^2 - 4x + 3 = 0$

2. a. $3x^2 + 6x - 2 = 0$

b. $3x^2 - 6x - 2 = 0$

c. $3x^2 + 6x + 2 = 0$

d. $2x^2 + 6x + 3 = 0$

e. $2x^2 + 6x - 3 = 0$

3. a. $x^2 - 2x + 2 = 0$

b. $x^2 - 2x + 5 = 0$

c. $x^2 + 2x + 2 = 0$

4. a. $x^2 - 4x + 5 = 0$

b. $x^2 + 4x + 5 = 0$

c. $a^2 + 4a + 1 = 0$

Solve each equation. Use factoring or the quadratic formula, whichever is appropriate. (Try factoring first. If you have any difficulty factoring, then go right to the quadratic formula.)

5. $\frac{1}{6}x^2 - \frac{1}{2}x + \frac{1}{3} = 0$

6. $\frac{1}{4}x^2 + \frac{1}{4}x - \frac{1}{2} = 0$

7. $\frac{x^2}{2} + 1 = \frac{2x}{3}$

8. $\frac{x^2}{2} + \frac{2}{3} = -\frac{2x}{3}$

9. $y^2 - 5y = 0$

10. $2y^2 + 10y = 0$

11. $30x^2 + 40x = 0$

12. $50x^2 - 20x = 0$

13. $\frac{2t^2}{3} - t = -\frac{1}{6}$

14. $\frac{t^2}{3} - \frac{t}{2} = -\frac{3}{2}$

15. $0.01x^2 + 0.06x - 0.08 = 0$

16. $0.02x^2 - 0.03x + 0.05 = 0$

17. $2x + 3 = -2x^2$

18. $2x - 3 = 3x^2$

▒ = Videos available by instructor request

▶ = Online student support materials available at academic.cengage.com/login

19. $100x^2 - 200x + 100 = 0$

20. $100x^2 - 600x + 900 = 0$

21. $\frac{1}{2}r^2 = \frac{1}{6}r - \frac{2}{3}$

22. $\frac{1}{4}r^2 = \frac{2}{5}r + \frac{1}{10}$

23. $(x - 3)(x - 5) = 1$

24. $(x - 3)(x + 1) = -6$

25. $(x + 3)^2 + (x - 8)(x - 1) = 16$

26. $(x - 4)^2 + (x + 2)(x + 1) = 9$

27. $\frac{x^2}{3} - \frac{5x}{6} = \frac{1}{2}$

28. $\frac{x^2}{6} + \frac{5}{6} = -\frac{x}{3}$

Multiply both sides of each equation by its LCD. Then solve the resulting equation.

29. $\frac{1}{x + 1} - \frac{1}{x} = \frac{1}{2}$

30. $\frac{1}{x + 1} + \frac{1}{x} = \frac{1}{3}$

31. $\frac{1}{y - 1} + \frac{1}{y + 1} = 1$

32. $\frac{2}{y + 2} + \frac{3}{y - 2} = 1$

33. $\frac{1}{x + 2} + \frac{1}{x + 3} = 1$

34. $\frac{1}{x + 3} + \frac{1}{x + 4} = 1$

35. $\frac{6}{r^2 - 1} - \frac{1}{2} = \frac{1}{r + 1}$

36. $2 + \frac{5}{r - 1} = \frac{12}{(r - 1)^2}$

Solve each equation. In each case you will have three solutions.

37. $x^3 - 8 = 0$

38. $x^3 - 27 = 0$

39. $8a^3 + 27 = 0$

40. $27a^3 + 8 = 0$

41. $125t^3 - 1 = 0$

42. $64t^3 + 1 = 0$

Each of the following equations has three solutions. Look for the greatest common factor, then use the quadratic formula to find all solutions.

43. $2x^3 + 2x^2 + 3x = 0$

44. $6x^3 - 4x^2 + 6x = 0$

45. $3y^4 = 6y^3 - 6y^2$

46. $4y^4 = 16y^3 - 20y^2$

47. $6t^5 + 4t^4 = -2t^3$

48. $8t^5 + 2t^4 = -10t^3$

49. Which two of the expressions below are equivalent?

   a. $\frac{6 + 2\sqrt{3}}{4}$

   b. $\frac{3 + \sqrt{3}}{2}$

   c. $6 + \frac{\sqrt{3}}{2}$

50. Which two of the expressions below are equivalent?

   a. $\frac{8 - 4\sqrt{2}}{4}$

   b. $2 - 4\sqrt{3}$

   c. $2 - \sqrt{2}$

▶ **51.** Solve $3x^2 - 5x = 0$
   a. By factoring

   b. By the quadratic formula

▶ **52.** Solve $3x^2 + 23x - 70 = 0$
   a. By factoring

   b. By the quadratic formula

53. Can the equation $x^2 - 4x + 7 = 0$ be solved by factoring? Solve the equation.

54. Can the equation $x^2 = 5$ be solved by factoring? Solve the equation.

▶ **55.** Is $x = -1 + i$ a solution to $x^2 + 2x = -2$?

▶ **56.** Is $x = 2 + 2i$ a solution to $(x - 2)^2 = -4$?

57. Let $f(x) = x^2 - 2x - 3$. Find all values for the variable x, which produce the following values of $f(x)$.
a. $f(x) = 0$
b. $f(x) = -11$
c. $f(x) = -2x + 1$
d. $f(x) = 2x + 1$

58. Let $g(x) = x^2 + 16$. Find all values for the variable x, which produce the following values of $g(x)$.
a. $g(x) = 0$
b. $g(x) = 20$
c. $g(x) = 8x$
d. $g(x) = -8x$

59. Let $f(x) = \dfrac{10}{x^2}$. Find all values for the variable x, for which $f(x) = g(x)$.
a. $g(x) = 3 + \dfrac{1}{x}$
b. $g(x) = 8x - \dfrac{17}{x^2}$
c. $g(x) = 0$
d. $g(x) = 10$

60. Let $h(x) = \dfrac{x + 2}{x}$. Find all values for the variable x, for which $h(x) = f(x)$.
a. $f(x) = 2$
b. $f(x) = x + 2$
c. $f(x) = x - 2$
d. $f(x) = \dfrac{x - 2}{4}$

Applying the Concepts

61. Falling Object An object is thrown downward with an initial velocity of 5 feet per second. The relationship between the distance s it travels and time t is given by $s = 5t + 16t^2$. How long does it take the object to fall 74 feet?

62. Coin Toss A coin is tossed upward with an initial velocity of 32 feet per second from a height of 16 feet above the ground. The equation giving the object's height h at any time t is $h = 16 + 32t - 16t^2$. Does the object ever reach a height of 32 feet?

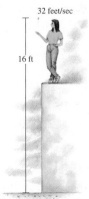
32 feet/sec
16 ft

63. Profit The total cost (in dollars) for a company to manufacture and sell x items per week is $C = 60x + 300$, whereas the revenue brought in by selling all x items is $R = 100x - 0.5x^2$. How many items must be sold to obtain a weekly profit of $300?

64. Profit Suppose a company manufactures and sells x picture frames each month with a total cost of $C = 1,200 + 3.5x$ dollars. If the revenue obtained by selling x frames is $R = 9x - 0.002x^2$, find the number of frames it must sell each month if its monthly profit is to be $2,300.

Kathleen Olson

65. Photograph Cropping The following figure shows a photographic image on a 10.5-centimeter by 8.2-centimeter background. The overall area of the background is to be reduced to 80% of its original area by cutting off (cropping) equal strips on all four sides. What is the width of the strip that is cut from each side?

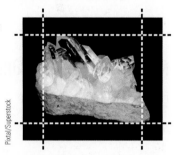
Pixtal/Superstock

66. Area of a Garden A garden measures 20.3 meters by 16.4 meters. To double the area of the garden, strips of equal width are added to all four sides.
a. Draw a diagram that illustrates these conditions.
b. What are the new overall dimensions of the garden?

67. Area and Perimeter A rectangle has a perimeter of 20 yards and an area of 15 square yards.
a. Write two equations that state these facts in terms of the rectangle's length, l, and its width, w.
b. Solve the two equations from part (a) to determine the actual length and width of the rectangle.
c. Explain why two answers are possible to part (b).

68. Population Size Writing in 1829, former President James Madison made some predictions about the growth of the population of the United States. The populations he predicted fit the equation

$$y = 0.029x^2 - 1.39x + 42$$

where y is the population in millions of people x years from 1829.

Library of Congress

a. Use the equation to determine the approximate year President Madison would have predicted that the U.S. population would reach 100,000,000.

b. If the U.S. population in 2006 was approximately 300 million, were President Madison's predictions accurate in the long term? Explain why or why not.

Maintaining Your Skills

Divide, using long division.

69. $\dfrac{8y^2 - 26y - 9}{2y - 7}$

70. $\dfrac{6y^2 + 7y - 18}{3y - 4}$

71. $\dfrac{x^3 + 9x^2 + 26x + 24}{x + 2}$

72. $\dfrac{x^3 + 6x^2 + 11x + 6}{x + 3}$

Simplify each expression. (Assume $x, y > 0$.)

73. $25^{1/2}$

74. $8^{1/3}$

75. $\left(\dfrac{9}{25}\right)^{3/2}$

76. $\left(\dfrac{16}{81}\right)^{3/4}$

77. $8^{-2/3}$

78. $4^{-3/2}$

79. $\dfrac{(49x^8y^{-4})^{1/2}}{(27x^{-3}y^9)^{-1/3}}$

80. $\dfrac{(x^{-2}y^{1/3})^6}{x^{-10}y^{3/2}}$

Getting Ready for the Next Section

Find the value of $b^2 - 4ac$ when

81. $a = 1, b = -3, c = -40$

82. $a = 2, b = 3, c = 4$

83. $a = 4, b = 12, c = 9$

84. $a = -3, b = 8, c = -1$

Solve.

85. $k^2 - 144 = 0$

86. $36 - 20k = 0$

Multiply.

87. $(x - 3)(x + 2)$

88. $(t - 5)(t + 5)$

89. $(x - 3)(x - 3)(x + 2)$

90. $(t - 5)(t + 5)(t - 3)$

Extending the Concepts

So far, all the equations we have solved have had coefficients that were rational numbers. Here are some equations that have irrational coefficients and some that have complex coefficients. Solve each equation. (Remember, $i^2 = -1$.)

91. $x^2 + \sqrt{3}x - 6 = 0$

92. $x^2 - \sqrt{5}x - 5 = 0$

93. $\sqrt{2}x^2 + 2x - \sqrt{2} = 0$

94. $\sqrt{7}x^2 + 2\sqrt{2}x - \sqrt{7} = 0$

95. $x^2 + ix + 2 = 0$

96. $x^2 + 3ix - 2 = 0$

Additional Items Involving Solutions to Equations

OBJECTIVES

A Find the number and kind of solutions to a quadratic equation by using the discriminant.

B Find an unknown constant in quadratic equation so that there is exactly one solution.

C Find an equation from its solutions.

In this section we will do two things. First, we will define the discriminant and use it to find the kind of solutions a quadratic equation has without solving the equation. Second, we will use the zero-factor property to build equations from their solutions.

The Discriminant

The quadratic formula

$$x = \frac{-b \pm \sqrt{b^2 - 4ac}}{2a}$$

gives the solutions to any quadratic equation in standard form. When working with quadratic equations, there are times when it is important only to know what kind of solutions the equation has.

> **DEFINITION** The expression under the radical in the quadratic formula is called the **discriminant**:
>
> $$\text{Discriminant} = D = b^2 - 4ac$$

The discriminant indicates the number and type of solutions to a quadratic equation, when the original equation has integer coefficients. For example, if we were to use the quadratic formula to solve the equation $2x^2 + 2x + 3 = 0$, we would find the discriminant to be

$$b^2 - 4ac = 2^2 - 4(2)(3) = -20$$

Since the discriminant appears under a square root symbol, we have the square root of a negative number in the quadratic formula. Our solutions therefore would be complex numbers. Similarly, if the discriminant were 0, the quadratic formula would yield

$$x = \frac{-b \pm \sqrt{0}}{2a} = \frac{-b \pm 0}{2a} = \frac{-b}{2a}$$

and the equation would have one rational solution, the number $\dfrac{-b}{2a}$.

The following table gives the relationship between the discriminant and the type of solutions to the equation.

For the equation $ax^2 + bx + c = 0$ where a, b, and c are integers and $a \neq 0$:

If the Discriminant $b^2 - 4ac$ is	Then the Equation Will Have
Negative	Two complex solutions containing i
Zero	One rational solution
A positive number that is also a perfect square	Two rational solutions
A positive number that is not a perfect square	Two irrational solutions

In the second and third cases, when the discriminant is 0 or a positive perfect square, the solutions are rational numbers. The quadratic equations in these two cases are the ones that can be factored.

EXAMPLES For each equation, give the number and kind of solutions.

1. $x^2 - 3x - 40 = 0$

SOLUTION Using $a = 1$, $b = -3$, and $c = -40$ in $b^2 - 4ac$, we have $(-3)^2 - 4(1)(-40) = 9 + 160 = 169$.

The discriminant is a perfect square. The equation therefore has two rational solutions.

2. $2x^2 - 3x + 4 = 0$

SOLUTION Using $a = 2$, $b = -3$, and $c = 4$, we have

$$b^2 - 4ac = (-3)^2 - 4(2)(4) = 9 - 32 = -23$$

The discriminant is negative, implying the equation has two complex solutions that contain i.

3. $4x^2 - 12x + 9 = 0$

SOLUTION Using $a = 4$, $b = -12$, and $c = 9$, the discriminant is

$$b^2 - 4ac = (-12)^2 - 4(4)(9) = 144 - 144 = 0$$

Since the discriminant is 0, the equation will have one rational solution.

4. $x^2 + 6x = 8$

SOLUTION We first must put the equation in standard form by adding -8 to each side. If we do so, the resulting equation is

$$x^2 + 6x - 8 = 0$$

Now we identify a, b, and c as 1, 6, and -8, respectively:

$$b^2 - 4ac = 6^2 - 4(1)(-8) = 36 + 32 = 68$$

The discriminant is a positive number but not a perfect square. Therefore, the equation will have two irrational solutions.

 EXAMPLE 5 Find an appropriate k so that the equation $4x^2 - kx = -9$ has exactly one rational solution.

SOLUTION We begin by writing the equation in standard form:

$$4x^2 - kx + 9 = 0$$

Using $a = 4$, $b = -k$, and $c = 9$, we have

$$b^2 - 4ac = (-k)^2 - 4(4)(9)$$
$$= k^2 - 144$$

An equation has exactly one rational solution when the discriminant is 0. We set the discriminant equal to 0 and solve:

$$k^2 - 144 = 0$$
$$k^2 = 144$$
$$k = \pm 12$$

Choosing k to be 12 or -12 will result in an equation with one rational solution.

32. $t = -\dfrac{4}{5}, t = 2$

33. $x = 3, x = -3, x = \dfrac{5}{6}$

34. $x = 5, x = -5, x = \dfrac{2}{3}$

35. $a = -\dfrac{1}{2}, a = \dfrac{3}{5}$

36. $a = -\dfrac{1}{3}, a = \dfrac{4}{7}$

37. $x = -\dfrac{2}{3}, x = \dfrac{2}{3}, x = 1$

▶ **38.** $x = -\dfrac{4}{5}, x = \dfrac{4}{5}, x = -1$

39. $x = 2, x = -2, x = 3, x = -3$

40. $x = 1, x = -1, x = 5, x = -5$

▶ **41.** $x = \sqrt{5}, x = -\sqrt{5}$

▶ **42.** $x = \sqrt{13}, x = -\sqrt{13}$

▶ **43.** $x = \sqrt{3}, x = -\sqrt{7}$

▶ **44.** $x = \sqrt{2}, x = -\sqrt{5}$

▶ **45.** $x = 4i, x = -4i$

▶ **46.** $x = 9i, x = -9i$

▶ **47.** $x = 2i, x = 3i$

▶ **48.** $x = -2i, x = -3i$

▶ **49.** $x = 2 + i, x = 2 - i$

▶ **50.** $x = 3 + 2i, x = 3 - 2i$

▶ **51.** $x = \dfrac{1 + \sqrt{5}}{2}, x = \dfrac{1 - \sqrt{5}}{2}$

▶ **52.** $x = \dfrac{-1 + \sqrt{5}}{2}, x = \dfrac{-1 - \sqrt{5}}{2}$

53. The graphs of three quadratic functions are shown below. Match each function with the number of solutions to equation $y = 0$, that it contains.
   a. One real solution
   b. No real solutions

c. Two real solutions

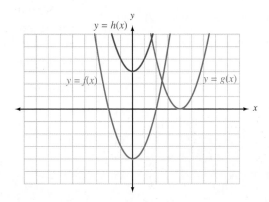

Maintaining Your Skills

Multiply. (Assume all variables represent positive numbers for the rest of the problems in this section.)

54. $a^4(a^{3/2} - a^{1/2})$

55. $(a^{1/2} - 5)(a^{1/2} + 3)$

56. $(x^{3/2} - 3)^2$

57. $(x^{1/2} - 8)(x^{1/2} + 8)$

Divide.

58. $\dfrac{30x^{3/4} - 25x^{5/4}}{5x^{1/4}}$

59. $\dfrac{45x^{5/3}y^{7/3} - 36x^{8/3}y^{4/3}}{9x^{2/3}y^{1/3}}$

60. Factor $5(x - 3)^{1/2}$ from $10(x - 3)^{3/2} - 15(x - 3)^{1/2}$.

61. Factor $2(x + 1)^{1/3}$ from $8(x + 1)^{4/3} - 2(x + 1)^{1/3}$.

Factor each of the following.

62. $2x^{2/3} - 11x^{1/3} + 12$

63. $9x^{2/3} + 12x^{1/3} + 4$

Getting Ready for the Next Section

Simplify.

64. $(x + 3)^2 - 2(x + 3) - 8$

65. $(x - 2)^2 - 3(x - 2) - 10$

66. $(2a - 3)^2 - 9(2a - 3) + 20$

67. $(3a - 2)^2 + 2(3a - 2) - 3$

68. $2(4a + 2)^2 - 3(4a + 2) - 20$

69. $6(2a + 4)^2 - (2a + 4) - 2$

Solve.

70. $x^2 = \dfrac{1}{4}$

71. $x^2 = -2$

72. $\sqrt{x} = -3$

73. $\sqrt{x} = 2$

74. $x + 3 = 4$

75. $x + 3 = -2$

76. $y^2 - 2y - 8 = 0$

77. $y^2 + y - 6 = 0$

78. $4y^2 + 7y - 2 = 0$

79. $6x^2 - 13x - 5 = 0$

Extending the Concepts

Find all solutions to the following equations. Solve using algebra and by graphing. If rounding is necessary, round to the nearest hundredth. A calculator can be used in these problems.

80. $x^2 = 4x + 5$

81. $4x^2 = 8x + 5$

82. $x^2 - 1 = 2x$

83. $4x^2 - 1 = 4x$

Find all solutions to each equation. If rounding is necessary, round to the nearest hundredth.

84. $2x^3 - x^2 - 2x + 1 = 0$

85. $3x^3 - 2x^2 - 3x + 2 = 0$

86. $2x^3 + 2 = x^2 + 4x$

87. $3x^3 - 9x = 2x^2 - 6$

10.4 Equations Quadratic in Form

OBJECTIVES

A Solve equations that are reducible to a quadratic equation.

B Solve application problems using equations quadratic in form.

We are now in a position to put our knowledge of quadratic equations to work to solve a variety of equations.

EXAMPLE 1   Solve $(x + 3)^2 - 2(x + 3) - 8 = 0$.

▶ **SOLUTION**   We can see that this equation is quadratic in form by replacing $x + 3$ with another variable, say y. Replacing $x + 3$ with y we have

$$y^2 - 2y - 8 = 0$$

We can solve this equation by factoring the left side and then setting each factor equal to 0.

$$y^2 - 2y - 8 = 0$$

$$(y - 4)(y + 2) = 0 \qquad \textbf{Factor}$$

$$y - 4 = 0 \quad \text{or} \quad y + 2 = 0 \qquad \textbf{Set factors to 0}$$

$$y = 4 \quad \text{or} \qquad y = -2$$

Since our original equation was written in terms of the variable x, we would like our solutions in terms of x also. Replacing y with $x + 3$ and then solving for x we have

$$x + 3 = 4 \quad \text{or} \quad x + 3 = -2$$

$$x = 1 \quad \text{or} \qquad x = -5$$

The solutions to our original equation are 1 and -5.

The method we have just shown lends itself well to other types of equations that are quadratic in form, as we will see. In this example, however, there is another method that works just as well. Let's solve our original equation again, but this time, let's begin by expanding $(x + 3)^2$ and $2(x + 3)$.

$$(x + 3)^2 - 2(x + 3) - 8 = 0$$

$$x^2 + 6x + 9 - 2x - 6 - 8 = 0 \qquad \textbf{Multiply}$$

$$x^2 + 4x - 5 = 0 \qquad \textbf{Combine similar terms}$$

$$(x - 1)(x + 5) = 0 \qquad \textbf{Factor}$$

$$x - 1 = 0 \qquad \text{or} \qquad x + 5 = 0 \qquad \textbf{Set factors to 0}$$

$$x = 1 \qquad \text{or} \qquad x = -5$$

As you can see, either method produces the same result.

 EXAMPLE 2   Solve $4x^4 + 7x^2 = 2$.

SOLUTION This equation is quadratic in x^2. We can make it easier to look at by using the substitution $y = x^2$. (The choice of the letter y is arbitrary. We could just as easily use the substitution $m = x^2$.) Making the substitution $y = x^2$ and then solving the resulting equation we have

$$4y^2 + 7y = 2$$

$$4y^2 + 7y - 2 = 0 \qquad \textbf{Standard form}$$

$$(4y - 1)(y + 2) = 0 \qquad \textbf{Factor}$$

$$4y - 1 = 0 \qquad \text{or} \qquad y + 2 = 0 \qquad \textbf{Set factors to 0}$$

$$y = \frac{1}{4} \qquad \text{or} \qquad y = -2$$

Now we replace y with x^2 to solve for x:

$$x^2 = \frac{1}{4} \qquad \text{or} \qquad x^2 = -2$$

$$x = \pm\sqrt{\frac{1}{4}} \qquad \text{or} \qquad x = \pm\sqrt{-2} \qquad \textbf{Theorem 1}$$

$$x = \pm\frac{1}{2} \qquad \text{or} \qquad x = \pm i\sqrt{2}$$

The solution set is $\{\frac{1}{2}, -\frac{1}{2}, i\sqrt{2}, -i\sqrt{2}\}$.

 EXAMPLE 3   Solve for x: $x + \sqrt{x} - 6 = 0$.

SOLUTION   To see that this equation is quadratic in form, we have to notice that $(\sqrt{x})^2 = x$; that is, the equation can be rewritten as

$$(\sqrt{x})^2 + \sqrt{x} - 6 = 0$$

Replacing $\sqrt{x}$ with y and solving as usual, we have

$$y^2 + y - 6 = 0$$

$$(y + 3)(y - 2) = 0$$

$$y + 3 = 0 \qquad \text{or} \qquad y - 2 = 0$$

$$y = -3 \qquad \text{or} \qquad y = 2$$

Again, to find x, we replace with $\sqrt{x}$ and solve:

$$\sqrt{x} = -3 \qquad \text{or} \qquad \sqrt{x} = 2$$

$$x = 9 \qquad\qquad\qquad x = 4 \qquad \textbf{Square both sides of each equation}$$

Since we squared both sides of each equation, we have the possibility of obtaining extraneous solutions. We have to check both solutions in our original equation.

When $x = 9$ When $x = 4$

the equation $x + \sqrt{x} - 6 = 0$ the equation $x + \sqrt{x} - 6 = 0$

becomes $9 + \sqrt{9} - 6 \overset{?}{=} 0$ becomes $4 + \sqrt{4} - 6 \overset{?}{=} 0$

$9 + 3 - 6 \overset{?}{=} 0$ $4 + 2 - 6 \overset{?}{=} 0$

$6 \neq 0$ $0 = 0$

This means 9 is   This means 4 is

extraneous.   a solution.

The only solution to the equation $x + \sqrt{x} - 6 = 0$ is $x = 4$.

We should note here that the two possible solutions, 9 and 4, to the equation in Example 3 can be obtained by another method. Instead of substituting for $\sqrt{x}$, we can isolate it on one side of the equation and then square both sides to clear the equation of radicals.

$$x + \sqrt{x} - 6 = 0$$

$\sqrt{x} = -x + 6$ **Isolate x**

$x = x^2 - 12x + 36$ **Square both sides**

$0 = x^2 - 13x + 36$ **Add $-x$ to both sides**

$0 = (x - 4)(x - 9)$ **Factor**

$x - 4 = 0$ or $x - 9 = 0$

$x = 4$ $x = 9$

We obtain the same two possible solutions. Since we squared both sides of the equation to find them, we would have to check each one in the original equation. As was the case in Example 3, only $x = 4$ is a solution; $x = 9$ is extraneous.

EXAMPLE 4   If an object is tossed into the air with an upward velocity of 12 feet per second from the top of a building h feet high, the time it takes for the object to hit the ground below is given by the formula

$$16t^2 - 12t - h = 0$$

Solve this formula for t.

SOLUTION The formula is in standard form and is quadratic in t. The coefficients a, b, and c that we need to apply to the quadratic formula are $a = 16$, $b = -12$, and $c = -h$. Substituting these quantities into the quadratic formula, we have

$$t = \frac{12 \pm \sqrt{144 - 4(16)(-h)}}{2(16)}$$

$$= \frac{12 \pm \sqrt{144 + 64h}}{32}$$

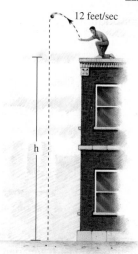

12 feet/sec

h

We can factor the perfect square 16 from the two terms under the radical and simplify our radical somewhat:

$$t = \frac{12 \pm \sqrt{16(9 + 4h)}}{32}$$

$$= \frac{12 \pm 4\sqrt{9 + 4h}}{32}$$

Now we can reduce to lowest terms by factoring a 4 from the numerator and denominator.

$$t = \frac{\cancel{4}(3 \pm \sqrt{9 + 4h})}{\cancel{4} \cdot 8}$$

$$= \frac{3 \pm \sqrt{9 + 4h}}{8}$$

If we were given a value of h, we would find that one of the solutions to this last formula would be a negative number. Since time is always measured in positive units, we wouldn't use that solution.

More About the Golden Ratio

Previously, we derived the golden ratio $\dfrac{1 + \sqrt{5}}{2}$ by finding the ratio of length to width for a golden rectangle. The golden ratio was actually discovered before the golden rectangle by the Greeks who lived before Euclid. The early Greeks found the golden ratio by dividing a line segment into two parts so that the ratio of the shorter part to the longer part was the same as the ratio of the longer part to the whole segment. When they divided a line segment in this manner, they said it was divided in "extreme and mean ratio." Figure 1 illustrates a line segment divided this way.

A   B   C

FIGURE 1

If point B divides segment AC in "extreme and mean ratio," then

$$\frac{\text{Length of shorter segment}}{\text{Length of longer segment}} = \frac{\text{Length of longer segment}}{\text{Length of whole segment}}$$

$$\frac{AB}{BC} = \frac{BC}{AC}$$

EXAMPLE 5   If the length of segment AB in Figure 1 is 1 inch, find the length of BC so that the whole segment AC is divided in "extreme and mean ratio."

SOLUTION   Using Figure 1 as a guide, if we let x = the length of segment BC, then the length of AC is $x + 1$. If B divides AC into "extreme and mean ratio," then the ratio of AB to BC must equal the ratio of BC to AC. Writing this relationship using the variable x, we have

$$\frac{1}{x} = \frac{x}{x+1}$$

If we multiply both sides of this equation by the LCD $x(x+1)$ we have

$$x + 1 = x^2$$

$$0 = x^2 - x - 1 \qquad \textbf{Write equation in standard form}$$

Since this last equation is not factorable, we apply the quadratic formula.

$$x = \frac{1 \pm \sqrt{(-1)^2 - 4(1)(-1)}}{2}$$

$$= \frac{1 \pm \sqrt{5}}{2}$$

Our equation has two solutions, which we approximate using decimals:

$$\frac{1 + \sqrt{5}}{2} \approx 1.618 \qquad \frac{1 - \sqrt{5}}{2} \approx -0.618$$

Since we originally let x equal the length of segment BC, we use only the positive solution to our equation. As you can see, the positive solution is the golden ratio.

USING TECHNOLOGY

Graphing Calculators

More About Example 1

As we mentioned earlier, algebraic expressions entered into a graphing calculator do not have to be simplified to be evaluated. This fact applies to equations as well. We can graph the equation $y = (x + 3)^2 - 2(x + 3) - 8$ to assist us in solving the equation in Example 1. The graph is shown in Figure 2. Using the Zoom and Trace features at the x-intercepts gives us $x = 1$ and $x = -5$ as the solutions to the equation $0 = (x + 3)^2 - 2(x + 3) - 8$.

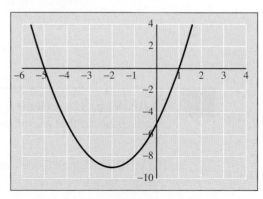

FIGURE 2

More About Example 2

Figure 3 shows the graph of $y = 4x^4 + 7x^2 - 2$. As we expect, the x-intercepts give the real number solutions to the equation $0 = 4x^4 + 7x^2 - 2$. The complex solutions do not appear on the graph.

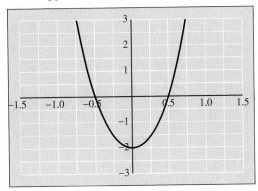

FIGURE 3

More About Example 3

In solving the equation in Example 3, we found that one of the possible solutions was an extraneous solution. If we solve the equation $x + \sqrt{x} - 6 = 0$ by graphing the function $y = x + \sqrt{x} - 6$, we find that the extraneous solution, 9, is not an x-intercept. Figure 4 shows that the only solution to the equation occurs at the x-intercept 4.

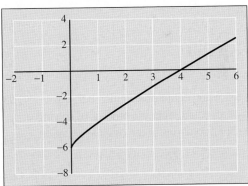

FIGURE 4

GETTING READY FOR CLASS

After reading through the preceding section, respond in your own words and in complete sentences.

1. What does it mean for an equation to be quadratic in form?
2. What are all the circumstances in solving equations (that we have studied) in which it is necessary to check for extraneous solutions?
3. How would you start to solve the equation $x + \sqrt{x} - 6 = 0$?
4. What does it mean for a line segment to be divided in "extreme and mean ratio"?

LINKING OBJECTIVES AND EXAMPLES

Next to each **objective** we have listed the examples that are best described by that objective.

A   1–3

B   4, 5

Solve each equation.

1. $(x - 3)^2 + 3(x - 3) + 2 = 0$

2. $(x + 4)^2 - (x + 4) - 6 = 0$

3. $2(x + 4)^2 + 5(x + 4) - 12 = 0$

4. $3(x - 5)^2 + 14(x - 5) - 5 = 0$

▶ **5.** $x^4 - 10x^2 + 9 = 0$

▶ **6.** $x^4 - 29x^2 + 100 = 0$

▶ **7.** $x^4 - 7x^2 + 12 = 0$

▶ **8.** $x^4 - 14x^2 + 45 = 0$

9. $x^4 - 6x^2 - 27 = 0$

10. $x^4 + 2x^2 - 8 = 0$

11. $x^4 + 9x^2 = -20$

12. $x^4 - 11x^2 = -30$

13. $(2a - 3)^2 - 9(2a - 3) = -20$

14. $(3a - 2)^2 + 2(3a - 2) = 3$

15. $2(4a + 2)^2 = 3(4a + 2) + 20$

16. $6(2a + 4)^2 = (2a + 4) + 2$

17. $6t^4 = -t^2 + 5$

18. $3t^4 = -2t^2 + 8$

19. $9x^4 - 49 = 0$

20. $25x^4 - 9 = 0$

Solve each of the following equations. Remember, if you square both sides of an equation in the process of solving it, you have to check all solutions in the original equation.

21. $x - 7\sqrt{x} + 10 = 0$ **22.** $x - 6\sqrt{x} + 8 = 0$

23. $t - 2\sqrt{t} - 15 = 0$

24. $t - 3\sqrt{t} - 10 = 0$

25. $6x + 11\sqrt{x} = 35$

26. $2x + \sqrt{x} = 15$

27. $x - 2\sqrt{x} - 8 = 0$

28. $x + 2\sqrt{x} - 3 = 0$

29. $x + 3\sqrt{x} - 18 = 0$

30. $x - 5\sqrt{x} - 14 = 0$

31. $2x + 9\sqrt{x} - 5 = 0$

32. $2x - \sqrt{x} - 6 = 0$

33. $(a - 2) - 11\sqrt{a - 2} + 30 = 0$

34. $(a - 3) - 9\sqrt{a - 3} + 20 = 0$

35. $(2x + 1) - 8\sqrt{2x + 1} + 15 = 0$

36. $(2x - 3) - 7\sqrt{2x - 3} + 12 = 0$

▶ **37.** $(x^2 + 1)^2 - 2(x^2 + 1) - 15 = 0$

▶ **38.** $(x^2 - 3)^2 - 4(x^2 - 3) - 12 = 0$

▶ **39.** $(x^2 + 5)^2 - 6(x^2 + 5) - 27 = 0$

▶ **40.** $(x^2 - 2)^2 - 6(x^2 - 2) - 7 = 0$

▶ **41.** $x^{-2} - 3x^{-1} + 2 = 0$

▶ **42.** $x^{-2} + x^{-1} - 2 = 0$

▶ **43.** $y^{-4} - 5y^{-2} = 0$

▶ **44.** $2y^{-4} + 10y^{-2} = 0$

▶ **45.** $x^{2/3} + 4x^{1/3} - 12 = 0$

▶ **46.** $x^{2/3} - 2x^{1/3} - 8 = 0$

▶ **47.** $x^{4/3} + 6x^{2/3} + 9 = 0$

▶ **48.** $x^{4/3} - 10x^{2/3} + 25 = 0$

49. Solve the formula $16t^2 - vt - h = 0$ for t.

50. Solve the formula $16t^2 + vt + h = 0$ for t.

51. Solve the formula $kx^2 + 8x + 4 = 0$ for x.

52. Solve the formula $k^2x^2 + kx + 4 = 0$ for x.

53. Solve $x^2 + 2xy + y^2 = 0$ for x by using the quadratic formula with $a = 1$, $b = 2y$, and $c = y^2$.

54. Solve $x^2 - 2xy + y^2 = 0$ for x by using the quadratic formula, with $a = 1$, $b = -2y$, and $c = y^2$.

Applying the Concepts

For Problems 55–56, t is in seconds.

55. Falling Object An object is tossed into the air with an upward velocity of 8 feet per second from the top of a building h feet high. The time it takes for the object to hit the ground below is given by the formula $16t^2 - 8t - h = 0$. Solve this formula for t.

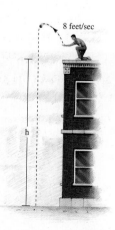

8 feet/sec

h

= Videos available by instructor request

▶ = Online student support materials available at academic.cengage.com/login

56. Falling Object An object is tossed into the air with an upward velocity of 6 feet per second from the top of a building h feet high. The time it takes for the object to hit the ground below is given by the formula $16t^2 - 6t - h = 0$. Solve this formula for t.

57. Saint Louis Arch The shape of the famous "Gateway to the West" arch in Saint Louis can be modeled by a parabola. The equation for one such parabola is:

$$y = -\frac{1}{150}x^2 + \frac{21}{5}x$$

where x and y are in feet.

a. Sketch the graph of the arch's equation on a coordinate axis.

b. Approximately how far do you have to walk to get from one side of the arch to the other?

58. Area and Perimeter A total of 160 yards of fencing is to be used to enclose part of a lot that borders on a river. This situation is shown in the following diagram.

a. Write an equation that gives the relationship between the length and width and the 160 yards of fencing.

b. The formula for the area that is enclosed by the fencing and the river is $A = lw$. Solve the equation in part (a) for l, and then use the result to write the area in terms of w only.

c. Make a table that gives at least five possible values of w and associated area A.

d. From the pattern in your table shown in part (c), what is the largest area that can be enclosed by the 160 yards of fencing? (Try some other table values if necessary.)

Golden Ratio Use Figure 1 from this section as a guide to working Problems 59–60.

59. If AB in Figure 1 is 4 inches, and B divides AC in "extreme and mean ratio," find BC, and then show that BC is 4 times the golden ratio.

60. If AB in Figure 1 is $\frac{1}{2}$ inch, and B divides AC in "extreme and mean ratio," find BC, and then show that the ratio of BC to AB is the golden ratio.

Maintaining Your Skills

Combine, if possible.

61. $5\sqrt{7} - 2\sqrt{7}$

62. $6\sqrt{2} - 9\sqrt{2}$

63. $\sqrt{18} - \sqrt{8} + \sqrt{32}$

64. $\sqrt{50} + \sqrt{72} - \sqrt{8}$

65. $9x\sqrt{20x^3y^2} + 7y\sqrt{45x^5}$

66. $5x^2\sqrt{27xy^3} - 6y\sqrt{12x^5y}$

Multiply.

67. $(\sqrt{5} - 2)(\sqrt{5} + 8)$

68. $(2\sqrt{3} - 7)(2\sqrt{3} + 7)$

69. $(\sqrt{x} + 2)^2$

70. $(3 - \sqrt{x})(3 + \sqrt{x})$

Rationalize the denominator.

71. $\dfrac{\sqrt{7}}{\sqrt{7} - 2}$

72. $\dfrac{\sqrt{5} - \sqrt{2}}{\sqrt{5} + \sqrt{2}}$

Getting Ready for the Next Section

73. Evaluate $y = 3x^2 - 6x + 1$ for $x = 1$.

74. Evaluate $y = -2x^2 + 6x - 5$ for $x = \dfrac{3}{2}$.

75. Let $P(x) = -0.1x^2 + 27x - 500$ and find $P(135)$.

76. Let $P(x) = -0.1x^2 + 12x - 400$ and find $P(600)$.

Solve.

77. $0 = a(80)^2 + 70$

78. $0 = a(80)^2 + 90$

79. $x^2 - 6x + 5 = 0$

80. $x^2 - 3x - 4 = 0$

81. $-x^2 - 2x + 3 = 0$

82. $-x^2 + 4x + 12 = 0$

83. $2x^2 - 6x + 5 = 0$

84. $x^2 - 4x + 5 = 0$

Fill in the blanks to complete the square.

85. $x^2 - 6x + \boxed{} = (x - \boxed{})^2$

86. $x^2 + 10x + \boxed{} = (x + \boxed{})^2$

87. $y^2 + 2y + \boxed{} = (y + \boxed{})^2$

88. $y^2 - 12y + \boxed{} = (y - \boxed{})^2$

Extending the Concepts

Find the x- and y-intercepts.

89. $y = x^3 - 4x$

90. $y = x^4 - 10x^2 + 9$

91. $y = 3x^3 + x^2 - 27x - 9$

92. $y = 2x^3 + x^2 - 8x - 4$

93. The graph of $y = 2x^3 - 7x^2 - 5x + 4$ crosses the x-axis at $x = 4$. Where else does it cross the x-axis?

94. The graph of $y = 6x^3 + x^2 - 12x + 5$ crosses the x-axis at $x = 1$. Where else does it cross the x-axis?

10.5   Graphing Parabolas

OBJECTIVES

A   Graph a parabola.

B   Solve application problems using information from a graph.

C   Find an equation from its graph.

The solution set to the equation

$$y = x^2 - 3$$

consists of ordered pairs. One method of graphing the solution set is to find a number of ordered pairs that satisfy the equation and to graph them. We can obtain some ordered pairs that are solutions to $y = x^2 - 3$ by use of a table as follows:

x	$y = x^2 - 3$	y	Solutions
−3	$y = (-3)^2 - 3 = 9 - 3 = 6$	6	(−3, 6)
−2	$y = (-2)^2 - 3 = 4 - 3 = 1$	1	(−2, 1)
−1	$y = (-1)^2 - 3 = 1 - 3 = -2$	−2	(−1, −2)
0	$y = 0^2 \quad - 3 = 0 - 3 = -3$	−3	(0, −3)
1	$y = 1^2 \quad - 3 = 1 - 3 = -2$	−2	(1, −2)
2	$y = 2^2 \quad - 3 = 4 - 3 = 1$	1	(2, 1)
3	$y = 3^2 \quad - 3 = 9 - 3 = 6$	6	(3, 6)

Graphing these solutions and then connecting them with a smooth curve, we have the graph of $y = x^2 - 3$ (Figure 1).

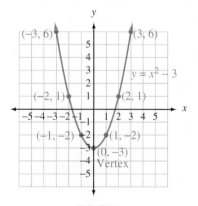

FIGURE 1

This graph is an example of a *parabola*. All equations of the form $y = ax^2 + bx + c$, $a \neq 0$ have parabolas for graphs.

the vertex is at $(3, -4)$. As a matter of fact, this is the same kind of reasoning we used when we derived the formula $x = \dfrac{-b}{2a}$ for the x-coordinate of the vertex.

 EXAMPLE 2   Graph $y = -x^2 - 2x + 3$.

SOLUTION   To find the x-intercepts, we let $y = 0$:

$$0 = -x^2 - 2x + 3$$

$$0 = x^2 + 2x - 3 \qquad \textbf{Multiply each side by } -1$$

$$0 = (x + 3)(x - 1)$$

$$x = -3 \quad \text{or} \quad x = 1$$

The x-coordinate of the vertex is given by

$$x = \frac{-b}{2a} = \frac{-(-2)}{2(-1)} = \frac{2}{-2} = -1$$

To find the y-coordinate of the vertex, we substitute -1 for x in our original equation to get

$$y = -(-1)^2 - 2(-1) + 3 = -1 + 2 + 3 = 4$$

Our parabola has x-intercepts at -3 and 1 and a vertex at $(-1, 4)$. Figure 3 shows the graph. We say the graph is *concave down* since it opens downward.

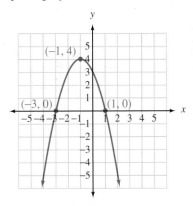

FIGURE 3

Again, we could have obtained the coordinates of the vertex by completing the square on the first two terms on the right side of our equation. To do so, we must first factor -1 from the first two terms. (Remember, the leading coefficients must be 1 to complete the square.) When we complete the square, we add 1 inside the parentheses, which actually decreases the right side of the equation by -1 since everything in the parentheses is multiplied by -1. To make up for it, we add 1 outside the parentheses.

$$y = -1(x^2 + 2x) + 3$$

$$y = -1(x^2 + 2x + \textbf{1}) + 3 + \textbf{1}$$

$$y = -1(x + 1)^2 + 4$$

The last line tells us that the *largest* value of y will be 4, and that will occur when $x = -1$.

EXAMPLE 3   Graph $y = 3x^2 - 6x + 1$.

SOLUTION   To find the x-intercepts, we let $y = 0$ and solve for x:

$$0 = 3x^2 - 6x + 1$$

Since the right side of this equation does not factor, we can look at the discriminant to see what kind of solutions are possible. The discriminant for this equation is

$$b^2 - 4ac = 36 - 4(3)(1) = 24$$

Since the discriminant is a positive number but not a perfect square, the equation will have irrational solutions. This means that the x-intercepts are irrational numbers and will have to be approximated with decimals using the quadratic formula. Rather than use the quadratic formula, we will find some other points on the graph, but first let's find the vertex.

Here are both methods of finding the vertex:

Using the formula that gives us the x-coordinate of the vertex, we have:	To complete the square on the right side of the equation, we factor 3 from the first two terms, add 1 inside the parentheses, and add -3 outside the parentheses (this amounts to adding 0 to the right side):

$$x = \frac{-b}{2a} = \frac{-(-6)}{2(3)} = 1$$

Substituting 1 for x in the equation gives us the y-coordinate of the vertex:

$$y = 3 \cdot 1^2 - 6 \cdot 1 + 1 = -2$$

$$y = 3(x^2 - 2x \qquad) + 1$$

$$y = 3(x^2 - 2x + \mathbf{1}) + 1 - \mathbf{3}$$

$$y = 3(x - 1)^2 - 2$$

In either case, the vertex is $(1, -2)$.

If we can find two points, one on each side of the vertex, we can sketch the graph. Let's let $x = 0$ and $x = 2$ since each of these numbers is the same distance from $x = 1$ and $x = 0$ will give us the y-intercept.

When $x = 0$	When $x = 2$
$y = 3(0)^2 - 6(0) + 1$	$y = 3(2)^2 - 6(2) + 1$
$= 0 - 0 + 1$	$= 12 - 12 + 1$
$= 1$	$= 1$

The two points just found are $(0, 1)$ and $(2, 1)$. Plotting these two points along with the vertex $(1, -2)$, we have the graph shown in Figure 4.

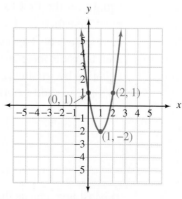

FIGURE 4

Looking back at the diagram, we see the regions that satisfy these conditions are between 2 and 3 or above 5. Here is our solution set:

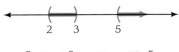

$$2 < x < 3 \quad \text{or} \quad x > 5$$

GETTING READY FOR CLASS

After reading through the preceding section, respond in your own words and in complete sentences.

1. What is the first step in solving a quadratic inequality?

2. How do you show that the endpoint of a line segment is not part of the graph of a quadratic inequality?

3. How would you use the graph of $y = ax^2 + bx + c$ to help you find the graph of $ax^2 + bx + c < 0$?

4. Can a quadratic inequality have exactly one solution? Give an example.

LINKING OBJECTIVES AND EXAMPLES

Next to each **objective** we have listed the examples that are best described by that objective.

A   1–3

Problem Set 10.6

Online support materials can be found at academic.cengage.com/login

Solve each of the following inequalities and graph the solution set.

▶ **1.** $x^2 + x - 6 > 0$

2. $x^2 + x - 6 < 0$

3. $x^2 - x - 12 \leq 0$

4. $x^2 - x - 12 \geq 0$

5. $x^2 + 5x \geq -6$

6. $x^2 - 5x > 6$

▶ **7.** $6x^2 < 5x - 1$

8. $4x^2 \geq -5x + 6$

▶ **9.** $x^2 - 9 < 0$

10. $x^2 - 16 \geq 0$

11. $4x^2 - 9 \geq 0$

12. $9x^2 - 4 < 0$

13. $2x^2 - x - 3 < 0$

14. $3x^2 + x - 10 \geq 0$

15. $x^2 - 4x + 4 \geq 0$

16. $x^2 - 4x + 4 < 0$

17. $x^2 - 10x + 25 < 0$

18. $x^2 - 10x + 25 > 0$

▶ **19.** $(x - 2)(x - 3)(x - 4) > 0$

20. $(x - 2)(x - 3)(x - 4) < 0$

21. $(x + 1)(x + 2)(x + 3) \leq 0$

22. $(x + 1)(x + 2)(x + 3) \geq 0$

▶ **23.** $\dfrac{x - 1}{x + 4} \leq 0$

▶ **24.** $\dfrac{x + 4}{x - 1} \leq 0$

▶ **25.** $\dfrac{3x - 8}{x + 6} < 0$

▶ **26.** $\dfrac{5x - 3}{x + 1} < 0$

▶ **27.** $\dfrac{x - 2}{x - 6} > 0$

▶ **28.** $\dfrac{x - 1}{x - 3} \geq 0$

▶ **29.** $\dfrac{x - 2}{(x + 3)(x - 4)} < 0$

▶ **30.** $\dfrac{x - 1}{(x + 2)(x - 5)} < 0$

▶ **31.** $\dfrac{2}{x - 4} - \dfrac{1}{x - 3} > 0$

32. $\dfrac{4}{x + 3} - \dfrac{3}{x + 2} > 0$

33. Write each statement using inequality notation.

    a. $x - 1$ is always positive.

    b. $x - 1$ is never negative.

    c. $x - 1$ is greater than or equal to 0.

34. Match each expression on the left with a phrase on the right.

    a. $(x - 1)^2 \geq 0$ **i.** Never true

    b. $(x - 1)^2 < 0$ **ii.** Sometimes true

    c. $(x - 1)^2 \leq 0$ **iii.** Always true

Solve each inequality by inspection without showing any work.

▶ **35.** $(x - 1)^2 < 0$

▶ **36.** $(x + 2)^2 < 0$

▶ **37.** $(x - 1)^2 \leq 0$

▶ **38.** $(x + 2)^2 \leq 0$

39. $(x - 1)^2 \geq 0$

40. $(x + 2)^2 \geq 0$

41. $\dfrac{1}{(x - 1)^2} \geq 0$

42. $\dfrac{1}{(x + 2)^2} > 0$

▢ = Videos available by instructor request

▶ = Online student support materials available at academic.cengage.com/login

43. $x^2 - 6x + 9 < 0$ **44.** $x^2 - 6x + 9 \leq 0$

45. $x^2 - 6x + 9 > 0$

46. $\dfrac{1}{x^2 - 6x + 9} > 0$

47. The graph of $y = x^2 - 4$ is shown in Figure 6. Use the graph to write the solution set for each of the following:

   a. $x^2 - 4 < 0$
   b. $x^2 - 4 > 0$
   c. $x^2 - 4 = 0$

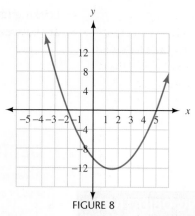

FIGURE 8

50. The graph of $y = x^2 + x - 12$ is shown in Figure 9. Use the graph to write the solution set for each of the following:

   a. $x^2 + x - 12 < 0$
   b. $x^2 - x - 12 > 0$
   c. $x^2 + x - 12 = 0$

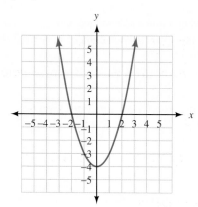

FIGURE 6

48. The graph of $y = 4 - x^2$ is shown in Figure 7. Use the graph to write the solution set for each of the following:

   a. $4 - x^2 < 0$
   b. $4 - x^2 > 0$
   c. $4 - x^2 = 0$

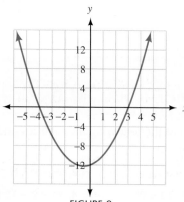

FIGURE 9

51. The graph of $y = x^3 - 3x^2 - x + 3$ is shown in Figure 10. Use the graph to write the solution set for each of the following:

   a. $x^3 - 3x^2 - x + 3 < 0$
   b. $x^3 - 3x^2 - x + 3 > 0$
   c. $x^3 - 3x^2 - x + 3 = 0$

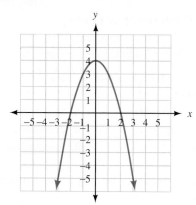

FIGURE 7

49. The graph of $y = x^2 - 3x - 10$ is shown in Figure 8. Use the graph to write the solution set for each of the following:

   a. $x^2 - 3x - 10 < 0$
   b. $x^2 - 3x - 10 > 0$
   c. $x^2 - 3x - 10 = 0$

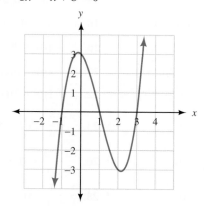

FIGURE 10

52. The graph of $y = x^3 + 4x^2 - 4x - 16$ is shown in Figure 11. Use the graph to write the solution set for each of the following:

a. $x^3 + 4x^2 - 4x - 16 < 0$
b. $x^3 + 4x^2 - 4x - 16 > 0$
c. $x^3 + 4x^2 - 4x - 16 = 0$

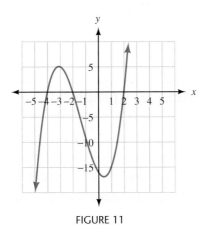

FIGURE 11

Applying the Concepts

53. Dimensions of a Rectangle The length of a rectangle is 3 inches more than twice the width. If the area is to be at least 44 square inches, what are the possibilities for the width?

54. Dimensions of a Rectangle The length of a rectangle is 5 inches less than 3 times the width. If the area is to be less than 12 square inches, what are the possibilities for the width?

55. Revenue A manufacturer of portable radios knows that the weekly revenue produced by selling x radios is given by the equation $R = 1,300p - 100p^2$, where p is the price of each radio. What price should she charge for each radio if she wants her weekly revenue to be at least $4,000?

56. Revenue A manufacturer of small calculators knows that the weekly revenue produced by selling x calculators is given by the equation $R = 1,700p - 100p^2$, where p is the price of each calculator. What price should be charged for each calculator if the revenue is to be at least $7,000 each week?

Maintaining Your Skills

Use a calculator to evaluate. Give answers to 4 decimal places.

57. $\dfrac{50,000}{32,000}$

58. $\dfrac{2.4362}{1.9758} - 1$

59. $\dfrac{1}{2}\left(\dfrac{4.5926}{1.3876} - 2\right)$

60. $1 + \dfrac{0.06}{12}$

Solve each equation.

61. $\sqrt{3t - 1} = 2$

62. $\sqrt{4t + 5} + 7 = 3$

63. $\sqrt{x + 3} = x - 3$

64. $\sqrt{x + 3} = \sqrt{x} - 3$

Graph each equation.

65. $y = \sqrt[3]{x - 1}$

66. $y = \sqrt[3]{x} - 1$

Extending the Concepts

Graph the solution set for each inequality.

67. $x^2 - 2x - 1 < 0$

68. $x^2 - 6x + 7 < 0$

69. $x^2 - 8x + 13 > 0$

70. $x^2 - 10x + 18 > 0$

Chapter 10 SUMMARY

Theorem 1 [10.1]

EXAMPLES

1. If $(x - 3)^2 = 25$
then $x - 3 = \pm 5$
$x = 3 \pm 5$
$x = 8$ or $x = -2$

If $a^2 = b$, where b is a real number, then

$$a = \sqrt{b} \quad \text{or} \quad a = -\sqrt{b}$$

which can be written as $a = \pm\sqrt{b}$.

To Solve a Quadratic Equation by Completing the Square [10.1]

2. Solve $x^2 - 6x - 6 = 0$
$x^2 - 6x = 6$
$x^2 - 6x + \mathbf{9} = 6 + \mathbf{9}$
$(x - 3)^2 = 15$
$x - 3 = \pm\sqrt{15}$
$x = 3 \pm \sqrt{15}$

Step 1: Write the equation in the form $ax^2 + bx = c$.

Step 2: If $a \neq 1$, divide through by the constant a so the coefficient of x^2 is 1.

Step 3: Complete the square on the left side by adding the square of $\frac{1}{2}$ the coefficient of x to both sides.

Step 4: Write the left side of the equation as the square of a binomial. Simplify the right side if possible.

Step 5: Apply Theorem 1, and solve as usual.

The Quadratic Theorem [10.2]

3. If $2x^2 + 3x - 4 = 0$, then
$x = \dfrac{-3 \pm \sqrt{9 - 4\,(2)(-4)}}{2(2)}$

$= \dfrac{-3 \pm \sqrt{41}}{4}$

For any quadratic equation in the form $ax^2 + bx + c = 0$, $a \neq 0$, the two solutions are

$$x = \frac{-b \pm \sqrt{b^2 - 4ac}}{2a}$$

This last expression is known as the *quadratic formula*.

The Discriminant [10.3]

4. The discriminant for
$x^2 + 6x + 9 = 0$
is $D = 36 - 4(1)(9) = 0$,
which means the equation
has one rational solution.

The expression $b^2 - 4ac$ that appears under the radical sign in the quadratic formula is known as the *discriminant*.

We can classify the solutions to $ax^2 + bx + c = 0$:

The Solutions Are	When the Discriminant Is
Two complex numbers containing i	Negative
One rational number	Zero
Two rational numbers	A positive perfect square
Two irrational numbers	A positive number, but not a perfect square

Exponential and Logarithmic Functions

11

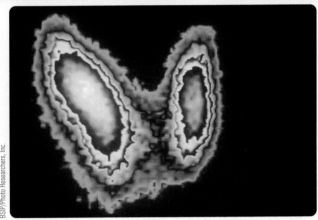

BSIP/Photo Researchers, Inc.

I f you have had any problems with or had testing done on your thyroid gland, then you may have come in contact with radioactive iodine-131. Like all radioactive elements, iodine-131 decays naturally. The half-life of iodine-131 is 8 days, which means that every 8 days a sample of iodine-131 will decrease to half of its original amount. The following table and graph show what happens to a 1,600-microgram sample of iodine-131 over time.

Iodine-131 as a Function of Time	
t (days)	A (micrograms)
0	1,600
8	800
16	400
24	200
32	100

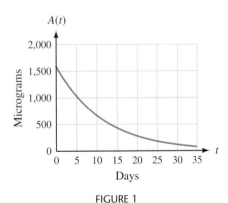

FIGURE 1

The function represented by the information in the table and Figure 1 is

$$A(t) = 1,600 \cdot 2^{-t/8}$$

It is one of the types of functions we will study in this chapter.

▶ Improve your grade and save time!
Go online to **academic.cengage.com/login** where you can
- Watch videos of instructors working through the in-text examples
- Follow step-by-step online tutorials of in-text examples and review questions
- Work practice problems
- Check your readiness for an exam by taking a pre-test and exploring the modules recommended in your Personalized Study plan
- Receive help from a live tutor online through vMentor™

Try it out! Log in with an access code or purchase access at **www.ichapters.com**.

11.1 Exponential Functions

OBJECTIVES

A Find function values for exponential functions.

B Graph exponential functions.

C Work problems involving exponential growth and decay.

To obtain an intuitive idea of how exponential functions behave, we can consider the heights attained by a bouncing ball. When a ball used in the game of racquetball is dropped from any height, the first bounce will reach a height that is $\frac{2}{3}$ of the original height. The second bounce will reach $\frac{2}{3}$ of the height of the first bounce, and so on, as shown in Figure 1.

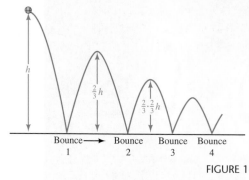

FIGURE 1

If the ball is dropped initially from a height of 1 meter, then during the first bounce it will reach a height of $\frac{2}{3}$ meter. The height of the second bounce will reach $\frac{2}{3}$ of the height reached on the first bounce. The maximum height of any bounce is $\frac{2}{3}$ of the height of the previous bounce.

$$\text{Initial height:} \quad h = 1$$

$$\text{Bounce 1:} \quad h = \frac{2}{3}(1) = \frac{2}{3}$$

$$\text{Bounce 2:} \quad h = \frac{2}{3}\left(\frac{2}{3}\right) = \left(\frac{2}{3}\right)^2$$

$$\text{Bounce 3:} \quad h = \frac{2}{3}\left(\frac{2}{3}\right)^2 = \left(\frac{2}{3}\right)^3$$

$$\text{Bounce 4:} \quad h = \frac{2}{3}\left(\frac{2}{3}\right)^3 = \left(\frac{2}{3}\right)^4$$

$$\vdots$$

$$\text{Bounce } n: \quad h = \frac{2}{3}\left(\frac{2}{3}\right)^{n-1} = \left(\frac{2}{3}\right)^n$$

This last equation is exponential in form. We classify all exponential functions together with the following definition.

DEFINITION An **exponential function** is any function that can be written in the form

$$f(x) = b^x$$

where b is a positive real number other than 1.

Each of the following is an exponential function:

$$f(x) = 2^x \qquad y = 3^x \qquad f(x) = \left(\frac{1}{4}\right)^x$$

LINKING OBJECTIVES AND EXAMPLES

Next to each objective we have listed the examples that are best described by that objective.

A	1–4
B	1–3
C	4
D	5–10

GETTING READY FOR CLASS

After reading through the preceding section, respond in your own words and in complete sentences.

1. What is a logarithm?
2. What is the relationship between $y = 2^x$ and $y = \log_2 x$? How are their graphs related?
3. Will the graph of $y = \log_b x$ ever appear in the second or third quadrants? Explain why or why not.
4. Explain why $\log_2 0 = x$ has no solution for x.

Problem Set 11.3

Online support materials can be found at academic.cengage.com/login

Write each of the following equations in logarithmic form.

1. $2^4 = 16$

2. $3^2 = 9$

3. $125 = 5^3$

4. $16 = 4^2$

5. $0.01 = 10^{-2}$

6. $0.001 = 10^{-3}$

▶ 7. $2^{-5} = \dfrac{1}{32}$

8. $4^{-2} = \dfrac{1}{16}$

9. $\left(\dfrac{1}{2}\right)^{-3} = 8$

10. $\left(\dfrac{1}{3}\right)^{-2} = 9$

11. $27 = 3^3$

12. $81 = 3^4$

Write each of the following equations in exponential form.

13. $\log_{10} 100 = 2$

14. $\log_2 8 = 3$

15. $\log_2 64 = 6$

16. $\log_2 32 = 5$

17. $\log_8 1 = 0$

18. $\log_9 9 = 1$

19. $\log_{10} 0.001 = -3$

20. $\log_{10} 0.0001 = -4$

21. $\log_6 36 = 2$

22. $\log_7 49 = 2$

23. $\log_5 \dfrac{1}{25} = -2$

24. $\log_3 \dfrac{1}{81} = -4$

Solve each of the following equations for x.

▶ 25. $\log_3 x = 2$

26. $\log_4 x = 3$

27. $\log_5 x = -3$

28. $\log_2 x = -4$

29. $\log_2 16 = x$

30. $\log_3 27 = x$

▶ 31. $\log_8 2 = x$

32. $\log_{25} 5 = x$

33. $\log_x 4 = 2$

34. $\log_x 16 = 4$

35. $\log_x 5 = 3$

36. $\log_x 8 = 2$

■ = Videos available by instructor request

▶ = Online student support materials available at academic.cengage.com/login

Sketch the graph of each of the following logarithmic equations.

37. $y = \log_3 x$

38. $y = \log_{1/2} x$

39. $y = \log_{1/3} x$

40. $y = \log_4 x$

41. $y = \log_5 x$

42. $y = \log_{1/5} x$

43. $y = \log_{10} x$

44. $y = \log_{1/4} x$

Simplify each of the following.

▶ **45.** $\log_2 16$

46. $\log_3 9$

47. $\log_{25} 125$

48. $\log_9 27$

49. $\log_{10} 1,000$

50. $\log_{10} 10,000$

▶ **51.** $\log_3 3$

52. $\log_4 4$

▶ **53.** $\log_5 1$

54. $\log_{10} 1$

▶ **55.** $\log_3(\log_6 6)$

56. $\log_5(\log_3 3)$

57. $\log_4[\log_2(\log_2 16)]$

58. $\log_4[\log_3(\log_2 8)]$

Applying the Concepts

$pH = -\log_{10}[H^+]$, where $[H^+]$ is the concentration of the hydrogen ions in solution. An acid solution has a pH below 7, and a basic solution has a pH higher than 7.

59. In distilled water, the concentration of hydrogen ions is $[H^+] = 10^{-7}$. What is the pH?

60. Find the pH of a bottle of vinegar if the concentration of hydrogen ions is $[H^+] = 10^{-3}$.

61. A hair conditioner has a pH of 6. Find the concentration of hydrogen ions, $[H^+]$, in the conditioner.

62. If a glass of orange juice has a pH of 4, what is the concentration of hydrogen ions, $[H^+]$, in the juice?

63. **Magnitude of an Earthquake** Find the magnitude M of an earthquake with a shock wave that measures $T = 100$ on a seismograph.

64. **Magnitude of an Earthquake** Find the magnitude M of an earthquake with a shock wave that measures $T = 100,000$ on a seismograph.

65. **Shock Wave** If an earthquake has a magnitude of 8 on the Richter scale, how many times greater is its shock wave than the smallest shock wave measurable on a seismograph?

66. **Shock Wave** If an earthquake has a magnitude of 6 on the Richter scale, how many times greater is its shock wave than the smallest shock wave measurable on a seismograph?

Maintaining Your Skills

Fill in the blanks to complete the square.

67. $x^2 + 10x + \boxed{} = (x + \boxed{})^2$

68. $x^2 + 4x + \boxed{} = (x + \boxed{})^2$

69. $y^2 - 2y + \boxed{} = (y - \boxed{})^2$

70. $y^2 + 3y + \boxed{} = \left(y + \boxed{} \right)^2$

Solve.

71. $-y^2 = 9$

72. $7 + y^2 = 11$

73. $-x^2 - 8 = -4$

74. $10x^2 = 100$

75. $2x^2 + 4x - 3 = 0$

76. $3x^2 + 4x - 2 = 0$

77. $(2y - 3)(2y - 1) = -4$

78. $(y - 1)(3y - 3) = 10$

79. $t^3 - 125 = 0$

80. $8t^3 + 1 = 0$

81. $4x^5 - 16x^4 = 20x^3$

82. $3x^4 + 6x^2 = 6x^3$

83. $\dfrac{1}{x - 3} + \dfrac{1}{x + 2} = 1$

84. $\dfrac{1}{x + 3} + \dfrac{1}{x - 2} = 1$

Getting Ready for the Next Section

Simplify.

85. $8^{2/3}$

86. $27^{2/3}$

Solve.

87. $(x + 2)(x) = 2^3$

88. $(x + 3)(x) = 2^2$

89. $\dfrac{x - 2}{x + 1} = 9$

90. $\dfrac{x + 1}{x - 4} = 25$

Write in exponential form.

91. $\log_2 [(x + 2)(x)] = 3$

92. $\log_4 [x(x - 6)] = 2$

93. $\log_3\left(\dfrac{x - 2}{x + 1} \right) = 4$

94. $\log_5\left(\dfrac{x - 1}{x - 4} \right) = 2$

Extending the Concepts

95. The graph of the exponential function $y = f(x) = b^x$ is shown here. Use the graph to complete parts a through d.

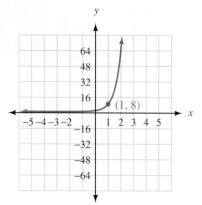

a. Fill in the table.

x	f(x)
−1	
0	
1	
2	

b. Fill in the table.

x	$f^{-1}(x)$
	−1
	0
	1
	2

c. Find the equation for $f(x)$.
d. Find the equation for $f^{-1}(x)$.

96. The graph of the exponential function $y = f(x) = b^x$ is shown here. Use the graph to complete parts a through d.

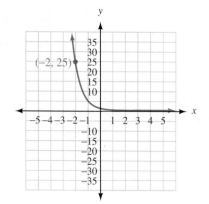

a. Fill in the table.

x	f(x)
−1	
0	
1	
2	

b. Fill in the table.

x	$f^{-1}(x)$
	−1
	0
	1
	2

c. Find the equation for $f(x)$.
d. Find the equation for $f^{-1}(x)$.

11.4 Properties of Logarithms

OBJECTIVES

A Use the properties of logarithms to convert between expanded form and single logarithms.

B Use the properties of logarithms to solve equations that contain logarithms.

If we search for the word *decibel* in *Microsoft Bookshelf,* we find the following definition:

> A unit used to express relative difference in power or intensity, usually between two acoustic or electric signals, equal to ten times the common logarithm of the ratio of the two levels.

Decibels	Comparable to
10	A light whisper
20	Quiet conversation
30	Normal conversation
40	Light traffic
50	Typewriter, loud conversation
60	Noisy office
70	Normal traffic, quiet train
80	Rock music, subway
90	Heavy traffic, thunder
100	Jet plane at takeoff

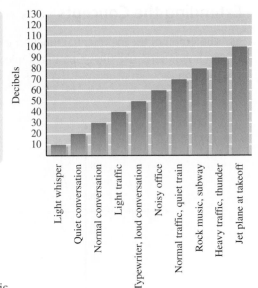

The precise definition for a *decibel* is

$$D = 10 \log_{10}\left(\frac{I}{I_0}\right)$$

where I is the intensity of the sound being measured, and I_0 is the intensity of the least audible sound. (Sound intensity is related to the amplitude of the sound wave that models the sound and is given in units of watts per meter².) In this section, we will see that the preceding formula can also be written as

$$D = 10(\log_{10} I - \log_{10} I_0)$$

The rules we use to rewrite expressions containing logarithms are called the *properties of logarithms*. There are three of them.

For the following three properties, x, y, and b are all positive real numbers, $b \neq 1$, and r is any real number.

Property 1

$$\log_b (xy) = \log_b x + \log_b y$$

In words: The logarithm of a *product* is the *sum* of the logarithms.

Property 2

$$\log_b \left(\frac{x}{y}\right) = \log_b x - \log_b y$$

In words: The logarithm of a *quotient* is the *difference* of the logarithms.

Property 3

$$\log_b x^r = r \log_b x$$

In words: The logarithm of a number raised to a *power* is the *product* of the power and the logarithm of the number.

Proof of Property 1 To prove property 1, we simply apply the first identity for logarithms given in the preceding section:

$$b^{\log_b xy} = xy = (b^{\log_b x})(b^{\log_b y}) = b^{\log_b x + \log_b y}$$

Since the first and last expressions are equal and the bases are the same, the exponents $\log_b xy$ and $\log_b x + \log_b y$ must be equal. Therefore,

$$\log_b xy = \log_b x + \log_b y$$

The proofs of properties 2 and 3 proceed in much the same manner, so we will omit them here. The examples that follow show how the three properties can be used.

 EXAMPLE 1   Expand, using the properties of logarithms: $\log_5 \dfrac{3xy}{z}$

SOLUTION   Applying property 2, we can write the quotient of $3xy$ and z in terms of a difference:

$$\log_5 \frac{3xy}{z} = \log_5 3xy - \log_5 z$$

Applying property 1 to the product $3xy$, we write it in terms of addition:

$$\log_5 \frac{3xy}{z} = \log_5 3 + \log_5 x + \log_5 y - \log_5 z$$

 EXAMPLE 2   Expand, using the properties of logarithms:

$$\log_2 \frac{x^4}{\sqrt{y} \cdot z^3}$$

SOLUTION   We write $\sqrt{y}$ as $y^{1/2}$ and apply the properties:

$$\log_2 \frac{x^4}{\sqrt{y} \cdot z^3} = \log_2 \frac{x^4}{y^{1/2} z^3} \qquad \sqrt{y} = y^{1/2}$$

$$= \log_2 x^4 - \log_2(y^{1/2} \cdot z^3) \qquad \textbf{Property 2}$$

$$= \log_2 x^4 - (\log_2 y^{1/2} + \log_2 z^3) \qquad \textbf{Property 1}$$

$$= \log_2 x^4 - \log_2 y^{1/2} - \log_2 z^3 \qquad \textbf{Remove parentheses}$$

$$= 4 \log_2 x - \frac{1}{2} \log_2 y - 3 \log_2 z \qquad \textbf{Property 3}$$

We can also use the three properties to write an expression in expanded form as just one logarithm.

 EXAMPLE 3   Write as a single logarithm:

$$2 \log_{10} a + 3 \log_{10} b - \frac{1}{3} \log_{10} c$$

SOLUTION   We begin by applying property 3:

$$2 \log_{10} a + 3 \log_{10} b - \frac{1}{3} \log_{10} c$$

$$= \log_{10} a^2 + \log_{10} b^3 - \log_{10} c^{1/3} \qquad \textbf{Property 3}$$

$$= \log_{10} (a^2 \cdot b^3) - \log_{10} c^{1/3} \qquad \textbf{Property 1}$$

$$= \log_{10} \frac{a^2 b^3}{c^{1/3}} \qquad \textbf{Property 2}$$

$$= \log_{10} \frac{a^2 b^3}{\sqrt[3]{c}} \qquad c^{1/3} = \sqrt[3]{c}$$

The properties of logarithms along with the definition of logarithms are useful in solving equations that involve logarithms.

 EXAMPLE 4   Solve for x: $\log_2(x + 2) + \log_2 x = 3$

SOLUTION   Applying property 1 to the left side of the equation allows us to write it as a single logarithm:

$$\log_2(x + 2) + \log_2 x = 3$$

$$\log_2[(x + 2)(x)] = 3$$

The last line can be written in exponential form using the definition of logarithms:

$$(x + 2)(x) = 2^3$$

Solve as usual:

$$x^2 + 2x = 8$$

$$x^2 + 2x - 8 = 0$$

$$(x + 4)(x - 2) = 0$$

$$x + 4 = 0 \quad \text{or} \quad x - 2 = 0$$

$$x = -4 \quad \text{or} \quad x = 2$$

In the previous section we noted the fact that x in the expression $y = \log_b x$ cannot be a negative number. Since substitution of $x = -4$ into the original equation gives

$$\log_2(-2) + \log_2(-4) = 3$$

which contains logarithms of negative numbers, we cannot use -4 as a solution. The solution set is $\{2\}$.

LINKING OBJECTIVES AND EXAMPLES

Next to each **objective** we have listed the examples that are best described by that objective.

A 1–3

B 4

GETTING READY FOR CLASS

After reading through the preceding section, respond in your own words and in complete sentences.

1. Explain why the following statement is false: "The logarithm of a product is the product of the logarithms."
2. Explain why the following statement is false: "The logarithm of a quotient is the quotient of the logarithms."
3. Explain the difference between $\log_b m + \log_b n$ and $\log_b(m + n)$. Are they equivalent?
4. Explain the difference between $\log_b(mn)$ and $(\log_b m)(\log_b n)$. Are they equivalent?

Problem Set 11.4

Online support materials can be found at academic.cengage.com/login

Use the three properties of logarithms given in this section to expand each expression as much as possible.

▶ **1.** $\log_3 4x$

2. $\log_2 5x$

▶ **3.** $\log_6 \dfrac{5}{x}$

4. $\log_3 \dfrac{x}{5}$

5. $\log_2 y^5$

6. $\log_7 y^3$

▶ **7.** $\log_9 \sqrt[3]{z}$

8. $\log_8 \sqrt{z}$

9. $\log_6 x^2 y^4$

10. $\log_{10} x^2 y^4$

11. $\log_5 \sqrt{x} \cdot y^4$

12. $\log_8 \sqrt[3]{xy^6}$

13. $\log_b \dfrac{xy}{z}$

14. $\log_b \dfrac{3x}{y}$

15. $\log_{10} \dfrac{4}{xy}$

16. $\log_{10} \dfrac{5}{4y}$

▶ **17.** $\log_{10} \dfrac{x^2 y}{\sqrt{z}}$

18. $\log_{10} \dfrac{\sqrt{x} \cdot y}{z^3}$

19. $\log_{10} \dfrac{x^3 \sqrt{y}}{z^4}$

20. $\log_{10} \dfrac{x^4 \sqrt[3]{y}}{\sqrt{z}}$

21. $\log_b \sqrt[3]{\dfrac{x^2 y}{z^4}}$

22. $\log_b \sqrt[4]{\dfrac{x^4 y^3}{z^5}}$

Write each expression as a single logarithm.

▶ **23.** $\log_b x + \log_b z$

24. $\log_b x - \log_b z$

25. $2 \log_3 x - 3 \log_3 y$

26. $4 \log_2 x + 5 \log_2 y$

27. $\dfrac{1}{2} \log_{10} x + \dfrac{1}{3} \log_{10} y$

28. $\dfrac{1}{3} \log_{10} x - \dfrac{1}{4} \log_{10} y$

▶ **29.** $3 \log_2 x + \dfrac{1}{2} \log_2 y - \log_2 z$

30. $2 \log_3 x + 3 \log_3 y - \log_3 z$

31. $\dfrac{1}{2} \log_2 x - 3 \log_2 y - 4 \log_2 z$

▨ = Videos available by instructor request

▶ = Online student support materials available at academic.cengage.com/login

32. $3 \log_{10} x - \log_{10} y - \log_{10} z$

33. $\frac{3}{2} \log_{10} x - \frac{3}{4} \log_{10} y - \frac{4}{5} \log_{10} z$

34. $3 \log_{10} x - \frac{4}{3} \log_{10} y - 5 \log_{10} z$

Solve each of the following equations.

35. $\log_2 x + \log_2 3 = 1$

36. $\log_3 x + \log_3 3 = 1$

37. $\log_3 x - \log_3 2 = 2$

38. $\log_3 x + \log_3 2 = 2$

▶ **39.** $\log_3 x + \log_3(x - 2) = 1$

40. $\log_6 x + \log_6(x - 1) = 1$

41. $\log_3(x + 3) - \log_3(x - 1) = 1$

42. $\log_4(x - 2) - \log_4(x + 1) = 1$

43. $\log_2 x + \log_2(x - 2) = 3$

44. $\log_4 x + \log_4(x + 6) = 2$

45. $\log_8 x + \log_8(x - 3) = \frac{2}{3}$

46. $\log_{27} x + \log_{27}(x + 8) = \frac{2}{3}$

47. $\log_5 \sqrt{x} + \log_5 \sqrt{6x + 5} = 1$

48. $\log_2 \sqrt{x} + \log_2 \sqrt{6x + 5} = 1$

Applying the Concepts

49. Decibel Formula Use the properties of logarithms to rewrite the decibel formula $D = 10 \log_{10}(\frac{I}{I_0})$ as

$$D = 10(\log_{10} I - \log_{10} I_0).$$

50. Decibel Formula In the decibel formula $D = 10 \log_{10}(\frac{I}{I_0})$, the threshold of hearing, I_0, is

$$I_0 = 10^{-12} \text{ watts/meter}^2$$

Substitute 10^{-12} for I_0 in the decibel formula, and then show that it simplifies to

$$D = 10(\log_{10} I + 12)$$

51. Acoustic Powers The formula $N = \log_{10} \frac{P_1}{P_2}$ is used in radio electronics to find the ratio of the acoustic powers of two electric circuits in terms of their electric powers. Find N if P_1 is 100 and P_2 is 1. Then use the same two values of P_1 and P_2 to find N in the formula $N = \log_{10} P_1 - \log_{10} P_2$.

52. Henderson–Hasselbalch Formula Doctors use the Henderson–Hasselbalch formula to calculate the pH of a person's blood. pH is a measure of the acidity and/or the alkalinity of a solution. This formula is represented as

$$\text{pH} = 6.1 + \log_{10}\left(\frac{x}{y}\right)$$

where x is the base concentration and y is the acidic concentration. Rewrite the Henderson–Hasselbalch formula so that the logarithm of a quotient is not involved.

Maintaining Your Skills

Divide.

53. $\dfrac{12x^2 + y^2}{36}$

54. $\dfrac{x^2 + 4y^2}{16}$

55. Divide $25x^2 + 4y^2$ by 100

56. Divide $4x^2 + 9y^2$ by 36

Use the discriminant to find the number and kind of solutions to the following equations.

57. $2x^2 - 5x + 4 = 0$

58. $4x^2 - 12x = -9$

For each of the following problems, find an equation with the given solutions.

59. $x = -3, x = 5$

60. $x = 2, x = -2, x = 1$

61. $y = \dfrac{2}{3}, y = 3$

62. $y = -\dfrac{3}{5}, y = 2$

Getting Ready for the Next Section

Simplify.

63. 5^0

64. 4^1

65. $\log_3 3$

66. $\log_5 5$

67. $\log_b b^4$

68. $\log_a a^k$

Use a calculator to find each of the following. Write your answer in scientific notation with the first number in each answer rounded to the nearest tenth.

69. $10^{-5.6}$

70. $10^{-4.1}$

Divide and round to the nearest whole number.

71. $\dfrac{2.00 \times 10^8}{3.96 \times 10^6}$

72. $\dfrac{3.25 \times 10^{12}}{1.72 \times 10^{10}}$

11.5 Common Logarithms and Natural Logarithms

OBJECTIVES

A Use a calculator to find common logarithms.

B Use a calculator to find a number given its common logarithm.

C Simplify expressions containing natural logarithms.

Acid rain was first discovered in the 1960s by Gene Likens and his research team, who studied the damage caused by acid rain to Hubbard Brook in New Hampshire. Acid rain is rain with a pH of 5.6 and below. As you will see as you work your way through this section, pH is defined in terms of common logarithms—one of the topics we present in this section. So, when you are finished with this section, you will have a more detailed knowledge of pH and acid rain.

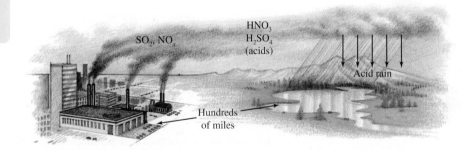

There are two kinds of logarithms that occur more frequently than other logarithms. They are logarithms with base 10 and natural logarithms, or logarithms with base e. Logarithms with a base of 10 are very common because our number system is a base-10 number system. For this reason, we call base-10 logarithms *common logarithms*.

> **DEFINITION** A **common logarithm** is a logarithm with a base of 10. Because common logarithms are used so frequently, it is customary, in order to save time, to omit notating the base; that is,
>
> $$\log_{10} x = \log x$$
>
> When the base is not shown, it is assumed to be 10.

Common Logarithms

Common logarithms of powers of 10 are simple to evaluate. We need only recognize that $\log 10 = \log_{10} 10 = 1$ and apply the third property of logarithms: $\log_b x^r = r \log_b x$.

$$\log 1{,}000 = \log 10^3 \ = 3 \log 10 \ = 3(1) \ = 3$$
$$\log 100 \ = \log 10^2 \ = 2 \log 10 \ = 2(1) \ = 2$$
$$\log 10 \ \ = \log 10^1 \ = 1 \log 10 \ = 1(1) \ = 1$$
$$\log 1 \ \ \ = \log 10^0 \ = 0 \log 10 \ = 0(1) \ = 0$$
$$\log 0.1 \ \ = \log 10^{-1} = -1 \log 10 = -1(1) = -1$$
$$\log 0.01 \ = \log 10^{-2} = -2 \log 10 = -2(1) = -2$$
$$\log 0.001 = \log 10^{-3} = -3 \log 10 = -3(1) = -3$$

Note

Remember, when the base is not written it is assumed to be 10.

To find common logarithms of numbers that are not powers of 10, we use a calculator with a $\boxed{\log}$ key.

Check the following logarithms to be sure you know how to use your calculator. (These answers have been rounded to the nearest ten-thousandth.)

$$\log 7.02 \approx 0.8463$$
$$\log 1.39 \approx 0.1430$$
$$\log 6.00 \approx 0.7782$$
$$\log 9.99 \approx 0.9996$$

 EXAMPLE 1   Use a calculator to find log 2,760.

SOLUTION $\log 2{,}760 \approx 3.4409$

To work this problem on a scientific calculator, we simply enter the number 2,760 and press the key labeled $\boxed{\log}$. On a graphing calculator we press the $\boxed{\log}$ key first, then 2,760.

The 3 in the answer is called the *characteristic,* and the decimal part of the logarithm is called the *mantissa.*

 EXAMPLE 2   Find log 0.0391.

SOLUTION $\log 0.0391 \approx -1.4078$

 EXAMPLE 3   Find log 0.00523.

SOLUTION $\log 0.00523 \approx -2.2815$

EXAMPLE 4   Find *x* if log *x* = 3.8774.

SOLUTION We are looking for the number whose logarithm is 3.8774. On a scientific calculator, we enter 3.8774 and press the key labeled $\boxed{10^x}$. On a graphing calculator we press $\boxed{10^x}$ first, then 3.8774. The result is 7,540 to four significant digits. Here's why:

$$\text{If} \qquad \log x = 3.8774$$

$$\text{then} \qquad x = 10^{3.8774}$$

$$\approx 7,540$$

The number 7,540 is called the *antilogarithm* or just *antilog* of 3.8774, that is, 7,540 is the number whose logarithm is 3.8774.

 EXAMPLE 5   Find x if $\log x = -2.4179$.

SOLUTION   Using the $\boxed{10^x}$ key, the result is 0.00382.

$$\text{If} \qquad \log x = -2.4179$$

$$\text{then} \qquad x = 10^{-2.4179}$$

$$\approx 0.00382$$

The antilog of -2.4179 is 0.00382; that is, the logarithm of 0.00382 is -2.4179.

Applications

Previously, we found that the magnitude M of an earthquake that produces a shock wave T times larger than the smallest shock wave that can be measured on a seismograph is given by the formula

$$M = \log_{10} T$$

We can rewrite this formula using our shorthand notation for common logarithms as

$$M = \log T$$

 EXAMPLE 6   The San Francisco earthquake of 1906 is estimated to have measured 8.3 on the Richter scale. The San Fernando earthquake of 1971 measured 6.6 on the Richter scale. Find T for each earthquake, and then give some indication of how much stronger the 1906 earthquake was than the 1971 earthquake.

SOLUTION   For the 1906 earthquake:

If $\log T = 8.3$, then $T = 2.00 \times 10^8$.

For the 1971 earthquake:

If $\log T = 6.6$, then $T = 3.98 \times 10^6$.

Dividing the two values of T and rounding our answer to the nearest whole number, we have

$$\frac{2.00 \times 10^8}{3.98 \times 10^6} \approx 50$$

The shock wave for the 1906 earthquake was approximately 50 times stronger than the shock wave for the 1971 earthquake.

8.3

In chemistry, the pH of a solution is the measure of the acidity of the solution. The definition for pH involves common logarithms. Here it is:

$$pH = -\log[H^+]$$

where $[H^+]$ is the concentration of the hydrogen ion in moles per liter. The range for pH is from 0 to 14. Pure water, a neutral solution, has a pH of 7. An acidic solution, such as vinegar, will have a pH less than 7, and an alkaline solution, such as ammonia, has a pH above 7.

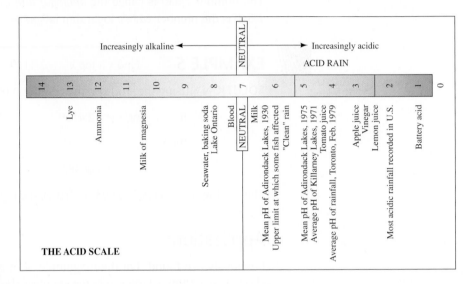

THE ACID SCALE

EXAMPLE 7

Normal rainwater has a pH of 5.6. What is the concentration of the hydrogen ion in normal rainwater?

SOLUTION Substituting 5.6 for pH in the formula $pH = -\log[H^+]$, we have

$5.6 = -\log[H^+]$	**Substitution**
$\log[H^+] = -5.6$	**Isolate the logarithm**
$[H^+] = 10^{-5.6}$	**Write in exponential form**
$\approx 2.5 \times 10^{-6}$ mole per liter	**Answer in scientific notation**

EXAMPLE 8

The concentration of the hydrogen ion in a sample of acid rain known to kill fish is 3.2×10^{-5} mole per liter. Find the pH of this acid rain to the nearest tenth.

SOLUTION Substituting 3.2×10^{-5} for $[H^+]$ in the formula $pH = -\log[H^+]$, we have

$pH = -\log[3.2 \times 10^{-5}]$	**Substitution**
$\approx -(-4.5)$	**Evaluate the logarithm**
$= 4.5$	**Simplify**

Natural Logarithms

> **DEFINITION** A **natural logarithm** is a logarithm with a base of e. The natural logarithm of x is denoted by $\ln x$; that is,
>
> $$\ln x = \log_e x$$

The postage stamp shown here contains one of the two special identities we mentioned previously in this chapter, but stated in terms of natural logarithms.

We can assume that all our properties of exponents and logarithms hold for expressions with a base of e since e is a real number. Here are some examples intended to make you more familiar with the number e and natural logarithms.

 EXAMPLE 9   Simplify each of the following expressions.

   a. $e^0 = 1$

   b. $e^1 = e$

   c. $\ln e = 1$ **In exponential form, $e^1 = e$**

   d. $\ln 1 = 0$ **In exponential form, $e^0 = 1$**

   e. $\ln e^3 = 3$

   f. $\ln e^{-4} = -4$

   g. $\ln e^t = t$

 EXAMPLE 10   Use the properties of logarithms to expand the expression $\ln Ae^{5t}$.

SOLUTION Since the properties of logarithms hold for natural logarithms, we have

$$\ln Ae^{5t} = \ln A + \ln e^{5t}$$

$$= \ln A + 5t \ln e$$

$$= \ln A + 5t \qquad \text{**Because $\ln e = 1$**}$$

EXAMPLE 11   If ln 2 = 0.6931 and ln 3 = 1.0986, find

a. ln 6   **b.** ln 0.5   **c.** ln 8

SOLUTION

a. Since $6 = 2 \cdot 3$, we have

$$\ln 6 = \ln (2 \cdot 3)$$

$$= \ln 2 + \ln 3$$

$$= 0.6931 + 1.0986$$

$$= 1.7917$$

b. Writing 0.5 as $\frac{1}{2}$ and applying property 2 for logarithms gives us

$$\ln 0.5 = \ln \frac{1}{2}$$

$$= \ln 1 - \ln 2$$

$$= 0 - 0.6931$$

$$= -0.6931$$

c. Writing 8 as 2^3 and applying property 3 for logarithms, we have

$$\ln 8 = \ln 2^3$$

$$= 3 \ln 2$$

$$= 3(0.6931)$$

$$= 2.0793$$

LINKING OBJECTIVES AND EXAMPLES

Next to each objective we have listed the examples that are best described by that objective.

A	1–3
B	4–8
C	9–11

GETTING READY FOR CLASS

After reading through the preceding section, respond in your own words and in complete sentences.

1. What is a common logarithm?
2. What is a natural logarithm?
3. Is *e* a rational number? Explain.
4. Find ln *e*, and explain how you arrived at your answer.

We can see from Example 4 that the equation of any circle with its center at the origin and radius r will be

$$x^2 + y^2 = r^2$$

EXAMPLE 5   Find the center and radius, and sketch the graph, of the circle whose equation is

$$(x - 1)^2 + (y + 3)^2 = 4$$

SOLUTION   Writing the equation in the form

$$(x - a)^2 + (y - b)^2 = r^2$$

we have

$$(x - 1)^2 + [y - (-3)]^2 = 2^2$$

The center is at $(1, -3)$ and the radius is 2. The graph is shown in Figure 5.

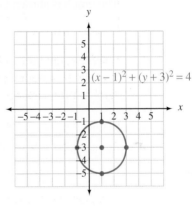

FIGURE 5

EXAMPLE 6   Sketch the graph of $x^2 + y^2 = 9$.

SOLUTION   Since the equation can be written in the form

$$(x - 0)^2 + (y - 0)^2 = 3^2$$

it must have its center at $(0, 0)$ and a radius of 3. The graph is shown in Figure 6.

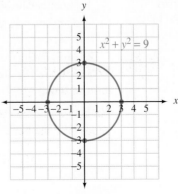

FIGURE 6

EXAMPLE 7 Sketch the graph of $x^2 + y^2 + 6x - 4y - 12 = 0$.

SOLUTION To sketch the graph, we must find the center and radius. The center and radius can be identified if the equation has the form

$$(x - a)^2 + (y - b)^2 = r^2$$

The original equation can be written in this form by completing the squares on x and y:

$$x^2 + y^2 + 6x - 4y - 12 = 0$$

$$x^2 + 6x + y^2 - 4y = 12$$

$$x^2 + 6x + \mathbf{9} + y^2 - 4y + \mathbf{4} = 12 + \mathbf{9} + \mathbf{4}$$

$$(x + 3)^2 + (y - 2)^2 = 25$$

$$(x + 3)^2 + (y - 2)^2 = 5^2$$

From the last line it is apparent that the center is at $(-3, 2)$ and the radius is 5. The graph is shown in Figure 7.

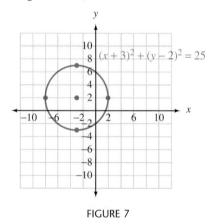

FIGURE 7

LINKING OBJECTIVES AND EXAMPLES

Next to each **objective** we have listed the examples that are best described by that objective.

A	1, 2
B	3, 4
C	5–7

GETTING READY FOR CLASS

After reading through the preceding section, respond in your own words and in complete sentences.

1. Describe the distance formula in words, as if you were explaining to someone how they should go about finding the distance between two points.
2. What is the mathematical definition of a circle?
3. How are the distance formula and the equation of a circle related?
4. When graphing a circle from its equation, why is completing the square sometimes useful?

Find the distance between the following points.

1. (3, 7) and (6, 3)

2. (4, 7) and (8, 1)

3. (0, 9) and (5, 0)

4. (−3, 0) and (0, 4)

▶ **5.** (3, −5) and (−2, 1)

6. (−8, 9) and (−3, −2)

7. (−1, −2) and (−10, 5)

8. (−3, −8) and (−1, 6)

▶ **9.** Find x so the distance between $(x, 2)$ and $(1, 5)$ is $\sqrt{13}$.

10. Find x so the distance between $(−2, 3)$ and $(x, 1)$ is 3.

11. Find y so the distance between $(7, y)$ and $(8, 3)$ is 1.

12. Find y so the distance between $(3, −5)$ and $(3, y)$ is 9.

Write the equation of the circle with the given center and radius.

13. Center (2, 3); $r = 4$

14. Center (3, −1); $r = 5$

▶ **15.** Center (3, −2); $r = 3$

16. Center (−2, 4); $r = 1$

17. Center (−5, −1); $r = \sqrt{5}$

18. Center (−7, −6); $r = \sqrt{3}$

19. Center (0, −5); $r = 1$

20. Center (0, −1); $r = 7$

21. Center (0, 0); $r = 2$

22. Center (0, 0); $r = 5$

Give the center and radius, and sketch the graph of each of the following circles.

▶ **23.** $x^2 + y^2 = 4$

24. $x^2 + y^2 = 16$

25. $(x − 1)^2 + (y − 3)^2 = 25$

26. $(x − 4)^2 + (y − 1)^2 = 36$

27. $(x + 2)^2 + (y − 4)^2 = 8$

28. $(x − 3)^2 + (y + 1)^2 = 12$

29. $(x + 1)^2 + (y + 1)^2 = 1$

30. $(x + 3)^2 + (y + 2)^2 = 9$

▶ **31.** $x^2 + y^2 − 6y = 7$

32. $x^2 + y^2 + 10x = 0$

33. $x^2 + y^2 − 4x − 6y = −4$

34. $x^2 + y^2 − 4x + 2y = 4$

▶ **35.** $x^2 + y^2 + 2x + y = \dfrac{11}{4}$

36. $x^2 + y^2 − 6x − y = −\dfrac{1}{4}$

Both of the following circles pass through the origin. In each case, find the equation.

37.

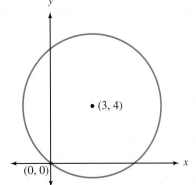

38.

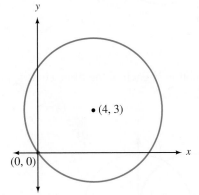

39. Find the equations of circles *A*, *B*, and *C* in the following diagram. The three points are the centers of the three circles.

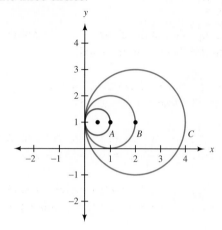

40. Each of the following circles passes through the origin. The centers are as shown. Find the equation of each circle.

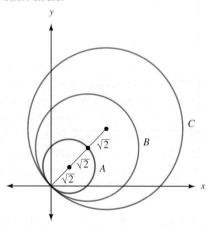

41. Find the equation of each of the three circles shown here.

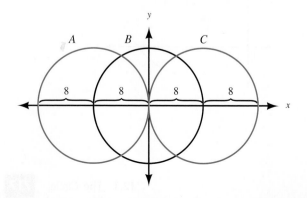

42. A parabola and a circle each contain the points (−4, 0), (0, 4), and (4, 0). Sketch the graph of each curve on the same coordinate system, then write an equation for each curve.

43. Ferris Wheel A giant Ferris wheel has a diameter of 240 feet and sits 12 feet above the ground. As shown in the diagram below, the wheel is 500 feet from the entrance to the park. The *xy*-coordinate system containing the wheel has its origin on the ground at the center of the entrance. Write an equation that models the shape of the wheel.

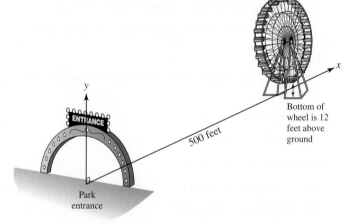

Diameter is 240 feet

Bottom of wheel is 12 feet above ground

500 feet

ENTRANCE

Park entrance

44. Magic Rings A magician is holding two rings that seem to lie in the same plane and intersect in two points. Each ring is 10 inches in diameter.

a. Find the equation of each ring if a coordinate system is placed with its origin at the center of the first ring and the *x*-axis contains the center of the second ring.

b. Find the equation of each ring if a coordinate system is placed with its origin at the center of the second ring and the *x*-axis contains the center of the first ring.

Maintaining Your Skills

Find the equation of the inverse of each of the following functions. Write the inverse using the notation $f^{-1}(x)$, if the inverse is itself a function.

45. $f(x) = 3^x$ **46.** $f(x) = 5^x$

47. $f(x) = 2x + 3$ **48.** $f(x) = 3x - 2$

49. $f(x) = \dfrac{x + 3}{5}$ **50.** $f(x) = \dfrac{x - 2}{6}$

Getting Ready for the Next Section

Solve.

51. $y^2 = 9$ **52.** $x^2 = 25$

53. $-y^2 = 4$

54. $-x^2 = 16$

55. $\dfrac{-x^2}{9} = 1$

56. $\dfrac{y^2}{100} = 1$

57. Divide $4x^2 + 9y^2$ by 36.

58. Divide $25x^2 + 4y^2$ by 100.

Find the x-intercepts and the y-intercepts.

59. $3x - 4y = 12$

60. $y = 3x^2 + 5x - 2$

61. If $\dfrac{x^2}{25} + \dfrac{y^2}{9} = 1$, find y when x is 3.

62. If $\dfrac{x^2}{25} + \dfrac{y^2}{9} = 1$, find y when x is -4.

Extending the Concepts

A circle is *tangent to* a line if it touches, but does not cross, the line.

63. Find the equation of the circle with center at (2, 3) if the circle is tangent to the y-axis.

64. Find the equation of the circle with center at (3, 2) if the circle is tangent to the x-axis.

65. Find the equation of the circle with center at (2, 3) if the circle is tangent to the vertical line $x = 4$.

66. Find the equation of the circle with center at (3, 2) if the circle is tangent to the horizontal line $y = 6$.

Find the distance from the origin to the center of each of the following circles.

67. $x^2 + y^2 - 6x + 8y = 144$

68. $x^2 + y^2 - 8x + 6y = 144$

69. $x^2 + y^2 - 6x - 8y = 144$

70. $x^2 + y^2 + 8x + 6y = 144$

71. If we were to solve the equation $x^2 + y^2 = 9$ for y, we would obtain the equation $y = \pm\sqrt{9 - x^2}$. This last equation is equivalent to the two equations $y = \sqrt{9 - x^2}$, in which y is always positive, and $y = -\sqrt{9 - x^2}$, in which y is always negative. Look at the graph of $x^2 + y^2 = 9$ in Example 6 of this section and indicate what part of the graph each of the two equations corresponds to.

72. Solve the equation $x^2 + y^2 = 9$ for x, and then indicate what part of the graph in Example 6 each of the two resulting equations corresponds to.

OBJECTIVES

A Graph an ellipse.

B Graph a hyperbola.

The photograph in Figure 1 shows Halley's comet as it passed close to earth in 1986. Like the planets in our solar system, it orbits the sun in an elliptical path (Figure 2). While it takes the earth 1 year to complete one orbit around the sun, it takes Halley's comet 76 years. The first known sighting of Halley's comet was in 239 B.C. Its most famous appearance occurred in 1066 A.D., when it was seen at the Battle of Hastings.

The Orbit of Halley's Comet

FIGURE 1   FIGURE 2

We begin this section with an introductory look at equations that produce ellipses. To simplify our work, we consider only ellipses that are centered about the origin.

Suppose we want to graph the equation

$$\frac{x^2}{25} + \frac{y^2}{9} = 1$$

We can find the y-intercepts by letting $x = 0$, and we can find the x-intercepts by letting $y = 0$:

When $x = 0$

$$\frac{0^2}{25} + \frac{y^2}{9} = 1$$

$$y^2 = 9$$

$$y = \pm 3$$

When $y = 0$

$$\frac{x^2}{25} + \frac{0^2}{9} = 1$$

$$x^2 = 25$$

$$x = \pm 5$$

The graph crosses the y-axis at $(0, 3)$ and $(0, -3)$ and the x-axis at $(5, 0)$ and $(-5, 0)$. Graphing these points and then connecting them with a smooth curve gives the graph shown in Figure 3.

Note

We can find other ordered pairs on the graph by substituting values for x (or y) and then solving for y (or x). For example, if we let $x = 3$, then

$$\frac{3^2}{25} + \frac{y^2}{9} = 1$$

$$\frac{9}{25} + \frac{y^2}{9} = 1$$

$$0.36 + \frac{y^2}{9} = 1$$

$$\frac{y^2}{9} = 0.64$$

$$y^2 = 5.76$$

$$y = \pm 2.4$$

This would give us the two ordered pairs $(3, -2.4)$ and $(3, 2.4)$.

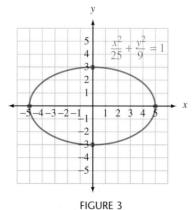

FIGURE 3

A graph of the type shown in Figure 3 is called an *ellipse*. If we were to find some other ordered pairs that satisfy our original equation, we would find that their graphs lie on the ellipse. Also, the coordinates of any point on the ellipse will satisfy the equation. We can generalize these results as follows.

The Ellipse The graph of any equation of the form

$$\frac{x^2}{a^2} + \frac{y^2}{b^2} = 1 \qquad \textbf{Standard form}$$

will be an *ellipse* centered at the origin. The ellipse will cross the x-axis at $(a, 0)$ and $(-a, 0)$. It will cross the y-axis at $(0, b)$ and $(0, -b)$. When a and b are equal, the ellipse will be a circle.

The most convenient way to graph an ellipse is to locate the intercepts.

 EXAMPLE 1   Sketch the graph of $4x^2 + 9y^2 = 36$.

SOLUTION   To write the equation in the form

$$\frac{x^2}{a^2} + \frac{y^2}{b^2} = 1$$

we must divide both sides by 36:

$$\frac{4x^2}{36} + \frac{9y^2}{36} = \frac{36}{36}$$

$$\frac{x^2}{9} + \frac{y^2}{4} = 1$$

The graph crosses the x-axis at $(3, 0)$, $(-3, 0)$ and the y-axis at $(0, 2)$, $(0, -2)$, as shown in Figure 4.

Note

When the equation is written in standard form, the x-intercepts are the positive and negative square roots of the number below x^2. The y-intercepts are the square roots of the number below y^2.

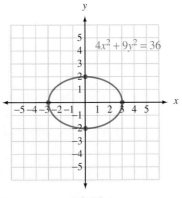

FIGURE 4

Hyperbolas

Figure 5 shows Europa, one of Jupiter's moons, as it was photographed by the Galileo space probe in the late 1990s. To speed up the trip from Earth to Jupiter—nearly a billion miles—Galileo made use of the *slingshot effect*. This involves flying a hyperbolic path very close to a planet, so that gravity can be used to gain velocity as the space probe hooks around the planet (Figure 6).

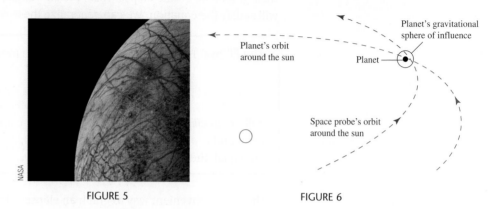

FIGURE 5   FIGURE 6

We use the rest of this section to consider equations that produce hyperbolas. Consider the equation

$$\frac{x^2}{9} - \frac{y^2}{4} = 1$$

If we were to find a number of ordered pairs that are solutions to the equation and connect their graphs with a smooth curve, we would have Figure 7.

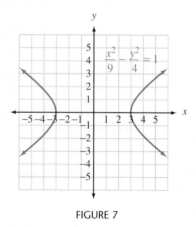

FIGURE 7

This graph is an example of a *hyperbola*. Notice that the graph has x-intercepts at $(3, 0)$ and $(-3, 0)$. The graph has no y-intercepts and hence does not cross the y-axis since substituting $x = 0$ into the equation yields

$$\frac{0^2}{9} - \frac{y^2}{4} = 1$$

$$-y^2 = 4$$

$$y^2 = -4$$

for which there is no real solution. We can, however, use the number below y^2 to help sketch the graph. If we draw a rectangle that has its sides parallel to the x- and y-axes and that passes through the x-intercepts and the points on the y-axis corresponding to the square roots of the number below y^2, $+2$ and -2, it

54. Health Expenditures The total national health expenditures, in billions of dollars, increased by approximately $50 billion from 1995 to 1996 and from 1996 to 1997. The total national health expenditures for 1997 was approximately $1,092 billion (U.S. Census Bureau, National Health Expenditures).

  a. If the same approximate increase is assumed to continue, what would be the estimated total expenditures for 1998, 1999, and 2000?

  b. Write the general term for the sequence, where n represents the number of years since 1997.

  c. If health expenditures continue to increase at this approximate rate, what would the expected expenditures be in 2010?

55. Polygons The formula for the sum of the interior angles of a polygon with n sides is $a_n = 180°(n - 2)$.

  a. Write a sequence to represent the sum of the interior angles of a polygon with 3, 4, 5, and 6 sides.

  b. What would be the sum of the interior angles of a polygon with 20 sides?

  c. What happens when $n = 2$ to indicate that a polygon cannot be formed with only two sides?

56. Pendulum A pendulum swings 10 feet left to right on its first swing. On each swing following the first, the pendulum swings $\frac{4}{5}$ of the previous swing.

  a. Write a sequence for the distance traveled by the pendulum on the first, second, and third swing.

  b. Write a general term for the sequence, where n represents the number of the swing.

  c. How far will the pendulum swing on its tenth swing? (Round to the nearest hundredth.)

Maintaining Your Skills

Find x in each of the following.

57. $\log_9 x = \dfrac{3}{2}$ **58.** $\log_x \dfrac{1}{4} = -2$

Simplify each expression.

59. $\log_2 32$ **60.** $\log_{10} 10{,}000$

61. $\log_3[\log_2 8]$ **62.** $\log_5[\log_6 6]$

Getting Ready for the Next Section

Simplify.

63. $-2 + 6 + 4 + 22$

64. $9 - 27 + 81 - 243$

65. $-8 + 16 - 32 + 64$

66. $-4 + 8 - 16 + 32 - 64$

67. $(1 - 3) + (4 - 3) + (9 - 3) + (16 - 3)$

68. $(1 - 3) + (9 + 1) + (16 + 1) + (25 + 1) + (36 + 1)$

69. $-\dfrac{1}{3} + \dfrac{1}{9} - \dfrac{1}{27} + \dfrac{1}{81}$

70. $\dfrac{1}{2} + \dfrac{2}{3} + \dfrac{3}{4} + \dfrac{4}{5}$

71. $\dfrac{1}{3} + \dfrac{1}{2} + \dfrac{3}{5} + \dfrac{2}{3}$

72. $\dfrac{1}{16} + \dfrac{1}{32} + \dfrac{1}{64}$

Extending the Concepts

73. As n increases, the terms in the sequence

$$a_n = \left(1 + \frac{1}{n}\right)^n$$

get closer and closer to the number e (that's the same e we used in defining natural logarithms). It takes some fairly large values of n, however, before we can see this happening. Use a calculator to find a_{100}, $a_{1,000}$, $a_{10,000}$, and $a_{100,000}$, and compare them to the decimal approximation we gave for the number e.

74. The sequence

$$a_n = \left(1 + \frac{1}{n}\right)^{-n}$$

gets close to the number $\frac{1}{e}$ as n becomes large. Use a calculator to find approximations for a_{100} and $a_{1,000}$, and then compare them to $\frac{1}{2.7183}$.

75. Write the first ten terms of the sequence defined by the recursion formula

$$a_1 = 1,\ a_2 = 1,\ a_n = a_{n-1} + a_{n-2} \qquad n > 2$$

76. Write the first ten terms of the sequence defined by the recursion formula

$$a_1 = 2,\ a_2 = 2,\ a_n = a_{n-1} + a_{n-2} \qquad n > 2$$

13.2 Series

OBJECTIVES

A Expand and simplify an expression written with summation notation.

B Write a series using summation notation.

There is an interesting relationship between the sequence of odd numbers and the sequence of squares that is found by adding the terms in the sequence of odd numbers.

$$1 \qquad\qquad = 1$$

$$1 + 3 \qquad\quad = 4$$

$$1 + 3 + 5 \qquad = 9$$

$$1 + 3 + 5 + 7 = 16$$

When we add the terms of a sequence the result is called a series.

> **DEFINITION** The sum of a number of terms in a sequence is called a **series.**

A sequence can be finite or infinite, depending on whether the sequence ends at the nth term. For example,

$$1, 3, 5, 7, 9$$

is a finite sequence, but

$$1, 3, 5, \ldots$$

is an infinite sequence. Associated with each of the preceding sequences is a series found by adding the terms of the sequence:

$$1 + 3 + 5 + 7 + 9 \qquad \text{Finite series}$$

$$1 + 3 + 5 + \ldots \qquad \text{Infinite series}$$

In this section, we will consider only finite series. We can introduce a new kind of notation here that is a compact way of indicating a finite series. The notation is called *summation notation,* or *sigma notation* because it is written using the Greek letter sigma. The expression

$$\sum_{i=1}^{4} (8i - 10)$$

is an example of an expression that uses summation notation. The summation notation in this expression is used to indicate the sum of all the expressions $8i - 10$ from $i = 1$ up to and including $i = 4$. That is,

$$\sum_{i=1}^{4} (8i - 10) = (8 \cdot 1 - 10) + (8 \cdot 2 - 10) + (8 \cdot 3 - 10) + (8 \cdot 4 - 10)$$

$$= -2 + 6 + 14 + 22$$

$$= 40$$

The letter i as used here is called the *index of summation,* or just *index* for short. Here are some examples illustrating the use of summation notation.

EXAMPLE 1   Expand and simplify $\displaystyle\sum_{i=1}^{5} (i^2 - 1)$.

SOLUTION We replace i in the expression $i^2 - 1$ with all consecutive integers from 1 up to 5, including 1 and 5:

$$\sum_{i=1}^{5} (i^2 - 1) = (1^2 - 1) + (2^2 - 1) + (3^2 - 1) + (4^2 - 1) + (5^2 - 1)$$

$$= 0 + 3 + 8 + 15 + 24$$

$$= 50$$

 EXAMPLE 2   Expand and simplify $\displaystyle\sum_{i=3}^{6} (-2)^i$.

SOLUTION We replace i in the expression $(-2)^i$ with the consecutive integers beginning at 3 and ending at 6:

$$\sum_{i=3}^{6} (-2)^i = (-2)^3 + (-2)^4 + (-2)^5 + (-2)^6$$

$$= -8 + 16 + (-32) + 64$$

$$= 40$$

USING TECHNOLOGY

Summing Series on a Graphing Calculator

A TI-83 graphing calculator has a built-in sum(function that, when used with the seq(function, allows us to add the terms of a series. Let's repeat Example 1 using our graphing calculator. First, we go to LIST and select MATH. The fifth option in that list is sum(, which we select. Then we go to LIST again and select OPS. From that list we select seq(. Next we enter X^2 − 1, X, 1, 5, and then we close both sets of parentheses. Our screen shows the following:

$$\text{sum(seq(X\textasciicircum2} - 1, \text{X, 1, 5))} \quad \text{which will give us} \sum_{i=1}^{5} (i^2 - 1)$$

When we press $\boxed{\text{ENTER}}$ the calculator displays 50, which is the same result we obtained in Example 1.

 EXAMPLE 3   Expand $\displaystyle\sum_{i=2}^{5} (x^i - 3)$.

SOLUTION We must be careful not to confuse the letter x with i. The index i is the quantity we replace by the consecutive integers from 2 to 5, not x:

$$\sum_{i=2}^{5} (x^i - 3) = (x^2 - 3) + (x^3 - 3) + (x^4 - 3) + (x^5 - 3)$$

In the first three examples, we were given an expression with summation notation and asked to expand it. The next examples in this section illustrate how we can write an expression in expanded form as an expression involving summation notation.

EXAMPLE 4 Write with summation notation: $1 + 3 + 5 + 7 + 9$.

SOLUTION A formula that gives us the terms of this sum is

$$a_i = 2i - 1$$

where i ranges from 1 up to and including 5. Notice we are using the subscript i in exactly the same way we used the subscript n in the previous section—to indicate the general term. Writing the sum

$$1 + 3 + 5 + 7 + 9$$

with summation notation looks like this:

$$\sum_{i=1}^{5} (2i - 1)$$

 EXAMPLE 5 Write with summation notation: $3 + 12 + 27 + 48$.

SOLUTION We need a formula, in terms of i, that will give each term in the sum. Writing the sum as

$$3 \cdot 1^2 + 3 \cdot 2^2 + 3 \cdot 3^2 + 3 \cdot 4^2$$

we see the formula

$$a_i = 3 \cdot i^2$$

where i ranges from 1 up to and including 4. Using this formula and summation notation, we can represent the sum

$$3 + 12 + 27 + 48$$

as

$$\sum_{i=1}^{4} 3i^2$$

 EXAMPLE 6 Write with summation notation:

$$\frac{x + 3}{x^3} + \frac{x + 4}{x^4} + \frac{x + 5}{x^5} + \frac{x + 6}{x^6}$$

SOLUTION A formula that gives each of these terms is

$$a_i = \frac{x + i}{x^i}$$

where i assumes all integer values between 3 and 6, including 3 and 6. The sum can be written as

$$\sum_{i=3}^{6} \frac{x + i}{x^i}$$

Determine which of the following sequences are arithmetic progressions. For those that are arithmetic progressions, identify the common difference d.

1. 1, 2, 3, 4, . . .

2. 4, 6, 8, 10, . . .

3. 1, 2, 4, 7, . . .

4. 1, 2, 4, 8, . . .

▶ **5.** 50, 45, 40, . . .

6. 1, $\dfrac{1}{2}$, $\dfrac{1}{4}$, $\dfrac{1}{8}$, . . .

▶ **7.** 1, 4, 9, 16, . . .

8. 5, 7, 9, 11, . . .

9. $\dfrac{1}{3}$, 1, $\dfrac{5}{3}$, $\dfrac{7}{3}$, . . .

10. 5, 11, 17, . . .

Each of the following problems refers to arithmetic sequences.

▶ **11.** If $a_1 = 3$ and $d = 4$, find a_n and a_{24}.

12. If $a_1 = 5$ and $d = 10$, find a_n and a_{100}.

13. If $a_1 = 6$ and $d = -2$, find a_{10} and S_{10}.

14. If $a_1 = 7$ and $d = -1$, find a_{24} and S_{24}.

▶ **15.** If $a_6 = 17$ and $a_{12} = 29$, find the term a_1, the common difference d, and then find a_{30}.

16. If $a_5 = 23$ and $a_{10} = 48$, find the first term a_1, the common difference d, and then find a_{40}.

17. If the third term is 16 and the eighth term is 26, find the first term, the common difference, and then find a_{20} and S_{20}.

18. If the third term is 16 and the eighth term is 51, find the first term, the common difference, and then find a_{50} and S_{50}.

19. If $a_1 = 3$ and $d = 4$, find a_{20} and S_{20}.

20. If $a_1 = 40$ and $d = -5$, find a_{25} and S_{25}.

21. If $a_4 = 14$ and $a_{10} = 32$, find a_{40} and S_{40}.

22. If $a_7 = 0$ and $a_{11} = -\dfrac{8}{3}$, find a_{61} and S_{61}.

23. If $a_6 = -17$ and $S_6 = -12$, find a_1 and d.

24. If $a_{10} = 12$ and $S_{10} = 40$, find a_1 and d.

25. Find a_{85} for the sequence 14, 11, 8, 5, . . .

26. Find S_{100} for the sequence $-32, -25, -18, -11, \ldots$

27. If $S_{20} = 80$ and $a_1 = -4$, find d and a_{39}.

28. If $S_{24} = 60$ and $a_1 = 4$, find d and a_{116}.

▶ **29.** Find the sum of the first 100 terms of the sequence 5, 9, 13, 17, ...

30. Find the sum of the first 50 terms of the sequence 8, 11, 14, 17, ...

31. Find a_{35} for the sequence 12, 7, 2, -3, ...

32. Find a_{45} for the sequence 25, 20, 15, 10, ...

33. Find the tenth term and the sum of the first 10 terms of the sequence $\dfrac{1}{2}$, 1, $\dfrac{3}{2}$, 2, ...

34. Find the 15th term and the sum of the first 15 terms of the sequence $-\dfrac{1}{3}$, 0, $\dfrac{1}{3}$, $\dfrac{2}{3}$, ...

Applying the Concepts

Straight-Line Depreciation Straight-line depreciation is an accounting method used to help spread the cost of new equipment over a number of years. The value at any time during the life of the machine can be found with a linear equation in two variables. For income tax purposes, however, it is the value at the end of the year that is most important, and for this reason sequences can be used.

35. Value of a Copy Machine A large copy machine sells for $18,000 when it is new. Its value decreases

$3,300 each year after that. We can use an arithmetic sequence to find the value of the machine at the end of each year. If we let a_0 represent the value when it is purchased, then a_1 is the value after 1 year, a_2 is the value after 2 years, and so on.

a. Write the first 5 terms of the sequence.

b. What is the common difference?

c. Construct a line graph for the first 5 terms of the sequence.

d. Use the line graph to estimate the value of the copy machine 2.5 years after it is purchased.

e. Write the sequence from part a using a recursive formula.

36. **Value of a Forklift** An electric forklift sells for $125,000 when new. Each year after that, it decreases $16,500 in value.
a. Write an arithmetic sequence that gives the value of the forklift at the end of each of the first 5 years after it is purchased.

b. What is the common difference for this sequence?

c. Construct a line graph for this sequence.

d. Use the line graph to estimate the value of the forklift 3.5 years after it is purchased.

e. Write the sequence from part (a) using a recursive formula.

37. **Distance** A rocket travels vertically 1,500 feet in its first second of flight, and then about 40 feet less each succeeding second. Use these estimates to answer the following questions.
a. Write a sequence of the vertical distance traveled by a rocket in each of its first 6 seconds.

b. Is the sequence in part (a) an arithmetic sequence? Explain why or why not.

c. What is the general term of the sequence in part (a)?

38. **Depreciation** Suppose an automobile sells for N dollars new, and then depreciates 40% each year.
a. Write a sequence for the value of this automobile (in terms of N) for each year.

b. What is the general term of the sequence in part (a)?

c. Is the sequence in part (a) an arithmetic sequence? Explain why it is or is not.

39. **Triangular Numbers** The first four triangular numbers are $\{1, 3, 6, 10, \ldots\}$, and are illustrated in the following diagram.

a. Write a sequence of the first 15 triangular numbers.

b. Write the recursive general term for the sequence of triangular numbers.

c. Is the sequence of triangular numbers an arithmetic sequence? Explain why it is or is not.

40. **Arithmetic Means** Three (or more) arithmetic means between two numbers may be found by forming an arithmetic sequence using the original two numbers and the arithmetic means. For example, three arithmetic means between 10 and 34 may be found by examining the sequence $\{10, a, b, c, 34\}$. For the sequence to be arithmetic, the common difference must be 6; therefore, $a = 16$, $b = 22$, and $c = 28$. Use this idea to answer the following questions.
a. Find four arithmetic means between 10 and 35.

b. Find three arithmetic means between 2 and 62.

c. Find five arithmetic means between 4 and 28.

33. $(x - y)^{18}$ **34.** $(x - 2y)^{65}$

Write the first three terms in the expansion of each of the following.

35. $(x + 1)^{15}$ **36.** $(x - 1)^{15}$

37. $(x - y)^{12}$ **38.** $(x + y)^{12}$

39. $(x + 2)^{20}$ **40.** $(x - 2)^{20}$

Write the first two terms in the expansion of each of the following.

▶ **41.** $(x + 2)^{100}$

42. $(x - 2)^{50}$

43. $(x + y)^{50}$

44. $(x - y)^{100}$

45. Find the ninth term in the expansion of $(2x + 3y)^{12}$.

46. Find the sixth term in the expansion of $(2x + 3y)^{12}$.

47. Find the fifth term of $(x - 2)^{10}$.

48. Find the fifth term of $(2x - 1)^{10}$.

49. Find the sixth term in the expansion of $(x - 2)^{12}$.

50. Find the ninth term in the expansion of $(7x - 1)^{10}$.

51. Find the third term in the expansion of $(x - 3y)^{25}$.

52. Find the 24th term in the expansion of $(2x - y)^{26}$.

53. Write the formula for the 12th term of $(2x + 5y)^{20}$. Do not simplify.

54. Write the formula for the eighth term of $(2x + 5y)^{20}$. Do not simplify.

55. Write the first three terms of the expansion of $(x^2y - 3)^{10}$.

56. Write the first three terms of the expansion of $(x - \frac{1}{x})^{50}$.

Applying the Concepts

57. **Probability** The third term in the expansion of $\left(\frac{1}{2} + \frac{1}{2}\right)^7$ will give the probability that in a family with 7 children, 5 will be boys and 2 will be girls. Find the third term.

58. **Probability** The fourth term in the expansion of $\left(\frac{1}{2} + \frac{1}{2}\right)^8$ will give the probability that in a family with 8 children, 3 will be boys and 5 will be girls. Find the fourth term.

Maintaining Your Skills

Solve each equation. Write your answers to the nearest hundreth.

59. $5^x = 7$ **60.** $10^x = 15$

61. $8^{2x+1} = 16$ **62.** $9^{3x-1} = 27$

63. **Compound Interest** How long will it take $400 to double if it is invested in an account with an annual interest rate of 10% compounded four times a year?

64. **Compound Interest** How long will it take $200 to become $800 if it is invested in an account with an annual interest rate of 8% compounded four times a year?

Extending the Concepts

65. Calculate both $\binom{8}{5}$ and $\binom{8}{3}$ to show that they are equal.

66. Calculate both $\binom{10}{8}$ and $\binom{10}{2}$ to show that they are equal.

67. Simplify $\binom{20}{12}$ and $\binom{20}{8}$.

68. Simplify $\binom{15}{10}$ and $\binom{15}{5}$.

69. **Pascal's Triangle** Copy the first eight rows of Pascal's triangle into the eight rows of the triangular array to the right. (Each number in Pascal's triangle will go into one of the hexagons in the array.) Next, color in each hexagon that contains an odd number. What pattern begins to emerge from this coloring process?

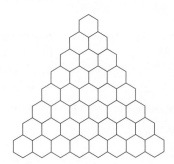

Sequences [13.1]

EXAMPLES

1. In the sequence
$1, 3, 5, \ldots, 2n - 1, \ldots$
$a_1 = 1$
$a_2 = 3$
$a_3 = 5$
and
$a_n = 2n - 1$

A *sequence* is a function whose domain is the set of positive integers. The terms of a sequence are denoted by

$$a_1, a_2, a_3, \ldots, a_n, \ldots$$

where a_1 (read "a sub 1") is the first term, a_2 the second term, and a_n the nth or *general term*.

Summation Notation [13.2]

2. $\displaystyle\sum_{i=3}^{6} (-2)^i$
$= (-2)^3 + (-2)^4 + (-2)^5 + (-2)^6$
$= -8 + 16 + (-32) + 64$
$= 40$

The notation

$$\sum_{i=1}^{n} a_i = a_1 + a_2 + a_3 + \cdots + a_n$$

is called *summation notation* or *sigma notation*. The letter i as used here is called the *index of summation* or just *index*.

Arithmetic Sequences [13.3]

3. For the sequence
$3, 7, 11, 15, \ldots,$
$a_1 = 3$ and $d = 4$. The general term is
$a_n = 3 + (n - 1)4$
$\quad = 4n - 1$
Using this formula to find the tenth term, we have
$a_{10} = 4(10) - 1 = 39$
The sum of the first 10 terms is
$S_{10} = \dfrac{10}{2}(3 + 39) = 210$

An *arithmetic sequence* is a sequence in which each term comes from the preceding term by adding a constant amount each time. If the first term of an arithmetic sequence is a_1 and the amount we add each time (called the *common difference*) is d, then the nth term of the progression is given by

$$a_n = a_1 + (n - 1)d$$

The sum of the first n terms of an arithmetic sequence is

$$S_n = \frac{n}{2}(a_1 + a_n)$$

S_n is called the nth *partial sum*.

Geometric Sequences [13.4]

4. For the geometric progression
$3, 6, 12, 24, \ldots,$
$a_1 = 3$ and $r = 2$. The general term is
$a_n = 3 \cdot 2^{n-1}$
The sum of the first 10 terms is
$S_{10} = \dfrac{3(2^{10} - 1)}{2 - 1} = 3{,}069$

A *geometric sequence* is a sequence of numbers in which each term comes from the previous term by multiplying by a constant amount each time. The constant by which we multiply each term to get the next term is called the *common ratio*. If the first term of a geometric sequence is a_1 and the common ratio is r, then the formula that gives the general term a_n is

$$a_n = a_1 r^{n-1}$$

The sum of the first n terms of a geometric sequence is given by the formula

$$S_n = \frac{a_1(r^n - 1)}{r - 1}$$

The Sum of an Infinite Geometric Series [13.4]

5. The sum of the series

$$\frac{1}{3} + \frac{1}{6} + \frac{1}{12} + \cdots$$

is

$$S = \frac{\frac{1}{3}}{1 - \frac{1}{2}} = \frac{\frac{1}{3}}{\frac{1}{2}} = \frac{2}{3}$$

If a geometric sequence has first term a_1 and common ratio r such that $|r| < 1$, then the following is called an *infinite geometric series:*

$$S = \sum_{i=0}^{\infty} a_1 r^i = a_1 + a_1 r + a_1 r^2 + a_1 r^3 + \cdots$$

Its sum is given by the formula

$$S = \frac{a_1}{1 - r}$$

Factorials [13.5]

The notation $n!$ is called *n factorial* and is defined to be the product of each consecutive integer from n down to 1. That is,

$$0! = 1 \qquad \textbf{(By definition)}$$

$$1! = 1$$

$$2! = 2 \cdot 1$$

$$3! = 3 \cdot 2 \cdot 1$$

$$4! = 4 \cdot 3 \cdot 2 \cdot 1$$

and so on.

Binomial Coefficients [13.5]

6. $\binom{7}{3} = \frac{7!}{3!(7-3)!}$

$$= \frac{7!}{3! \cdot 4!}$$

$$= \frac{7 \cdot 6 \cdot 5 \cdot 4 \cdot 3 \cdot 2 \cdot 1}{(3 \cdot 2 \cdot 1)(4 \cdot 3 \cdot 2 \cdot 1)}$$

$$= 35$$

The notation $\binom{n}{r}$ is called a *binomial coefficient* and is defined by

$$\binom{n}{r} = \frac{n!}{r!(n-r)!}$$

Binomial coefficients can be found by using the formula above or by *Pascal's triangle,* which is

```
              1
           1     1
        1     2     1
     1     3     3     1
  1     4     6     4     1
1    5    10    10    5    1
```

and so on.

Binomial Expansion [13.5]

7. $(x+2)^4$

$$= x^4 + 4x^3 \cdot 2 + 6x^2 \cdot 2^2 + 4x \cdot 2^3 + 2^4$$

$$= x^4 + 8x^3 + 24x^2 + 32x + 16$$

If n is a positive integer, then the formula for expanding $(x + y)^n$ is given by

$$(x+y)^n = \binom{n}{0}x^n y^0 + \binom{n}{1}x^{n-1}y^1 + \binom{n}{2}x^{n-2}y^2 + \cdots + \binom{n}{n}x^0 y^n$$

Chapter 13 Review Test

The problems below form a comprehensive review of the material in this chapter. They can be used to study for exams. If you would like to take a practice test on this chapter, you can use the odd-numbered problems. Give yourself an hour and work as many of the odd-numbered problems as possible. When you are finished, or when an hour has passed, check your answers with the answers in the back of the book. You can use the even-numbered problems for a second practice test.

Write the first four terms of the sequence with the following general terms. [13.1]

1. $a_n = 2n + 5$

2. $a_n = 3n - 2$

3. $a_n = n^2 - 1$

4. $a_n = \dfrac{n + 3}{n + 2}$

5. $a_1 = 4, a_n = 4a_{n-1}, n > 1$

6. $a_1 = \dfrac{1}{4}, a_n = \dfrac{1}{4}a_{n-1}, n > 1$

Determine the general term for each of the following sequences. [13.1]

7. $2, 5, 8, 11, \ldots$

8. $-3, -1, 1, 3, 5, \ldots$

9. $1, 16, 81, 256, \ldots$

10. $2, 5, 10, 17, \ldots$

11. $\dfrac{1}{2}, \dfrac{1}{4}, \dfrac{1}{8}, \dfrac{1}{16}, \ldots$

12. $2, \dfrac{3}{4}, \dfrac{4}{9}, \dfrac{5}{16}, \dfrac{6}{25}, \ldots$

Expand and simplify each of the following. [13.2]

13. $\displaystyle\sum_{i=1}^{4} (2i + 3)$

14. $\displaystyle\sum_{i=1}^{3} (2i^2 - 1)$

15. $\displaystyle\sum_{i=2}^{3} \dfrac{i^2}{i + 2}$

16. $\displaystyle\sum_{i=1}^{4} (-2)^{i-1}$

17. $\displaystyle\sum_{i=3}^{5} (4i + i^2)$

18. $\displaystyle\sum_{i=4}^{6} \dfrac{i + 2}{i}$

Write each of the following sums with summation notation. [13.2]

19. $3 + 6 + 9 + 12$

20. $3 + 7 + 11 + 15$

21. $5 + 7 + 9 + 11 + 13$

22. $4 + 9 + 16$

23. $\dfrac{1}{3} + \dfrac{1}{4} + \dfrac{1}{5} + \dfrac{1}{6}$

24. $\dfrac{1}{3} + \dfrac{2}{9} + \dfrac{3}{27} + \dfrac{4}{81} + \dfrac{5}{243}$

25. $(x - 2) + (x - 4) + (x - 6)$

26. $\dfrac{x}{x + 1} + \dfrac{x}{x + 2} + \dfrac{x}{x + 3} + \dfrac{x}{x + 4}$

Determine which of the following sequences are arithmetic progressions, geometric progressions, or neither. [13.3, 13.4]

27. $1, -3, 9, -27, \ldots$

28. $7, 9, 11, 13, \ldots$

29. $5, 11, 17, 23, \ldots$

30. $\dfrac{1}{2}, \dfrac{1}{3}, \dfrac{1}{4}, \dfrac{1}{5}, \ldots$

31. $4, 8, 16, 32, \ldots$

32. $\dfrac{1}{2}, \dfrac{1}{4}, \dfrac{1}{8}, \dfrac{1}{16}, \ldots$

33. $12, 9, 6, 3, \ldots$

34. $2, 5, 9, 14, \ldots$

Each of the following problems refers to arithmetic progressions. [13.3]

35. If $a_1 = 2$ and $d = 3$, find a_n and a_{20}.

36. If $a_1 = 5$ and $d = -3$, find a_n and a_{16}.

37. If $a_1 = -2$ and $d = 4$, find a_{10} and S_{10}.

38. If $a_1 = 3$ and $d = 5$, find a_{16} and S_{16}.

39. If $a_5 = 21$ and $a_8 = 33$, find the first term a_1, the common difference d, and then find a_{10}.

40. If $a_3 = 14$ and $a_7 = 26$, find the first term a_1, the common difference d, and then find a_9 and S_9.

41. If $a_4 = -10$ and $a_8 = -18$, find the first term a_1, the common difference d, and then find a_{20} and S_{20}.

42. Find the sum of the first 100 terms of the sequence $3, 7, 11, 15, 19, \ldots$

43. Find a_{40} for the sequence $100, 95, 90, 85, 80, \ldots$

Each of the following problems refers to infinite geometric progressions. [13.4]

44. If $a_1 = 3$ and $r = 2$, find a_n and a_{20}.

45. If $a_1 = 5$ and $r = -2$, find a_n and a_{16}.

46. If $a_1 = 4$ and $r = \dfrac{1}{2}$, find a_n and a_{10}.

47. If $a_1 = -2$ and $r = \dfrac{1}{3}$, find the sum.

48. If $a_1 = 4$ and $r = \dfrac{1}{2}$, find the sum.

49. If $a_3 = 12$ and $a_4 = 24$, find the first term a_1, the common ratio r, and then find a_6.

50. Find the tenth term of the sequence $3, 3\sqrt{3}, 9, 9\sqrt{3}, \ldots$

CHAPTER 7 REVIEW TEST

1. -3 **2.** $\frac{10}{13}$ **3.** $-\frac{5}{11}$ **4.** $-\frac{4}{9}$ **5.** $-\frac{3}{2}, 4$ **6.** $-\frac{1}{9}, \frac{1}{9}$ **7.** $1, 4$ **8.** $-2, -\frac{2}{3}, \frac{2}{3}$ **9.** 4 feet by 12 feet

10. $3, 4, 5$ **11.** $2, 4$ **12.** $-1, 5$ **13.** $-1, 4$ **14.** $-\frac{5}{3}, 3$ **15.** $-\frac{3}{2}, 3$ **16.** $-\frac{5}{3}, \frac{7}{3}$ **17.** $(-\infty, 12]$ **18.** $(-1, \infty)$

19. $[2, 6]$ **20.** $[-3, 2]$ **21.** $(-\infty, -\frac{3}{2}] \cup [3, \infty)$ **22.** $(-\infty, \frac{4}{5}] \cup [2, \infty)$

23.

24.

25.

26.

27. $\varnothing$ **28.** All real numbers   **29.** $(0, 1)$ **30.** Lines coincide   **31.** $(3, 5)$ **32.** $(4, 8)$ **33.** $(\frac{3}{2}, \frac{1}{2})$ **34.** $(4, 1)$

35. $(-5, 3)$ **36.** Parallel lines   **37.** $(3, -1, 4)$ **38.** $(2, \frac{1}{2}, -3)$ **39.** Dependent system   **40.** Dependent system

41. $(2, -1, 4)$ **42.** Dependent system

43.

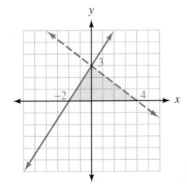

44.

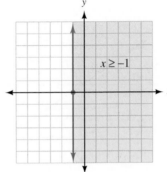

45.

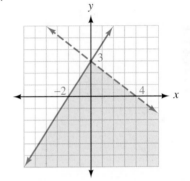

46.

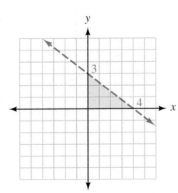

47.

48.

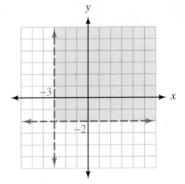

CHAPTER 8

Problem Set 8.1

1. $\frac{3}{2}$ **3.** No slope **5.** $\frac{2}{3}$

7.

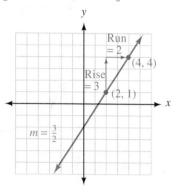

9.

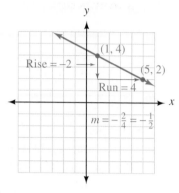

11.

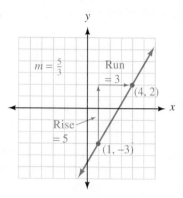

13.

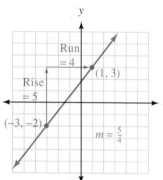

15.

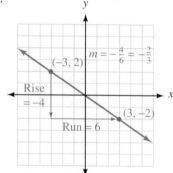

17.

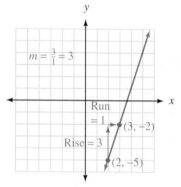

19. $x = 3$ **21.** $a = 5$ **23.** $b = 2$

25.

x	y
0	2
3	0

Slope $= -\frac{2}{3}$

27.

x	y
0	−5
3	−3

Slope $= \frac{2}{3}$

29.

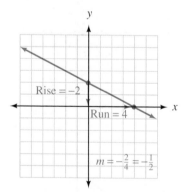

31. $\frac{1}{5}$ **33.** 0 **35.** − 1 **37.** $-\frac{3}{2}$ **39. a.** yes **b.** no **41. a.** 17.5 mph **b.** 40 km/h **c.** 120 ft/sec **d.** 28 m/min

43. Slopes: A, −50; B −75; C, −25 **45. a.** 15,800 feet **b.** $-\frac{7}{100}$

47. 6.5 percent/year. Over the 10 years from 1995 to 2005, the percent of adults that access the Internet has increased at an average rate of 6.5% each year.

49. a. −4,768.6 cases/year. New cases of rubella decreased at an average of 4,768.6 cases per year from 1969 to 1979.
 b. −999.1 cases/year. New cases of rubella decreased at an average of 999.1 cases per year from 1979 to 1989.
 c. The slope is 0, meaning that the number of new cases of German Measles is constant at 9 cases/year.

51. 0 **53.** $y = -\frac{3}{2}x + 6$ **55.** $t = \frac{A - P}{Pr}$ **57.** −1 **59.** −2 **61.** $y = mx + b$ **63.** $y = -2x - 5$ **65.** 5 **67.** $\left(\frac{5}{4}, \frac{7}{4}\right)$

Problem Set 8.2

1. $y = 2x + 3$ **3.** $y = x - 5$ **5.** $y = \frac{1}{2}x + \frac{3}{2}$ **7.** $y = 4$ **9. a.** 3 **b.** $-\frac{1}{3}$ **11. a.** -3 **b.** $\frac{1}{3}$ **13. a.** $-\frac{2}{5}$ **b.** $\frac{5}{2}$

15. Slope = 3,
y-intercept = -2,
perpendicular slope = $-\frac{1}{3}$

17. Slope = $\frac{2}{3}$,
y-intercept = -4,
perpendicular slope = $-\frac{3}{2}$

19. Slope = $-\frac{4}{5}$,
y-intercept = 4,
perpendicular slope = $\frac{5}{4}$

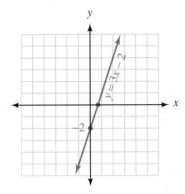

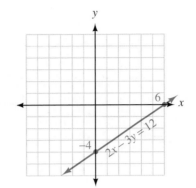

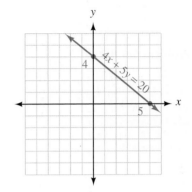

21. Slope = $\frac{1}{2}$, y-intercept = -4, $y = \frac{1}{2}x - 4$ **23.** Slope = $-\frac{2}{3}$, y-intercept = 3, $y = -\frac{2}{3}x + 3$ **25.** $y = 2x - 1$ **27.** $y = -\frac{1}{2}x - 1$

29. $y = -3x + 1$ **31.** $y = \frac{2}{3}x + \frac{2}{3}$ **33.** $y = -\frac{1}{4}x + \frac{3}{4}$ **35.** $x - y = 2$ **37.** $2x - y = 3$ **39.** $6x - 5y = 3$

41. a. x-intercept = $\frac{10}{3}$; y-intercept = -5 **b.** (4, 1) **c.** $y = \frac{3}{2}x - 5$ **d.** No

43. a. x-intercept = $\frac{4}{3}$; y-intercept = -2 **b.** (2, 1) **c.** $y = \frac{3}{2}x - 2$ **d.** No

45. a. 2 **b.** $y = 2x - 3$ **c.** $y = -3$ **d.** 2 **e.**

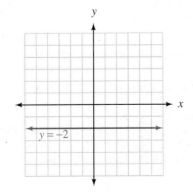

47. (0, -4), (2, 0); $y = 2x - 4$ **49.** (-2, 0) (0, 4); $y = 2x + 4$

51. Slope = 0,
y-intercept = -2

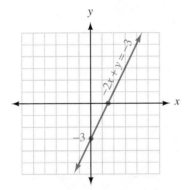

53. $y = 3x + 7$ **55.** $y = -\frac{5}{2}x - 13$ **57.** $y = \frac{1}{4}x + \frac{1}{4}$ **59.** $y = -\frac{2}{3}x + 2$ **61. b.** $86°$ F   **63. a.** \$190,000   **b.** \$19   **c.** \$6.50

65. a. $y = \frac{69}{2}x - 68{,}994$ **b.** 213 books **67. a.** $y = 6.5x - 12{,}958.5$ **b.** In 2006 80.5% of adults will be online.

c. When we try to estimate the number of adults online in 2010 we get a number larger than 100%, which is impossible.

69. 23 inches, 5 inches **71.** $46.50 **73.**

x	y
0	0
10	75
20	150

75.

x	y
0	0
1	1
1	−1

77. $y = -\frac{3}{2}x + 12$ **79.** $y = -\frac{4}{3}x - \frac{3}{2}$

Problem Set 8.3

1. Domain = {1, 2, 4}; Range = {3, 5, 1}; a function **3.** Domain = {−1, 1, 2}; range = {3, −5}; a function
5. Domain = {7, 3}; Range = {−1, 4}; not a function **7.** Domain = {a, b, c, d}; range = {3, 4, 5}; a function
9. Domain = {a}; Range = {1, 2, 3, 4}; not a function **11.** Yes **13.** No **15.** No **17.** Yes **19.** Yes
21. Domain = $\{x \mid -5 \le x \le 5\}$; Range = $\{y \mid 0 \le y \le 5\}$ **23.** Domain = $\{x \mid -5 \le x \le 3\}$; Range = $\{y \mid y = 3\}$
25. Domain = All real numbers; **27.** Domain = All real numbers; **29.** Domain = $\{x \mid x \ge -1\}$;
Range = $\{y \mid y \ge -1\}$; Range = $\{y \mid y \ge 4\}$; Range = All real numbers;
a function a function not a function

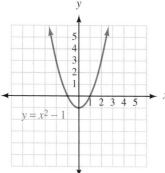

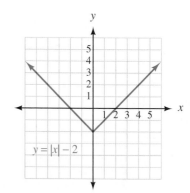

31. Domain = $\{x \mid x \ge 4\}$; **33.** Domain = All real numbers; **35.** Domain = All real numbers;
Range = All real numbers; Range = $\{y \mid y \ge 0\}$; Range = $\{y \mid y \ge -2\}$;
not a function a function a function

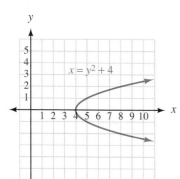

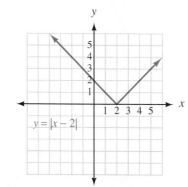

37. a. $y = 8.5x$ for $10 \le x \le 40$
b.

Hours Worked	Function Rule	Gross Pay ($)
x	y = 8.5x	y
10	y = 8.5(10) = 85	85
20	y = 8.5(20) = 170	170
30	y = 8.5(30) = 255	255
40	y = 8.5(40) = 340	340

c.

d. Domain = $\{x \mid 10 \le x \le 40\}$; Range = $\{y \mid 85 \le y \le 340\}$ **e.** Minimum = $85; Maximum = $340

39. a.

Time (sec) t	Function Rule $h = 16t - 16t^2$	Distance (ft) h
0	$h = 16(0) - 16(0)^2$	0
0.1	$h = 16(0.1) - 16(0.1)^2$	1.44
0.2	$h = 16(0.2) - 16(0.2)^2$	2.56
0.3	$h = 16(0.3) - 16(0.3)^2$	3.36
0.4	$h = 16(0.4) - 16(0.4)^2$	3.84
0.5	$h = 16(0.5) - 16(0.5)^2$	4
0.6	$h = 16(0.6) - 16(0.6)^2$	3.84
0.7	$h = 16(0.7) - 16(0.7)^2$	3.36
0.8	$h = 16(0.8) - 16(0.8)^2$	2.56
0.9	$h = 16(0.9) - 16(0.9)^2$	1.44
1	$h = 16(1) - 16(1)^2$	0

b. Domain = $\{t \mid 0 \le t \le 1\}$; Range = $\{h \mid 0 \le h \le 4\}$

c.

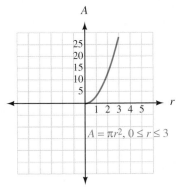

41. a.

$A = \pi r^2, 0 \le r \le 3$

b. Domain = $\{r \mid 0 \le r \le 3\}$
Range = $\{A \mid 0 \le A \le 9\pi\}$

43. a. Yes **b.** Domain = $\{t \mid 0 \le t \le 6\}$; Range = $\{h \mid 0 \le h \le 60\}$ **c.** $t = 3$ **d.** $h = 60$ **e.** $t = 6$ **45. a.** III **b.** I **c.** II **d.** IV   **47.** 10

49. -14 **51.** 1   **53.** -3 **55.** $-\frac{6}{5}$ **57.** $-\frac{7}{640}$ **59.** 150   **61.** 113   **63.** -9 **65. a.** 6 **b.** 7.5   **67. a.** 27 **b.** 6

69. Domain = All real numbers;
Range = $\{y \mid y \le 5\}$;
a function

71. Domain = $\{x \mid x \ge 3\}$;
Range = All real numbers;
not a function

73. Domain = $\{x \mid -4 \le x \le 4\}$;
Range = $\{y \mid -4 \le y \le 4\}$;
not a function

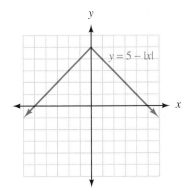

$y = 5 - |x|$

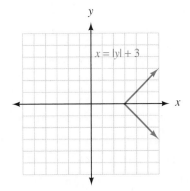

$x = |y| + 3$

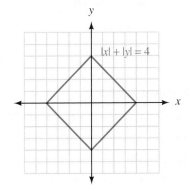

$|x| + |y| = 4$

Problem Set 8.4

1. -1 **3.** -11 **5.** 2 **7.** 4 **9.** 35 **11.** -13 **13.** 1 **15.** -9 **17.** 8 **19.** 19 **21.** 16 **23.** 0 **25.** $3a^2 - 4a + 1$
27. 4 **29.** 0 **31.** 2 **33.** -8 **35.** -1 **37.** $2a^2 - 8$ **39.** $2b^2 - 8$ **41.** 0 **43.** -2 **45.** -3
47.   **49.** $x = 4$ **51.**

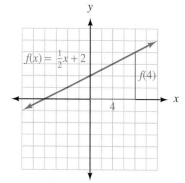
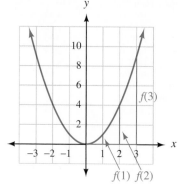

53. $V(3) = 300$, the painting is worth \$300 in 3 years; $V(6) = 600$, the painting is worth \$600 in 6 years.
55. a. True   **b.** True   **c.** True   **d.** False   **e.** True
57. a. \$5,625   **b.** \$1,500   **c.** $\{t \mid 0 \le t \le 5\}$ **d.**
    e. $\{V(t) \mid 1,500 \le V(t) \le 18,000\}$ **f.** 2.42 years

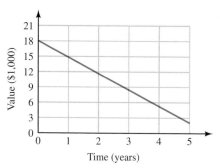

59. $-\frac{2}{3}, 4$ **61.** $-2, 1$ **63.** $\varnothing$ **65.** $-500 + 27x - 0.1x^2$ **67.** $6x^2 - 2x - 4$ **69.** $2x^2 + 8x + 8$ **71.** $0.6m - 42$
73. $4x^2 - 7x + 3$ **75. a.** 2 **b.** 0
77. a.

Weight (ounces)	0.6	1.0	1.1	2.5	3.0	4.8	5.0	5.3
Cost (cents)	39	39	63	87	87	135	135	159

b. More than 2 ounces, but not more than 3 ounces; $2 < x \le 3$

   c. domain: $\{x \mid 0 < x \le 6\}$ **d.** range: $\{c \mid c = 39, 63, 87, 111, 135, 159\}$

Problem Set 8.5

1. $6x + 2$ **3.** $-2x + 8$ **5.** $4x - 7$ **7.** $3x^2 + x - 2$ **9.** $-2x + 3$ **11.** $9x^3 - 15x^2$ **13.** $\dfrac{3x^2}{3x - 5}$ **15.** $\dfrac{3x - 5}{3x^2}$
17. $3x^2 + 4x - 7$ **19.** 15 **21.** 98 **23.** $\frac{3}{2}$ **25.** 1 **27.** 40 **29.** 147 **31. a.** 81 **b.** 29 **c.** $(x + 4)^2$ **d.** $x^2 + 4$
33. a. -2 **b.** -1 **c.** $4x^2 + 12x - 1$ **37.** 6 **39.** 2 **41.** 3 **43.** -8 **45.** 6 **47.** 5 **49.** 3 **51.** -6
53. a. $R(x) = 11.5x - 0.05x^2$ **b.** $C(x) = 2x + 200$ **c.** $P(x) = -0.05x^2 + 9.5x - 200$ **d.** $\overline{C}(x) = 2 + \dfrac{200}{x}$
55. a. $M(x) = 220 - x$ **b.** 196 **c.** 142 **d.** 135 **e.** 128 **57.** $2 < x < 4$ **59.** $x < 4$ or $x > 8$ **61.** $-\frac{5}{7} \le x \le 1$ **63.** 196 **65.** 4
67. 1.6 **69.** 3 **71.** $2,400$

Problem Set 8.6

1. Direct   **3.** Direct   **5.** Direct   **7.** Direct   **9.** Direct   **11.** Inverse   **13.** 30 **15.** 5 **17.** -6 **19.** $\frac{1}{2}$ **21.** 40
23. 225 **25.** 2 **27.** 10 **29.** $\frac{81}{5}$ **31.** 40.5 **33.** 64 **35.** 8 **37.** $\frac{50}{3}$ pounds

39. a. $T = 4P$ **c.** 70 pounds per square inch   **41.** 12 pounds per square inch   **43. a.** $f = \frac{80}{d}$ **c.** An f-stop of 8

b.

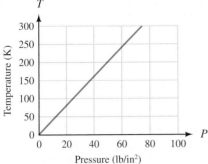

b.

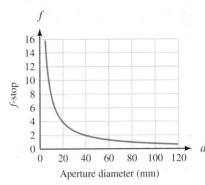

45. $\frac{1{,}504}{15}$ square inches   **47.** 1.5 ohms   **49.** 1.28 meters   **51.** $F = G\,\frac{m_1 m_2}{d^2}$ **53.** 12   **55.** 28   **57.** $-\frac{7}{4}$

59. $w = \frac{P - 2l}{2}$ **61.** $t \geq -6$ **63.** $x < 6$ **65.** 6, 2   **67.** $\varnothing$ **69.** $x \leq -9$ or $x \geq -1$

71. All real numbers   **73. a.** Square of speed **b.** $d = 0.0675s^2$ **c.** About 204 feet from the cannon **d.** About 7.5 feet further away

CHAPTER 8 REVIEW TEST

1.

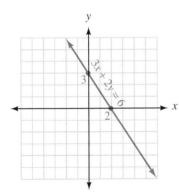

2.

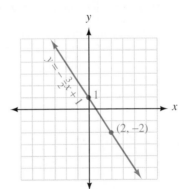

3.

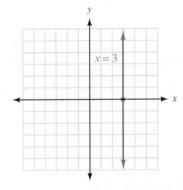

4. -2 **5.** 0   **6.** 3   **7.** 5   **8.** -5 **9.** 6   **10.** $y = 3x + 5$ **11.** $y = -2x$ **12.** $m = 3, b = -6$ **13.** $m = \frac{2}{3}, b = -3$

14. $y = 2x$ **15.** $y = -\frac{1}{3}x$ **16.** $y = 2x + 1$ **17.** $y = 7$ **18.** $y = -\frac{3}{2}x - \frac{17}{2}$ **19.** $y = 2x - 7$ **20.** $y = \frac{1}{3}x - \frac{2}{3}$

21. Domain = {2, 3, 4}; Range = {4, 3, 2}; a function   **22.** Domain = {6, -4, -2}; Range = {3, 0}; a function   **23.** 2   **24.** 4

25. 0   **26.** 1   **27.** 1   **28.** $3a + 2$ **29.** 1   **30.** 31   **31.** 24   **32.** 6   **33.** 4   **34.** 25   **35.** 84 pounds

36. 16 footcandles

CHAPTER 9

Problem Set 9.1

1. 12   **3.** Not a real number   **5.** -7 **7.** -3 **9.** 2   **11.** Not a real number   **13.** 0.2   **15.** 0.2   **17.** 5   **19.** -6

21. $\frac{1}{6}$ **23.** $\frac{2}{5}$ **25.** $6a^4$ **27.** $3a^4$ **29.** $2x^2y$ **31.** $2a^3b^5$ **33.** 6   **35.** -3 **37.** 2   **39.** -2 **41.** 2   **43.** $\frac{9}{5}$ **45.** 9

47. 125   **49.** $\frac{1}{3}$ **51.** $\frac{1}{27}$ **53.** $\frac{6}{5}$ **55.** $\frac{8}{27}$ **57.** 7   **59.** $\frac{3}{4}$ **61.** $x^{4/5}$ **63.** a **65.** $\frac{1}{x^{2/5}}$ **67.** $x^{1/6}$ **69.** $x^{9/25}y^{1/2}z^{1/5}$

71. $\frac{b^{7/4}}{a^{1/8}}$ **73.** $y^{3/10}$ **75.** $\frac{1}{a^2b^4}$ **77.** $\frac{s^{1/2}}{r^{20}}$ **79.** $10b^3$ **81.** 25 mph

83. a. 424 picometers **b.** 600 picometers **c.** 6×10^{-10} meters   **85. a.** B **b.** A **c.** C **d.** (0, 0) and (1, 1)   **87.** $x^6 - x^3$

89. $x^2 + 2x - 15$ **91.** $x^4 - 10x^2 + 25$ **93.** $x^3 - 27$ **95.** $x^6 - x^5$ **97.** $12a^2 - 11ab + 2b^2$ **99.** $x^6 - 4$ **101.** $3x - 4$

103. $(x - 5)(x + 2)$ **105.** $(3x - 2)(2x + 5)$ **107.** x^2 **109.** t **111.** $x^{1/3}$ **113.** $(9^{1/2} + 4^{1/2})^2 = (3 + 2)^2 = 5^2 = 25 \neq 9 +$

115. $\sqrt{\sqrt{a}} = (a^{1/2})^{1/2} = a^{1/4} = \sqrt[4]{a}$

Problem Set 9.2

1. $x + x^2$ **3.** $a^2 - a$ **5.** $6x^3 - 8x^2 + 10x$ **7.** $12x^2 - 36y^2$ **9.** $x^{4/3} - 2x^{2/3} - 8$ **11.** $a - 10a^{1/2} + 21$ **13.** $20y^{2/3} - 7y^{1/3} - 6$
15. $10x^{4/3} + 21x^{2/3}y^{1/2} + 9y$ **17.** $t + 10t^{1/2} + 25$ **19.** $x^3 + 8x^{3/2} + 16$ **21.** $a - 2a^{1/2}b^{1/2} + b$ **23.** $4x - 12x^{1/2}y^{1/2} + 9y$ **25.** $a - 3$
27. $x^3 - y^3$ **29.** $t - 8$ **31.** $4x^3 - 3$ **33.** $x + y$ **35.** $a - 8$ **37.** $8x + 1$ **39.** $t - 1$ **41.** $2x^{1/2} + 3$ **43.** $3x^{1/3} - 4y^{1/3}$
45. $3a - 2b$ **47.** $3(x-2)^{1/2}(4x-11)$ **49.** $5(x-3)^{7/5}(x-6)$ **51.** $3(x+1)^{1/2}(3x^2 + 3x + 2)$ **53.** $(x^{1/3} - 2)(x^{1/3} - 3)$
55. $(a^{1/5} - 4)(a^{1/5} + 2)$ **57.** $(2y^{1/3} + 1)(y^{1/3} - 3)$ **59.** $(3t^{1/5} + 5)(3t^{1/5} - 5)$ **61.** $(2x^{1/7} + 5)^2$ **63.** -8 **65.** 90 **67.** -3 **69.** 0
71. $\dfrac{3+x}{x^{1/2}}$ **73.** $\dfrac{x+5}{x^{1/3}}$ **75.** $\dfrac{x^3 + 3x^2 + 1}{(x^3 + 1)^{1/2}}$ **77.** $\dfrac{-4}{(x^2 + 4)^{1/2}}$ **79.** 15.8% **81.** 3 **83.** $m = \frac{2}{3}, b = -2$ **85.** $y = \frac{2}{3}x + 6$
87. $y = \frac{2}{3}x - 2$ **89.** 5 **91.** 6 **93.** $4x^2y$ **95.** $5y$ **97.** 3 **99.** 2 **101.** $2ab$ **103.** 25 **105.** $48x^4y^2$ **107.** $4x^6y^6$

Problem Set 9.3

1. $2\sqrt{2}$ **3.** $7\sqrt{2}$ **5.** $12\sqrt{2}$ **7.** $4\sqrt{5}$ **9.** $4\sqrt{3}$ **11.** $15\sqrt{3}$ **13.** $3\sqrt[3]{2}$ **15.** $4\sqrt[3]{2}$ **17.** $6\sqrt[3]{2}$ **19.** $2\sqrt[5]{2}$ **21.** $3x\sqrt{2x}$
23. $2y\sqrt[4]{2y^3}$ **25.** $2xy^2\sqrt[3]{5xy}$ **27.** $4abc^2\sqrt[3]{3b}$ **29.** $2bc\sqrt[3]{6a^2c}$ **31.** $2xy^2\sqrt[5]{2x^3y^2}$ **33.** $3xy^2z\sqrt[5]{x^2}$ **35.** $2\sqrt{3}$
37. $\sqrt{-20}$, which is not a real number **39.** $\dfrac{\sqrt{11}}{2}$ **41.** $\dfrac{2\sqrt{3}}{3}$ **43.** $\dfrac{5\sqrt{6}}{6}$ **45.** $\dfrac{\sqrt{2}}{2}$ **47.** $\dfrac{\sqrt{5}}{5}$ **49.** $2\sqrt[3]{4}$ **51.** $\dfrac{2\sqrt[3]{3}}{3}$
53. $\dfrac{\sqrt[4]{24x^2}}{2x}$ **55.** $\dfrac{\sqrt[4]{8y^3}}{y}$ **57.** $\dfrac{\sqrt[3]{36xy^2}}{3y}$ **59.** $\dfrac{\sqrt[3]{6xy^2}}{3y}$ **61.** $\dfrac{\sqrt[4]{2x}}{2x}$ **63.** $\dfrac{3x\sqrt{15xy}}{5y}$ **65.** $\dfrac{5xy\sqrt{6xz}}{2z}$ **67.** $\dfrac{2ab\sqrt{6ac^2}}{3c}$
69. $\dfrac{2xy^2\sqrt[3]{3z^2}}{3z}$ **71.** $\sqrt{x}$ **73.** $\sqrt[6]{xy}$ **75.** $\sqrt[12]{a}$ **77.** $x\sqrt[9]{6x}$ **79.** $ab^2c\sqrt[6]{c}$ **81.** $abc^2\sqrt[15]{3a^2b}$ **83.** $2b^2\sqrt[3]{a}$ **85.** $5|x|$
87. $3|xy|\sqrt{3x}$ **89.** $|x-5|$ **91.** $|2x+3|$ **93.** $2|a(a+2)|$ **95.** $2|x|\sqrt{x-2}$ **97.** $\sqrt{9+16} = \sqrt{25} = 5; \sqrt{9} + \sqrt{16} = 3 + 4 = 7$
99. $5\sqrt{13}$ feet **101. a.** 13 feet **b.** $2\sqrt{14} \approx 7.5$ feet **105.** 3 **107.** 0 **109.** 1 **111.** 3 **113.** $27xy^2$ **115.** $\dfrac{5}{6}x$
117. $3\sqrt{2}$ **119.** $5y\sqrt{3xy}$ **121.** $2a\sqrt[3]{ab^2}$ **123.** $12\sqrt[3]{5}$ **125.** $6\sqrt[3]{49}$ **127.** $\dfrac{\sqrt[10]{a^7}}{a}$ **129.** $\dfrac{\sqrt[20]{a^9}}{a}$
131.

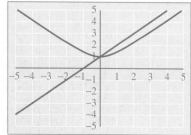

133. About $\frac{3}{4}$ of a unit when $x = 2$ **135.** $x = 0$

Problem Set 9.4

1. $7\sqrt{5}$ **3.** $-x\sqrt{7}$ **5.** $\sqrt[3]{10}$ **7.** $9\sqrt[5]{6}$ **9.** 0 **11.** $\sqrt{5}$ **13.** $-32\sqrt{2}$ **15.** $-3x\sqrt{2}$ **17.** $-2\sqrt[3]{2}$ **19.** $8x\sqrt[3]{xy^2}$
21. $3a^2b\sqrt{3ab}$ **23.** $11ab\sqrt[3]{3a^2b}$ **25.** $10xy\sqrt[4]{3y}$ **27.** $\sqrt{2}$ **29.** $\dfrac{8\sqrt{5}}{15}$ **31.** $\dfrac{(x-1)\sqrt{x}}{x}$ **33.** $\dfrac{3\sqrt{2}}{2}$ **35.** $\dfrac{5\sqrt{6}}{6}$ **37.** $\dfrac{8\sqrt[3]{25}}{5}$
39. a. $8\sqrt{2x}$ **b.** $-4\sqrt{2x}$ **41. a.** $4\sqrt{2x}$ **b.** $2\sqrt{2x}$ **43. a.** $3x\sqrt{2}$ **b.** $-x\sqrt{2}$ **45. a.** $3\sqrt{2x} + 3$ **b.** $-\sqrt{2x} - 7$
47. $\sqrt{12} \approx 3.464; 2\sqrt{3} = 2(1.732) \approx 3.464$ **49.** $\sqrt{8} + \sqrt{18} \approx 2.828 + 4.243 = 7.071; \sqrt{50} \approx 7.071; \sqrt{26} \approx 5.099$ **55.** $\dfrac{\sqrt{3}}{2}$
57.

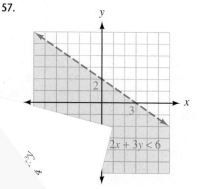

59.

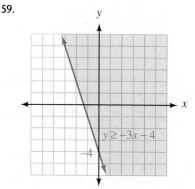

61.

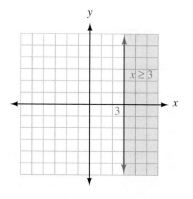

63. $-\frac{1}{5} \le x \le \frac{3}{5}$ **65.** 6   **67.** $4x^2 + 3xy - y^2$ **69.** $x^2 + 6x + 9$ **71.** $x^2 - 4$ **73.** $6\sqrt{2}$ **75.** 6   **77.** $9x$ **79.** $\dfrac{\sqrt{6}}{2}$
81. $-xy$ **83.** $-xy^2z^3$ **85.** $5b^2c\sqrt[4]{2a^3b}$ **87.** $-3ab\sqrt[3]{2a}$

Problem Set 9.5

1. $3\sqrt{2}$ **3.** $10\sqrt{21}$ **5.** 720   **7.** 54   **9.** $\sqrt{6} - 9$ **11.** $24 + 6\sqrt[3]{4}$ **13.** $2 + 2\sqrt[3]{3}$ **15.** $xy\sqrt[3]{y} + x^2\sqrt[3]{y}$ **17.** $2x^2\sqrt[4]{x} + 2x^3\sqrt[4]{2}$
19. $7 + 2\sqrt{6}$ **21.** $x + 2\sqrt{x} - 15$ **23.** $34 + 20\sqrt{3}$ **25.** $19 + 8\sqrt{3}$ **27.** $x - 6\sqrt{x} + 9$ **29.** $4a - 12\sqrt{ab} + 9b$ **31.** $x + 4\sqrt{x - 4}$
33. $x - 6\sqrt{x - 5} + 4$ **35.** 1   **37.** $a - 49$ **39.** $25 - x$ **41.** $x - 8$ **43.** $10 + 6\sqrt{3}$ **45.** $5 + \sqrt[3]{12} + \sqrt[3]{18}$
47. $x^2 + x\sqrt[3]{x^2y^2} + \sqrt[3]{xy} + y$ **49.** $\dfrac{\sqrt{2}}{2}$ **51.** $\dfrac{\sqrt{x}}{x}$ **53.** $\dfrac{4\sqrt{3}}{3}$ **55.** $\dfrac{x\sqrt{6}}{3}$ **57.** $\dfrac{2\sqrt{10x}}{5x}$ **59.** $\dfrac{\sqrt{2}}{2}$ **61.** $\dfrac{2x\sqrt{2y}}{y}$
63. $\dfrac{a\sqrt{6c}}{2bc}$ **65.** $\dfrac{2ac\sqrt[3]{b^2}}{b}$ **67.** $\dfrac{\sqrt{3} + 1}{2}$ **69.** $\dfrac{5 - \sqrt{5}}{4}$ **71.** $\dfrac{x + 3\sqrt{x}}{x - 9}$ **73.** $\dfrac{10 + 3\sqrt{5}}{11}$ **75.** $\dfrac{3\sqrt{x} + 3\sqrt{y}}{x - y}$
77. $2 + \sqrt{3}$ **79.** $\dfrac{11 - 4\sqrt{7}}{3}$ **81.** $\dfrac{a + 2\sqrt{ab} + b}{a - b}$ **83.** $\dfrac{x + 4\sqrt{x} + 4}{x - 4}$ **85.** $\dfrac{5 - \sqrt{21}}{4}$ **87.** $\dfrac{\sqrt{x} - 3x + 2}{1 - x}$ **89.** $\dfrac{8\sqrt{3}}{3}$
91. $\dfrac{11\sqrt{5}}{5}$ **93.** $-4\sqrt{3}$ **95.** $5\sqrt{3}$
97. $(\sqrt[3]{2} + \sqrt[3]{3})(\sqrt[3]{4} - \sqrt[3]{6} + \sqrt[3]{9})$
$= \sqrt[3]{8} - \sqrt[3]{12} + \sqrt[3]{18} + \sqrt[3]{12} - \sqrt[3]{18} + \sqrt[3]{27}$
$= 2 + 3 = 5$
99. $10\sqrt{3}$ **101.** $x + 6\sqrt{x} + 9$ **103.** 75   **105.** $\dfrac{5\sqrt{2}}{4}$ seconds; $\dfrac{5}{2}$ seconds   **111.** 147   **113.** 50   **115.** 100
117. $t^2 + 10t + 25$ **119.** x **121.** 7   **123.** $-4, -3$ **125.** $-6, -3$ **127.** $-5, -2$ **129.** $\dfrac{x(\sqrt{x - 2} - 4)}{x - 18}$
131. $\dfrac{x(\sqrt{x + 5} + 5)}{x - 20}$ **133.** $\dfrac{3(\sqrt{5x} - x)}{5 - x}$

Problem Set 9.6

1. 4   **3.** ∅   **5.** 5   **7.** ∅   **9.** $\dfrac{39}{2}$ **11.** ∅   **13.** 5   **15.** 3   **17.** $-\dfrac{32}{3}$ **19.** 3, 4   **21.** $-1, -2$
23. Possible solutions $\frac{1}{2}$ and 11; only 11 checks; 11   **25.** 3, 7   **27.** $-\dfrac{1}{4}$, 14   **29.** -1 **31.** ∅   **33.** 7   **35.** 0, 3   **37.** -4
39. 8   **41.** 0   **43.** 9   **45.** 0   **47.** 8   **49.** Possible solution 9, which does not check; ∅
51. Possible solutions 0 and 32; only 0 checks; 0   **53.** Possible solutions -2 and 6; only 6 checks; 6   **55.** $\dfrac{1}{2}$ **57.** $\dfrac{1}{2}$, 1
59. 5, 13   **61.** $-\dfrac{3}{2}$ **63.** Possible solution 1, which does not check; ∅   **65.** $\dfrac{1}{2}$
67. Possible solution 2, which does not check; ∅   **69.** 1   **71.** -11
73.

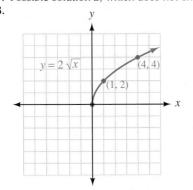

75.

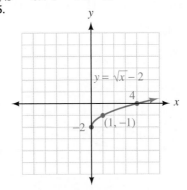

77.

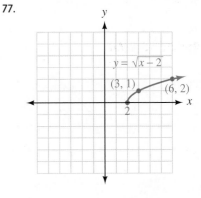

79.

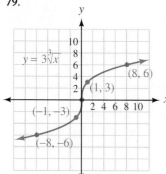

81.

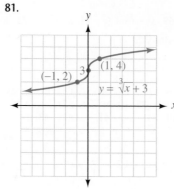

83.

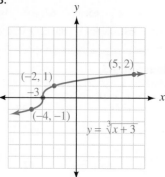

85. $h = 100 - 16t^2$ **87.** $\frac{392}{121} \approx 3.24$ feet **89.** 5 meters **91.** 2,500 meters **93.** $\sqrt{6} - 2$ **95.** $x + 10\sqrt{x} + 25$ **97.** $\frac{x - 3\sqrt{x}}{x - 9}$
99. 5 **101.** $2\sqrt{3}$ **103.** -1 **105.** 1 **107.** 4 **109.** 2 **111.** $10 - 2x$ **113.** $2 - 3x$ **115.** $6 + 7x - 20x^2$ **117.** $8x - 12x^2$
119. $4 + 12x + 9x^2$ **121.** $4 - 9x^2$

Problem Set 9.7
1. $6i$ **3.** $-5i$ **5.** $6i\sqrt{2}$ **7.** $-2i\sqrt{3}$ **9.** 1 **11.** -1 **13.** $-i$ **15.** $x = 3, y = -1$ **17.** $x = -2, y = -\frac{1}{2}$ **19.** $x = -8, y = -5$
21. $x = 7, y = \frac{1}{2}$ **23.** $x = \frac{3}{7}, y = \frac{2}{5}$ **25.** $5 + 9i$ **27.** $5 - i$ **29.** $2 - 4i$ **31.** $1 - 6i$ **33.** $2 + 2i$ **35.** $-1 - 7i$ **37.** $6 + 8i$
39. $2 - 24i$ **41.** $-15 + 12i$ **43.** $18 + 24i$ **45.** $10 + 11i$ **47.** $21 + 23i$ **49.** $-2 + 2i$ **51.** $2 - 11i$ **53.** $-21 + 20i$ **55.** $-2i$
57. $-7 - 24i$ **59.** 5 **61.** 40 **63.** 13 **65.** 164 **67.** $-3 - 2i$ **69.** $-2 + 5i$ **71.** $\frac{8}{13} + \frac{12}{13}i$ **73.** $-\frac{18}{13} - \frac{12}{13}i$ **75.** $-\frac{5}{13} + \frac{12}{13}i$
77. $\frac{13}{15} - \frac{2}{5}i$ **79.** $R = -11 - 7i$ ohms **81.** domain = {1, 3, 4}; range = {2, 4}; is a function
83. domain = {1, 2, 3}; range = {1, 2, 3}; is a function **85.** is a function **87.** not a function

CHAPTER 9 REVIEW TEST

1. 7 **2.** -3 **3.** 2 **4.** 27 **5.** $2x^3y^2$ **6.** $\frac{1}{16}$ **7.** x^2 **8.** a^2b^4 **9.** $a^{7/20}$ **10.** $a^{5/12}b^{8/3}$ **11.** $12x + 11x^{1/2}y^{1/2} - 15y$

12. $a^{2/3} - 10a^{1/3} + 25$ **13.** $4x^{1/2} + 2x^{5/6}$ **14.** $2(x - 3)^{1/4}(4x - 13)$ **15.** $\frac{x + 5}{x^{1/4}}$ **16.** $2\sqrt{3}$ **17.** $5\sqrt{2}$ **18.** $2\sqrt[3]{2}$ **19.** $3x\sqrt{2}$

20. $4ab^2c\sqrt{5a}$ **21.** $2abc\sqrt[4]{2bc^2}$ **22.** $\frac{3\sqrt{2}}{2}$ **23.** $3\sqrt[3]{4}$ **24.** $\frac{4x\sqrt{21xy}}{7y}$ **25.** $\frac{2y\sqrt[3]{45x^2z^2}}{3z}$ **26.** $-2x\sqrt{6}$ **27.** $3\sqrt{3}$

28. $\frac{8\sqrt{5}}{5}$ **29.** $7\sqrt{2}$ **30.** $11a^2b\sqrt{3ab}$ **31.** $-4xy\sqrt[3]{xz^2}$ **32.** $\sqrt{6} - 4$ **33.** $x - 5\sqrt{x} + 6$ **34.** $3\sqrt{5} + 6$ **35.** $6 + \sqrt{35}$

36. $\frac{63 + 12\sqrt{7}}{47}$ **37.** 0 **38.** 3 **39.** 5 **40.** $\varnothing$

41.

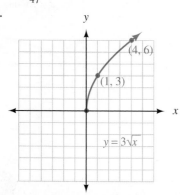

42.

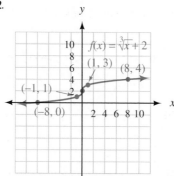

43. 1 **44.** $-i$ **45.** $x = -\frac{3}{2}, y = -\frac{1}{2}$ **46.** $x = -2, y = -4$ **47.** $9 + 3i$ **48.** $-7 + 4i$ **49.** $-6 + 12i$ **50.** $5 + 14i$
51. $12 + 16i$ **52.** 25 **53.** $1 - 3i$ **54.** $-\frac{6}{5} + \frac{3}{5}i$ **55.** 22.5 feet **56.** 65 yards

CHAPTER 10

Problem Set 10.1

1. ± 5 **3.** $\pm \dfrac{\sqrt{3}}{2}$ **5.** $\pm 2i\sqrt{3}$ **7.** $\pm \dfrac{3\sqrt{5}}{2}$ **9.** $-2, 3$ **11.** $\dfrac{-3 \pm 3i}{2}$ **13.** $-4 \pm 3i\sqrt{3}$ **15.** $\dfrac{3 \pm 2i}{2}$

17. $x^2 + 12x + 36 = (x + 6)^2$ **19.** $x^2 - 4x + 4 = (x - 2)^2$ **21.** $a^2 - 10a + 25 = (a - 5)^2$ **23.** $x^2 + 5x + \dfrac{25}{4} = (x + \dfrac{5}{2})^2$

25. $y^2 - 7y + \dfrac{49}{4} = (y - \dfrac{7}{2})^2$ **27.** $x^2 + \dfrac{1}{2}x + \dfrac{1}{16} = (x + \dfrac{1}{4})^2$ **29.** $x^2 + \dfrac{2}{3}x + \dfrac{1}{9} = (x + \dfrac{1}{3})^2$ **31.** $-3, -9$ **33.** $1 \pm 2i$

35. $4 \pm \sqrt{15}$ **37.** $\dfrac{5 \pm \sqrt{37}}{2}$ **39.** $1 \pm \sqrt{5}$ **41.** $\dfrac{4 \pm \sqrt{13}}{3}$ **43.** $\dfrac{3 \pm i\sqrt{71}}{8}$ **45. a.** No. **b.** $\pm 3i$ **47. a.** 0, 6 **b.** 0, 6

49. a. $-7, 5$ **b.** $-7, 5$ **51.** No   **53. a.** $\dfrac{7}{5}$ **b.** 3 **c.** $\dfrac{7 \pm 2\sqrt{2}}{5}$ **d.** $\dfrac{71}{5}$ **e.** 3   **55.** ± 1 **57.** ± 1 **59.** ± 1

61.

x	$f(x)$	$g(x)$	$h(x)$
-2	49	49	25
-1	25	25	13
0	9	9	9
1	1	1	13
2	1	1	25

63. 3

65. a. $-1, 6$ **b.** $-1, 6$ **c.** -6 **d.** -10 **67.** $\dfrac{\sqrt{3}}{2}$ inch, 1 inch   **69.** $\sqrt{2}$ inches   **71.** 781 feet   **73.** 7.3% to the nearest tenth

75. $20\sqrt{2} \approx 28$ feet   **77.** $3\sqrt{5}$ **79.** $3y^2\sqrt{3y}$ **81.** $3x^2y\sqrt[3]{2y^2}$ **83.** $\dfrac{3\sqrt{2}}{2}$ **85.** $\sqrt[3]{2}$ **87.** 13   **89.** 7   **91.** $\dfrac{1}{2}$ **93.** 13

95. $(3t - 2)(9t^2 + 6t + 4)$ **97.** $x = \pm 2a$ **99.** $x = \dfrac{-p \pm \sqrt{p^2 - 4q}}{2}$ **101.** $x = \dfrac{-p \pm \sqrt{p^2 - 12q}}{6}$ **103.** $(x - 5)^2 + (y - 3)^2 = 2^2$

Problem Set 10.2

1. a. $\dfrac{-2 \pm \sqrt{10}}{3}$ **b.** $\dfrac{2 \pm \sqrt{10}}{3}$ **c.** $\dfrac{-2 \pm i\sqrt{2}}{3}$ **d.** $\dfrac{-2 \pm \sqrt{10}}{2}$ **e.** $\dfrac{2 \pm i\sqrt{2}}{2}$ **3. a.** $1 \pm i$ **b.** $1 \pm 2i$ **c.** $-1 \pm i$ **5.** 1, 2

7. $\dfrac{2 \pm i\sqrt{14}}{3}$ **9.** 0, 5   **11.** 0, $-\dfrac{4}{3}$ **13.** $\dfrac{3 \pm \sqrt{5}}{4}$ **15.** $-3 \pm \sqrt{17}$ **17.** $\dfrac{-1 \pm i\sqrt{5}}{2}$ **19.** 1   **21.** $\dfrac{1 \pm i\sqrt{47}}{6}$ **23.** $4 \pm \sqrt{2}$

25. $\dfrac{1}{2}, 1$ **27.** $-\dfrac{1}{2}, 3$ **29.** $\dfrac{-1 \pm i\sqrt{7}}{2}$ **31.** $1 \pm \sqrt{2}$ **33.** $\dfrac{-3 \pm \sqrt{5}}{2}$ **35.** $-5, 3$ **37.** $2, -1 \pm i\sqrt{3}$ **39.** $-\dfrac{3}{2}, \dfrac{3 \pm 3i\sqrt{3}}{4}$

41. $\dfrac{1}{5}, \dfrac{-1 \pm i\sqrt{3}}{10}$ **43.** $0, \dfrac{-1 \pm i\sqrt{5}}{2}$ **45.** $0, 1 \pm i$ **47.** $0, \dfrac{-1 \pm i\sqrt{2}}{3}$ **49.** a and b **51. a.** $\dfrac{5}{3}, 0$ **b.** $\dfrac{5}{3}, 0$ **53.** No, $2 \pm i\sqrt{3}$

55. Yes   **57. a.** $-1, 3$ **b.** $1 \pm i\sqrt{7}$ **c.** $-2, 2$ **d.** $2 \pm 2\sqrt{2}$ **59. a.** $-2, \dfrac{5}{3}$ **b.** $\dfrac{3}{2}, \dfrac{-3 \pm 3i\sqrt{3}}{4}$ **c.** $\varnothing$ **d.** $1, -1$ **61.** 2 seconds

63. $40 \pm 20 = 20$ or 60 items   **65.** 0.49 centimeter (8.86 cm is impossible)

67. a. $\ell + w = 10, \ell w = 15$
b. 8.16 yards, 1.84 yards
c. Two answers are possible because either dimension (long or short) may be considered the length.

69. $4y + 1 + \dfrac{-2}{2y - 7}$ **71.** $x^2 + 7x + 12$ **73.** 5   **75.** $\dfrac{27}{125}$ **77.** $\dfrac{1}{4}$ **79.** $21x^3y$ **81.** 169   **83.** 0   **85.** ± 12 **87.** $x^2 - x - 6$

89. $x^3 - 4x^2 - 3x + 18$ **91.** $-2\sqrt{3}, \sqrt{3}$ **93.** $\dfrac{-1 \pm \sqrt{3}}{\sqrt{2}} = \dfrac{-\sqrt{2} \pm \sqrt{6}}{2}$ **95.** $-2i, i$

Problem Set 10.3

1. $D = 16$, two rational   **3.** $D = 0$, one rational   **5.** $D = 5$, two irrational   **7.** $D = 17$, two irrational   **9.** $D = 36$, two rational
11. $D = 116$, two irrational   **13.** ± 10 **15.** ± 12 **17.** 9   **19.** -16 **21.** $\pm 2\sqrt{6}$ **23.** $x^2 - 7x + 10 = 0$ **25.** $t^2 - 3t - 18 = 0$
27. $y^3 - 4y^2 - 4y + 16 = 0$ **29.** $2x^2 - 7x + 3 = 0$ **31.** $4t^2 - 9t - 9 = 0$ **33.** $6x^3 - 5x^2 - 54x + 45 = 0$ **35.** $10a^2 - a - 3 = 0$
37. $9x^3 - 9x^2 - 4x + 4 = 0$ **39.** $x^4 - 13x^2 + 36 = 0$ **41.** $x^2 - 5 = 0$ **43.** $x^2 + (\sqrt{7} - \sqrt{3})x - \sqrt{21} = 0$ **45.** $x^2 + 16 = 0$
47. $x^2 - 5ix - 6 = 0$ **49.** $x^2 - 4x + 5 = 0$ **51.** $x^2 - x - 1 = 0$ **53. a.** $y = g(x)$ **b.** $y = h(x)$ **c.** $y = f(x)$ **55.** $a - 2a^{1/2} - 15$
57. $x - 64$ **59.** $5xy^2 - 4x^2y$ **61.** $2(x + 1)^{1/3}(4x + 3)$ **63.** $(3x^{1/3} + 2)^2$ **65.** $x^2 - 7x$ **67.** $9a^2 - 6a - 3$
69. $24a^2 + 94a + 90$ **71.** $\pm i\sqrt{2}$ **73.** 4   **75.** -5 **77.** $-3, 2$ **79.** $-\dfrac{1}{3}, \dfrac{5}{2}$ **81.** $-\dfrac{1}{2}, \dfrac{5}{2}$

83. $\dfrac{1 + \sqrt{2}}{2} \approx 1.21, \dfrac{1 - \sqrt{2}}{2} \approx -0.21$ **85.** $\dfrac{2}{3}, 1, -1$ **87.** $\dfrac{2}{3}, \sqrt{3}, \approx 1.73, -\sqrt{3}, \approx -1.73$

Problem Set 10.4

1. $1, 2$ **3.** $-8, -\frac{5}{2}$ **5.** $\pm 3, \pm 1$ **7.** $\pm 2, \pm\sqrt{3}$ **9.** $\pm 3, \pm i\sqrt{3}$ **11.** $\pm 2i, \pm i\sqrt{5}$ **13.** $\frac{7}{2}, 4$ **15.** $-\frac{9}{8}, \frac{1}{2}$ **17.** $\pm\frac{\sqrt{30}}{6}, \pm i$

19. $\pm\frac{\sqrt{21}}{3}, \pm\frac{i\sqrt{21}}{3}$ **21.** $4, 25$ **23.** Possible solutions 25 and 9; only 25 checks; 25

25. Possible solutions $\frac{25}{9}$ and $\frac{49}{4}$; only $\frac{25}{9}$ checks; $\frac{25}{9}$ **27.** Possible solutions 4 and 16; only 16 checks; 16

29. Possible solutions 9 and 36; only 9 checks; 9 **31.** Possible solutions $\frac{1}{4}$ and 25; only $\frac{1}{4}$ checks; $\frac{1}{4}$ **33.** $27, 38$ **35.** $4, 12$

37. $\pm 2, \pm 2i$ **39.** $\pm 2, \pm 2i\sqrt{2}$ **41.** $\frac{1}{2}, 1$ **43.** $\pm\frac{\sqrt{5}}{5}$ **45.** $-216, 8$ **47.** $\pm 3i\sqrt{3}$ **49.** $t = \frac{v \pm \sqrt{v^2 + 64h}}{32}$

51. $x = \frac{-4 \pm 2\sqrt{4 - k}}{k}$ **53.** $x = -y$ **55.** $t = \frac{1 \pm \sqrt{1 + h}}{4}$

57. a. **b.** 630 feet

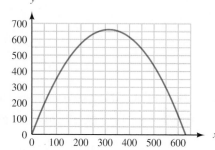

59. Let $x = BC$. Then $\frac{4}{x} = \frac{x}{x + 4}$, or $4(x + 4) = x^2$.

$$x = \frac{4 + 4\sqrt{5}}{2} = 4\left(\frac{1 + \sqrt{5}}{2}\right)$$

61. $3\sqrt{7}$ **63.** $5\sqrt{2}$ **65.** $39x^2y\sqrt{5x}$ **67.** $-11 + 6\sqrt{5}$ **69.** $x + 4\sqrt{x} + 4$ **71.** $\frac{7 + 2\sqrt{7}}{3}$ **73.** -2 **75.** $1,322.5$

77. $-\frac{7}{640}$ **79.** $1, 5$ **81.** $-3, 1$ **83.** $\frac{3}{2} \pm \frac{1}{2}i$ **85.** $9; 3$ **87.** $1; 1$ **89.** x-intercepts $= -2, 0, 2$; y-intercept $= 0$

91. x-intercepts $= -3, -\frac{1}{3}, 3$; y-intercept $= -9$ **93.** $\frac{1}{2}$ and -1

Problem Set 10.5

1. x-intercepts $= -3, 1$;
Vertex $= (-1, -4)$

3. x-intercepts $= -5, 1$;
Vertex $= (-2, 9)$

5. x-intercepts $= -1, 1$;
Vertex $= (0, -1)$

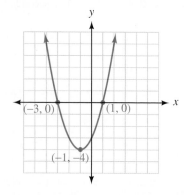

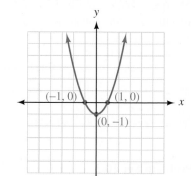

7. x-intercepts = 3, −3;
Vertex = (0, 9)

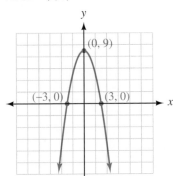

9. x-intercepts = −1, 3;
Vertex = (1, −8)

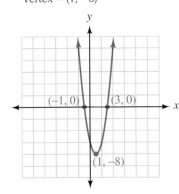

11. x-intercepts = $1 + \sqrt{5}$, $1 - \sqrt{5}$;
Vertex = (1, −5)

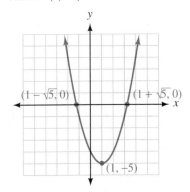

13. Vertex = (2, −8)

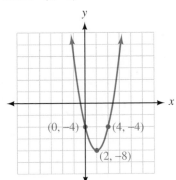

15. Vertex = (1, −4)

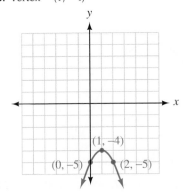

17. Vertex = (0, 1)

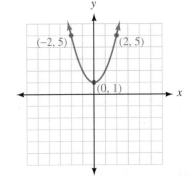

19. Vertex = (0, −3)

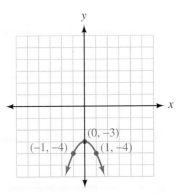

21. (3, −4) lowest   **23.** (1, 9) highest

25. (2, 16) highest   **27.** (−4, 16) highest   **29.** 875 patterns; maximum profit $731.25
31. The ball is in her hand when $h(t) = 0$, which means $t = 0$ or $t = 2$ seconds. Maximum height is $h(1) = 16$ feet.

33. Maximum $R = \$3{,}600$ when $p = \$6.00$ **35.** Maximum $R = \$7{,}225$ when $p = \$8.50$

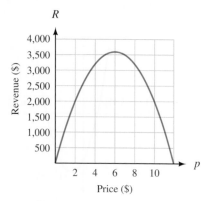

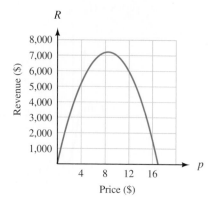

37. $y = -\frac{1}{135}(x - 90)^2 + 60$

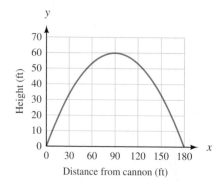

39. $1 - i$ **41.** $27 + 5i$ **43.** $\frac{1}{10} + \frac{3}{10}i$ **45.** $-2, 4$ **47.** $-\frac{1}{2}, \frac{2}{3}$ **49.** 3 **51.** $y = (x - 2)^2 - 4$

Problem Set 10.6

1.

3.

5.

7.

9.

11.

13.

15. All real numbers   **17.** No solution; $\varnothing$ **19.**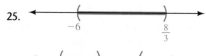

21.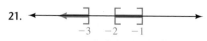

23.

25.

27.

29.

31.

33. a. $x - 1 > 0$ **b.** $x - 1 \geq 0$ **c.** $x - 1 \geq 0$ **35.** $\varnothing$ **37.** $x = 1$ **39.** all real numbers   **41.** $x > 1$ or $x < 1$ **43.** $\varnothing$
45. $x > 3$ or $x < 3$; all real numbers except 3   **47. a.** $-2 < x < 2$ **b.** $x < -2$ or $x > 2$ **c.** $x = -2$ or $x = 2$
49. a. $-2 < x < 5$ **b.** $x < -2$ or $x > 5$ **c.** $x = -2$ or $x = 5$
51. a. $x < -1$ or $1 < x < 3$ **b.** $-1 < x < 1$ or $x > 3$ **c.** $x = -1$ or $x = 1$ or $x = 3$ **53.** $x \geq 4$; the width is at least 4 inches
55. $5 \leq p \leq 8$; charge at least \$5 but no more than \$8 for each radio   **57.** 1.5625   **59.** 0.6549   **61.** $\frac{5}{3}$
63. Possible solutions 1 and 6; only 6 checks; 6

65.

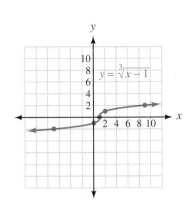

67.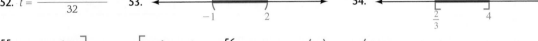

$1 - \sqrt{2} \qquad 1 + \sqrt{2}$

69.

$4 - \sqrt{3} \qquad 4 + \sqrt{3}$

CHAPTER 10 REVIEW TEST

1. $0, 5$　**2.** $0, \frac{4}{3}$　**3.** $\frac{4 \pm 7i}{3}$　**4.** $-3 \pm \sqrt{3}$　**5.** $-5, 2$　**6.** $-3, -2$　**7.** 3　**8.** 2　**9.** $\frac{-3 \pm \sqrt{3}}{2}$　**10.** $\frac{3 \pm \sqrt{5}}{2}$　**11.** $-5, 2$

12. $0, \frac{9}{4}$　**13.** $\frac{2 \pm i\sqrt{15}}{2}$　**14.** $1 \pm \sqrt{2}$　**15.** $-\frac{4}{3}, \frac{1}{2}$　**16.** 3　**17.** $5 \pm \sqrt{7}$　**18.** $0, \frac{5 \pm \sqrt{21}}{2}$　**19.** $-1, \frac{3}{5}$　**20.** $3, \frac{-3 \pm 3i\sqrt{3}}{2}$

21. $\frac{1 \pm i\sqrt{2}}{3}$　**22.** $\frac{3 \pm \sqrt{29}}{2}$　**23.** 100 or 170 items　**24.** 20 or 60 items　**25.** $D = 0$; 1 rational　**26.** $D = 0$; 1 rational

27. $D = 25$; 2 rational　**28.** $D = 361$; 2 rational　**29.** $D = 5$; 2 irrational　**30.** $D = 5$; 2 irrational　**31.** $D = -23$; 2 complex

32. $D = -87$; 2 complex　**33.** ± 20　**34.** ± 20　**35.** 4　**36.** 4　**37.** 25　**38.** 49　**39.** $x^2 - 8x + 15 = 0$　**40.** $x^2 - 2x - 8 = 0$

41. $2y^2 + 7y - 4 = 0$　**42.** $t^3 - 5t^2 - 9t + 45 = 0$　**43.** $-4, 12$　**44.** $-\frac{3}{4}, -\frac{1}{6}$　**45.** $\pm 2, \pm i\sqrt{3}$

46. Possible solutions 4 and 1, only 4 checks; 4　**47.** $\frac{9}{4}, 16$　**48.** 4　**49.** 4　**50.** 7　**51.** $t = \frac{5 \pm \sqrt{25 + 16h}}{16}$

52. $t = \frac{v \pm \sqrt{v^2 + 640}}{32}$　**53.**

$-1 \qquad 2$

54.

$\frac{2}{3} \qquad 4$

55.

$-4 \qquad \frac{3}{2}$

56.

$-4 -2 \qquad 3$

57.

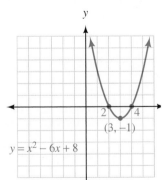

$(3, -1)$

$y = x^2 - 6x + 8$

58.

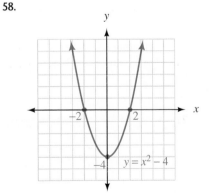

$y = x^2 - 4$

CHAPTER 11

Problem Set 11.1

1. 1   **3.** 2   **5.** $\frac{1}{27}$ **7.** 13

9.

11.

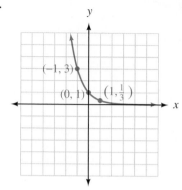

13.

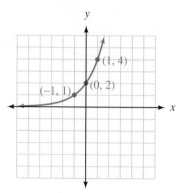

15.

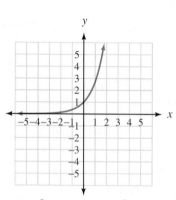

17.

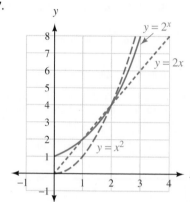

19.

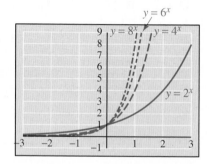

21. $h = 6 \cdot \left(\frac{2}{3}\right)^n$; 5th bounce: $6\left(\frac{2}{3}\right)^5 \approx 0.79$ feet   **23.** After 8 days, 700 micrograms; After 11 days, $1{,}400 \cdot 2^{-11/8} \approx 539.8$ micrograms

25. a. $A(t) = 1{,}200\left(1 + \dfrac{.06}{4}\right)^{4t}$ **b.** \$1,932.39 **c.** About 11.64 years **d.** \$1,939.29

27. a. The function underestimated the expenditures by \$69 billion.
   b. \$4,123 billion
      \$4,577 billion
      \$5,080 billion

29.

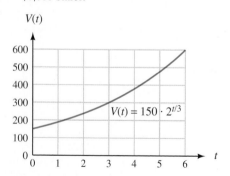

31. After 6 years

Problem Set 13.2

1. 36   **3.** 11   **5.** 18   **7.** $\frac{163}{60}$ **9.** 60   **11.** 40   **13.** 44   **15.** $-\frac{11}{32}$ **17.** $\frac{21}{10}$ **19.** $5x + 15$

21. $(x - 2) + (x - 2)^2 + (x - 2)^3 + (x - 2)^4$ **23.** $\frac{x+1}{x-1} + \frac{x+2}{x-1} + \frac{x+3}{x-1} + \frac{x+4}{x-1} + \frac{x+5}{x-1}$

25. $(x + 3)^3 + (x + 4)^4 + (x + 5)^5 + (x + 6)^6 + (x + 7)^7 + (x + 8)^8$ **27.** $(x - 6)^6 + (x - 8)^7 + (x - 10)^8 + (x - 12)^9$

29. $\sum\limits_{i=1}^{4} 2^i$ **31.** $\sum\limits_{i=2}^{6} 2^i$ **33.** $\sum\limits_{i=1}^{5} (4i + 1)$ **35.** $\sum\limits_{i=2}^{5} -(-2)^i$ **37.** $\sum\limits_{i=3}^{7} \frac{i}{i+1}$ **39.** $\sum\limits_{i=1}^{4} \frac{i}{2i+1}$ **41.** $\sum\limits_{i=6}^{9} (x - 2)^i$ **43.** $\sum\limits_{i=1}^{4} \left(1 + \frac{i}{x}\right)^{i+1}$

45. $\sum\limits_{i=3}^{5} \frac{x}{x+i}$ **47.** $\sum\limits_{i=2}^{4} x^i(x + i)$

49. a. $0.3 + 0.03 + 0.003 + 0.0003 + \ldots$ **b.** $0.2 + 0.02 + 0.002 + 0.0002 + \ldots$ **c.** $0.27 + 0.0027 + 0.000027 + \ldots$

51. seventh second: 208 feet; total: 784 feet   **53. a.** $16 + 48 + 80 + 112 + 144$ **b.** $\sum\limits_{i=1}^{5} (32i - 16)$ **55.** $3\log_2 x + \log_2 y$

57. $\frac{1}{3}\log_{10} x - 2\log_{10} y$ **59.** $\log_{10} \frac{x}{y^2}$ **61.** $\log_3 \frac{x^2}{y^3 z^4}$ **63.** 80   **65.** Possible solutions -1 and 8; only 8.   **67.** 74   **69.** $\frac{55}{2}$

71. $2n + 1$ **73.** $(3, 2)$ **75.** $\frac{13}{2}$

Problem Set 13.3

1. arithmetic; $d = 1$ **3.** not arithmetic   **5.** arithmetic; $d = -5$ **7.** not arithmetic   **9.** arithmetic; $d = \frac{2}{3}$

11. $a_n = 4n - 1$; $a_{24} = 95$ **13.** $a_{10} = -12$; $S_{10} = -30$ **15.** $a_1 = 7$; $d = 2$; $a_{30} = 65$ **17.** $a_1 = 12$; $d = 2$; $a_{20} = 50$; $S_{20} = 620$

19. $a_{20} = 79$; $S_{20} = 820$ **21.** $a_{40} = 122$; $S_{40} = 2,540$ **23.** $a_1 = 13$; $d = -6$ **25.** $a_{85} = -238$ **27.** $d = \frac{16}{19}$; $a_{85} = 28$

29. 20,300   **31.** $a_{35} = -158$ **33.** $a_{10} = 5$; $S_{10} = \frac{55}{2}$

35. a. $18,000, $14,700, $11,400, $8,100, $4,800 **b.** $-$3,300 **c.**

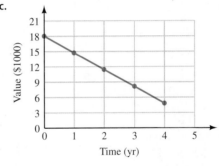

d. $9,750 **e.** $a_0 = 18000$; $a_n = a_{n-1} - 3300$ for $n \geq 1$ **37. a.** 1500 ft, 1460 ft, 1420 ft, 1380 ft, 1340 ft, 1300 ft
b. It is arithmetic because the same amount is subtracted from each succeeding term. **c.** $a_n = 1,540 - 40n$
39. a. 1, 3, 6, 10, 15, 21, 28, 36, 45, 55, 66, 78, 91, 105, 120 **b.** $a_1 = 1$; $a_n = n + a_{n-1}$ for $n \geq 2$
c. It is not arithmetic because the same amount is not added to each term.
41. 2.7604   **43.** -1.2396 **45.** 445   **47.** 4.45×10^{-8} **49.** $\frac{1}{16}$ **51.** $\sqrt{3}$ **53.** 2^n **55.** r^3 **57.** $\frac{2}{5}$

59. 255   **61.** $a_1 = 9$, $d = 4$ **63.** $d = -3$ **65.** $S_{50} = 50 + 1225\sqrt{2}$

Problem Set 13.4

1. geometric; $r = 5$ **3.** geometric; $r = \frac{1}{3}$ **5.** not geometric   **7.** geometric; $r = -2$ **9.** not geometric

11. $a_n = 4 \cdot 3^{n-1}$ **13.** $a_6 = \frac{1}{16}$ **15.** $a_{20} = -3$ **17.** $S_{10} = 10,230$ **19.** $S_{20} = 0$ **21.** $a_8 = \frac{1}{640}$ **23.** $S_5 = -\frac{31}{32}$

25. $a_{10} = 32$; $S_{10} = 62 + 31\sqrt{2}$ **27.** $a_6 = \frac{1}{1000}$; $S_6 = 111.111$ **29.** ± 2 **31.** $a_8 = 384$; $S_8 = 255$ **33.** $r = 2$

35. 1   **37.** 8   **39.** 4   **41.** $\frac{8}{9}$ **43.** $\frac{2}{3}$ **45.** $\frac{9}{8}$ **51. a.** \$450,000, \$315,000, \$220,500, \$154,350, \$108,045 **b.** 0.7

c.

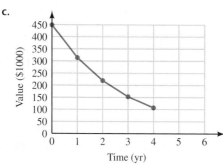

d. Approximately \$90,000 **e.** $a_0 = 450000$; $a_n = 0.7a_{n-1}$ for $n \geq 1$

53. a. $\frac{1}{2}$ **b.** $\frac{364}{729}$ **c.** $\frac{1}{1458}$ **55.** 300 feet   **57. a.** $a_n = 15(\frac{4}{5})^{n-1}$ **b.** 75 feet   **59.** 2.16   **61.** 6.36   **63.** $t = \frac{1}{5}\ln\frac{A}{10}$ **65.** 1

67. $x^3 + 2xy + y^2$ **69.** 15   **71.** $\frac{7}{11}$ **73.** $r = \frac{1}{3}$

75. a. stage 1:1; stage 2: $\frac{3}{4}$; stage 3: $\frac{9}{16}$; stage 4: $\frac{27}{64}$ **b.** geometric sequence **c.** 0 **d.** increasing sequence

Problem Set 13.5

1. $x^4 + 8x^3 + 24x^2 + 32x + 16$ **3.** $x^6 + 6x^5y + 15x^4y^2 + 20x^3y^3 + 15x^2y^4 + 6xy^5 + y^6$
5. $32x^5 + 80x^4 + 80x^3 + 40x^2 + 10x + 1$ **7.** $x^5 - 10x^4y + 40x^3y^2 - 80x^2y^3 + 80xy^4 - 32y^5$
9. $81x^4 - 216x^3 + 216x^2 - 96x + 16$ **11.** $64x^3 - 144x^2y + 108xy^2 - 27y^3$ **13.** $x^8 + 8x^6 + 24x^4 + 32x^2 + 16$
15. $x^6 + 3x^4y^2 + 3x^2y^4 + y^6$ **17.** $16x^4 + 96x^3y + 216x^2y^2 + 216xy^3 + 81y^4$ **19.** $\frac{x^3}{8} + \frac{x^2y}{4} + \frac{xy^2}{6} + \frac{y^3}{27}$
21. $\frac{x^3}{8} - 3x^2 + 24x - 64$ **23.** $\frac{x^4}{81} + \frac{2x^3y}{27} + \frac{x^2y^2}{6} + \frac{xy^3}{6} + \frac{y^4}{16}$ **25.** $x^9 + 18x^8 + 144x^7 + 672x^6$
27. $x^{10} - 10x^9y + 45x^8y^2 - 120x^7y^3$ **29.** $x^{25} + 75x^{24} + 2,700x^{23} + 62,100x^{22}$ **31.** $x^{60} - 120x^{59} + 7,080x^{58} - 273,760x^{57}$
33. $x^{18} - 18x^{17}y + 153x^{16}y^2 - 816x^{15}y^3$ **35.** $x^{15} + 15x^{14} + 105x^{13}$ **37.** $x^{12} - 12x^{11}y + 66x^{10}y^2$ **39.** $x^{20} + 40x^{19} + 760x^{18}$
41. $x^{100} + 200x^{99}$ **43.** $x^{50} + 50x^{49}y$ **45.** $51,963,120x^4y^8$ **47.** $3,360x^6$ **49.** $-25,344x^7$ **51.** $2,700x^{23}y^2$
53. $\binom{20}{11}(2x)^9(5y)^{11} = \frac{20!}{11!9!}(2x)^9(5y)^{11}$ **55.** $x^{20}y^{10} - 30x^{18}y^9 + 405x^{16}y^8$ **57.** $\frac{21}{128}$ **59.** $x \approx 1.21$ **61.** $\frac{1}{6}$ **63.** ≈ 17.5 years
65. 56   **67.** 125,970

CHAPTER 13 REVIEW TEST

1. 7, 9, 11, 13   **2.** 1, 4, 7, 10   **3.** 0, 3, 8, 15   **4.** $\frac{4}{3}, \frac{5}{4}, \frac{6}{5}, \frac{7}{6}$ **5.** 4, 16, 64, 256   **6.** $\frac{1}{4}, \frac{1}{16}, \frac{1}{64}, \frac{1}{256}$ **7.** $a_n = 3n - 1$
8. $a_n = 2n - 5$ **9.** $a_n = n^4$ **10.** $a_n = n^2 + 1$ **11.** $a_n = (\frac{1}{2})^n = 2^{-n}$ **12.** $a_n = \frac{n+1}{n^2}$ **13.** 32   **14.** 25   **15.** $\frac{14}{5}$
16. -5 **17.** 98   **18.** $\frac{127}{30}$ **19.** $\sum_{i=1}^{4} 3i$ **20.** $\sum_{i=1}^{4}(4i - 1)$ **21.** $\sum_{i=1}^{5}(2i + 3)$ **22.** $\sum_{i=2}^{4} i^2$ **23.** $\sum_{i=1}^{4}\frac{1}{i+2}$ **24.** $\sum_{i=1}^{5}\frac{i}{3^i}$
25. $\sum_{i=1}^{3}(x - 2i)$ **26.** $\sum_{i=1}^{4}\frac{x}{x+i}$ **27.** geometric   **28.** arithmetic   **29.** arithmetic   **30.** neither   **31.** geometric
32. geometric   **33.** arithmetic   **34.** neither   **35.** $a_n = 3n - 1$; $a_{20} = 59$ **36.** $a_n = 8 - 3n$; $a_{16} = -40$
37. $a_{10} = 34$; $S_{10} = 160$ **38.** $a_{16} = 78$; $S_{16} = 648$ **39.** $a_1 = 5$; $d = 4$; $a_{10} = 41$ **40.** $a_1 = 8$; $d = 3$; $a_9 = 32$; $S_9 = 180$
41. $a_1 = -4$; $d = -2$; $a_{20} = -42$; $S_{20} = -460$ **42.** 20,100   **43.** $a_{40} = -95$ **44.** $a_n = 3(2)^{n-1}$; $a_{20} = 1,572,864$
45. $a_n = 5(-2)^{n-1}$; $a_{16} = -163,840$ **46.** $a_n = 4(\frac{1}{2})^{n-1}$; $a_{10} = \frac{1}{128}$ **47.** -3 **48.** 8   **49.** $a_1 = 3$; $r = 2$; $a_6 = 96$ **50.** $243\sqrt{3}$
51. $x^4 - 8x^3 + 24x^2 - 32x + 16$ **52.** $16x^4 + 96x^3 + 216x^2 + 216x + 81$ **53.** $27x^3 + 54x^2y + 36xy^2 + 8y^3$
54. $x^{10} - 10x^8 + 40x^6 - 80x^4 + 80x^2 - 32$ **55.** $\frac{1}{16}x^4 + \frac{3}{2}x^3 + \frac{27}{2}x^2 + 54x + 81$ **56.** $\frac{1}{27}x^3 - \frac{1}{6}x^2y + \frac{1}{4}xy^2 - \frac{1}{8}y^3$
57. $x^{10} + 30x^9y + 405x^8y^2$ **58.** $x^9 - 27x^8y + 324x^7y^2$ **59.** $x^{11} + 11x^{10}y + 55x^9y^2$ **60.** $x^{12} - 24x^{11}y + 264x^{10}y^2$
61. $x^{16} - 32x^{15}y$ **62.** $x^{32} + 64x^{31}y$ **63.** $x^{50} - 50x^{49}$ **64.** $x^{150} + 150x^{149}y$ **65.** $-61,236x^5$ **66.** $5376x^6$

Index